FUNDAMENTALS AND NEW FRONTIERS OF BOSE-EINSTEIN CONDENSATION

FUNDAMENTALS AND NEW FRONTIERS OF BOSE-EINSTEIN CONDENSATION

Masahito Ueda
University of Tokyo, Japan

World Scientific

NEW JERSEY • LONDON • SINGAPORE • BEIJING • SHANGHAI • HONG KONG • TAIPEI • CHENNAI

Published by

World Scientific Publishing Co. Pte. Ltd.

5 Toh Tuck Link, Singapore 596224

USA office: 27 Warren Street, Suite 401-402, Hackensack, NJ 07601

UK office: 57 Shelton Street, Covent Garden, London WC2H 9HE

British Library Cataloguing-in-Publication Data
A catalogue record for this book is available from the British Library.

FUNDAMENTALS AND NEW FRONTIERS OF BOSE–EINSTEIN CONDENSATION

ISBN-13 978-981-283-959-6
ISBN-10 981-283-959-3

Desk Editor: Ryan Bong

Printed in Singapore.

Preface

Experimental realization of Bose–Einstein condensation (BEC) of dilute atomic gases [Anderson, *et al.* (1995); Davis, *et al.* (1995); Bradley, *et al.* (1995, 1997)] has ignited a virtual explosion of research. The unique feature of the atomic gas BEC is its unprecedented controllability, which makes the previously unthinkable possible. Almost all parameters of the system such as the temperature, number of atoms, and even strength and sign (attractive or repulsive) of interaction can be varied by several orders of magnitude. The interaction between atoms is usually considered to be an immutable, inherent property of individual atomic species. In alkali and some other Bose–Einstein condensates, we can not only control the strength of interaction but also switch the sign of interaction from repulsive to attractive and vice versa [Inouye, *et al.* (1998); Cornish (2000)]. The atomic-gas BEC may thus be regarded as an artificial macroscopic matter wave that act as an ideal testing ground for the investigation of quantum many-body physics. The atomic-gas BEC may also be regarded as an atom laser because the condensate provides a phase-coherent, intense atomic source with potential applications for precision measurement, lithography, and quantum computation. Fermionic species may also undergo BEC by forming molecules or Cooper pairs. Both molecular condensates [Greiner, *et al.* (2003); Zwierlein, *et al.* (2003)] and Bardeen–Cooper–Schrieffer-type resonant superfluids [Regal; *et al.* (2004); Zwierlein, *et al.* (2004)] have been realized using alkali fermions, opening up the new research field of strongly correlated gaseous superfluidity. This book is intended as an introduction to this rapidly developing, interdisciplinary field of research.

Most phase transitions occur due to interactions between constituent particles. For example, superconductivity occurs due to effective interactions between electrons, and ferromagnetism is caused by the exchange interaction between spins. In contrast, BEC is a genuinely quantum-statistical phase transition in that it occurs without the help of interaction (Einstein called it "condensation without interaction" [Einstein (1925)]). The fundamentals of noninteracting BECs are reviewed in Chapter 1.

In a real BEC system, interactions between atoms play a crucial role in determining the basic properties of the system. Neutral atoms have a hard core that is short-ranged (~ 1 Å) and strongly repulsive. At a longer distance (~ 100 Å), the atoms are attracted to each other because of the van der Waals force. When two atoms collide, they experience both these forces, and the net interaction can be either repulsive or attractive depending on the hyperfine and translational states of the colliding atoms. Under normal conditions, a dilute-gas BEC system can be treated as a weakly interacting Bose gas. The Bogoliubov theory of a weakly interacting Bose gas and related topics are described in Chapter 2.

One of the remarkable aspects of a dilute gas BEC system is the great success of the mean-field theory governed by the Gross–Pitaevskii (GP) equation [Gross (1961); Pitaevskii (1961)]. The GP equation describes the mean-field ground state as well as the linear and nonlinear response of the system. Various nonlinear matter-wave phenomena including four-wave mixing [Deng, *et al.* (1999); Rolston and Phillips (2002)] and topological excitations such as solitons [Denschlag, *et al.* (2000)] and vortices [Matthews, *et al.* (1999); Madison, *et al.* (2000)], have been successfully described by the GP equation. This remarkable success of the mean-field theory is due to the high ($> 99\%$) degree of condensation of bosons into a single-particle state, which in turn originates in an extremely low density ($\sim 10^{11} - 10^{15}$ cm^{-3}) of the system operating at ultralow temperatures ($\lesssim 10^{-6}$ K). The Gross–Pitaevskii theory together with its various applications is discussed in Chapter 3.

The linear response theory provides a general theoretical framework to investigate collective modes of Bose–Einstein condensates and superfluids. A sum-rule approach is also very useful for this purpose because the ground state for a dilute-gas Bose–Einstein condensate can be obtained very accurately. These subjects are discussed in Chapter 4.

Superfluidity manifests itself as a response of the system to its moving container. A statistical-mechanical theory to tackle such problems and some basic properties of superfluidity are described in Chapter 5.

Alkali atoms have both electronic spin $\mathbf{s}$ and nuclear spin $\mathbf{i}$, and these two spins interact with each other via the hyperfine interaction. When the energy of the hyperfine coupling exceeds the electronic and nuclear Zeeman energies as well as the thermal energy, the total spin $\mathbf{f} = \mathbf{s} + \mathbf{i}$, which is called the hyperfine spin, is a conserved quantum number. When atoms are confined in a magnetic potential, the spin of each atom points in the direction of an external magnetic field. The spin degrees of freedom are therefore frozen and the mean-field properties of the system are described by a scalar order parameter. When the system is confined in an optical trap, the frozen degrees of freedom are liberated, yielding a rich variety of phenomena arising from the magnetic moment of the atom. Since the magnetic moments of alkali atoms originate primarily from the electronic spin, this system's response to an external magnetic field is much greater than that of superfluid helium-3. We can expect interesting interplay between superfluidity and magnetism with the possibility of new ground states, spin domains, and vortex structures. Spinor condensates are discussed in Chapter 6.

When the rotational speed of the container of the system is faster than the critical frequency, vortices enter the system and form a vortex lattice. The direct observation of vortex lattice formation [Madison, *et al.* (2000); Abo-Shaeer, *et al.* (2001)] has attracted considerable interest in the equilibrium and nonequilibrium dynamics of condensates. The effect of rotation on neutral particles is equivalent to that of a magnetic field on charged particles. Therefore, the properties of a vortex lattice of neutral particles are similar to those of superconductors. Furthermore, it is pointed out that in systems containing neutral bosons that are subject to very fast rotation, the vortex lattice melts, and a new vortex liquid state similar to the Laughlin state in the fractional quantum Hall system may be realized. A brief overview of these subjects is presented in Chapter 7.

Almost every bosonic atom has its fermionic counterpart. Fermions and bosons of the same species exhibit the same properties at high temperature, but they exhibit remarkably different behavior when quantum degeneracy sets in. Bosons undergo BEC below the transition temperature; in contrast, fermions become degenerate below the Fermi temperature, where almost every quantum state below the Fermi energy is occupied by one fermion and most quantum states above the Fermi energy are empty. At even lower temperatures, fermionic systems may exhibit superfluidity by forming Cooper pairs via the Bardeen–Cooper–Schrieffer transition. This is a rapidly developing field that has relevance to high-temperature superconductivity. We describe the basics and some of the recent developments of ultracold fermionic systems in Chapter 8.

It is known that BEC does not occur at finite temperature in one- or two-dimensional infinite systems because thermal fluctuations destroy the off-diagonal long-range order (ODLRO). In one-dimensional systems, BEC does not occur even at absolute zero because quantum fluctuations wash out the ODLRO. However, confined low dimensional systems can exhibit BEC because long-wavelength fluctuations are cut off by confinement. We may thus investigate interesting phenomena associated with low-dimensional BEC, such as solitons and the Berezinskii–Kosterlitz–Thouless transition. These subjects are discussed in Chapter 9.

Atoms with magnetic moments and polar molecules undergo dipole–dipole interactions, which are long-ranged and anisotropic and yield a wealth of novel phenomena. The magnetic dipole–dipole interaction is by far the weakest of the relevant interactions in cold atom systems; yet it plays a dominant role in forming spin textures and magnetic ordering and produces a spectacular effect in the course of the collapsing dynamics. The electric dipole–dipole interaction between polar molecules, in contrast, is very strong and may cause instabilities of the system; at the same time, it has the potential to yield several exotic phases and for use in quantum information processing. Some basic properties of the dipolar condensates are reviewed in Chapter 10.

An optical lattice is a periodic potential created by interference between two counterpropagating laser beams. Atoms in an optical lattice behave like electrons in a crystal. An optical lattice can host bosons as well as fermions, and it offers an ideal testbed to simulate quantum many-body physics and quantum information processing. Chapter 11 provides a brief overview of some basic properties of this artificial condensed matter system.

Superfluids host a rich variety of topological defects such as vortices, monopoles, and skyrmions. Those topological excitations are best described by the homotopy theory. Chapter 12 is devoted to an introduction of the homotopy theory, classfication of topological excitations, and an account of am how to calculate various topological charges.

Fifteen years after its first experimental realization, the field of ultracold atomic gases is still growing at a remarkable speed, such that coverage of every topic of importance far exceeds the range of this or perhaps any book. Rather, I have chosen a small number of important issues and tried to discuss their physical aspects as engagingly as possible. Many of the phenomena that have been observed in the past decade and those that will possibly be observed in the near future are of fundamental importance because of the very fact that they are being "seen" on a macroscopic scale.

If this book succeeds in conveying even a portion of the fascination inherent in this field, it will have well served its intended purpose.

This book derives from a set of lecture notes delivered at several universities over the past decade or so. I have benefited greatly from students and colleagues who actively participated in the class and collaboration. Special thanks are due to Rina Kanamoto, Yuki Kawaguchi, Michikazu Kobayashi, Tony Leggett, Hiroki Saito, and Masaki Tezuka. I would like to thank all of them for their questions, comments, and criticisms that helped me clarify my thoughts and improve the presentation of the material in this book. I am grateful to A. Koda, Y. Ookawara, and A. Yoshida for their efficient editing and preparation of the figures.

March 2010
Tokyo
Masahito Ueda

Revisions and corrections will be posted on:
`http://cat.phys.s.u-tokyo.ac.jp/~ueda/E_kyokasyo.html/`

Contents

Chapter 1

Fundamentals of Bose–Einstein Condensation

1.1 Indistinguishability of Identical Particles

Quantum statistics is governed by the principle of indistinguishability of identical particles. Particles with integer (half-integer) spin (in multiples of $\hbar$, where $\hbar$ is the Planck constant divided by 2π) are called bosons (fermions). Bosons obey Bose–Einstein statistics in which there is no restriction on the occupation number of any single-particle state. Fermions obey Fermi–Dirac statistics in which not more than one particle can occupy any single-particle state. The many-body wave function of identical bosons (fermions) must be symmetric (antisymmetric) under the exchange of any two particles. This symmetry requirement drastically reduces the number of available quantum states of the system, resulting in highly nonclassical phenomena at low temperature.

To understand this, let us suppose that we obtain a wave function $\Phi(\xi_1, \xi_2)$ of a two-particle system by solving the Schrödinger equation, where ξ_1 and ξ_2 represent the space and possibly spin coordinates of the two particles. For identical bosons (fermions), the symmetrized (antisymmetrized) wave function is given by

$$\Psi(\xi_1, \xi_2) = \frac{1}{\sqrt{2}}[\Phi(\xi_1, \xi_2) \pm \Phi(\xi_2, \xi_1)], \tag{1.1}$$

where the plus (minus) sign indicates bosons (fermions). The joint probability of finding the two particles at ξ_1 and ξ_2 is given by

$$\begin{aligned} |\Psi(\xi_1, \xi_2)|^2 = &\frac{1}{2}\{|\Phi(\xi_1, \xi_2)|^2 + |\Phi(\xi_2, \xi_1)|^2 \\ &\pm 2\mathrm{Re}[\Phi^*(\xi_1, \xi_2)\Phi(\xi_2, \xi_1)]\}, \end{aligned} \tag{1.2}$$

where Re denotes the real part. Because of the last interference term in Eq. (1.2), the probability of finding the two identical bosons at the same coordinate, $|\Psi(\xi,\xi)|^2$, is twice as high as $|\Phi(\xi,\xi)|^2$, which gives the corresponding probability for distinguishable particles. In contrast, for fermions, $|\Psi(\xi,\xi)|^2$ vanishes in accordance with Pauli's exclusion principle.

Such a bunching effect of bosons becomes increasingly pronounced when the number of bosons is large. For N number of bosons, the symmetrized wave function is given by

$$\Psi(\xi_1,\xi_2,\cdots,\xi_N)=\frac{1}{\sqrt{N!}}\sum_{(i_1,i_2,\cdots,i_N)}\Phi(\xi_{i_1},\xi_{i_2},\cdots,\xi_{i_N}), \tag{1.3}$$

where the summation over $i_1,i_2,\cdots,i_N$ is to be taken over all $N!$ permutations of $1,2,\cdots,N$. The joint probability of finding all N bosons at the same coordinate is thus $N!$ times that for distinguishable bosons, $|\Phi(\xi,\xi,\cdots,\xi)|^2$, due to the constructive interference of the permuted probability amplitudes:

$$|\Psi(\xi,\xi,\cdots,\xi)|^2=N!|\Phi(\xi,\xi,\cdots,\xi)|^2. \tag{1.4}$$

The constructive interference of the probability amplitudes is effective only when the wave packets of bosons overlap each other. At temperature T, each wave packet has a spatial extent of $\Delta x\sim\hbar/\sqrt{Mk_{\rm B}T}$, where M is the mass of the boson and $k_{\rm B}$ is the Boltzmann constant. By setting Δx equal to the average interparticle distance $n^{-\frac{1}{3}}$, where n is the particle number density, we can estimate the transition temperature T_0 of Bose–Einstein condensation (BEC) to be

$$k_{\rm B}T_0\sim\frac{\hbar^2}{M}n^{\frac{2}{3}}. \tag{1.5}$$

Because of the large enhancement factor of $N!$ in Eq. (1.4), a large number of particles suddenly begin to condense into a single-particle state below T_0. When N is macroscopic, the onset of this condensation becomes prominent, endowing BEC with a conspicuous trait of quantum phase transition. Substituting $n=N/V$, where V is the volume of the system, in Eq. (1.5) gives

$$k_{\rm B}T_0\sim\frac{\hbar^2}{MV^{\frac{2}{3}}}N^{\frac{2}{3}}. \tag{1.6}$$

Here, $\hbar^2/(MV^{\frac{2}{3}})$ gives an estimate of the energy gap between the ground state and the first excited state. Classical particles would condense into the ground state below the corresponding temperature $T_{\rm cl}\sim\hbar^2(k_{\rm B}MV^{\frac{2}{3}})$. Equation (1.6) shows that BEC occurs at a considerably higher temperature; further, the large enhancement factor $N^{\frac{2}{3}}$ can be attributed to the interference effect as discussed above. A more quantitative treatment described in Sec. 1.2 will validate Eq. (1.5).

1.2 Ideal Bose Gas in a Uniform System

The grand partition function Ξ of a system of particles with the Hamiltonian $\hat{H}$ and particle-number operator $\hat{N}$ is given by

$$\Xi = \mathrm{Tr} e^{-\beta(\hat{H}-\mu\hat{N})}, \tag{1.7}$$

where $\beta \equiv (k_{\mathrm{B}}T)^{-1}$, Tr denotes a trace operation, and the chemical potential μ serves as a Lagrange multiplier that is to be determined so as to fix the average number of particles to a prescribed value.

For ideal (*i.e.*, noninteracting) identical bosons with the dispersion relation $\epsilon_{\mathbf{k}} = \hbar^2\mathbf{k}^2/2M$, $\hat{H} - \mu\hat{N}$ is given by

$$\hat{H} - \mu\hat{N} = \sum_{\mathbf{k}} (\epsilon_{\mathbf{k}} - \mu)\hat{n}_{\mathbf{k}}, \tag{1.8}$$

where $\hat{n}_{\mathbf{k}}$ denotes the number operator of particles with wave vector $\mathbf{k}$. Substituting Eq. (1.8) in Eq. (1.7) gives

$$\Xi = \prod_{\mathbf{k}} \sum_{n_{\mathbf{k}}=0}^{\infty} (e^{\beta(\mu-\epsilon_{\mathbf{k}})})^{n_{\mathbf{k}}}. \tag{1.9}$$

For the geometric series in Eq. (1.9) to converge, $e^{\beta(\mu-\epsilon_{\mathbf{k}})}$ must be less than one. It follows from $\epsilon_k \geq 0$ that

$$\mu < 0. \tag{1.10}$$

Then, Eq. (1.9) gives

$$\Xi = \prod_{\mathbf{k}} \frac{1}{1 - e^{\beta(\mu-\epsilon_{\mathbf{k}})}}.$$

The thermodynamic potential Ω is defined in terms of Ξ as

$$\Omega \equiv -\frac{1}{\beta}\ln\Xi = \frac{1}{\beta}\sum_{\mathbf{k}} \ln(1 - e^{\beta(\mu-\epsilon_{\mathbf{k}})}) = \sum_{\mathbf{k}} \Omega_{\mathbf{k}}, \tag{1.11}$$

where

$$\Omega_{\mathbf{k}} = \frac{1}{\beta}\ln(1 - e^{\beta(\mu-\epsilon_{\mathbf{k}})}). \tag{1.12}$$

The average number of particles with wave vector $\mathbf{k}$ is given by

$$\bar{n}_{\mathbf{k}} = -\frac{\partial\Omega_{\mathbf{k}}}{\partial\mu} = \frac{1}{e^{\beta(\epsilon_{\mathbf{k}}-\mu)} - 1}, \tag{1.13}$$

which is referred to as the Bose–Einstein distribution function. The average total number of bosons is expressed in terms of the chemical potential μ as

$$N = \sum_{\mathbf{k}} \frac{1}{e^{\beta(\epsilon_{\mathbf{k}}-\mu)} - 1}. \tag{1.14}$$

For a given N, μ is determined such that it satisfies Eq. (1.14).

In the thermodynamic limit in which both N and V are made infinite with the particle number density N/V held constant, the sum over $\mathbf{k}$ may be replaced with the following integral[1]:

$$\sum_{\mathbf{k}} \to \frac{V}{(2\pi)^3} \int d^3k. \tag{1.15}$$

Equation (1.14) then gives

$$\frac{N}{V} = \frac{1}{(2\pi)^3} \int d^3k \frac{1}{e^{\beta(\epsilon_{\mathbf{k}}-\mu)} - 1}. \tag{1.16}$$

When the temperature is reduced while maintaining N/V constant, μ increases and eventually becomes zero at a certain temperature T_0. Substituting $\mu = 0$ and $\epsilon_{\mathbf{k}} = \hbar^2\mathbf{k}^2/2M$ in Eq. (1.16) yields

$$\begin{aligned}\frac{N}{V} &= \frac{(Mk_{\mathrm{B}}T_0)^{3/2}}{\sqrt{2}\pi^2\hbar^3} \int_0^\infty \frac{\sqrt{x}}{e^x - 1} dx \\ &= \zeta\left(\frac{3}{2}\right)\left(\frac{Mk_{\mathrm{B}}T_0}{2\pi\hbar^2}\right)^{\frac{3}{2}} = 2.612 \left(\frac{Mk_{\mathrm{B}}T_0}{2\pi\hbar^2}\right)^{\frac{3}{2}},\end{aligned} \tag{1.17}$$

where $\zeta(x)$ is the Riemann zeta function, and the following formulae are used:

$$\int_0^\infty \frac{x^{a-1}}{e^x - 1} dx = \Gamma(a)\zeta(a),\ \zeta\left(\frac{3}{2}\right) = 2.612,\ \Gamma\left(\frac{3}{2}\right) = \frac{\sqrt{\pi}}{2}. \tag{1.18}$$

From Eq. (1.17), the transition temperature T_0 of BEC is given by

$$k_{\mathrm{B}}T_0 = \frac{2\pi}{[\zeta(3/2)]^{2/3}} \frac{\hbar^2}{M} \left(\frac{N}{V}\right)^{\frac{2}{3}} = 3.31 \frac{\hbar^2}{M} \left(\frac{N}{V}\right)^{\frac{2}{3}}, \tag{1.19}$$

which is in agreement with Eq. (1.5). For $T < T_0$, a nonzero fraction of bosons should therefore remain in the ground state; *i.e.*, they condense into the lowest-energy state. For $T < T_0$, the replacement of the sum with the integral (Eq. 1.15) is applicable only for particles with positive energy $\epsilon > 0$, since the particles with $\epsilon = 0$ cannot contribute to the integral in Eq. (1.17) because of the factor $\sqrt{x}$ in the integrand. From Eq. (1.17), we find that T_0 is related to the particle-number density N/V through the relation

$$\frac{N}{V} = \zeta\left(\frac{3}{2}\right)\left(\frac{Mk_{\mathrm{B}}T_0}{2\pi\hbar^2}\right)^{\frac{3}{2}}. \tag{1.20}$$

[1] In the presence of spin multiplicity $g = 2S + 1$, where $\hbar S$ is the spin of a boson, the following results hold true if we replace V by gV.

For $T < T_0$, we have

$$\frac{N_{\epsilon>0}}{V} = \zeta\left(\frac{3}{2}\right)\left(\frac{Mk_BT}{2\pi\hbar^2}\right)^{\frac{3}{2}}. \tag{1.21}$$

From Eqs. (1.20) and (1.21), it follows that

$$\frac{N_{\epsilon>0}}{N} = \left(\frac{T}{T_0}\right)^{\frac{3}{2}}. \tag{1.22}$$

This quantity is referred to as the normal fraction. Hence, the condensate fraction is given by

$$\frac{N_{\epsilon=0}}{N} = 1 - \left(\frac{T}{T_0}\right)^{\frac{3}{2}}. \tag{1.23}$$

BEC occurs when the de Broglie waves of individual bosons begin to overlap, *i.e.*, when the quantum degeneracy sets in. The thermal de Broglie length $\lambda_{\rm th}$ is conventionally defined as

$$k_BT = \frac{1}{2\pi M}\left(\frac{h}{\lambda_{\rm th}}\right)^2 \to \lambda_{\rm th} = \frac{h}{\sqrt{2\pi Mk_BT}}. \tag{1.24}$$

Here, $\lambda_{\rm th}$ characterizes the spatial extension of a wave packet of an individual boson. Substituting this in Eq. (1.20) gives

$$n\lambda_{\rm th}^3 = \zeta\left(\frac{3}{2}\right) \simeq 2.612 \quad \text{at } T = T_0. \tag{1.25}$$

Thus, the thermal de Broglie length at the transition temperature is on the order of the average interparticle distance $n^{-\frac{1}{3}}$. The quantity $n\lambda_{\rm th}^3$ is referred to as *phase-space density*. Equation (1.25) shows that an ideal Bose gas undergoes BEC at a phase-space density of 2.612.

At the transition point, the specific heat of an ideal Bose gas at constant volume is continuous and its derivative is discontinuous. Therefore, the BEC of an ideal Bose gas at constant volume is a third-order phase transition[2]. The specific heat of liquid ^{4}He shows a discontinuous jump at the lambda point (T = 2.17 K), which indicates that the superfluid transition of liquid ^{4}He is a second-order phase transition. By studying the similarity between the behavior of the specific heat of liquid ^{4}He near the lambda point and that of an ideal Bose gas, Fritz London found that BEC plays an essential role in both superfluidity and superconductivity [London (1938)]. In the several decades after London's seminal work, the physics community has gradually acknowledged the special role of BEC in superfluidity.

[2]If we consider the state of the system as a function of pressure P and volume at constant temperature, *i.e.*, the isotherm, P becomes constant below a certain volume, where the Bose–Einstein condensate coexists with the normal component in a manner analogous to gas-liquid transition. In this situation, BEC may be considered to be a first-order phase transition. Refer to K. Huang, *Statistical Mechanics*, 2nd edition (John Wiley & Sons, New York, 1987).

1.3 Off-Diagonal Long-Range Order: Bose System

The essence of BEC is the off-diagonal long-range order (ODLRO). We first explain the concept of the ODLRO using a simple example and then provide its general definition.

A system is said to possess an ODLRO if the single-particle density matrix

$$\rho_1(\mathbf{r},\mathbf{r}') \equiv \mathrm{Tr}\{\hat{\rho}\hat{\psi}^\dagger(\mathbf{r})\hat{\psi}(\mathbf{r}')\} \equiv \langle\hat{\psi}^\dagger(\mathbf{r})\hat{\psi}(\mathbf{r}')\rangle \tag{1.26}$$

has a large eigenvalue, *i.e.*, an eigenvalue proportional to the total number of particles N, where $\hat{\rho}$ is the density operator of the system and $\hat{\psi}^\dagger(\mathbf{r})$ ($\hat{\psi}(\mathbf{r}')$) is the field operator that creates (annihilates) a particle at $\mathbf{r}$ ($\mathbf{r}'$). Since $\hat{\rho}$ is Hermitian, $\rho_1(\mathbf{r},\mathbf{r}')$ is a Hermitian matrix. When the system is in a pure state $|\Phi\rangle$, Eq. (1.26) reduces to

$$\rho_1(\mathbf{r},\mathbf{r}') = \langle\Phi|\hat{\psi}^\dagger(\mathbf{r})\hat{\psi}(\mathbf{r}')|\Phi\rangle. \tag{1.27}$$

This expression implies that the single-particle density matrix gives the probability amplitude that the quantum state of the system remains unperturbed if a particle is removed from the system at $\mathbf{r}'$ and added to it at $\mathbf{r}$. Under normal conditions, $\rho_1(\mathbf{r},\mathbf{r}')$ decreases exponentially with increasing $|\mathbf{r}-\mathbf{r}'|$ [see Eq. (1.48)]. When the system undergoes BEC, the de Broglie waves of individual bosons overlap, and thus a particle at $\mathbf{r}'$ becomes indistinguishable from a particle at $\mathbf{r}$. As a consequence, $\rho_1(\mathbf{r},\mathbf{r}')$ does not vanish over a long distance $|\mathbf{r}-\mathbf{r}'|$. If this condition holds, the system is said to maintain spatial coherence over a long distance. As shown below, the system possesses an ODLRO when $\rho_1(\mathbf{r},\mathbf{r}')$ remains on the order of N/V as $|\mathbf{r}-\mathbf{r}'|$ increases. This shows that a particle can travel a long distance without disturbing the state of the system. In this respect, the ODLRO bears a close relation with superfluidity. When $\mathbf{r}=\mathbf{r}'$, Eq. (1.26) gives the particle number density.

Let us first consider a spatially uniform system. In this case, it is convenient to expand the field operator $\hat{\psi}(\mathbf{x})$ in terms of plane waves as follows:

$$\hat{\psi}(\mathbf{x}) = \frac{1}{\sqrt{V}}\sum_{\mathbf{k}}\hat{a}_{\mathbf{k}}e^{i\mathbf{k}\mathbf{x}}, \tag{1.28}$$

where $\hat{a}_{\mathbf{k}}$ is the annihilation operator of bosons with wave vector $\mathbf{k}$. We assume that $\hat{a}_{\mathbf{k}}$ satisfies the boson commutation relation

$$\left[\hat{a}_{\mathbf{k}},\hat{a}_{\mathbf{q}}^\dagger\right] = \delta_{\mathbf{kq}}. \tag{1.29}$$

The single-particle density matrix is then given by

$$\rho_1(\mathbf{x},\mathbf{y}) = \frac{1}{V}\sum_{\mathbf{k},\mathbf{q}}\langle\hat{a}^\dagger_{\mathbf{k}}\hat{a}_{\mathbf{q}}\rangle e^{-i(\mathbf{kx}-\mathbf{qy})}. \tag{1.30}$$

Here, we note that $\langle\hat{a}^\dagger_{\mathbf{k}}\hat{a}_{\mathbf{q}}\rangle = \delta_{\mathbf{kq}}\langle\hat{a}^\dagger_{\mathbf{k}}\hat{a}_{\mathbf{k}}\rangle$ holds for a translationally invariant system. To prove this, let us calculate the commutation relation between $\hat{a}^\dagger_{\mathbf{k}}\hat{a}_{\mathbf{q}}$ and the momentum operator $\hat{\mathbf{P}} = \sum_{\mathbf{p}}\hbar\mathbf{p}\hat{a}^\dagger_{\mathbf{p}}\hat{a}_{\mathbf{p}}$, given by

$$\left[\hat{\mathbf{P}},\hat{a}^\dagger_{\mathbf{k}}\hat{a}_{\mathbf{q}}\right] = \sum_{\mathbf{p}}\hbar\mathbf{p}\left[\hat{a}^\dagger_{\mathbf{p}}\hat{a}_{\mathbf{p}},\hat{a}^\dagger_{\mathbf{k}}\hat{a}_{\mathbf{q}}\right] = \hbar(\mathbf{k}-\mathbf{q})\hat{a}^\dagger_{\mathbf{k}}\hat{a}_{\mathbf{q}}. \tag{1.31}$$

Taking the thermal average of this quantity, we obtain

$$\begin{aligned}\left\langle\left[\hat{\mathbf{P}},\hat{a}^\dagger_{\mathbf{k}}\hat{a}_{\mathbf{q}}\right]\right\rangle &= Z^{-1}\mathrm{Tr}\left\{e^{-\beta\hat{H}}\left(\hat{\mathbf{P}}\hat{a}^\dagger_{\mathbf{k}}\hat{a}_{\mathbf{q}} - \hat{a}^\dagger_{\mathbf{k}}\hat{a}_{\mathbf{q}}\hat{\mathbf{P}}\right)\right\}\\ &= Z^{-1}\mathrm{Tr}\left\{e^{-\beta\hat{H}}\hat{\mathbf{P}}\hat{a}^\dagger_{\mathbf{k}}\hat{a}_{\mathbf{q}} - \hat{\mathbf{P}}e^{-\beta\hat{H}}\hat{a}^\dagger_{\mathbf{k}}\hat{a}_{\mathbf{q}}\right\},\end{aligned} \tag{1.32}$$

where $Z \equiv \mathrm{Tr}e^{-\beta\hat{H}}$ and the cyclic property of the trace,

$$\mathrm{Tr}(\hat{A}\hat{B}) = \mathrm{Tr}(\hat{B}\hat{A}),$$

is used in deriving the last equality. Since $[\hat{H},\hat{\mathbf{P}}] = 0$ for a spatially uniform system, the last term in Eq. (1.32) vanishes; thus,

$$\left\langle\left[\hat{\mathbf{P}},\hat{a}^\dagger_{\mathbf{k}}\hat{a}_{\mathbf{q}}\right]\right\rangle = 0.$$

Hence,

$$(\mathbf{k}-\mathbf{q})\langle\hat{a}^\dagger_{\mathbf{k}}\hat{a}_{\mathbf{q}}\rangle = 0, \tag{1.33}$$

which implies that

$$\langle\hat{a}^\dagger_{\mathbf{k}}\hat{a}_{\mathbf{q}}\rangle = \delta_{\mathbf{kq}}\langle\hat{a}^\dagger_{\mathbf{k}}\hat{a}_{\mathbf{k}}\rangle = \delta_{\mathbf{kq}}\langle\hat{n}_{\mathbf{k}}\rangle, \tag{1.34}$$

where $\hat{n}_{\mathbf{k}} \equiv \hat{a}^\dagger_{\mathbf{k}}\hat{a}_{\mathbf{k}}$ is the number operator.

Substituting Eq. (1.34) in Eq. (1.30) gives

$$\rho_1(\mathbf{x},\mathbf{y}) = \frac{1}{V}\sum_{\mathbf{k}}\langle\hat{n}_{\mathbf{k}}\rangle e^{-i\mathbf{k}(\mathbf{x}-\mathbf{y})} = \frac{\langle\hat{n}_0\rangle}{V} + \int\frac{d^3k}{(2\pi)^3}\langle\hat{n}_{\mathbf{k}}\rangle e^{-i\mathbf{k}(\mathbf{x}-\mathbf{y})}. \tag{1.35}$$

The last term vanishes in the limit $|\mathbf{x}-\mathbf{y}| \to \infty$ because of the rapidly oscillating term $e^{-i\mathbf{k}(\mathbf{x}-\mathbf{y})}$ (Riemann–Lebesgue lemma). Consequently,

$$\rho_1(\mathbf{x},\mathbf{y}) \to \frac{\langle\hat{n}_0\rangle}{V} \quad \text{as} \quad |\mathbf{x}-\mathbf{y}| \to \infty. \tag{1.36}$$

This result shows that the system exhibits the off-diagonal (*i.e.*, $\mathbf{x} \neq \mathbf{y}$) long-range order in the thermodynamic limit if and only if an extensive number of bosons (proportional to the volume) condense into a state with zero momentum. This shows that the ODLRO implies BEC.

Single-particle energy levels are not well defined in the presence of an interparticle interaction. However, the following reduced single-particle density operator is well defined in this case:

$$\hat{\rho}_1 \equiv \mathrm{Tr}_{2,3,\cdots,N}\hat{\rho}, \tag{1.37}$$

where $\mathrm{Tr}_{2,3,\cdots,N}$ denotes the trace over particles $2, 3, \cdots, N$; it should be noted that because bosons are identical, we can choose arbitrary $N-1$ particles without loss of generality. Let n_{M} be the maximum eigenvalue of

$$\hat{\sigma}_1 \equiv N\hat{\rho}_1. \tag{1.38}$$

The condition for a Bose–Einstein condensate to exist can be formulated as follows [Penrose and Onsager (1956)][3]:

$$\frac{n_{\mathrm{M}}}{N} = e^{O(1)}. \tag{1.39}$$

This definition is applicable irrespective of the presence or absence of interactions. It is also applicable when the system is not uniform. The single-particle density matrix can also be expressed in terms of the reduced single-particle density operator $\hat{\rho}_1$ as follows:

$$\rho_1(\mathbf{x},\mathbf{y}) = \langle\hat{\psi}^\dagger(\mathbf{x})\hat{\psi}(\mathbf{y})\rangle = \int d\mathbf{z}\langle\mathbf{z}|\hat{\rho}_1\hat{\psi}^\dagger(\mathbf{x})\hat{\psi}(\mathbf{y})|\mathbf{z}\rangle \tag{1.40}$$

$$= \langle\mathbf{y}|\hat{\rho}_1\hat{\psi}^\dagger(\mathbf{x})|0\rangle = \langle\mathbf{y}|\hat{\rho}_1|\mathbf{x}\rangle, \tag{1.41}$$

where $\hat{\psi}(\mathbf{y})|\mathbf{z}\rangle = \delta(\mathbf{y}-\mathbf{z})|0\rangle$ was used in deriving the third equality.

When the system is spatially uniform, the momentum is a good quantum number. Therefore,

$$\langle\mathbf{p}|\hat{\rho}_1|\mathbf{p}'\rangle \propto \delta(\mathbf{p}-\mathbf{p}'). \tag{1.42}$$

In this case, the condition of BEC is that a macroscopic number of particles occupy the same single-particle momentum state. When the system is not spatially uniform, the condition is given by

$$\rho_1(\mathbf{x},\mathbf{y}) \longrightarrow \psi^*(\mathbf{x})\psi(\mathbf{y}) \;\; \text{as} \;\; |\mathbf{x}-\mathbf{y}| \to \infty, \tag{1.43}$$

where $\psi(\mathbf{x})$, to a very good approximation, is an eigenfunction of the single-particle reduced density matrix $\rho_1(\mathbf{x},\mathbf{y})$:

$$\int d\mathbf{x}\hat{\rho}_1(\mathbf{x},\mathbf{y})\psi(\mathbf{x}) \simeq n_{\mathrm{M}}\psi(\mathbf{y}), \quad n_{\mathrm{M}} = \int d\mathbf{x}|\psi(\mathbf{x})|^2. \tag{1.44}$$

[3] $e^{O(1)}$ is a *positive* number of the order of unity.

Here, $\psi(\mathbf{x})$ in Eq. (1.43) is often referred to as the *condensate wave function* or the *order parameter* and $n_{\rm M}$ is the number of condensed bosons. The ratio $n_{\rm M}/N$ is referred to as the condensate fraction. We note that if $\psi(\mathbf{x})$ is a solution to the eigenvalue equation (1.44), $\psi(\mathbf{x})e^{i\phi}$ is also a solution to it, where ϕ is an arbitrary real number. The global phase of the condensate wave function is therefore arbitrary.

Comparing Eq. (1.43) with Eq. (1.36), we find that the condensate wave function is a thermodynamic quantity that appears in the thermodynamic limit. The significance of BEC is thus to generate a new thermodynamic variable that is a macroscopic wave function representing the ODLRO. With the macroscopic wave function, it is possible to describe coherent properties of a many-body system without referring to the microscopic details of the system.

For comparison, let us consider the classical limit of Eq. (1.35). In the thermodynamic limit at $T > T_0$, we have $\langle \hat{n}_0 \rangle / V = 0$. Hence, Eq. (1.35) reduces to

$$\rho_1(\mathbf{x}, \mathbf{y}) = \int \frac{d^3k}{(2\pi)^3} \langle \hat{n}_{\mathbf{k}} \rangle e^{-i\mathbf{k}(\mathbf{x}-\mathbf{y})}, \tag{1.45}$$

where

$$\langle \hat{n}_{\mathbf{k}} \rangle \simeq e^{\beta(\mu - \hbar^2 \mathbf{k}^2/2M)}.$$

The chemical potential μ is determined such that $\rho_1(\mathbf{x}, \mathbf{x})$ gives the total particle-number density n:

$$n = \int \frac{d^3k}{(2\pi)^3} \langle \hat{n}_{\mathbf{k}} \rangle = \frac{e^{\beta\mu}}{(2\pi)^3} \left(\int dk e^{-\beta \frac{\hbar^2 k^2}{2M}} \right)^3 = e^{\beta\mu} \left(\frac{M}{2\pi\beta\hbar^2} \right)^{\frac{3}{2}}. \tag{1.46}$$

Hence,

$$\mu = \frac{1}{\beta} \ln \left(n \lambda_{\rm th}^3 \right). \tag{1.47}$$

Substituting this in Eq. (1.46), we obtain

$$\rho_1(\mathbf{x}, \mathbf{y}) = \frac{e^{\beta\mu}}{(2\pi)^3} \int d^3k e^{-\frac{\beta \mathbf{k}^2}{2M} - i\mathbf{k}(\mathbf{x}-\mathbf{y})} = n e^{-\frac{M|\mathbf{x}-\mathbf{y}|^2}{2\beta\hbar^2}}; \tag{1.48}$$

we thus find that only the diagonal (*i.e.*, $\mathbf{x} = \mathbf{y}$) order remains nonvanishing in the high-temperature limit ($\beta \to 0$).

1.4 Off-Diagonal Long-Range Order: Fermi System

In the case of a Fermi system, the commutator on the left-hand side of (1.29) is replaced with the anti-commutator as follows:

$$\{\hat{c}_{\mathbf{k}\sigma}, \hat{c}^{\dagger}_{\mathbf{q}\sigma'}\} \equiv \hat{c}_{\mathbf{k}\sigma}\hat{c}^{\dagger}_{\mathbf{q}\sigma'} + \hat{c}^{\dagger}_{\mathbf{q}\sigma'}\hat{c}_{\mathbf{k}\sigma} = \delta_{\mathbf{kq}}\delta_{\sigma\sigma'}, \tag{1.49}$$

where $\hat{c}^{\dagger}_{\mathbf{q}\sigma'}$ $(\hat{c}_{\mathbf{k}\sigma})$ is the creation (annihilation) operator of a fermion with wave number $\mathbf{q}(\mathbf{k})$ and spin $\sigma'(\sigma)$. As in the case of bosons [see Eq. (1.34)], it can be shown that for a translation-invariant system

$$\langle \hat{c}^{\dagger}_{\mathbf{k}\sigma}\hat{c}_{\mathbf{q}\sigma'}\rangle = \delta_{\mathbf{kq}}\delta_{\sigma\sigma'}\langle \hat{c}^{\dagger}_{\mathbf{k}\sigma}\hat{c}_{\mathbf{k}\sigma}\rangle. \tag{1.50}$$

The single-particle reduced density matrix of a Fermi system is defined as

$$\rho_1(\mathbf{r}\sigma, \mathbf{r}'\sigma') = \langle \hat{\psi}^{\dagger}_{\sigma}(\mathbf{r})\hat{\psi}_{\sigma'}(\mathbf{r}')\rangle, \tag{1.51}$$

where $\hat{\psi}_{\sigma}(\mathbf{r})$ is the field operator of a fermion with spin σ at position $\mathbf{r}$. Substituting the Fourier expansion of $\hat{\psi}_{\sigma}(\mathbf{r})$,

$$\hat{\psi}_{\sigma}(\mathbf{r}) = \frac{1}{\sqrt{V}}\sum_{\mathbf{k}} \hat{c}_{\mathbf{k}\sigma}e^{i\mathbf{kr}}, \tag{1.52}$$

into (1.51) and using (1.50), we obtain

$$\rho_1(\mathbf{r}\sigma, \mathbf{r}'\sigma') = \int \frac{d^3k}{(2\pi)^3}\langle \hat{c}^{\dagger}_{\mathbf{k}\sigma}\hat{c}_{\mathbf{k}\sigma'}\rangle e^{-i\mathbf{k}(\mathbf{r}-\mathbf{r}')}. \tag{1.53}$$

Due to the Pauli exclusion principle, the occupation number of any single-particle state cannot exceed unity, and therefore, the term corresponding to the first term on the right-hand side of (1.35) does not appear in (1.53). At absolute zero,

$$\langle \hat{c}^{\dagger}_{\mathbf{k}\sigma}\hat{c}_{\mathbf{k}\sigma'}\rangle = \delta_{\sigma\sigma'}\theta(k_{\mathrm{F}} - |\mathbf{k}|), \tag{1.54}$$

where k_{F} is the Fermi wave number. Substituting (1.54) into (1.53), we obtain

$$\rho_1(\mathbf{r}\sigma, \mathbf{r}'\sigma') = \frac{\delta_{\sigma\sigma'}}{2\pi^2 r^3}(\sin k_{\mathrm{F}}r - k_{\mathrm{F}}r\cos k_{\mathrm{F}}r), \tag{1.55}$$

where $r \equiv |\mathbf{r} - \mathbf{r}'|$. Since ρ_1 vanishes for $r \to \infty$, the single-particle density matrix does not show ODLRO but decays algebraically.

The Fermi system may show ODLRO at the two-particle level. To verify this, let us consider the two-particle reduced density matrix of a Fermi system:

$$\rho_2(\mathbf{r}_1\sigma_1, \mathbf{r}_2\sigma_2; \mathbf{r}'_1\sigma'_1, \mathbf{r}'_2\sigma'_2) \equiv \langle \hat{\psi}^{\dagger}_{\sigma_1}(\mathbf{r}_1)\hat{\psi}^{\dagger}_{\sigma_2}(\mathbf{r}_2)\hat{\psi}_{\sigma'_2}(\mathbf{r}'_2)\hat{\psi}_{\sigma'_1}(\mathbf{r}'_1)\rangle. \tag{1.56}$$

Substituting (1.52) into (1.56), we have

$$\rho_2(\mathbf{r}_1\sigma_1, \mathbf{r}_2\sigma_2; \mathbf{r}_1'\sigma_1', \mathbf{r}_2'\sigma_2') = \frac{1}{V^2}\sum_{\mathbf{k},\mathbf{k}'\mathbf{K}} \langle \hat{c}^\dagger_{\mathbf{k}+\frac{\mathbf{K}}{2},\sigma_1} \hat{c}^\dagger_{-\mathbf{k}+\frac{\mathbf{K}}{2},\sigma_2} \hat{c}_{-\mathbf{k}'+\frac{\mathbf{K}}{2},\sigma_2'} \hat{c}_{\mathbf{k}'+\frac{\mathbf{K}}{2},\sigma_1'} \rangle \times e^{i[\mathbf{k}'\mathbf{r}'-\mathbf{k}\mathbf{r}+\mathbf{K}(\mathbf{R}-\mathbf{R}')]}, \tag{1.57}$$

where $\mathbf{r} \equiv \mathbf{r}_2 - \mathbf{r}_1$, $\mathbf{r}' \equiv \mathbf{r}_2' - \mathbf{r}_1'$, $\mathbf{R} \equiv \frac{\mathbf{r}_1+\mathbf{r}_2}{2}$, and $\mathbf{R}' \equiv \frac{\mathbf{r}_1'+\mathbf{r}_2'}{2}$. In the limit of $|\mathbf{R}-\mathbf{R}'| \to \infty$, all terms except for the $\mathbf{K} = \mathbf{0}$ term vanish due to the rapidly oscillating factor $e^{i\mathbf{K}(\mathbf{R}-\mathbf{R}')}$. Thus,

$$\lim_{|\mathbf{R}-\mathbf{R}'|\to\infty} \rho_2(\mathbf{r}_1\sigma_1, \mathbf{r}_2\sigma_2; \mathbf{r}_1'\sigma_1', \mathbf{r}_2'\sigma_2') = \frac{1}{V^2}\sum_{\mathbf{k},\mathbf{k}'} \langle \hat{c}^\dagger_{\mathbf{k},\sigma_1} \hat{c}^\dagger_{-\mathbf{k},\sigma_2} \hat{c}_{-\mathbf{k}',\sigma_2'} \hat{c}_{\mathbf{k}',\sigma_1'} \rangle \times e^{i(\mathbf{k}'\mathbf{r}'-\mathbf{k}\mathbf{r})}. \tag{1.58}$$

In terms of the Cooper-pair operator defined as

$$\hat{\Psi}_{\sigma_2\sigma_1}(\mathbf{r}) \equiv \frac{1}{V}\sum_{\mathbf{k}} \hat{c}_{-\mathbf{k},\sigma_2}\hat{c}_{\mathbf{k},\sigma_1} e^{i\mathbf{k}\mathbf{r}}, \tag{1.59}$$

Eq. (1.58) is expressed as

$$\lim_{|\mathbf{R}-\mathbf{R}'|\to\infty} \rho_2(\mathbf{r}_1\sigma_1, \mathbf{r}_2\sigma_2; \mathbf{r}_1'\sigma_1', \mathbf{r}_2'\sigma_2') = \langle \hat{\Psi}^\dagger_{\sigma_2\sigma_1}(\mathbf{r}) \hat{\Psi}_{\sigma_2'\sigma_1'}(\mathbf{r}') \rangle. \tag{1.60}$$

If the left-hand side of Eq. (1.60) does not vanish, its right-hand side may be written as

$$\langle \hat{\Psi}^\dagger_{\sigma_2\sigma_1}(\mathbf{r}) \hat{\Psi}_{\sigma_2'\sigma_1'}(\mathbf{r}') \rangle = \Psi^*_{\sigma_2\sigma_1}(\mathbf{r}) \Psi_{\sigma_2'\sigma_1'}(\mathbf{r}'). \tag{1.61}$$

Thus, two-particle correlations are essential for a Fermi system to show ODLRO [Gor'kov (1958); Yang (1962)], and if ODLRO occurs, the quantity $\Psi_{\sigma_2\sigma_1}(\mathbf{r})$ defined in Eq. (1.61) is referred to as the order parameter or the macroscopic wave function of Cooper pairs.

1.5 $U(1)$ Gauge Symmetry

The concept of order parameter plays a key role in our understanding of the second-order phase transition, which is accompanied by a change in symmetry. As a typical example, let us consider the case of a ferromagnet, where the order parameter is magnetization, which is an observable representing a collective order of microscopic spins, and is coupled to another observable — the magnetic field. In contrast, BEC is unique in that the order parameter is a macroscopic wave function that is complex and is not an observable per se, because the phase of the wave function is arbitrary.

The arbitrariness of the phase reflects the symmetry, called the $U(1)$ gauge symmetry, that results from the conservation of the particle number.

In the mean field theory, it is convenient to break the $U(1)$ symmetry and let the order parameter have a definite phase ϕ. A mathematical trick to implement this is to add to the Hamiltonian a term H' that establishes correlations among phases of states having different particle numbers:

$$H' = \epsilon \int \left[e^{-i\phi}\hat{\psi}(\mathbf{r}) + e^{i\phi}\hat{\psi}^\dagger(\mathbf{r}) \right] d\mathbf{r}. \tag{1.62}$$

The phase of the order parameter is thus coupled to the non-Hermitian field operators $\hat{\psi}^\dagger$ and $\hat{\psi}$. If we operate the perturbation given by the integrand of H' at a local point in a normal fluid, the phase at that point is fixed, but the phase at a point located at a distance greater than the correlation length becomes random. However, when the temperature is lowered below the transition temperature, the phase of the system becomes spatially uniform. If we change the phase over space, the energy of the system increases by an amount $\kappa(\nabla\phi)^2$, where the positive coefficient κ is enhanced by a repulsive interaction. We may consider that the stability of a superfluid is a consequence of the rigidity of the macroscopic wave function due to the repulsive interaction.

While breaking of the $U(1)$ gauge symmetry greatly simplifies calculations of physical quantities, there is one conceptual difficulty here. The symmetry-breaking perturbation (1.62) brings the system in a superposition of states having different particle numbers. However, for massive particles, such a superposition state is precluded by the superselection rule [Haag (1996)]. Fortunately, we can understand BEC and superfluidity without breaking the $U(1)$ symmetry [see Secs. 1.3 and 1.7]; moreover, theories with and without the $U(1)$ gauge theories virtually yield the same results in the thermodynamic limit. However, significant differences may arise in the mesoscopic regime in which the number of particles is finite.

The condition for BEC is often stated as follows:

$$\rho_1(\mathbf{x}, \mathbf{y}) \xrightarrow{|\mathbf{x}-\mathbf{y}|\to\infty} \langle\hat{\psi}^\dagger(\mathbf{x})\rangle\langle\hat{\psi}(\mathbf{y})\rangle. \tag{1.63}$$

Here, the symbol $\langle\cdots\rangle$ should be interpreted, to a good approximation, as the expectation value between states in which only the numbers of condensate particles, n_{M}, differ by one and all the other quantum numbers represented by $\{\xi_i\}$ remain unaltered:[4]

$$\begin{aligned} \langle\hat{\psi}^\dagger(\mathbf{x})\rangle &= \langle n_{\mathrm{M}}, \{\xi_i\}|\hat{\psi}^\dagger(\mathbf{x})|n_{\mathrm{M}} - 1, \{\xi_i\}\rangle, \\ \langle\hat{\psi}(\mathbf{y})\rangle &= \langle n_{\mathrm{M}} - 1, \{\xi_i\}|\hat{\psi}(\mathbf{y})|n_{\mathrm{M}}, \{\xi_i\}\rangle. \end{aligned} \tag{1.64}$$

[4]If we would interpret $\langle\hat{\psi}^\dagger(\mathbf{x})\rangle$ as an expectation value over a state with a definite particle number, we would get $\langle\hat{\psi}^\dagger(\mathbf{x})\rangle = 0$. A nonzero value of $\langle\hat{\psi}(\mathbf{x})\rangle$ might be obtained

To show this, let us expand the field operator as

$$\hat{\psi}(\mathbf{x}) = \hat{a}_0 u_0(\mathbf{x}) + \sum_{\xi} \hat{a}_x u_\xi(\mathbf{x}),$$

where $u_0(\mathbf{x})$ is the mode function of the condensate and $u_\xi(\mathbf{x})$'s are non-condensate modes. Then,

$$\hat{\psi}(\mathbf{x})|n_{\mathrm{M}}, \{\xi_i\}\rangle = \sqrt{n_{\mathrm{M}}} u_0(\mathrm{x})|n_{\mathrm{M}} - 1, \{\xi_i\}\rangle + \sum_{\xi} \sqrt{n_\xi} u_\xi(\mathrm{x})|n_{\mathrm{M}}, \cdots, \xi_i - 1, \cdots\rangle.$$

Since $n_{\mathrm{M}} \gg n_\xi$, Eqs. (1.64) hold to a good approximation. Physically, this result implies that when a single-particle state is macroscopically occupied, it plays the role of a particle reservoir that can absorb or emit a particle with negligible ($\sim N^{-1}$) influence upon itself.

Let us now consider the expectation value of the time-dependent field operator $\hat{\psi}(\mathbf{x}, t) = e^{i\hat{H}t/\hbar}\hat{\psi}(\mathbf{x})e^{-i\hat{H}t/\hbar}$ in the Heisenberg representation, where $\hat{H}|n_{\mathrm{M}}, \{\xi_i\}\rangle = E_N|n_{\mathrm{M}}, \{\xi_i\}\rangle$ with N being the total number of particles. Then,

$$\langle n_{\mathrm{M}} - 1, \{\xi_i\}|\hat{\psi}(\mathbf{x}, t)|n_{\mathrm{M}}, \{\xi_i\}\rangle = \langle\hat{\psi}(\mathbf{x})\rangle e^{-i(E_N - E_{N-1})t/\hbar}. \tag{1.65}$$

Since for $N \gg 1$

$$E_N - E_{N-1} \simeq \frac{\partial E_N}{\partial N} = \mu, \tag{1.66}$$

where μ is the chemical potential, the time dependence of the condensate wave function is governed by the chemical potential:

$$\psi(\mathbf{x}, t) = \psi(\mathbf{x})e^{-i\mu t/\hbar}. \tag{1.67}$$

1.6 Ground-State Wave Function of a Bose System

The most fundamental property of the ground-state wave function (GSWF) Ψ_0 of a Bose system is that it can be taken to be real, nodeless, and nondegenerate.

To show that Ψ_0 can be taken to be real, we decompose Ψ_0 into amplitude $|\Psi_0|$ and phase χ:

$$\Psi_0(\mathbf{x}_1, \mathbf{x}_2, \cdots, \mathbf{x}_N; t) = |\Psi_0(\mathbf{x}_1, \mathbf{x}_2, \cdots, \mathbf{x}_N; t)|e^{i\chi(\mathbf{x}_1, \mathbf{x}_2, \cdots, \mathbf{x}_N; t)}. \tag{1.68}$$

---

if we would assume that the system is in a superposition of states having different particle numbers. However, this assumption, which is often made in literature, runs counter to the superselection rule, as described in the preceding paragraph.

If χ is constant, we can consider it as zero since the overall phase can be chosen arbitrarily. If it varies in space, the system shows a mass flow. To verify this, let us express the density of particles at position $\mathbf{r}$ in terms of Ψ_0:

$$\rho(\mathbf{r},t) = \int d\mathbf{x}_1 \cdots d\mathbf{x}_N |\Psi_0(\mathbf{x}_1, \cdots, \mathbf{x}_N; t)|^2 \sum_{k=1}^{N} \delta(\mathbf{r} - \mathbf{x}_k). \quad (1.69)$$

The mass current density $\mathbf{j}(\mathbf{r},t)$, which, together with $\rho(\mathbf{r},t)$, satisfies the equation of continuity

$$\frac{\partial \rho}{\partial t} + \nabla \mathbf{j} = 0, \quad (1.70)$$

is given by

$$\mathbf{j}(\mathbf{r},t) = \frac{\hbar}{2Mi} \int d\mathbf{x}_1 \cdots d\mathbf{x}_N \sum_{k=1}^{N} (\Psi_0^* \nabla_k \Psi_0 - \Psi_0 \nabla_k \Psi_0^*) \delta(\mathbf{r} - \mathbf{x}_k), \quad (1.71)$$

where ∇_k denotes differentiation with respect to $\mathbf{x}_k$. Substitution of Eq. (1.68) into Eq. (1.71) yields

$$\mathbf{j}(\mathbf{r}) = \frac{\hbar}{M} \int d\mathbf{x}_1 \cdots d\mathbf{x}_N |\Psi_0|^2 \sum_{k=1}^{N} \delta(\mathbf{r} - \mathbf{r}_k) \nabla_k \chi. \quad (1.72)$$

When χ depends on coordinates, $\mathbf{j}$ is, in general, nonzero; therefore, Ψ_0 cannot be the ground state. Thus, the GSWF of a Bose system can be considered as real.

To show that Ψ_0 is nodeless, let us recall that the GSWF can be determined from the variational principle: Ψ_0 is determined so as to minimize the energy functional

$$F[\Psi] = \int d\mathbf{x}_1 \cdots d\mathbf{x}_N \left[\frac{\hbar^2}{2M} \sum_{k=1}^{N} (\nabla_k \Psi)^2 + V(\mathbf{x}_1, \cdots, \mathbf{x}_N) \Psi^2 \right], \quad (1.73)$$

where the potential function V is assumed to be finite everywhere, and Ψ is assumed to be real and subject to the normalization condition

$$\int d\mathbf{x}_1 \cdots d\mathbf{x}_N \Psi^2 = 1. \quad (1.74)$$

With a Lagrange multiplier E, the variational principle

$$\frac{\delta}{\delta \Psi} \left(F - E \int d\mathbf{x}_1 \cdots d\mathbf{x}_N \Psi^2 \right) = 0 \quad (1.75)$$

leads to the Schrödinger equation

$$\left(-\frac{\hbar^2}{2M}\sum_{k=1}^{N}\nabla_k^2 + V(\mathbf{x}_1,\cdots,\mathbf{x}_N)\right)\Psi = E\Psi. \tag{1.76}$$

It is noteworthy that Eqs. (1.73) and (1.74) are invariant under $\Psi \to -\Psi$; therefore, if Ψ is a solution to Eq. (1.76), $|\Psi|$ is also a solution that has the same energy E. Now suppose that the GSWF Ψ_0 has a node and changes its sign at a certain point; then, $|\Psi_0|$ must have a cusp at that point, and its derivative must be discontinuous there. We can then construct a new wave function Ψ' from $|\Psi_0|$ by smoothing out the cusp of $|\Psi_0|$ over an infinitesimal region. However, the energy of Ψ' would then be smaller than that of $|\Psi_0|$ because Ψ' has no such cusp that costs the kinetic energy but otherwise coincides with $|\Psi_0|$. By reductio ad absurdum, the GSWF of a Bose system has no node and can therefore be taken to be non-negative.

Let us assume that there are two such non-negative solutions Ψ_1 and Ψ_2. By the linearity of the Schrödiger equation, $\Psi_1 - \Psi_2$ is also a nodeless solution. This implies either $\Psi_1 \geq \Psi_2$ or $\Psi_1 \leq \Psi_2$, which, however, is compatible with the normalization condition (1.74) if and only if $\Psi_1 = \Psi_2$. Thus, the GSWF is nondegenerate.

A corollary of the uniqueness of the GSWF of a Bose system is that at absolute zero, the thermodynamic properties of a Bose system are the same as those of a Boltzmann gas.

1.7 BEC and Superfluidity

There is no unique relationship between BEC and superfluidity. An ideal-gas Bose–Einstein condensate shows no superfluidity and a two-dimensional superfluid shows no BEC. However, there are many cases in which BEC and superfluidity do occur simultaneously. Under such circumstances, a generic argument can be made, which offers insight into the interplay between BEC and superfluidity.

Consider a nonequilibrium situation in which the state of a system changes in time. Since ρ_1 is Hermitian, we can consider the representation in which the single-particle density operator is diagonal at all times:

$$\rho_1(\mathbf{r},\mathbf{r}';t) = \langle\hat{\psi}^\dagger(\mathbf{r},t)\hat{\psi}(\mathbf{r}',t)\rangle = \sum_\nu n_\nu(t)\psi_\nu^*(\mathbf{r},t)\psi_\nu(\mathbf{r}',t). \tag{1.77}$$

We denote the mode in which BEC occurs as $\nu = 0$, that is, $n_0 = O(N)$. If BEC occurs only in the $\nu = 0$ mode, $n_{\nu\neq 0} = O(1)$. In the limit $|\mathbf{r}-\mathbf{r}'| \to \infty$, contributions other than the BEC mode do not add up but rather cancel each other, since $\psi_{\nu\neq 0}$'s are orthogonal to each other. Thus, in the limit $|\mathbf{r}-\mathbf{r}'| \to \infty$, we have

$$\begin{aligned}\langle\hat{\psi}^\dagger(\mathbf{r},t)\hat{\psi}(\mathbf{r}',t)\rangle &\to n_0\psi_0^*(\mathbf{r},t)\psi_0(\mathbf{r}',t)\\ &\equiv \Psi^*(\mathbf{r},t)\Psi(\mathbf{r}',t),\end{aligned} \tag{1.78}$$

We may interpret

$$\Psi(\mathbf{r},t) \equiv \sqrt{n_0}\psi_0(\mathbf{r},t) \tag{1.79}$$

as the condensate wave function (or the order parameter) that is applicable to a nonequilibrium situation. The density of the condensed bosons is defined as

$$\rho(\mathbf{r},t) \equiv \Psi^*(\mathbf{r},t)\Psi(\mathbf{r},t). \tag{1.80}$$

It follows from Eq. (1.80) and the continuity equation

$$\frac{\partial}{\partial t}\rho(\mathbf{r},t) + \mathrm{div}\mathbf{j}(\mathbf{r},t) = 0 \tag{1.81}$$

that the current density of particles is given by

$$\mathbf{j}(\mathbf{r},t) = \frac{\hbar}{2Mi}[\Psi^*(\mathbf{r},t)\nabla\Psi(\mathbf{r},t) - \Psi(\mathbf{r},t)\nabla\Psi^*(\mathbf{r},t)]. \tag{1.82}$$

Let us decompose $\Psi(\mathbf{r},t)$ into the amplitude and phase as

$$\Psi(\mathbf{r},t) = A(\mathbf{r},t)e^{i\phi(\mathbf{r},t)}. \tag{1.83}$$

In terms of A and ϕ, ρ and $\mathbf{j}$ can be expressed as

$$\begin{aligned}\rho(\mathbf{r},t) &= A^2(\mathbf{r},t),\\ \mathbf{j}(\mathbf{r},t) &= A^2(\mathbf{r},t)\frac{\hbar}{M}\nabla\phi(\mathbf{r},t).\end{aligned}$$

The superfluid velocity $\mathbf{v}_\mathrm{s}(\mathbf{r},t)$ is defined as the ratio of $\mathbf{j}$ to ρ.

$$\mathbf{v}_\mathrm{s}(\mathbf{r},t) \equiv \frac{\mathbf{j}(\mathbf{r},t)}{\rho(\mathbf{r},t)} = \frac{\hbar}{M}\nabla\phi(\mathbf{r},t) \tag{1.84}$$

Thus, the phase of the condensate wave function plays the role of the velocity potential in superfluidity. The equation of motion for $\mathbf{v}_\mathrm{s}$ is given from Eq. (1.84) as

$$\frac{d}{dt}\mathbf{v}_\mathrm{s} = \nabla\Omega, \qquad \Omega \equiv \frac{\hbar}{M}\frac{\partial\phi}{\partial t}. \tag{1.85}$$

Because the superfluid velocity is the gradient of a scalar function, its rotation vanishes identically.

$$\mathrm{rot}\mathbf{v}_\mathrm{s} = 0. \tag{1.86}$$

Thus, a superfluid is irrotational. However, when we rotate a container holding a superfluid, the surface of the superfluid shows a parabolic meniscus on the periphery, as in the case of a normal fluid. This shows that the value of the surface integral of vorticity, $\int \mathrm{rot}\mathbf{v}_s dS$, is nonzero in apparent contradiction with Eq. (1.86). Onsager [Onsager (1949)] resolved this paradox by assuming that $\mathrm{rot}\mathbf{v}_s$ is nonzero only within microscopic regions where the liquid is not a superfluid. These singular regions are where vortex lines penetrate, as shown in Fig. 1.1. The integral of $\mathbf{v}_s$ along a closed

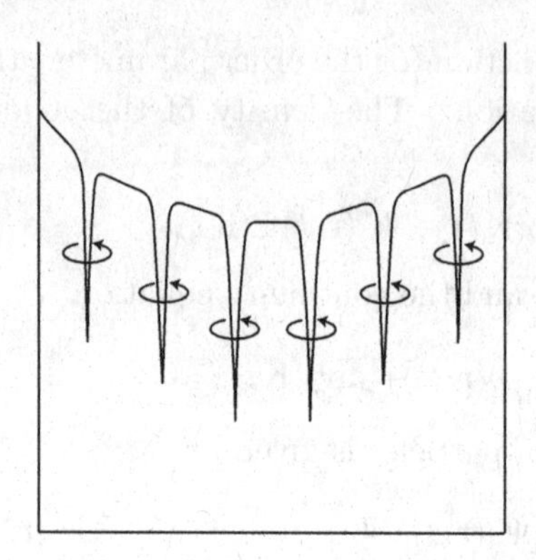

Fig. 1.1 Vortex lines. A rotating superfluid becomes normal within narrow regions where vortex lines penetrate. The size of each vortex line is on the order of the healing length, which is on the order of a few angstroms for ^{4}He and on the order of a few micrometers for alkali Bose–Einstein condensates.

contour gives

$$\oint \mathbf{v}_s d\mathbf{l} = \frac{\hbar}{M} \oint \nabla\phi(\mathbf{r}, t) d\mathbf{l}. \tag{1.87}$$

Because of the single-valuedness of the condensate wave function (or the order parameter), $\oint \nabla\phi(\mathbf{r}t)d\mathbf{l}$ is equal to an integer multiple of 2π, leading to the celebrated quantization of circulation

$$\oint \mathbf{v}_s d\mathbf{l} = \frac{h}{M} n, \tag{1.88}$$

where n is an integer. Thus, the line integral of the superfluid velocity $\mathbf{v}_s$ is quantized in units of

$$\kappa_0 \equiv \frac{h}{M} \simeq \begin{cases} 9.97 \times 10^{-4}\mathrm{cm}^2/\mathrm{s} & \text{for } ^4\mathrm{He}; \\ 4.59 \times 10^{-5}\mathrm{cm}^2/\mathrm{s} & \text{for } ^{87}\mathrm{Rb}, \end{cases}$$

where κ_0 is called the quantum of circulation. When the radius of a container is R and the angular frequency of rotation is ω, we have

$$\oint \mathbf{v}_s d\mathbf{l} = \omega R \cdot 2\pi R = 2\pi R^2 \omega = n\kappa_0.$$

The observed meniscus on the periphery can be explained if the vortex lines are distributed with density $n/(\pi R^2) = 2\omega/\kappa_0$. For the case of $\omega/2\pi =$ 100 Hz, we have $2\omega/\kappa_0 \simeq 0.013/\mu\mathrm{m}^2$ for ^{4}He and $0.3/\mu m^2$ for ^{87}Rb.

A persistent current in a ring geometry is attenuated if a quantized vortex having circulation with the same (opposite) sign crosses the ring from inside (outside) to outside (inside). The decay of a persistent current is caused by thermal excitations or quantum tunneling of quantized vortices. It is generally held that turbulent flow in a superfluid is induced by an entanglement of vortices. However, it is difficult to judge whether attenuation of a persistent current is caused by nucleation of a new vortex or by entanglement of already existing vortices. The study of turbulence using a gaseous Bose–Einstein condensate is expected to shed new light on this problem, because the dynamics of vortices can be visually tracked in real time.

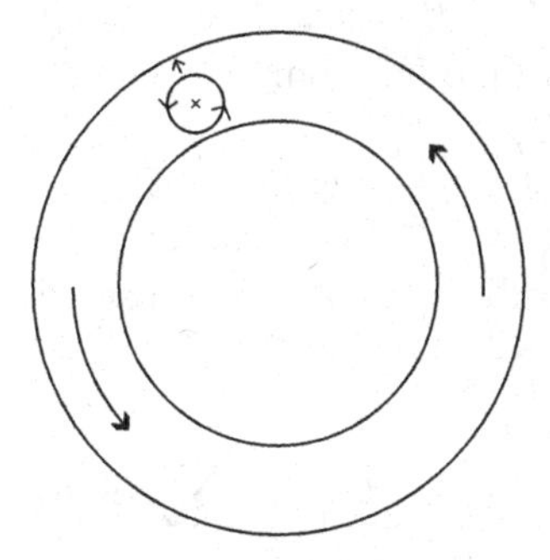

Fig. 1.2 Decay of persistent current in a torus. This decay occurs stepwise because a quantized vortex with the same (opposite) sense of circulation crosses the ring from inside (outside) to outside (inside).

The wave function of identical bosons must be symmetric under exchange of any two particles. The ground-state wave function of a Bose system does not have a node and can take only non-negative real numbers. The wave function of a low-lying excited state is expected to be similar to that of the ground state but modulated slightly depending on the nature of the excitation. One of the candidates that satisfy the requirement of Bose symmetry is given by

$$\psi(\mathbf{r}_1, \mathbf{r}_2, \cdots, \mathbf{r}_N; t) = \exp\left[i\sum_{j=1}^{N}\phi(\mathbf{r}_j, t)\right]\psi(\mathbf{r}_1, \mathbf{r}_2, \cdots, \mathbf{r}_N). \qquad (1.89)$$

Then, the eigenvector for $\rho_1(\mathbf{r}, \mathbf{r}')$ in Eq. (1.26) for a uniform system changes from a constant to a constant times $e^{i\phi(r,t)}$.

When ϕ is a real number, Eq. (1.89) can describe a state of superflow, because the superfluid velocity

$$\mathbf{v}_s(\mathbf{r}, t) = \frac{\hbar}{M}\nabla\phi(\mathbf{r}, t) \tag{1.90}$$

is nonzero if ϕ changes spatially. In particular, when the system is incompressible (*i.e.*, $|\psi| = \text{const.}$), the equation of continuity dictates that the potential function ϕ must satisfy Laplace's equation $\nabla^2\phi = 0$.

When ϕ is a complex number, Eq. (1.89) describes the state of a sound wave accompanied by density modulations. Let us consider, for simplicity, a one-dimensional case. Taking the velocity of the sound wave as $v_x = c\cos(kx - \omega t)$, the equation of continuity

$$\frac{\partial\rho}{\partial t} + \frac{\partial}{\partial x}(\rho_0 v_x) = 0$$

leads to the following density modulation:

$$\rho = \rho_0\left(1 + \frac{ck}{\omega}\cos(kx - \omega t)\right).$$

The wave function that describes such an excited state can be expressed by inspection as [Thouless (1998)]

$$\psi = \exp\left[i\frac{Mc}{\hbar k}\sum_j \sin(kx_j - \omega t) + \frac{ck}{2\omega}\sum_j \cos(kx_j - \omega t)\right]\psi_0. \tag{1.91}$$

It follows that the current density is given by

$$\begin{aligned} j_x &= \frac{\hbar}{2Mi}\left(\psi^*\frac{\partial}{\partial x_j}\psi - \psi\frac{\partial}{\partial x_j}\psi^*\right) \\ &= c\cos(kx_j - \omega t)e^{\frac{ck}{\omega}\sum_j\cos(kx_j - \omega t)}\psi_0^2. \end{aligned}$$

Since the density of particles is given by

$$\rho = |\psi|^2 = e^{\frac{ck}{\omega}\sum_j\cos(kx_j - \omega t)}\rho_0, \quad \rho_0 = \psi_0^2,$$

we have $j_x = v_x\rho$. We can obtain the many-body wave function for the system with one phonon excitation by linearizing the amplitude in Eq. (1.91) as

$$\psi = \frac{1}{\sqrt{N}}\sum_j e^{ikx_j}\psi_0.$$

The generalization of this argument to higher dimensions is straightforward.

1.8 Two-Fluid Model

Let us consider a metastable state in which a superfluid moves along a wall with velocity $\mathbf{v}_s$. As the mass current $\mathbf{j}$ vanishes for $\mathbf{v}_s = 0$, it is reasonable to assume a linear relation between $\mathbf{j}$ and $\mathbf{v}_s$ when $\mathbf{v}_s$ is small [Leggett (1991)]:

$$\mathbf{j}(\mathbf{r}) = \int K_s(\mathbf{r},\mathbf{r}')\mathbf{v}_s(\mathbf{r}')d\mathbf{r}', \tag{1.92}$$

where K_s depends on the microscopic details of the system. When the spatial variation of $\mathbf{v}_s(\mathbf{r})$ is smooth, we can assume the following local relation:

$$\mathbf{j}(\mathbf{r}) = \rho_s(\mathbf{r})\mathbf{v}_s(\mathbf{r}), \quad \rho_s(\mathbf{r}) \equiv \int K_s(\mathbf{r},\mathbf{r}')d\mathbf{r}'. \tag{1.93}$$

While the superfluid velocity $\mathbf{v}_s$ defined in Eq. (1.84) is a microscopic quantity, the superfluid density ρ_s is a hydrodynamic concept, as can be observed from the definition in Eq. (1.93). The normal fluid density ρ_n is defined as $\rho - \rho_s$.

A normal fluid may be considered as an assembly of quasiparticle excitations in a superfluid. When the entire system moves with velocity $\mathbf{v}_s$, the mass current is $\rho\mathbf{v}_s$. In addition, when there are quasiparticle excitations, they contribute to the mass current density by an additional amount

$$\rho_n(\mathbf{v}_n - \mathbf{v}_s), \tag{1.94}$$

where ρ_n is the normal density and $\mathbf{v}_n$ is the normal velocity. The total mass current density is then given by

$$\mathbf{j} = \rho\mathbf{v}_s + \rho_n(\mathbf{v}_n - \mathbf{v}_s) = (\rho - \rho_n)\mathbf{v}_s + \rho_n\mathbf{v}_n = \rho_s\mathbf{v}_s + \rho_n\mathbf{v}_n. \tag{1.95}$$

The superfluid component shows a vortex-free or irrotational potential flow, whereas the normal fluid component shows a viscous flow.

The field operators of bosons follow the canonical commutation relations

$$[\hat{\psi}(\mathbf{r}), \hat{\psi}^\dagger(\mathbf{r}')] = \delta(\mathbf{r}-\mathbf{r}'), \ [\hat{\psi}(\mathbf{r}), \hat{\psi}(\mathbf{r}')] = 0, \ [\hat{\psi}^\dagger(\mathbf{r}), \hat{\psi}^\dagger(\mathbf{r}')] = 0. \tag{1.96}$$

In analogy with Eq. (1.83), let us formally decompose $\hat{\psi}(\mathbf{r})$ as

$$\hat{\psi}(\mathbf{r}) = \sqrt{\hat{\rho}(\mathbf{r})}e^{i\hat{\phi}(\mathbf{r})}. \tag{1.97}$$

For Eq. (1.97) to be consistent with Eq. (1.96), it is sufficient to assume the following commutation relations between $\hat{\rho}$ and $\hat{\phi}$:

$$[\hat{\rho}(\mathbf{r}), \hat{\phi}(\mathbf{r}')] = i\delta(\mathbf{r}-\mathbf{r}'), \quad [\hat{\rho}(\mathbf{r}), \hat{\rho}(\mathbf{r}')] = 0, \quad [\hat{\phi}(\mathbf{r}), \hat{\phi}(\mathbf{r}')] = 0. \tag{1.98}$$

To show this, we first note that

$$\hat{\rho}(\mathbf{r}) = i\frac{\delta}{\delta\hat{\phi}(\mathbf{r})}, \quad \hat{\phi}(\mathbf{r}) = -i\frac{\delta}{\delta\hat{\rho}(\mathbf{r})}.$$

We can use these relations to obtain

$$[e^{i\hat{\phi}(\mathbf{r})}, \hat{\rho}(\mathbf{r}')] = e^{i\hat{\phi}(\mathbf{r})}\delta(\mathbf{r}-\mathbf{r}'), \quad [\hat{\rho}(\mathbf{r}), e^{i\hat{\phi}(\mathbf{r}')}] = -e^{i\hat{\phi}(\mathbf{r}')}\delta(\mathbf{r}-\mathbf{r}').$$

Furthermore, noting that (see the end of this section for details)

$$[e^{i\hat{\phi}(\mathbf{r})}, f[\hat{\rho}(\mathbf{r}')]] \simeq [e^{i\hat{\phi}(\mathbf{r})}, \hat{\rho}(\mathbf{r}')]f'[\hat{\rho}(\mathbf{r}')] = e^{i\hat{\phi}(\mathbf{r})}\delta(\mathbf{r}-\mathbf{r}')f'[\hat{\rho}(\mathbf{r}')],$$

we obtain the following relations:

$$\left[e^{i\hat{\phi}(\mathbf{r})}, \sqrt{\hat{\rho}(\mathbf{r}')}\right] \simeq e^{i\hat{\phi}(\mathbf{r})}\frac{1}{2\sqrt{\hat{\rho}(\mathbf{r}')}}\delta(\mathbf{r}-\mathbf{r}'),$$

$$\left[\sqrt{\hat{\rho}(\mathbf{r})}, e^{-i\hat{\phi}(\mathbf{r}')}\right] \simeq e^{-i\hat{\phi}(\mathbf{r}')}\frac{1}{2\sqrt{\hat{\rho}(\mathbf{r})}}\delta(\mathbf{r}-\mathbf{r}').$$

Thus,

$$\begin{aligned}
&\left[\hat{\psi}(\mathbf{r}), \hat{\psi}^\dagger(\mathbf{r}')\right] = \left[\sqrt{\hat{\rho}(\mathbf{r})}e^{i\hat{\phi}(\mathbf{r})}, e^{-i\hat{\phi}(\mathbf{r})}\sqrt{\hat{\rho}(\mathbf{r}')}\right] \\
&= \sqrt{\hat{\rho}(\mathbf{r})}e^{-i\hat{\phi}(\mathbf{r}')}\left[e^{i\hat{\phi}(\mathbf{r})}, \sqrt{\hat{\rho}(\mathbf{r}')}\right] + \left[\sqrt{\hat{\rho}(\mathbf{r})}, e^{-i\hat{\phi}(\mathbf{r}')}\right]\sqrt{\hat{\rho}(\mathbf{r}')}e^{i\hat{\phi}(\mathbf{r})} \\
&= \left[\sqrt{\hat{\rho}(\mathbf{r})}e^{-i\hat{\phi}(\mathbf{r}')}e^{i\hat{\phi}(\mathbf{r})}\frac{1}{2\sqrt{\hat{\rho}(\mathbf{r}')}} + e^{-i\hat{\phi}(\mathbf{r}')}\frac{1}{2\sqrt{\hat{\rho}(\mathbf{r})}}\sqrt{\hat{\rho}(\mathbf{r}')}e^{i\hat{\phi}(\mathbf{r})}\right]\delta(\mathbf{r}-\mathbf{r}') \\
&= \delta(\mathbf{r}-\mathbf{r}').
\end{aligned}$$

Let us define the phase operator $\hat{\phi}$ and the total number operator $\hat{N}$ as

$$\hat{\phi} \equiv \frac{1}{V}\int \hat{\phi}(\mathbf{r})d\mathbf{r}, \tag{1.99}$$

$$\hat{N} \equiv \int \hat{\rho}(\mathbf{r})d\mathbf{r}. \tag{1.100}$$

It is easy to show that

$$[\hat{N}, \hat{\phi}] = i. \tag{1.101}$$

From Eq. (1.101), it follows that

$$\hat{\phi} = -i\frac{\partial}{\partial\hat{N}}.$$

We can use this result to obtain

$$\frac{\partial\hat{\phi}}{\partial t} = \frac{i}{\hbar}[\hat{H}, \hat{\phi}] = -\frac{1}{\hbar}\frac{\partial\hat{H}}{\partial\hat{N}}. \tag{1.102}$$

Taking the expectation value on both sides of Eq. (1.102), we have

$$\frac{\partial \phi}{\partial t} = -\frac{1}{\hbar}\frac{\partial E}{\partial N} = -\frac{1}{\hbar}\mu, \tag{1.103}$$

where μ is the chemical potential of the system. We substitute this into Eq. (1.85) and find that $\mathbf{v}_s$ satisfies the following Euler equation:

$$\frac{d}{dt}\mathbf{v}_s = \frac{\partial \mathbf{v}_s}{\partial t} + (\mathbf{v}_s \cdot \nabla)\mathbf{v}_s = -\frac{1}{M}\nabla\mu. \tag{1.104}$$

Thus, the superfluid velocity is accelerated by the gradient of the chemical potential. On the other hand, the total mass current density $\mathbf{j}$ is driven by the pressure gradient,

$$\frac{d}{dt}\mathbf{j} = -\nabla P, \tag{1.105}$$

where P is the pressure. Substituting Eq. (1.95) into Eq. (1.105), we obtain

$$\rho_s \dot{\mathbf{v}}_s + \rho_n \dot{\mathbf{v}}_n = -\nabla P. \tag{1.106}$$

Substituting Eq. (1.104) for $\dot{\mathbf{v}}_s$ into Eq. (1.106) gives

$$\rho_n \dot{\mathbf{v}}_n = \frac{\rho_s}{M}\nabla\mu - \nabla P. \tag{1.107}$$

The Gibbs–Duhem relation [see Appendix A] holds in a system at thermal equilibrium:

$$SdT - VdP + Nd\mu = 0. \tag{1.108}$$

When there is no temperature variation in the system, *i.e.*, $dT = 0$, we have

$$\rho_n \dot{\mathbf{v}}_n = \frac{1}{M}(\rho_s - \rho)\nabla\mu. \tag{1.109}$$

When the system is stationary at absolute zero, we have $\dot{\mathbf{v}}_n = 0$. Therefore, we find that the superfluid density is equal to the total density ($\rho_s = \rho$) [Leggett (1991)]. (It must be noted, however, that $N_0 \neq N$.)

Finally, let us examine how the approximation

$$[e^{i\hat{\phi}(\mathbf{r})}, f[\hat{\rho}(\mathbf{r}')]] \simeq [e^{i\hat{\phi}(\mathbf{r})}, \hat{\rho}(\mathbf{r}')] f'[\hat{\rho}(\mathbf{r}')] \tag{1.110}$$

is justified. To analyze the validity of this approximation, let us start with the following operator equation:

$$\begin{aligned}[\hat{A}, \hat{B}^n] = \; & \hat{B}^{n-1}[\hat{A}, \hat{B}] + \hat{B}^{n-2}[\hat{A}, \hat{B}]\hat{B} + \hat{B}^{n-3}[\hat{A}, \hat{B}]\hat{B}^2 \\ & + \cdots + [\hat{A}, \hat{B}]\hat{B}^{n-1}.\end{aligned} \tag{1.111}$$

The approximation (1.110) is obtained if the right-hand side of Eq. (1.111) is approximated by $[\hat{A}, \hat{B}](\hat{B}^n)'$:

$$[\hat{A}, \hat{B}^n] \simeq [\hat{A}, \hat{B}](\hat{B}^n)'. \tag{1.112}$$

Then,

$$\begin{aligned}[\hat{A}, f(\hat{B})] &= \sum_{n=0}^{\infty} \frac{1}{n!}[\hat{A}, f^{(n)}(\hat{B}=0)\hat{B}^n] \simeq [\hat{A}, \hat{B}] \sum_{n=0}^{\infty} \frac{1}{n!} f^{(n)}(0)(\hat{B}^n)' \\ &= [\hat{A}, \hat{B}] f'(\hat{B});\end{aligned}$$

hence, Eq. (1.110) is proved. The following approximation is essential for deriving Eq. (1.112):

$$\hat{B}^{n-k}[\hat{A}, \hat{B}]\hat{B}^{k-1} = [\hat{A}, \hat{B}]\hat{B}^{n-1} + [\hat{B}^{n-k}, [\hat{A}, \hat{B}]]\hat{B}^{k-1}.$$

It is clear that if we ignore the last double commutator, Eq. (1.112) follows. As observed from Eq. (1.98), each commutation relation yields an extra unit density compared to $\hat{\rho}(\mathbf{r})$. Therefore, if the particle-number density is considerably higher than 1, we can ignore the double commutator. This requirement is satisfied when the system is Bose–Einstein condensed.

1.9 Fragmented Condensate[5]

When the one-particle density matrix has k (≥ 2) eigenvalues that are of the same order as the total number of particles, the system comprises k condensates whose relative phases are random, and the system as a whole is called a fragmented Bose–Einstein condensate (fragmented condensate). The concept of a fragmented BEC was first discussed by Nozières and Saint James [Nozières and James (1982); Nozières (1995)] and has witnessed a remarkable resurgence of interest, particularly in systems of gaseous Bose–Einstein condensates [for reviews, see Castin and Herzog (2001) and Mueller, *et al.* (2006)]. The fragmented condensate is obtained when the system possesses an exact symmetry. Here, we illustrate this concept using a few examples.

1.9.1 *Two-state model*

First, let us consider a model in which N spinless bosons occupy only two states with $\mathbf{k}$ and $-\mathbf{k}$. The corresponding field operator is given by

$$\hat{\psi}(\mathbf{r}) = \frac{1}{\sqrt{V}}\left(\hat{a}_{\mathbf{k}} e^{i\mathbf{kr}} + \hat{a}_{-\mathbf{k}} e^{-i\mathbf{kr}}\right), \tag{1.113}$$

[5]This section may be skipped on first reading.

and the second-quantized Hamiltonian is given by

$$\hat{H} = \int d\mathbf{r}\, \hat{\psi}(\mathbf{r})\left(-\frac{\hbar^2\nabla^2}{2M}\right)\hat{\psi}(\mathbf{r}) + \frac{gV}{2}\int d\mathbf{r}\, \hat{\psi}^\dagger(\mathbf{r})\hat{\psi}^\dagger(\mathbf{r})\hat{\psi}(\mathbf{r})\hat{\psi}(\mathbf{r})$$
$$= \epsilon_{\mathbf{k}}\hat{N} + \frac{g}{2}\hat{N}(\hat{N}-1) + g\hat{N}_{\mathbf{k}}\hat{N}_{-\mathbf{k}}, \tag{1.114}$$

where $\epsilon_{\mathbf{k}} \equiv \frac{\hbar^2\mathbf{k}^2}{2M}$, $\hat{N}_{\pm\mathbf{k}} \equiv \hat{a}^\dagger_{\pm\mathbf{k}}\hat{a}_{\pm\mathbf{k}}$, and $\hat{N} \equiv \hat{N}_{\mathbf{k}} + \hat{N}_{-\mathbf{k}}$. Since $\hat{N}$ is constant, the ground state of the system is determined by the last term in Eq. (1.114).

In the case of a repulsive interaction $g > 0$, the ground state is

$$|N_{\mathbf{k}} = N,\ N_{-\mathbf{k}} = 0\rangle \quad \text{or} \quad |N_{\mathbf{k}} = 0,\ N_{-\mathbf{k}} = N\rangle, \tag{1.115}$$

and it is doubly degenerate. The single-particle density matrix

$$\rho_1 = \begin{pmatrix} \langle\hat{a}^\dagger_{\mathbf{k}}\hat{a}_{\mathbf{k}}\rangle & \langle\hat{a}^\dagger_{\mathbf{k}}\hat{a}_{-\mathbf{k}}\rangle \\ \langle\hat{a}^\dagger_{-\mathbf{k}}\hat{a}_{\mathbf{k}}\rangle & \langle\hat{a}^\dagger_{-\mathbf{k}}\hat{a}_{-\mathbf{k}}\rangle \end{pmatrix} \tag{1.116}$$

for each ground state in Eq. (1.115) is

$$\begin{pmatrix} N & 0 \\ 0 & 0 \end{pmatrix} \quad \text{or} \quad \begin{pmatrix} 0 & 0 \\ 0 & N \end{pmatrix}. \tag{1.117}$$

Thus, the eigenvalues of ρ_1 are N and 0 and only one condensate exists for both cases.

In the case of an attractive interaction $g < 0$, the ground state of the Hamiltonian (1.114) is given by the Fock state:

$$|F\rangle = \left|N_{\mathbf{k}} = \frac{N}{2}, N_{-\mathbf{k}} = \frac{N}{2}\right\rangle, \tag{1.118}$$

where we assume that N is even. Then, ρ_1 becomes

$$\rho_1 = \begin{pmatrix} \frac{N}{2} & 0 \\ 0 & \frac{N}{2} \end{pmatrix}; \tag{1.119}$$

thus, the system is fragmented. From this example, we find that fragmentation occurs when the system undergoes BEC to form several degenerate single-particle states.

It is instructive to compare the Fock state (1.118) with the coherent state

$$|\phi\rangle = \frac{1}{\sqrt{2^N N!}}\left(\hat{a}^\dagger_{\mathbf{k}}e^{-\frac{i}{2}\phi} + \hat{a}^\dagger_{-\mathbf{k}}e^{\frac{i}{2}\phi}\right)^N |\text{vac}\rangle. \tag{1.120}$$

The single-particle density matrix (1.116) for this state is

$$\rho_1 = \begin{pmatrix} \frac{N}{2} & \frac{N}{2}e^{i\phi} \\ \frac{N}{2}e^{-i\phi} & \frac{N}{2} \end{pmatrix}, \tag{1.121}$$

and the eigenvalues are found to be N and 0. Thus, $|\phi\rangle$ describes a single condensate. The expectation value of $g\hat{N}_{\mathbf{k}}\hat{N}_{-\mathbf{k}}$ over $|\phi\rangle$ is calculated to give $gN(N-1)/4$. Thus, the energy of the fragmented state (1.118), which is $gN^2/4$, is lower than that of the coherent state (1.120) by $|g|N/4$. Interestingly, the average of $|\phi\rangle$ over ϕ gives $|F\rangle$:

$$\int_0^{2\pi} \frac{d\phi}{2\pi}|\phi\rangle = \sqrt{\frac{N!}{2^N\left[\left(\frac{N}{2}\right)!\right]^2}}\left|\frac{N}{2},\frac{N}{2}\right\rangle \propto |F\rangle. \tag{1.122}$$

Thus, $|F\rangle$ is the equal-weighted average of the family of degenerate states $|\phi\rangle$.

1.9.2 *Degenerate double-well model*

Next, let us consider a degenerate double-well model [see Fig. 1.3] [Leggett (2001)], where the Hamiltonian is given by

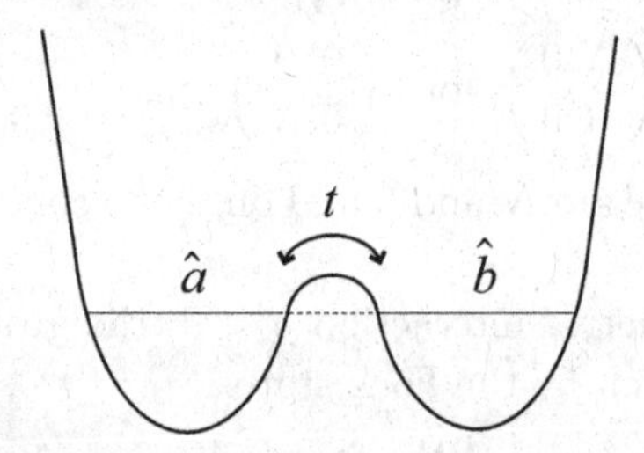

Fig. 1.3 A degenerate double-well model. Bosons can hop between the two wells with the transfer energy t. In each well, bosons undergo contact interactions.

$$\hat{H} = -t\left(\hat{a}^\dagger\hat{b} + \hat{b}^\dagger\hat{a}\right) + \frac{g}{2}\left(\hat{a}^{\dagger 2}\hat{a}^2 + \hat{b}^{\dagger 2}\hat{b}^2\right), \tag{1.123}$$

where t is the transfer energy and g is the strength of the contact interaction in each well. Assuming that the total number of particles is constant ($\hat{N} = \hat{N}_a + \hat{N}_b = N$), Eq. (1.123) can be rewritten as

$$\hat{H} = -t\left(\hat{a}^\dagger\hat{b} + \hat{b}^\dagger\hat{a}\right) + \frac{g}{4}N(N-2) + \frac{g}{4}\left(\hat{N}_a - \hat{N}_b\right)^2. \tag{1.124}$$

In the case of a repulsive interaction $g > 0$, a candidate for the ground state is the Fock state:

$$\left|N_a = \frac{N}{2}, N_b = \frac{N}{2}\right\rangle, \tag{1.125}$$

$$E^{\text{Fock}} = \frac{g}{4}N(N-2). \tag{1.126}$$

As shown in Eq. (1.119), this is a fragmented condensate. Another candidate is the coherent state in which all N bosons condense into a single-particle state described by the creation operator $(\hat{a}^\dagger e^{-\frac{i}{2}} + \hat{b}^\dagger e^{\frac{i}{2}\phi})/\sqrt{2}$:

$$|\phi\rangle = \frac{1}{\sqrt{2^N N!}} \left(\hat{a}^\dagger e^{-\frac{i}{2}\phi} + \hat{b}^\dagger e^{\frac{i}{2}\phi}\right)^N |0\rangle. \tag{1.127}$$

Calculating the expectation value of Eq. (1.123) over the state (1.127), we obtain

$$E^{\text{coherent}} = \frac{g}{4} N(N-1) - tN\cos\phi. \tag{1.128}$$

Hence, if $|t| > g/4$, the coherent state (1.127) can have lower energy than the Fock state (1.126). The single-particle density matrix for the coherent state is given by

$$\rho_1 = \begin{pmatrix} \langle \hat{a}^\dagger \hat{a} \rangle & \langle \hat{a}^\dagger \hat{b} \rangle \\ \langle \hat{b}^\dagger \hat{a} \rangle & \langle \hat{b}^\dagger \hat{b} \rangle \end{pmatrix} = \begin{pmatrix} \frac{N}{2} & \frac{N}{2} e^{i\phi} \\ \frac{N}{2} e^{-i\phi} & \frac{N}{2} \end{pmatrix}. \tag{1.129}$$

The eigenvalues of ρ_1 are determined from

$$\det \begin{pmatrix} \frac{N}{2} - \lambda & \frac{N}{2} e^{-i\phi} \\ \frac{N}{2} e^{-i\phi} & \frac{N}{2} - \lambda \end{pmatrix} = \lambda(\lambda - N) = 0 \tag{1.130}$$

to be N and 0. The state (1.127) is, therefore, a single condensate.

On the other hand, if $|t| < g/4$, the Fock state (1.125) has lower energy than the coherent state (1.128). Note that the critical value $t_c = g/4$ is smaller than the interaction energy per particle $\sim gN/4$ by a factor of N. Thus, a tiny symmetry-breaking perturbation, which is induced by the hopping term, makes the fragmented state unstable against the coherent state. The physical reason for the vulnerability of the fragmented condensate lies in the fact that the fragmented state (1.125) has two macroscopically occupied components that enhance the matrix element of the hopping term by a factor of N [see the last term in Eq. (1.128)]. Thus, for the fragmented state to survive, the symmetry-breaking perturbation must be suppressed to below $1/N$. This is why a single condensate is so robust in the thermodynamic limit; yet, it is possible to realize a fragmented condensate in a mesoscopic regime.

In the case of an attractive interaction $g < 0$, the last term in Eq. (1.124) attains the minimum energy when the number difference $\hat{N}_a - \hat{N}_b$ is maximal. When $t = 0$, the minimal energy is attained by a Schrödinger cat state:

$$|\text{cat}\rangle = \frac{1}{\sqrt{2}} \left(|N, 0\rangle + |0, N\rangle\right). \tag{1.131}$$

If the double-well potential has an asymmetry, the single condensate that has the lower energy, *i.e.*, $|N,0\rangle$ or $|0,N\rangle$, will be favored. When the two wells are degenerate, the cat state will be favored, because atoms tend to bunch up due to the attractive interaction; however, because of the degeneracy of the two possible states, the system evolves, resulting in a superposition state (1.131). The major symmetry-breaking perturbation for the cat state is the asymmetry between the two wells. If the energies of the wells are degenerate, the hopping term plays the role of the symmetry-breaking perturbation in favor of the coherent state (1.127).

The single-particle density matrix of the cat state is given by Eq. (1.119) and has two extensive eigenvalues. Thus, we find that two-state systems have two possible candidates for a fragmented condensate: the Fock state in the case of a repulsive interaction and the Schrödinger cat state in the case of an attractive interaction.

1.9.3 *Spin-1 antiferromagnetic BEC*

Finally, let us consider a system of N spin-1 bosons described by the Hamiltonian

$$\hat{H} = J\hat{\mathbf{S}} \cdot \hat{\mathbf{S}}, \tag{1.132}$$

where

$$\hat{\mathbf{S}} = \sum_{m,n=-1}^{1} (\mathbf{S})_{mn} \hat{a}_m^\dagger \hat{a}_n. \tag{1.133}$$

Here, $\mathbf{S} = (S_x, S_y, S_z)$ is a vector of spin-1 matrices, and $\hat{a}_m^\dagger$ and $\hat{a}_n$ are the creation and annihilation operators of the bosons with magnetic sublevels $m = 0, \pm 1$ [see Chap. 6 for details]. For simplicity, we assume that N is even. In the case of an antiferromagnetic interaction $J > 0$, the ground state is a spin-singlet state ($S = 0$), as described in Sec. 6.3.2. Introducing the spin operators in the Cartesian coordinates

$$\hat{\mathbf{A}} = \left(\hat{A}_x, \hat{A}_y, \hat{A}_z\right) \equiv \left(-\frac{\hat{a}_1 - \hat{a}_{-1}}{\sqrt{2}}, \frac{\hat{a}_1 + \hat{a}_{-1}}{\sqrt{2}}, \hat{a}_0\right), \tag{1.134}$$

the spin-singlet state is expressed as

$$|S=0\rangle = \frac{1}{\sqrt{(N+1)!}} \left(\hat{\mathbf{A}}^\dagger \cdot \hat{\mathbf{A}}^\dagger\right)^{\frac{N}{2}} |0\rangle. \tag{1.135}$$

In this example, ρ_1 is a 3×3 matrix, and its (i,j) component (*i.e.*, $i,j = x,y,z$) is given by $\langle \hat{A}_i^\dagger \hat{A}_j \rangle$. It can be shown [Koashi and Ueda (2000);

Ho and Yip (2000)] that ρ_1 has triply degenerate eigenvalues equal to $N/3$ and is therefore fragmented. The physics behind the fragmentation is the rotational invariance of the spin-singlet state, which transforms as a scalar under rotation; it follows from Schur's lemma that the irreducible representation of ρ_1 must be proportional to the identity matrix. Since $\mathrm{Tr}\rho_1 = N$, all the eigenvalues must be $N/3$.

The state that breaks the spherical symmetry is given by

$$|\mathbf{n}\rangle = \frac{\left(\mathbf{n}\cdot\hat{\mathbf{A}}^\dagger\right)^N}{\sqrt{N!}}|0\rangle. \tag{1.136}$$

The single-particle density matrix for this state has eigenvalues N, 0, and 0, and therefore, $|\mathbf{n}\rangle$ describes a single condensate. If we calculate the average of Eq. (1.136) over the direction of $\mathbf{n}$, we reproduce the fragmented state in Eq. (1.135) [Mueller, *et al.* (2006)]:

$$\begin{aligned}\int\frac{d\Omega}{4\pi}|\mathbf{n}\rangle &= \frac{1}{2}\int_{-1}^{1}d(\cos\theta)\frac{1}{\sqrt{N!}}\left(\sqrt{\hat{\mathbf{A}}^\dagger\cdot\hat{\mathbf{A}}^\dagger}\cos\theta\right)^N|0\rangle \\ &= \frac{1}{(N+1)\sqrt{N!}}\left(\hat{\mathbf{A}}^\dagger\cdot\hat{\mathbf{A}}^\dagger\right)^{\frac{N}{2}}|0\rangle,\end{aligned} \tag{1.137}$$

where $d\Omega$ is an element of the solid angle. Thus, a fragmented state is interpreted as the average of the symmetry-broken state over the direction of $\mathbf{n}$.

1.10 Interference Between Independent Condensates

An interference experiment involving two independent Bose–Einstein condensates that was demonstrated by the MIT group [Andrews, *et al.* (1997)] provided rather surprising results. Independently prepared condensates have no definite relative-phase relationship; yet, they exhibit interference patterns whose peak positions exhibit shot-to-shot fluctuations [Hadzibabic, *et al.* (2004)].

To understand this phenomenon, we begin by recalling Young's double-slit experiment, as shown in Fig. 1.4(a). An atomic wave emanating from source A passes through slits 1 and 2 and recombines on the screen to produce an interference pattern. It is important to recognize that independently and identically prepared single atoms can exhibit such interference. A single-atom state that passes through the double slits is expressed as

$$|\text{one atom}\rangle = \cos\theta|1,0\rangle + \sin\theta|0,1\rangle, \tag{1.138}$$

where $|1, 0\rangle$ ($|0, 1\rangle$) describes the state in which an atom passes through slit 1 (slit 2) at position $\mathbf{r}_1$ ($\mathbf{r}_2$). The probability $P(\mathbf{r})$ that the atom is detected at position $\mathbf{r}$ is given by

$$P(\mathbf{r}) = c\langle \text{one atom}|\hat{\Psi}^\dagger(\mathbf{r})\hat{\Psi}(\mathbf{r})|\text{one atom}\rangle, \tag{1.139}$$

where c is a constant and $\hat{\Psi}(\mathbf{r})$, the field operator at position $\mathbf{r}$; it can be expressed in terms of the annhilation operators of the atom at slits 1 and 2, $\hat{a}_1$ and $\hat{a}_2$, as

$$\hat{\Psi}(\mathbf{r}) = \hat{a}_1 u_1 e^{i\mathbf{k}_1(\mathbf{r}-\mathbf{r}_1)} + \hat{a}_2 u_2 e^{i\mathbf{k}_2(\mathbf{r}-\mathbf{r}_2)}, \tag{1.140}$$

where u_1 and u_2 are geometrical factors that depend on the experimental conditions. Substituting Eq. (1.140) into Eq. (1.139), we obtain

$$P(\mathbf{r}) = I_1 + I_2 + \sqrt{I_1 I_2}\cos(\mathbf{k}_1\mathbf{r}_1 - \mathbf{k}_2\mathbf{r}_2 + \phi), \tag{1.141}$$

where $I_1 \equiv c|u_1|^2\cos^2\theta$, $I_2 \equiv c|u_2|^2\sin^2\theta$, and $\phi \equiv \arg(u_1^* u_2)$.

This example shows that the necessary and sufficient conditions for single atoms to exhibit Young's interference pattern are that (i) one cannot tell, in principle, the slit (1 or 2) through which each atom passes and that (ii) every atom is prepared in an identical state (1.138). The first condition (i) leads to the superposition of the two possibilities (1.140) and the second condition (ii) allows the state of the atoms to be represented by the same state vector (1.138).

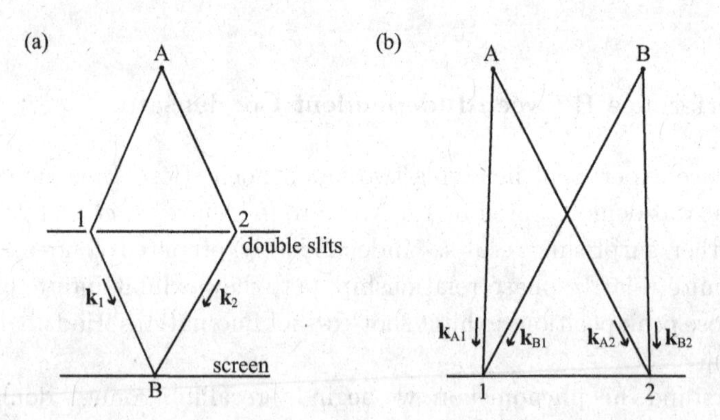

Fig. 1.4 (a) Young's double slit experiment, and (b) two-particle interference experiment.

Next, let us consider an interference experiment involving two condensates, as shown in Fig. 1.4(b). The initial atomic state is

$$|\text{AB}\rangle = |N_\text{A}\rangle|N_\text{B}\rangle, \tag{1.142}$$

where $N_{\rm A}$ and $N_{\rm B}$ are the atomic numbers of condensates A and B, respectively. The fact that they are Bose-Einstein condensed is implicit in the assumption that they are represented by single-mode Fock states. Because the two condensates are assumed to be independent, $|\Psi\rangle$ is the direct product of their state vectors with no relative-phase relationship between them.

Now, suppose that the two atomic waves emanating from condensates A and B arrive at various positions on the screen. The field operator at position $\mathbf{r}_i$ $(i = 1, 2)$ on the screen is expressed in terms of the annihilation operators $\hat{a}_{\rm A}$ and $\hat{a}_B$ of the two condensates as

$$\hat{\Psi}(\mathbf{r}_i) = \hat{a}_{\rm A} u_{\rm A} e^{i\mathbf{k}_{{\rm A}i}(\mathbf{r}_i - \mathbf{r}_{\rm A})} + \hat{a}_{\rm B} u_{\rm B} e^{i\mathbf{k}_{{\rm B}i}(\mathbf{r}_i - \mathbf{r}_{\rm B})}. \tag{1.143}$$

The joint probability distribution that one atom is detected at $\mathbf{r}_1$ and another atom is detected at $\mathbf{r}_2$ is given by

$$P(\mathbf{r}_1, \mathbf{r}_2) = c' \langle {\rm AB}|\hat{\Psi}^\dagger(\mathbf{r}_1)\hat{\Psi}^\dagger(\mathbf{r}_2)\hat{\Psi}(\mathbf{r}_2)\hat{\Psi}(\mathbf{r}_1)|{\rm AB}\rangle, \tag{1.144}$$

where c' is a constant. Substituting Eqs. (1.142) and (1.143) in this equation, we obtain

$$P(\mathbf{r}_1, \mathbf{r}_2) = 2c' N_{\rm A} N_{\rm B} |u_{\rm A} u_{\rm B}|^2 (1 + \cos\varphi) + \text{constant}, \tag{1.145}$$

where

$$\begin{aligned} \varphi &= \mathbf{k}_{\rm A1}(\mathbf{r}_1 - \mathbf{r}_{\rm A}) + \mathbf{k}_{\rm B2}(\mathbf{r}_2 - \mathbf{r}_{\rm B}) - \mathbf{k}_{\rm A2}(\mathbf{r}_2 - \mathbf{r}_{\rm A}) - \mathbf{k}_{\rm B1}(\mathbf{r}_1 - \mathbf{r}_{\rm B}) \\ &\simeq \frac{k}{L} |\mathbf{r}_1 - \mathbf{r}_2| |\mathbf{r}_{\rm A} - \mathbf{r}_{\rm B}|. \end{aligned} \tag{1.146}$$

Here, L is the distance between the source and screen, and we assume $|\mathbf{k}_{Ai}| = |\mathbf{k}_{Bi}| = k$ $(i = 1, 2)$ to obtain the last equality. Equations (1.145) and (1.146) show that the two independent condensates do exhibit an interference pattern. Because φ does not depend on the individual positions $\mathbf{r}_1$ and $\mathbf{r}_2$ but on the distance between them $|\mathbf{r}_1 - \mathbf{r}_2|$, the interference pattern exhibits shot-to-shot fluctuations with regard to where the peaks and valleys appear on the screen. Averaging over many events, the interference pattern disappears; in fact, the ensemble average of the one-particle density $\langle\hat{\Psi}^\dagger(\mathbf{r})\hat{\Psi}(\mathbf{r})\rangle$ over the state (1.142) shows no interference pattern. These results indicate that the interference pattern observed by the MIT group is a two-particle effect similar to the Hanbury Brown–Twiss experiment for photons [Hanbury Brown and Twiss (1956)]. The two-particle interference may also be interpreted as a conquence of the emergence of coherence by the action of a sequence of measurements [Javanainen and Yoo (1996); Naraschewski, *et al.* (1996); Castin and Dalibard (1997)].

1.11 Feshbach Resonance

Feshbach resonance [Feshbach (1958); Fano (1961)] is a resonance of a many-body system between a scattering state in one collision channel and a bound state in another. Suppose that two particles with total energy E in the center-of-mass frame interact via a molecular potential $V_{\rm o}(r)$ [see Fig. 1.5]. Such a collision path of a scattering state is referred to as an open channel. We assume that there exists another collision path, referred to as a closed channel, that supports a bound molecular state with energy $E_{\rm b}$. A Feshbach resonance occurs when the incident particles come into resonance with the bound state. The presence of such a bound state as an intermediate state in a scattering process alters the energy of the scattering state. According to the second-order perturbation theory, the energy shift ΔE is given by

$$\Delta E = \frac{|\langle {\rm b}|\hat{H}_{\rm int}|{\rm i}\rangle|^2}{E - E_{\rm b}}, \tag{1.147}$$

where $|{\rm i}\rangle$ and $|{\rm b}\rangle$ are the incident and bound states, respectively, and $\hat{H}_{\rm int}$ is the interaction Hamiltonian that couples the closed and open channels. For ultracold collisions, we may set $E = 0$ in Eq. (1.147) to obtain

$$\Delta E = -\frac{|\langle {\rm b}|\hat{H}_{\rm int}|{\rm i}\rangle|^2}{E_{\rm b}}. \tag{1.148}$$

A magnetic Feshbach resonance utilizes the fact that the magnetic moment of a pair of atoms changes from $\mu_{\rm o}$ in the open channel to $\mu_{\rm c}$ in the closed

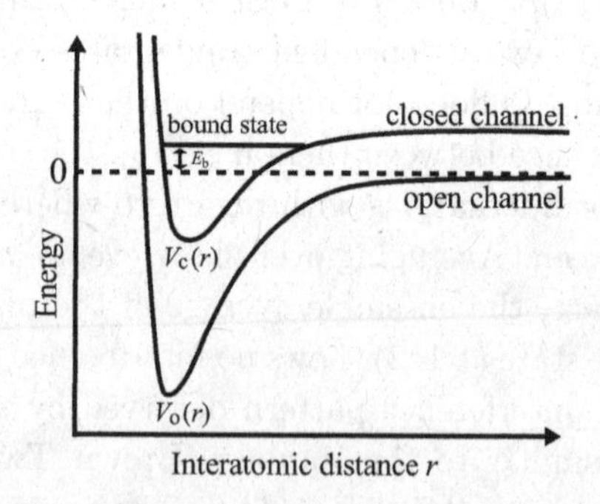

Fig. 1.5 Schematic illustration of a Feshbach resonance, where $V_{\rm c}(r)$ and $V_{\rm o}(r)$ are interatomic potentials for closed and open channels, respectively. The closed channel is assumed to support a bound state with energy $E_{\rm b}$.

channel. In the presence of an external magnetic field B, the energy of the bound state relative to the open-channel threshold changes as

$$E_{\mathrm{b}} = \delta\mu(B - B_0), \tag{1.149}$$

where $\delta\mu \equiv \mu_{\mathrm{o}} - \mu_{\mathrm{c}}$ and B_0 is the value of B at which E_{b} vanishes. Substituting Eq. (1.149) in Eq. (1.148), we obtain

$$\Delta E = -\frac{|\langle \mathrm{b}|\hat{H}_{\mathrm{int}}|\mathrm{i}\rangle|^2}{\delta\mu(B - B_0)}. \tag{1.150}$$

As discussed in the next section, the interaction between ultracold atoms is proportional to the scattering length a [see Eq. (2.25)]. Therefore, near a Feshbach resonance, the scattering length depends on B as

$$a = a_{\mathrm{bg}}\left(1 - \frac{\Delta}{B - B_0}\right), \tag{1.151}$$

where a_{bg} is the background scattering length and $\Delta \propto (\delta\mu)^{-1}$, the width of the resonance. The magnitude and sign of the scattering length can thus be tuned by an external magnetic field [Inouye, *et al.* (1998); Courteille, *et al.* (1998); Cornish (2000)].

A scattering state may be brought into resonance with a bound state by other means. If it is implemented optically, it is called an optical Feshbach resonance. If the resonance is caused by a quasibound state that is supported by an open-channel potential V_{o} rather than V_{c}, it is called a shape resonance. Comprehensive surveys of Feshbach resonances have been reported, *e.g.*, in [Köhler, *et al.* (2006); Chin, *et al.* (2008)].

Chapter 2

Weakly Interacting Bose Gas

2.1 Interactions Between Neutral Atoms

The interactions between neutral atoms are strongly repulsive when the interatomic distance is of the order of the size of the atoms. The origin of this strong repulsion is the Coulomb interaction between electrons; however, this interaction is effective only at atomic-scale distances because at longer distances, electric charges are screened and atoms tend to behave like neutral particles with electronic polarizability.

When the distance between atoms is of the order of 100 Å, an attractive force called the van der Waals force (or fluctuating dipole force) acts between the atoms. A neutral atom has an instantaneous electric dipole moment $\mathbf{p}_1$ that points randomly and averages to zero: $\overline{\mathbf{p}_1} = 0$. An instantaneous dipole moment $\mathbf{p}_1$ of an atom induces the dipole moment $\mathbf{p}_2$ of another atom at a distance of r as

$$\mathbf{p}_2 = \alpha \mathbf{E} \propto -\frac{\alpha \mathbf{p}_1}{r^3},$$

where α is the atomic polarizability. The interaction energy $V(r)$ between the dipole moments at distance r is given by

$$V(r) \propto \frac{\mathbf{p}_1 \mathbf{p}_2}{r^3} \propto -\frac{\alpha \mathbf{p}_1^2}{r^6};$$

this does not vanish upon time averaging $\left(\overline{\mathbf{p}_1^2} \neq 0\right)$ and causes a weak attractive force between the atoms.

Two colliding atoms experience both attractive and repulsive parts of the potential, and the net force can be attractive or repulsive. According to experimental results, ^{1}H, ^{4}He, ^{23}Na, ^{41}K, and ^{87}Rb undergo repulsive interactions, whereas ^{7}Li, ^{39}K, ^{85}Rb, and ^{133}Cs undergo attractive ones at zero magnetic field.

From a theoretical point of view, the presence of a singular short-ranged potential makes it difficult to calculate the physical properties of a system from first principles, because simple perturbative approaches such as the Born approximation break down. In fact, the scattering amplitude $\tilde{V}(\mathbf{k})$ for a potential $V(\mathbf{r})$ is given in the Born approximation by its Fourier transform

$$\tilde{V}(\mathbf{k}) = \int d\mathbf{r} V(\mathbf{r}) e^{-i\mathbf{k}\mathbf{r}} \tag{2.1}$$

that becomes infinite if $V(\mathbf{r})$ diverges. It is therefore of interest to determine whether it is possible to replace a real potential with an effective potential to which perturbation theory is applicable. This is indeed possible when the gas is dilute, *i.e.*, when the average interparticle distance is much longer than the range of the interaction, as discussed in the next section.

As shown above, the van der Waals potential, which acts between the electric dipoles of atoms, is of the form $-C/r^6$. Let r_0 be a characteristic length scale over which the interaction between atoms acts at ultralow temperatures. Since at ultralow temperatures, the total energy of two colliding atoms is nearly zero as compared to that at the other energy scales, we obtain

$$\text{total energy} = \frac{\hbar^2}{M r_0^2} - \frac{C}{r_0^6} = 0 \rightarrow r_0 = \left(\frac{CM}{\hbar^2}\right)^{\frac{1}{4}}, \tag{2.2}$$

where M is the mass of the atom. The coupling constant C is determined from a simple dimensional analysis; the relevant length scale is the Bohr radius

$$a_{\mathrm{B}} = \frac{4\pi\epsilon_0 \hbar^2}{e^2 m_{\mathrm{e}}} \simeq 5.29 \times 10^{-11} \text{ m}, \tag{2.3}$$

where m_{e} is the mass of the electron, and the relevant energy scale is the Coulomb energy $e^2/(4\pi\epsilon_0 a_{\mathrm{B}})$. Thus, C can be expressed in terms of these quantities as

$$C = C_6 \cdot \frac{e^2}{4\pi\epsilon_0 a_{\mathrm{B}}} \cdot a_{\mathrm{B}}^6 = \frac{C_6 e^2}{4\pi\epsilon_0} a_{\mathrm{B}}^5 = C_6 \frac{\hbar^2}{m_{\mathrm{e}}} a_{\mathrm{B}}^4, \tag{2.4}$$

where C_6 is a dimensionless constant that has a value of the order of a few thousands for alkali atoms. Substituting Eq. (2.4) in Eq. (2.2) gives

$$r_0 = \left(C_6 \frac{\mathrm{M}}{m_{\mathrm{e}}}\right)^{\frac{1}{4}} a_{\mathrm{B}}. \tag{2.5}$$

The mass of the atom M is typically 10,000 times that of the electron m_{e}. Considering this and using a typical value $C_6 \sim 10^4$, we obtain $r_0 \sim 100 a_{\mathrm{B}}$

as a rough estimate of the s-wave scattering length of an alkali atom or a typical extent of the least bound state.

A minimal model that takes into account the two important properties of the interatomic potential—strong repulsion at short distance and the van der Waals attraction at long distance—is given by

$$V(r) = \begin{cases} \infty & \text{if } r < R_0; \\ -\frac{C_6}{r^6} & \text{otherwise.} \end{cases} \tag{2.6}$$

The scattering problem with this potential can be solved exactly [Gribakin and Flambaum (1993)] with the result

$$a = \bar{a}\left[1 - \tan\left(\Phi - \frac{3}{8}\pi\right)\right], \tag{2.7}$$

where $\bar{a}$ is the average scattering length and $\Phi = \sqrt{2MC_6}/(2\hbar R_0)$ is the WKB phase shift that the incoming wave experiences due to the potential (2.6). Assuming that Φ distributes randomly from one atomic species to another, we find from Eq. (2.7) that the probability of finding the sign of the scattering length of a given atomic species to be positive is $3/4$.

Atoms are composite particles comprising electrons, protons, and neutrons. They have internal structures that are revealed by their discrete excitation spectra. It is noteworthy that atoms behave as structureless bosons or fermions at ultralow temperatures. This can be explained as follows. The ground-state and excited-state energies of particles confined to a sphere with radius r are of the order of

$$\epsilon \sim \frac{\hbar^2}{2Mr^2}.$$

When the temperature T of the system satisfies the condition

$$k_B T \ll \epsilon, \tag{2.8}$$

the system cannot get excited, and therefore, the internal structures of the composite particles do not manifest themselves. If we consider the transition temperature of BEC as a typical temperature of the system,

$$k_B T_0 \sim \frac{\hbar^2}{M}\left(\frac{N}{V}\right)^{2/3} = \frac{\hbar^2}{M}\frac{1}{R^2},$$

where R is the average interatomic distance, the condition $k_B T \ll \epsilon$ is replaced by

$$\frac{k_B T_0}{\epsilon} \sim \left(\frac{r}{R}\right)^2 \ll 1. \tag{2.9}$$

Condition (2.8) is thus satisfied if $R \gg r$, that is, if the gas is dilute. In the case of alkali BEC, typically, $R \sim 1000$ Å and $r \sim 1$ Å, and thus, $(r/R)^2 \sim 10^{-6}$. Condition (2.9) is thus well satisfied.

It is also noteworthy that gaseous BECs can be realized for alkali atoms whose ground states are known to be metals. The first step in the solidification of atomic gases is for these gases to form molecules. However, for a molecule to be formed, at least three atoms must collide for the laws of energy and momentum conservations to be obeyed; then, two of the three atoms form a molecule, while a third one carries away extra energy and momentum that are released upon molecular formation. In a tenuous gas, the probability of three atoms being found within the range of interaction is very small, and therefore, BECs can actually be formed in an excited many-body state with a sufficiently long lifetime (typically a minute). As long as three-body inelastic collisions are negligible, BECs with repulsive interactions can exist in a thermodynamically stable state. However, BECs with attractive interactions can only exist in a metastable state even if the inelastic collisions are negligible, as it is prone to collapse due to dynamical instability, as discussed in Sec. 3.8.

2.2 Pseudo-Potential Method

If the range of the interaction is much shorter than the mean interatomic distance, it is not necessary to know the details of the short-distance behavior of the interaction potential to describe the low-energy properties of the system. These properties will then be described only by the knowledge of the asymptotic behavior of the wave function before and after scattering characterized by the phase shift of the scattered wave.

Consider the scattering of two particles via a hard-sphere potential of radius a. The wave function of the two particles with respect to the relative coordinate $\mathbf{r}$ satisfies the Schrödinger equation@

$$(\nabla^2 + k^2)\psi(\mathbf{r}) = 0 \quad (r > a) \tag{2.10}$$

subject to the boundary condition

$$\psi(\mathbf{r}) = 0 \quad (r \leq a). \tag{2.11}$$

The hard-core boundary condition (2.11) makes it difficult to calculate the many-body wave function. Huang and Yang [Huang and Yang (1957)] pointed out that boundary condition (2.11) can be effectively replaced with a pseudo-potential. In particular, when only the s-wave scattering is important and as long as the solution with $r \geq a$ is concerned, Eq. (2.10) with boundary condition (2.11) is equivalent to the following equation:

$$(\nabla^2 + k^2)\psi(\mathbf{r}) = 4\pi \frac{\tan(ka)}{k}\delta(\mathbf{r})\frac{\partial}{\partial r}(r\psi(\mathbf{r})) \quad (r \geq 0). \tag{2.12}$$

That is, for $r \geq a$, the solution of Eq. (2.12) coincides with that of Eq. (2.10) with boundary condition (2.11).

To prove this, we first note that the general solution of Eq. (2.10) can be expanded as

$$\psi(\mathbf{r}) = \sum_{l=0}^{\infty}\sum_{m=-l}^{l} A_{lm}[j_l(kr) - \tan\eta_l \, n_l(kr)]Y_{lm}(\Omega), \tag{2.13}$$

where $j_l(x)$ and $n_l(x)$ are the spherical Bessel function and the spherical Neumann function, respectively, and $Y_{lm}(\Omega)$ is the spherical harmonic function. Here, l denotes the relative angular momentum of the two scattering atoms and m is the magnetic quantum number. The boundary condition $\psi(\mathbf{r}) = 0$ at $|r| = a$ is satisfied if

$$\tan\eta_l = \frac{j_l(ka)}{n_l(ka)}. \tag{2.14}$$

For s-wave scattering (*i.e.*, $l = 0$),

$$j_0(x) = \frac{\sin x}{x}, \quad n_0(x) = -\frac{\cos x}{x},$$

and hence, Eq. (2.14) leads to

$$\eta_0 = -ka.$$

Substituting these results in Eq. (2.13) gives

$$\psi(r) \propto j_0(kr) - \tan\eta_0 n_0(kr) = \frac{1}{\cos ka}\frac{\sin k(r-a)}{kr}. \tag{2.15}$$

This solution satisfies the Schrödinger equation (2.10) for $|r| \geq a$; however, it cannot formally be extrapolated to $|r| < a$ because it is singular at $r = 0$. This singularity can be removed if we add an appropriate term proportional to the delta function to the right-hand side of Eq. (2.10). In fact,

$$\begin{aligned}
\nabla^2 \frac{\sin k(r-a)}{kr} &= \frac{\sin k(r-a)}{k}\nabla^2\frac{1}{r} + \frac{1}{r}\nabla^2\frac{\sin k(r-a)}{k} - \frac{2}{r^2}\cos k(r-a) \\
&= 4\pi\delta(\mathbf{r})\frac{\sin ka}{k} - k^2\frac{\sin k(r-a)}{kr},
\end{aligned} \tag{2.16}$$

where

$$\nabla^2 \frac{1}{r} = -4\pi\delta(\mathbf{r}) \tag{2.17}$$

is used. Hence,

$$\begin{aligned}(\nabla^2 + k^2)\frac{\sin k(r-a)}{kr} &= 4\pi\delta(\mathbf{r})\frac{\sin ka}{k} = 4\pi\delta(\mathbf{r})\frac{\tan(ka)}{k}\cos k(r-a) \\ &= 4\pi\frac{\tan(ka)}{k}\delta(\mathbf{r})\frac{\partial}{\partial r}\left(r\frac{\sin k(r-a)}{kr}\right),\end{aligned}$$

and Eq. (2.12) is reproduced.

Equation (2.12) can be rewritten as

$$-\frac{\hbar^2\nabla^2}{2(M/2)}\psi(\mathbf{r}) + V^{\text{pseudo}}(\mathbf{r})\psi(\mathbf{r}) = E\psi(\mathbf{r}), \tag{2.18}$$

where $M/2$ is the reduced mass,

$$E = \frac{\hbar^2 k^2}{M}, \tag{2.19}$$

and

$$V^{\text{pseudo}}(\mathbf{r})\psi(\mathbf{r}) \equiv \frac{4\pi\hbar^2}{M}\frac{\tan(ka)}{k}\delta(\mathbf{r})\frac{\partial}{\partial r}(r\psi(\mathbf{r})). \tag{2.20}$$

We thus find that the hard-sphere boundary condition at $r = a$ can be replaced with the pseudo-potential defined in Eq. (2.20). We also see from Eq. (2.15) that the use of the pseudopotential is equivalent to imposing the following boundary condition

$$\left.\frac{1}{r\psi}\frac{d}{dr}(r\psi)\right|_{r=0} = -k\cot(ka) \tag{2.21}$$

which in the low-energy limit reduces to [Bethe and Peierls (1935)]

$$\lim_{k\to 0}\left.\frac{1}{r\psi}\frac{d}{dr}(r\psi)\right|_{r=0} = -\frac{1}{a}. \tag{2.22}$$

Note that $V^{\text{pseudo}}(\mathbf{r})$ does not satisfy hermiticity. To show this, we take a regular function $\phi(\mathbf{r})$ and a function $\psi(\mathbf{r})$ that has a $1/r$-singularity at $r = 0$ but is regular elsewhere, *i.e.*, $\psi(\mathbf{r}) = \tilde{\psi}(\mathbf{r}) + A/r$, where $\tilde{\psi}(\mathbf{r})$ is a regular part of $\psi(\mathbf{r})$. The hermiticity condition of V is

$$\int \phi^*(V^{\text{pseudo}}\psi)d\mathbf{r} \overset{?}{=} \int \psi(V^{\text{pseudo}}\phi)^* d\mathbf{r}.$$

However, it can be shown by a straightforward calculation that the integral on the left-hand side is proportional to $\phi^*(0)\tilde{\psi}(0)$, which is finite, whereas the one on the right-hand side is proportional to $(\tilde{\psi} + A/r)\phi^*|_{r=0}$, which is infinite. Thus, $V^{\text{pseudo}}(\mathbf{r})$ is not hermitian. Because of this non-hermiticity, $V^{\text{pseudo}}(\mathbf{r})$ introduced in Eq. (2.20) is called as a *pseudo-potential.*

When the energy of the system is so small or when the temperature is so low that $ka \ll 1$, we have

$$\frac{\tan(ka)}{k} = a + \frac{1}{2}(ka)^2 r_{\text{eff}} + \cdots = a + O(a^3), \tag{2.23}$$

where $r_{\text{eff}} = \frac{2}{3}a$ is an effective range of the interaction. When the terms of the order of $O(a^3)$ are ignored, pseudo-potential (2.20) reduces to

$$V^{\text{pseudo}}(\mathbf{r}) = U_0 \delta(\mathbf{r}) \frac{\partial}{\partial r} r, \tag{2.24}$$

where

$$U_0 = \frac{4\pi\hbar^2 a}{M}. \tag{2.25}$$

Moreover, if $\psi(\mathbf{r})$ does not diverge at the origin, the pseudo-potential becomes a simple delta-function potential

$$V(\mathbf{r}) = U_0 \delta(\mathbf{r}). \tag{2.26}$$

In fact, if $\psi(\mathbf{r})$ is regular at the origin, we have

$$\begin{aligned} V^{\text{pseudo}}(\mathbf{r})\psi(\mathbf{r}) &= U_0\delta(\mathbf{r})\frac{\partial}{\partial r}(r\psi(\mathbf{r})) = U_0\delta(\mathbf{r})\left(\psi(\mathbf{r}) + r\frac{\partial\psi(\mathbf{r})}{\partial r}\right) \\ &= U_0\delta(\mathbf{r})\psi(\mathbf{r}). \end{aligned}$$

However, if $\psi(\mathbf{r})$ has a $1/r$-singularity, the operator $\frac{\partial}{\partial r}r$ eliminates it.

It is straightforward to generalize the two-body problem to the N-body problem, where the corresponding Hamiltonian is given by

$$H = -\sum_{i=1}^{N} \frac{\hbar^2}{2M}\nabla_i^2 + \frac{4\pi\hbar^2 a}{M}\sum_{i<j}\delta(\mathbf{r}_i - \mathbf{r}_j)\frac{\partial}{\partial r_{ij}} r_{ij}, \tag{2.27}$$

where $r_{ij} \equiv |\mathbf{r}_i - \mathbf{r}_j|$. It is known that the δ-function potential in three dimensions does not scatter waves. Therefore, the eigenspectrum that is obtained by exactly diagonalizing Eq. (2.27) without the term $(\partial/\partial r_{ij})r_{ij}$ should exactly agree with that of the corresponding free-particle system. However, we can accurately calculate the ground-state energy up to terms of the order of a^3 by using the pseudo-potential as a perturbation, because (i) it is exact to replace $\tan(ka)/k$ with a because the wave number in the ground state is $k = 0$, and (ii) the symmetry requirement of bosons excludes the contribution of the p wave ($l = 1$) that is of the order of a^3.

The ground-state wave function does not diverge at $r = 0$. Therefore, when one is interested in the ground-state properties, Hamiltonian (2.27) can be simplified to

$$H = -\sum_{i=1}^{N} \frac{\hbar^2}{2M} \nabla_i^2 + \frac{4\pi a \hbar^2}{m} \sum_{i<j} \delta(\mathbf{r}_i - \mathbf{r}_j). \tag{2.28}$$

The pseudo-potential that describes a three-body collision can be described as

$$\frac{\hbar^2 K}{M} \sum_{i<j<k} \delta(\mathbf{r}_i - \mathbf{r}_j)\delta(\mathbf{r}_j - \mathbf{r}_k), \tag{2.29}$$

where by the dimensional analysis, K is of the order of a^4.

The wave function corresponding to the s-wave scattering is given from Eq. (2.15) as

$$\psi(\mathbf{r}) = \frac{A}{\cos ka} \frac{\sin k(r-a)}{kr}; \tag{2.30}$$

in the low-energy limit ($k \to 0$), this reduces to

$$\psi(\mathbf{r}) = A\left(1 - \frac{a}{r}\right). \tag{2.31}$$

For $a > 0$, the wave function is suppressed around $r = 0$ as compared to that in the non-interacting case with $a = 0$ [see Fig. 2.1 (a)]. This can be interpreted as being due to repulsive interaction. For $a < 0$, the wave function is enhanced around $r = 0$ as compared to that in the non-interacting case [see Fig. 2.1 (b)]. This can be interpreted as being due to attractive interaction.

2.3 Bogoliubov Theory

2.3.1 *Bogoliubov transformations*

The many-body Hamiltonian (2.27) is expressed in the second-quantized form as

$$\hat{H} = \int d\mathbf{r}\, \hat{\psi}^\dagger(\mathbf{r}) \left(-\frac{\hbar^2 \nabla^2}{2M}\right) \hat{\psi}(\mathbf{r}) + \frac{U_0}{2} \iint d\mathbf{r}_1 d\mathbf{r}_2 \hat{\psi}^\dagger(\mathbf{r}_1)\hat{\psi}^\dagger(\mathbf{r}_2)\delta(\mathbf{r}_1 - \mathbf{r}_2)\frac{\partial}{\partial r_{12}}\left[r_{12}\hat{\psi}(\mathbf{r}_1)\hat{\psi}(\mathbf{r}_2)\right], \tag{2.32}$$

where $r_{12} \equiv |\mathbf{r}_1 - \mathbf{r}_2|$, and the field operators $\hat{\psi}$ and $\hat{\psi}^\dagger$ satisfy the canonical commutation relations

$$\left[\hat{\psi}(\mathbf{r}), \hat{\psi}^\dagger(\mathbf{r}')\right] = \delta(\mathbf{r} - \mathbf{r}'), \left[\hat{\psi}(\mathbf{r}), \hat{\psi}(\mathbf{r}')\right] = 0, \left[\hat{\psi}^\dagger(\mathbf{r}), \hat{\psi}^\dagger(\mathbf{r}')\right] = 0. \tag{2.33}$$

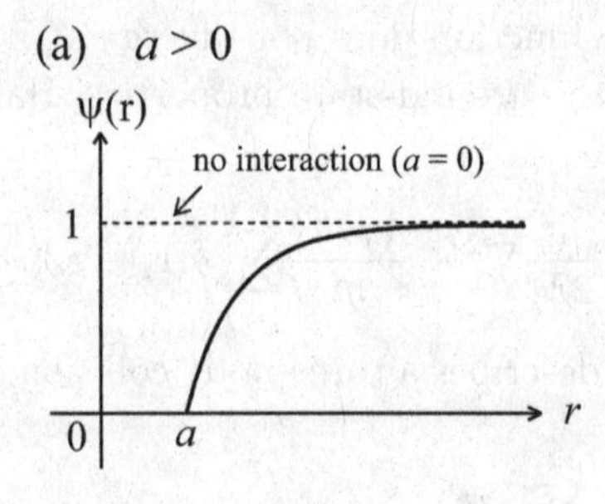

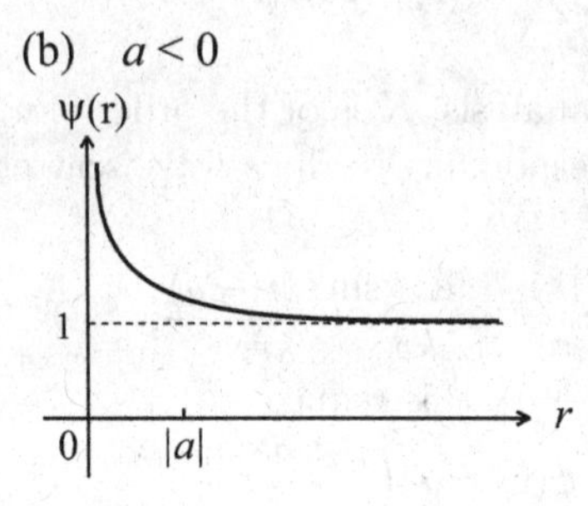

Fig. 2.1 Behavior of the wave function for (a) $a > 0$ (repulsive interaction) and (b) $a < 0$ (attractive interaction).

In Eq. (2.32), the operator $\delta(\mathbf{r}_1 - \mathbf{r}_2)(\partial/\partial r_{12})r_{12}$ can be replaced with $\delta(\mathbf{r}_1 - \mathbf{r}_2)$, unless the divergence of the $1/r$ form appears in the integral. With this proviso, Eq. (2.32) reduces to

$$\hat{H} = \int d^3r \hat{\psi}^\dagger(\mathbf{r}) \left(-\frac{\hbar^2 \nabla^2}{2M} \right) \hat{\psi}(\mathbf{r}) + \frac{U_0}{2} \int d\mathbf{r} \hat{\psi}^\dagger(\mathbf{r})^2 \hat{\psi}(\mathbf{r})^2. \quad (2.34)$$

In Eq. (2.34), by substituting the Fourier expansion of the field operator

$$\hat{\psi}(\mathbf{r}) = \frac{1}{\sqrt{V}} \sum_{\mathbf{k}} \hat{a}_{\mathbf{k}} e^{i\mathbf{k}\mathbf{r}}, \quad (2.35)$$

where V is the volume of the system, we find that

$$\hat{H} = \sum_{\mathbf{k}} \frac{\hbar^2 k^2}{2M} \hat{a}_{\mathbf{k}}^\dagger \hat{a}_{\mathbf{k}} + \frac{U_0}{2V} \sum_{\mathbf{pqk}} \hat{a}_{\mathbf{p}}^\dagger \hat{a}_{\mathbf{q}}^\dagger \hat{a}_{\mathbf{p}+\mathbf{k}} \hat{a}_{\mathbf{q}-\mathbf{k}}, \quad (2.36)$$

where the creation and annihilation operators are assumed to satisfy the canonical commutation relations:

$$[\hat{a}_{\mathbf{k}}, \hat{a}_{\mathbf{k}'}^\dagger] = \delta_{\mathbf{k}\mathbf{k}'}, \; [\hat{a}_{\mathbf{k}}, \hat{a}_{\mathbf{k}'}] = 0, \; [\hat{a}_{\mathbf{k}}^\dagger, \hat{a}_{\mathbf{k}'}^\dagger] = 0. \quad (2.37)$$

Suppose that most of the particles are Bose–Einstein condensed in the $\mathbf{k} = 0$ state, that is,

$$n_0 \approx N, \quad \sum_{\mathbf{k}}{}' n_{\mathbf{k}} \ll N, \tag{2.38}$$

where $n_{\mathbf{k}}$ is the number of particles in the $\mathbf{k} = 0$ state; N, the total number of particles; and $\sum_{\mathbf{k}}'$, the summation over $\mathbf{k}$ excluding $\mathbf{k} = \mathbf{0}$. In the Bogoliubov approximation, we retain terms up to the second order in $\hat{a}_0$ in Eq. (2.36):

$$\begin{aligned}
&\sum_{\mathbf{pqk}} \hat{a}^\dagger_{\mathbf{p}} \hat{a}^\dagger_{\mathbf{q}} \hat{a}_{\mathbf{p+k}} \hat{a}_{\mathbf{q-k}} \\
&\cong \hat{a}_0^{\dagger^2} \hat{a}_0^2 + \hat{a}_0^{\dagger^2} \sum_{\mathbf{k}}{}' \hat{a}_{\mathbf{k}} \hat{a}_{-\mathbf{k}} + \sum_{\mathbf{k}}{}' \hat{a}^\dagger_{\mathbf{k}} \hat{a}^\dagger_{-\mathbf{k}} \hat{a}_0^2 + 4\hat{a}_0^\dagger \hat{a}_0 \sum_{\mathbf{k}}{}' \hat{a}^\dagger_{\mathbf{k}} \hat{a}_{\mathbf{k}} \\
&\cong \left(\hat{a}_0^\dagger \hat{a}_0 + \sum_{\mathbf{k}}{}' \hat{a}^\dagger_{\mathbf{k}} \hat{a}_{\mathbf{k}}\right)^2 + 2\hat{a}_0^\dagger \hat{a}_0 \sum_{\mathbf{k}}{}' \hat{a}^\dagger_{\mathbf{k}} \hat{a}_{\mathbf{k}} + \hat{a}_0^{\dagger^2} \sum_{\mathbf{k}}{}' \hat{a}_{\mathbf{k}} \hat{a}_{-\mathbf{k}} + \hat{a}_0^2 \sum_{\mathbf{k}}{}' \hat{a}^\dagger_{\mathbf{k}} \hat{a}^\dagger_{-\mathbf{k}} \\
&= N^2 + 2\hat{a}_0^\dagger \hat{a}_0 \sum_{\mathbf{k}}{}' \hat{a}^\dagger_{\mathbf{k}} \hat{a}_{\mathbf{k}} + \hat{a}_0^{\dagger^2} \sum_{\mathbf{k}}{}' \hat{a}_{\mathbf{k}} \hat{a}_{-\mathbf{k}} + \hat{a}_0^2 \sum_{\mathbf{k}}{}' \hat{a}^\dagger_{\mathbf{k}} \hat{a}^\dagger_{-\mathbf{k}}.
\end{aligned}$$

For the same order of approximation, we may replace $\hat{a}_0^\dagger \hat{a}_0$, $\hat{a}_0^{\dagger^2}$, and $\hat{a}_0^2$ in the last line with N. Then, Eq. (2.36) reduces to

$$\begin{aligned}
\hat{H} = {} & \frac{U_0 n}{2} N \\
& + \sum_{\mathbf{k}}{}' \left[(\epsilon_k + U_0 n) \hat{a}^\dagger_{\mathbf{k}} \hat{a}_{\mathbf{k}} + \frac{1}{2} U_0 n \left(\hat{a}^\dagger_{\mathbf{k}} \hat{a}^\dagger_{-\mathbf{k}} + \hat{a}_{-\mathbf{k}} \hat{a}_{\mathbf{k}} \right) \right],
\end{aligned} \tag{2.39}$$

where $\epsilon_k = \hbar^2 k^2 / 2M$ and $n = N/V$.

To diagonalize this Hamiltonian, we perform the canonical transformations known as Bogoliubov transformations:

$$\hat{a}_{\mathbf{k}} = \frac{\hat{b}_{\mathbf{k}} - \alpha_k \hat{b}^\dagger_{-\mathbf{k}}}{\sqrt{1 - \alpha_k^2}}, \quad \hat{a}^\dagger_{\mathbf{k}} = \frac{\hat{b}^\dagger_{\mathbf{k}} - \alpha_k \hat{b}_{-\mathbf{k}}}{\sqrt{1 - \alpha_k^2}}, \tag{2.40}$$

where $\hat{b}_{\mathbf{k}}$ and $\hat{b}^\dagger_{\mathbf{k}}$ obey the canonical commutation relations

$$\left[\hat{b}_{\mathbf{k}}, \hat{b}^\dagger_{\mathbf{k}'}\right] = \delta_{\mathbf{k}\mathbf{k}'}, \quad \left[\hat{b}_{\mathbf{k}}, \hat{b}_{\mathbf{k}'}\right] = 0, \quad \left[\hat{b}^\dagger_{\mathbf{k}}, \hat{b}^\dagger_{\mathbf{k}'}\right] = 0. \tag{2.41}$$

We show that $\hat{b}^\dagger_{\mathbf{k}}$ and $\hat{b}_{\mathbf{k}}$ are the creation and annihiliation operators of a quasiparticle called Bogolon. Substituting Eq. (2.40) in Eq. (2.39) and putting

$$x = \sqrt{\frac{\epsilon_k}{U_0 n}}, \tag{2.42}$$

we have

$$
\begin{aligned}
&(\epsilon_k + U_0 n)\, \hat{a}_{\mathbf{k}}^\dagger \hat{a}_{\mathbf{k}} + \frac{1}{2} U_0 n \left(\hat{a}_{\mathbf{k}}^\dagger \hat{a}_{-\mathbf{k}}^\dagger + \hat{a}_{-\mathbf{k}} \hat{a}_{\mathbf{k}} \right) \\
&= \frac{U_0 n}{1-\alpha_k^2} \Big\{ (x^2 + 1 - \alpha_k) \hat{b}_{\mathbf{k}}^\dagger \hat{b}_{\mathbf{k}} + \left[(x^2+1)\alpha_k^2 - \alpha_k \right] \hat{b}_{-\mathbf{k}} \hat{b}_{-\mathbf{k}}^\dagger \\
&\quad - \left(\hat{b}_{\mathbf{k}}^\dagger \hat{b}_{-\mathbf{k}}^\dagger + \hat{b}_{-\mathbf{k}} \hat{b}_{\mathbf{k}} \right) \left[(x^2+1)\, \alpha_k - \frac{1+\alpha_k^2}{2} \right] \Big\}. \qquad (2.43)
\end{aligned}
$$

With the requirement that the coefficient of $\hat{b}_{\mathbf{k}}^\dagger \hat{b}_{-\mathbf{k}}^\dagger + \hat{b}_{-\mathbf{k}} \hat{b}_{\mathbf{k}}$ vanishes, we have

$$
\alpha_k = 1 + x^2 - x\sqrt{x^2+2}, \qquad (2.44)
$$

where the minus sign is taken because $\alpha_k < 1$ [see Eq. (2.40)]. Then, the Hamiltonian becomes

$$
\begin{aligned}
\hat{H} &= \frac{U_0 n}{2} N + \sum_{\mathbf{k}}{}' \sqrt{\epsilon_k(\epsilon_k + 2U_0 n)}\, \hat{b}_{\mathbf{k}}^\dagger \hat{b}_{\mathbf{k}} \\
&\quad - \frac{1}{2} \sum_{\mathbf{k}}{}' \left[U_0 n + \epsilon_k - \sqrt{\epsilon_k(\epsilon_k + 2U_0 n)} \right]. \qquad (2.45)
\end{aligned}
$$

The **k**-summation in the last term diverges. To examine the nature of this divergence, we expand the last term in Eq. (2.45):

$$
U_0 n + \epsilon_k - \sqrt{\epsilon_k(\epsilon_k + 2U_0 n)} = \frac{(U_0 n)^2}{2\epsilon_k} + \cdots. \qquad (2.46)
$$

Thus, the divergence is of the $1/k^2$ form which changes to the $1/r$ form upon Fourier transformation to a real space. This term should be ignored because it is to be eliminated by the action of the operator $\delta(\mathbf{r})(\partial/\partial r)r$ in the pseudo-potential. By subtracting the corresponding term, the last term in Eq. (2.45) converges, giving

$$
\begin{aligned}
&-\sum_{\mathbf{k}}{}' \left[U_0 n + \epsilon_k - \sqrt{\epsilon_k(\epsilon_k + 2U_0 n)} - \frac{(U_0 n)^2}{2\epsilon_k} \right] \\
&= -\frac{\hbar^2}{4M} \frac{V}{2\pi^2} g^{\frac{5}{2}} \int_0^\infty dx x^2 \left(1 + x^2 - x\sqrt{x^2+2} - \frac{1}{2x^2} \right) \\
&= \frac{U_0 n}{2} N \frac{128}{15} \sqrt{\frac{na^3}{\pi}}, \qquad (2.47)
\end{aligned}
$$

where $g \equiv 8\pi a n$ and

$$
\int_0^\infty dx\, x^2 \left(1 + x^2 - x\sqrt{x^2+2} - \frac{1}{2x^2} \right) = -\frac{8}{15}\sqrt{2}
$$

is used to obtain the last equality. Substituting Eq. (2.47) in Eq. (2.45), we obtain

$$\hat{H} = \frac{2\pi a\hbar^2 N^2}{MV}\left(1 + \frac{128}{15}\sqrt{\frac{na^3}{\pi}}\right) + \sum_{\mathbf{k}}{}' \frac{\hbar^2 k}{2M}\sqrt{k^2 + 16\pi na}\, \hat{b}^\dagger_{\mathbf{k}}\hat{b}_{\mathbf{k}}. \quad (2.48)$$

The first two terms on the right-hand side give the ground-state energy E_0. Including higher-order corrections [Wu (1959); Hugenholtz and Pines (1959); Sawada (1959); Braaten, *et al.* (2001)], we obtain

$$E_0 = \frac{2\pi a\hbar^2 N^2}{MV}\left[1 + \frac{128}{15}\sqrt{\frac{na^3}{\pi}} + 8\left(\frac{4\pi}{3} - \sqrt{3}\right) na^3 \ln(na^3) + Cna^3\right], \quad (2.49)$$

where $n = N/V$. Note that the expansion parameter of the theory is na^3. The coefficient C of the last term is contributed from a shape-dependent parameter of the s-wave scattering as well as the p-wave scattering, both of which are proportional to a^3. The last term in Eq. (2.49) involves nonuniversal effects that are sensitive to 3-body physics [Braaten, *et al.* (2001)]; other terms are universal in that they are insensitive to the short-range physics and characterized by the scattering length alone. The first two terms on the right-hand side were obtained by Lee and Yang [Lee and Yang (1957); Lee, *et al.* (1957)], and the third term was obtained by Wu [Wu (1959); Hugenholtz and Pines (1959); Sawada (1959)]. The appearance of the logarithmic term in the expansion implies that it is an asymptotic expansion.

The last term in Eq. (2.48) describes the Bogoliubov excitations called Bogoliubov phonons and the dispersion relation is given by

$$E_{\mathbf{k}} = \frac{\hbar^2 k}{2M}\sqrt{k^2 + 16\pi an}. \quad (2.50)$$

In the long-wavelength limit $k \to 0$, Eq. (2.50) reduces to

$$E_{\mathbf{k}} = \frac{\hbar^2 k}{2M}\sqrt{16\pi an} \propto k. \quad (2.51)$$

Thus, the excitation spectrum at the low-energy limit ($k \to 0$) is identical to that of a phonon $E_{\mathbf{k}} = \hbar ck$, where the sound velocity c is given by

$$c = \frac{\hbar}{M}\sqrt{4\pi an}. \quad (2.52)$$

At the high-energy limit ($k \to \infty$), on the other hand, Eq. (2.50) reduces to

$$E_{\mathbf{k}} = \frac{\hbar^2 k^2}{2M} + \frac{4\pi\hbar^2 a}{M}n, \quad (2.53)$$

implying that excitations move freely in a uniform Hartree potential $4\pi\hbar^2 an/M$. As Eq. (2.50) suggests, the crossover between the phonon spectrum (2.51) and the single-particle spectrum (2.53) occurs at $k = \xi^{-1}$, where

$$\xi = \frac{1}{\sqrt{8\pi an}} \tag{2.54}$$

is known as the healing length.

The sound velocity can also be calculated from the compressibility $\kappa = -(1/V)(\partial V/\partial P_0)$ of the ground state:

$$c = \frac{1}{\sqrt{Mn\kappa}} = \sqrt{-\frac{V}{M}\frac{\partial P_0}{\partial V}}, \tag{2.55}$$

where

$$P_0 = -\frac{\partial E_0}{\partial V} \tag{2.56}$$

is the pressure of the ground state. Considering the first two terms in Eq. (2.49), we obtain

$$c = \frac{\hbar}{M}\sqrt{4\pi an}\left(1 + 16\sqrt{\frac{na^3}{\pi}}\right). \tag{2.57}$$

The first term on the right-hand side is equivalent to Eq. (2.52), and the second term gives the Lee–Yang correction to it. Experiments on a zero-sound wave were performed for a sodium BEC [Andrews, *et al.* (1997,1998)].

2.3.2 *Bogoliubov ground state*

The eigenfunction of the diagonalized Hamiltonian (2.48) can be expanded in terms of the Fock state $|n_{\mathbf{k}_1}, n_{\mathbf{k}_2}, \cdots\rangle$ in which there are $n_{\mathbf{k}_1}$ quasiparticles with wave vector $\mathbf{k}_1$, $n_{\mathbf{k}_2}$ quasiparticles with wave vector $\mathbf{k}_2$, etc. The number operator of quasiparticles with wave vector $\mathbf{k}$ is given by $\hat{b}_{\mathbf{k}}^\dagger \hat{b}_{\mathbf{k}}$ and satisfies the eigenvalue equation

$$\hat{b}_{\mathbf{k}}^\dagger \hat{b}_{\mathbf{k}} |\cdots, n_{\mathbf{k}}, \cdots\rangle = n_{\mathbf{k}} |\cdots, n_{\mathbf{k}}, \cdots\rangle. \tag{2.58}$$

The Bogoliubov ground state $|\psi_0\rangle \equiv |0, 0, \cdots\rangle$ is the state in which no quasiparticles are present and therefore satisfies

$$\hat{b}_{\mathbf{k}} |\psi_0\rangle = 0 \tag{2.59}$$

for all $\mathbf{k}$. Since the Bogoliubov transformations (2.40) connect only the states with wave vectors $\mathbf{k}$ and $-\mathbf{k}$, $|\psi_0\rangle$ may be expanded in terms of the states $|n_1, m_1; n_2, m_2; \cdots\rangle$ in which n_i particles are present in the $\mathbf{k}_i$ state and m_i particles in the $-\mathbf{k}_i$ state:

$$|\psi_0\rangle = \sum_{n_1,m_1=0}^{\infty} \sum_{n_2,m_2=0}^{\infty} \cdots \left(C^{\mathbf{k}_1}_{n_1 m_1} C^{\mathbf{k}_2}_{n_2 m_2} \cdots\right) |n_1, m_1; n_2, m_2; \cdots\rangle. \quad (2.60)$$

Substituting this in Eq. (2.59) and noting that $\hat{b}_{\mathbf{k}}$ is related to the creation and annihiliation operators of original particles as

$$\hat{b}_{\mathbf{k}} = \frac{1}{\sqrt{1-\alpha_k^2}} \left(\hat{a}_{\mathbf{k}} + \alpha_k \hat{a}^\dagger_{-\mathbf{k}}\right), \quad (2.61)$$

we have

$$\begin{aligned} 0 &= \sum_{n,m=0}^{\infty} C^{\mathbf{k}}_{nm} \left[\sqrt{n}|n-1, m\rangle + \alpha_k \sqrt{m+1}|n, m+1\rangle\right] \\ &= \sum_{n,m=0}^{\infty} \left[C^{\mathbf{k}}_{n+1,m}\sqrt{n+1} + C^{\mathbf{k}}_{n,m-1}\alpha_k\sqrt{m}\right] |n, m\rangle, \end{aligned}$$

where we define $C^{\mathbf{k}}_{n,-1} \equiv 0$. From the orthogonality of the basis states, we obtain

$$C^{\mathbf{k}}_{n+1,m}\sqrt{n+1} + C^{\mathbf{k}}_{n,m-1}\alpha_k\sqrt{m} = 0. \quad (2.62)$$

Substituting $m = 0$, we have $C^{\mathbf{k}}_{n+1,0}\sqrt{n+1} = 0$; hence, $C^{\mathbf{k}}_{n+1,0} = 0$ for $n \geq 0$. We also obtain $C^{\mathbf{k}}_{0,n+1} = 0$ by following a similar procedure for $\hat{b}_{-\mathbf{k}}$. If we assume $C^{\mathbf{k}}_{n,m-1} = 0$ for $n \neq m-1$, we have $C^{\mathbf{k}}_{n+1,m} = 0$ from Eq. (2.62). By mathematical induction, all the off-diagonal components vanish. The diagonal component is determined from Eq. (2.62) as

$$C^{\mathbf{k}}_{mm} + \alpha_k C^{\mathbf{k}}_{m-1,m-1} = 0 \rightarrow C^{\mathbf{k}}_{mm} = (-\alpha_k)^m C^{\mathbf{k}}_{00}. \quad (2.63)$$

Here, $C^{\mathbf{k}}_{00}$ is determined from the normalization condition of the wave function. We thus find that the ground state is a state in which pairs of particles with wave vectors $\mathbf{k}$ and $-\mathbf{k}$ are excited. Let $|n_1, n_2, \cdots\rangle$ be the state in which there are n_1 pairs of particles with wave vectors $\mathbf{k}_1$ and $-\mathbf{k}_1$, n_2 pairs of particles with wave vectors $\mathbf{k}_2$ and $-\mathbf{k}_2$, etc. Then, $|\psi_0\rangle$ can be expressed as

$$|\psi_0\rangle = Z \sum_{n_1=0}^{\infty} \sum_{n_2=0}^{\infty} \cdots \left[(-\alpha_1)^{n_1} (-\alpha_2)^{n_2} \cdots\right] |n_1, n_2, \cdots\rangle |\phi_0\rangle, \quad (2.64)$$

where $|\phi_0\rangle$ is a state with $\mathbf{k} = 0$ [see Eq. (2.80)], and Z is the normalization factor, which is determined from

$$\langle\psi_0|\psi_0\rangle = 1 = Z^2 \sum_{n_1=0}^{\infty} \sum_{n_2=0}^{\infty} \cdots (-\alpha_1)^{2n_1} (-\alpha_2)^{2n_2} \cdots = Z^2 \prod_{k_x>0} \left(1-\alpha_k^2\right)^{-1},$$

where $\prod_{k_x>0}$ implies that the product is to be taken over all k_y and k_z and all positive k_x. Hence,

$$Z = \prod_{k_x>0} \sqrt{1-\alpha_k^2} = \exp\left[-\frac{4}{9}(3\pi - 8)N\sqrt{\frac{na^3}{\pi}}\right]. \tag{2.65}$$

Note that Z is the overlap integral between the ground states with and without interaction. Equation (2.65) shows that the interacting ground state becomes orthogonal to the noninteracting ground state for $N \to \infty$. To prove the last equality in Eq. (2.65), we note that

$$\prod_{k_x>0} \sqrt{1-\alpha_k^2} = \exp\left[\frac{1}{2}\sum_{k_x>0} \ln\left(1-\alpha_k^2\right)\right].$$

Here, the sum in the exponent can be converted into an integral as

$$\begin{aligned} \frac{1}{2}\frac{V}{(2\pi)^3}2\pi\int_0^\infty k^2 dk \ln\left(1-\alpha_k^2\right) &= \frac{V}{8\pi^2}\int_0^\infty dk \frac{k^3}{3}\frac{2\alpha_k}{1-\alpha_k^2}\frac{d\alpha_k}{dk} \\ = \frac{V}{12\pi^2}\int_0^\infty dk k^3 \frac{\alpha_k}{1-\alpha_k^2}\frac{d\alpha_k}{dk} &= -\frac{V}{12\pi^2}\int_0^1 d\alpha_k k^3 \frac{\alpha_k}{1-\alpha_k^2}. \end{aligned} \tag{2.66}$$

We then eliminate k in favor of α_k using the relation

$$k = \sqrt{8\pi a n}\frac{1-\alpha_k}{\sqrt{2\alpha_k}}. \tag{2.67}$$

The last term in Eq. (2.66) then becomes

$$\begin{aligned} &-\frac{V}{12\pi^2}(8\pi a n)^{\frac{3}{2}}\frac{1}{\sqrt{8}}\int_0^1 d\alpha \frac{(1-\alpha)^3}{\alpha^{3/2}}\frac{\alpha}{1-\alpha^2} \\ &= -\frac{2N}{3}\sqrt{\frac{na^3}{\pi}}\int_0^1 d\alpha \frac{1}{\sqrt{\alpha}}\frac{(1-\alpha)^2}{1+\alpha}. \end{aligned} \tag{2.68}$$

Substituting $\alpha = t^2$, the integral can be evaluated as

$$2\int_0^1 t dt \frac{1}{t}\frac{\left(1-t^2\right)^2}{1+t^2} = \frac{2}{3}(3\pi - 8).$$

We thus obtain Eq. (2.65).

The Bogoliubov ground state (2.64) can also be expressed in a second-quantized form as

$$|\psi_0\rangle = Z\exp\left[-\sum_{k_x>0}\alpha_k \hat{a}_{\mathbf{k}}^\dagger \hat{a}_{-\mathbf{k}}^\dagger\right]|\phi_0\rangle. \tag{2.69}$$

To prove this, it is sufficient to consider only the terms with $\mathbf{k} = \mathbf{k}_1$. Then, we have

$$\exp\left(-\alpha_{k_1}\hat{a}^\dagger_{\mathbf{k}_1}\hat{a}^\dagger_{-\mathbf{k}_1}\right)|\phi_0\rangle = \sum_{n_1=0}^{\infty}\frac{(-\alpha_{k_1})^{n_1}}{n_1!}\left(\hat{a}^\dagger_{\mathbf{k}_1}\hat{a}^\dagger_{-\mathbf{k}_1}\right)^{n_1}|\phi_0\rangle$$
$$= \sum_{n_1=0}^{\infty}(-\alpha_{k_1})^{n_1}|n_1\rangle|\phi_0\rangle,$$

which is the same as the corresponding term in Eq. (2.64). Equation (2.69) can also be proved more directly as follows. Substituting Eq. (2.61) in Eq. (2.59) gives

$$\hat{b}_{\mathbf{k}}|\psi_0\rangle = \frac{1}{\sqrt{1-\alpha_k^2}}\left(\hat{a}_{\mathbf{k}} + \alpha_k\hat{a}^\dagger_{-\mathbf{k}}\right)|\psi_0\rangle = 0. \tag{2.70}$$

From the commutation relation $\left[\hat{a}_{\mathbf{k}}, \hat{a}^\dagger_{\mathbf{k}}\right] = 1$, we may formally write $\hat{a}_{\mathbf{k}} = d/d\hat{a}^\dagger_{\mathbf{k}}$; then, Eq. (2.70) is expressed as

$$\left(\frac{d}{d\hat{a}^\dagger_{\mathbf{k}}} + \alpha_k\hat{a}^\dagger_{-\mathbf{k}}\right)|\psi_0\rangle = 0. \tag{2.71}$$

By inspection, we find that Eq. (2.69) is indeed a solution of Eq. (2.71).

2.3.3 *Low-lying excitations and condensate fraction*

The state in which one Bogoliubov phonon with momentum $\hbar\mathbf{k}$ is excited is given by

$$|\psi_{\mathbf{k}}\rangle = \hat{b}^\dagger_{\mathbf{k}}|\psi_0\rangle = \frac{\hat{a}^\dagger_{\mathbf{k}} + \alpha_k\hat{a}_{-\mathbf{k}}}{\sqrt{1-\alpha_k^2}}|\psi_0\rangle. \tag{2.72}$$

Substituting Eq. (2.64) on the right-hand side of this equation gives

$$\frac{\hat{a}^\dagger_{\mathbf{k}} + \alpha_k\hat{a}_{-\mathbf{k}}}{\sqrt{1-\alpha_k^2}} Z\sum\left[\cdots(-\alpha_k)^{n_{\mathbf{k}}}\cdots\right]|\cdots, n_{\mathbf{k}}, \cdots\rangle, \tag{2.73}$$

where $|n_{\mathbf{k}}\rangle$ represents the state that has $n_{\mathbf{k}}$ pairs of particles with wave vectors $\mathbf{k}$ and $-\mathbf{k}$, namely $|n_{\mathbf{k}}\rangle \equiv |n_{\mathbf{k}}, n_{-\mathbf{k}}\rangle$. Noting the fact that $n_{\mathbf{k}} = n_{-\mathbf{k}}$, Eq. (2.73) becomes

$$Z\sum\cdots\frac{(-\alpha_k)^{n_{\mathbf{k}}}}{\sqrt{1-\alpha_k^2}}\left[\sqrt{n_{\mathbf{k}}+1}|n_{\mathbf{k}}+1, n_{\mathbf{k}}\rangle + \alpha_k\sqrt{n_{\mathbf{k}}}|n_{\mathbf{k}}, n_{\mathbf{k}}-1\rangle\right]$$
$$= Z\sum\cdots\frac{(-\alpha_k)^{n_{\mathbf{k}}}}{\sqrt{1-\alpha_k^2}}\left(1-\alpha_k^2\right)\sqrt{n_{\mathbf{k}}+1}|n_{\mathbf{k}}+1, n_{\mathbf{k}}\rangle = \sqrt{1-\alpha_k^2}\hat{a}^\dagger_{\mathbf{k}}|\psi_0\rangle.$$

We thus obtain[1]

$$|\psi_{\mathbf{k}}\rangle = \hat{b}_{\mathbf{k}}^{\dagger}|\psi_0\rangle = \sqrt{1-\alpha_k^2}\hat{a}_{\mathbf{k}}^{\dagger}|\psi_0\rangle. \tag{2.74}$$

Note that the state (2.74) is normalized to unity for

$$\langle\psi_{\mathbf{k}}|\psi_{\mathbf{k}}\rangle = \langle\psi_0|\hat{b}_{\mathbf{k}}\hat{b}_{\mathbf{k}}^{\dagger}|\psi_0\rangle = \langle\psi_0|\hat{b}_{\mathbf{k}}^{\dagger}\hat{b}_{\mathbf{k}}+1|\psi_0\rangle = \langle\psi_0|\psi_0\rangle = 1, \tag{2.75}$$

where we used $\hat{b}_{\mathbf{k}}|\psi_0\rangle = 0$. The one-phonon state (2.74) is therefore formed when a particle with momentum $\hbar\mathbf{k}$ is added to the Bogoliubov ground state $|\psi_0\rangle$. The cordinate representation of the one-phonon state (2.74) is obtained from Eq. (B.20) in Appendix B as

$$\psi_{\mathbf{k}}(\mathbf{r}_1, \mathbf{r}_2, \cdots, \mathbf{r}_N) = \sum_{n=1}^{N} e^{i\mathbf{k}\mathbf{r}_n}\psi_0(\mathbf{r}_1, \mathbf{r}_2, \cdots, \mathbf{r}_N). \tag{2.76}$$

In the Bogoliubov ground state $|\psi_0\rangle$, bosons are virtually excited to states with $\mathbf{k} \neq 0$ even at $T = 0$ due to the interactions. The average number of virtually excited bosons with wave vector $\mathbf{k}$ is obtained from Eqs. (2.40) and (2.75) by

$$\langle n_{\mathbf{k}}\rangle = \langle\psi_0|\hat{a}_{\mathbf{k}}^{\dagger}\hat{a}_{\mathbf{k}}|\psi_0\rangle = \frac{\alpha_k^2}{1-\alpha_k^2} \quad (k \neq 0). \tag{2.77}$$

Summing Eq. (2.77) over $\mathbf{k} \neq \mathbf{0}$ gives the total number of virtually excited bosons:

$$\sum_{\mathbf{k}}{}' \langle n_{\mathbf{k}}\rangle = \sum_{k}{}' \frac{\alpha_k^2}{1-\alpha_k^2} = \frac{V}{(2\pi)^3}4\pi\int_0^{\infty} k^2 dk \frac{\alpha_k^2}{1-\alpha_k^2}.$$

---

[1]Another proof of Eq. (2.74): From Eq. (2.69) and

$$e^{\alpha_k \hat{a}_{\mathbf{k}}^{\dagger}\hat{a}_{-\mathbf{k}}^{\dagger}}\hat{a}_{-\mathbf{k}}e^{-\alpha\hat{a}_{\mathbf{k}}^{\dagger}\hat{a}_{-\mathbf{k}}^{\dagger}} = -\alpha_k\hat{a}_k^{\dagger} + \hat{a}_{-\mathbf{k}},$$

we have

$$\begin{aligned}\hat{a}_{-\mathbf{k}}|\psi_0\rangle &= Z\hat{a}_{-\mathbf{k}}e^{-\sum\alpha_k\hat{a}_{\mathbf{k}}^{\dagger}\hat{a}_{-\mathbf{k}}^{\dagger}}|\phi_0\rangle \\ &= Ze^{-\sum\alpha_k\hat{a}_{\mathbf{k}}^{\dagger}\hat{a}_{-\mathbf{k}}^{\dagger}}\left(-\alpha_k\hat{a}_k^{\dagger} + \hat{a}_{-\mathbf{k}}\right)|\phi_0\rangle = -\alpha_k\hat{a}_{\mathbf{k}}^{\dagger}|\psi_0\rangle.\end{aligned}$$

Therefore, we obtain

$$\left(\hat{a}_{\mathbf{k}}^{\dagger} + \alpha_k\hat{a}_{-\mathbf{k}}\right)|\psi_0\rangle = (1-\alpha_k^2)\,\hat{a}_{\mathbf{k}}^{\dagger}|\psi_0\rangle,$$

which proves Eq. (2.74).

By performing integrations in a manner similar to Eqs. (2.66)–(2.68), we obtain

$$\begin{aligned}\sum_{\mathbf{k}}{}' \langle n_{\mathbf{k}} \rangle &= \frac{V}{3\pi^2} \int_0^1 d\alpha_k k^3 \frac{\alpha_k}{(1-\alpha_k^2)^2} \\ &= \frac{V}{3\pi^2} \left(\frac{8\pi a N}{V} \right)^{\frac{3}{2}} \frac{1}{\sqrt{8}} \int_0^1 d\alpha \frac{(1-\alpha)^3}{\alpha^{3/2}} \frac{\alpha}{(1-\alpha^2)^2} \\ &= \frac{8N}{3} \sqrt{\frac{a^3 N}{\pi V}}.\end{aligned}$$

Hence, the total number of virtually excited bosons is given by

$$\sum_k{}' \langle n_{\mathbf{k}} \rangle = \frac{8}{3} \sqrt{\frac{na^3}{\pi}} N, \tag{2.78}$$

and the fraction of the bosons remaining in the $\mathbf{k} = \mathbf{0}$ state, which is referred to as the condensate fraction, is given by

$$\frac{\langle n_0 \rangle}{N} = 1 - \frac{8}{3} \sqrt{\frac{na^3}{\pi}}. \tag{2.79}$$

This result shows that for a dilute gas with $na^3 \ll 1$, the condition (2.38), which is assumed in deriving the effective Hamiltonian (2.39), is satisfied.

2.3.4 *Properties of Bogoliubov ground state*

To investigate the properties of the Bogoliubov ground state, it is computationally convenient to assume a coherent state for $|\phi_0\rangle$:

$$|\phi_0\rangle = e^{-\frac{|\phi_0|^2}{2}} e^{\phi_0 \hat{a}_0^\dagger} |\text{vac}\rangle = e^{-\frac{|\phi_0|^2}{2}} \sum_{n_0=0}^{\infty} \frac{\phi_0^{n_0}}{\sqrt{n_0!}} |n_0\rangle, \tag{2.80}$$

where $|\phi_0\rangle$ is an eigenstate of the annihilation operator

$$\hat{a}_0 |\phi_0\rangle = \phi_0 |\phi_0\rangle.$$

Substituting Eq. (2.80) in Eq. (2.69), we obtain the ground-state wave function in the Bogoliubov approximation:

$$|\psi_0\rangle = N \exp\left[\phi_0 \hat{a}_0^\dagger - \sum_{k_x > 0} \alpha_k \hat{a}_{\mathbf{k}}^\dagger \hat{a}_{-\mathbf{k}}^\dagger \right] |\text{vac}\rangle, \tag{2.81}$$

where

$$N = e^{-\frac{|\phi|^2}{2}} Z = e^{-\frac{|\phi|^2}{2}} \exp\left[-\frac{4}{9} (3\pi - 8) N \sqrt{\frac{na^3}{\pi}} \right]. \tag{2.82}$$

The expectation value of the annihilation operator over $|\psi_0\rangle$ does not vanish; it is given as

$$\begin{aligned}\langle \hat{a}_0 \rangle &= N^2 \langle \text{vac}| e^{\phi_0^* \hat{a}_0 - \sum_{k_x>0} \alpha_k^* \hat{a}_{\mathbf{k}} \hat{a}_{-\mathbf{k}}} \hat{a}_0 e^{\phi_0 \hat{a}_0^\dagger - \sum_{k_x>0} \alpha_k \hat{a}_{\mathbf{k}}^\dagger \hat{a}_{-\mathbf{k}}^\dagger} |\text{vac}\rangle \\ &= N^2 \frac{\partial}{\partial \phi_0^*} N^{-2} = \phi_0, \end{aligned} \tag{2.83}$$

where the last equality is obtained by using Eq. (2.82). The average number of particles in the $\mathbf{k} = 0$ state is given by

$$n_0 \equiv \langle \hat{a}_0^\dagger \hat{a}_0 \rangle = \langle \hat{a}_0 \hat{a}_0^\dagger - 1 \rangle = N^2 \left(\frac{\partial^2}{\partial \phi_0 \partial \phi_0^*} - 1 \right) N^{-2} = |\phi_0|^2. \tag{2.84}$$

On the other hand, the average number of particles in the $\mathbf{k} \neq 0$ state is given by

$$\begin{aligned} n_{\mathbf{k}} \equiv \langle \hat{a}_{\mathbf{k}}^\dagger \hat{a}_{\mathbf{k}} \rangle &= \left(1 - |\alpha_k|^2\right) \langle \text{vac}| e^{\alpha_k^* \hat{a}_{\mathbf{k}} \hat{a}_{-\mathbf{k}}} \hat{a}_{\mathbf{k}}^\dagger \hat{a}_{\mathbf{k}} e^{\alpha_k \hat{a}_{\mathbf{k}}^\dagger \hat{a}_{-\mathbf{k}}^\dagger} |\text{vac}\rangle \\ &= \left(1 - |\alpha_k|^2\right) \sum_{n,m=0}^{\infty} (-1)^{n+m} \alpha_k^{*n} \alpha_k^m \langle n_{\mathbf{k}}, n_{-\mathbf{k}} | \hat{a}_{\mathbf{k}}^\dagger \hat{a}_{\mathbf{k}} | m_{\mathbf{k}}, m_{-\mathbf{k}} \rangle \\ &= \sum_{n=0}^{\infty} n |\alpha_k|^{2n} = \frac{|\alpha_k|^2}{1 - |\alpha_k|^2}. \end{aligned} \tag{2.85}$$

Solving Eq. (2.85) for $|\alpha_k|$, we obtain

$$|\alpha_k| = \sqrt{\frac{n_{\mathbf{k}}}{1 + n_{\mathbf{k}}}}. \tag{2.86}$$

The expectation value of $\hat{a}_{\mathbf{k}} \hat{a}_{-\mathbf{k}}$, which is referred to as the anomalous average, is given by

$$\begin{aligned} \chi_k \equiv \langle \hat{a}_{\mathbf{k}} \hat{a}_{-\mathbf{k}} \rangle &= N^{-2} \left(-\frac{\partial}{\partial \alpha_k^*} \right) N^2 \\ &= \frac{\alpha_k}{1 - |\alpha_k|^2} = \frac{\alpha_k}{|\alpha_k|} \sqrt{n_{\mathbf{k}}(1 + n_{\mathbf{k}})}. \end{aligned} \tag{2.87}$$

Let us assume ϕ_0 and α_k to be variational parameters and determine their expression by minimizing the expectation value of $\hat{H} - \mu \hat{N}$, where

$$\hat{H} - \mu \hat{N} = \sum_{\mathbf{k}} (\epsilon_{\mathbf{k}} - \mu) \hat{a}_{\mathbf{k}}^\dagger \hat{a}_{\mathbf{k}} + \frac{1}{2} \sum_{\mathbf{qkk}'} V_{\mathbf{q}} \hat{a}_{\mathbf{k}+\mathbf{q}}^\dagger \hat{a}_{\mathbf{k}'-\mathbf{q}}^\dagger \hat{a}_{\mathbf{k}} \hat{a}_{\mathbf{k}'}. \tag{2.88}$$

The expectation value of the first term on the right-hand side is $\sum_{\mathbf{k}}(\epsilon_{\mathbf{k}} - \mu)n_{\mathbf{k}}$, and that of the second term is approximated to give

$$\begin{aligned}\frac{1}{2}\sum_{\mathbf{qkk}'} V_{\mathbf{q}}\left\langle \hat{a}^\dagger_{\mathbf{k}+\mathbf{q}}\hat{a}^\dagger_{\mathbf{k}'-\mathbf{q}}\hat{a}_{\mathbf{k}}\hat{a}_{\mathbf{k}'}\right\rangle \cong \frac{1}{2}V_0\sum_{\mathbf{kk}'}\left\langle \hat{a}^\dagger_{\mathbf{k}}\hat{a}^\dagger_{\mathbf{k}'}\hat{a}_{\mathbf{k}}\hat{a}_{\mathbf{k}'}\right\rangle \\ +\frac{1}{2}\sum_{\mathbf{q}(\neq 0),\mathbf{k}} V_{\mathbf{q}}\left\langle \hat{a}^\dagger_{\mathbf{k}+\mathbf{q}}\hat{a}^\dagger_{\mathbf{k}}\hat{a}_{\mathbf{k}}\hat{a}_{\mathbf{k}+\mathbf{q}}\right\rangle \\ +\frac{1}{2}\sum_{\mathbf{q}(\neq 0),\mathbf{k}} V_{\mathbf{q}}\left\langle \hat{a}^\dagger_{\mathbf{k}+\mathbf{q}}\hat{a}^\dagger_{-\mathbf{k}-\mathbf{q}}\hat{a}_{\mathbf{k}}\hat{a}_{-\mathbf{k}}\right\rangle,\end{aligned} \tag{2.89}$$

where each term on the right-hand side is obtained by substituting $\mathbf{q} = 0$ (Hartree), $\mathbf{k}' = \mathbf{k} + \mathbf{q}$ (Fock), or $\mathbf{k}' = -\mathbf{k}$ (pairing). The Hartree term gives

$$\begin{aligned}\sum_{\mathbf{kk}'}\left\langle \hat{a}^\dagger_{\mathbf{k}}\hat{a}^\dagger_{\mathbf{k}'}\hat{a}_{\mathbf{k}}\hat{a}_{\mathbf{k}'}\right\rangle &= \sum_{\mathbf{k}\neq\mathbf{k}'} n_{\mathbf{k}}n_{\mathbf{k}'} + \sum_{\mathbf{k}} n_{\mathbf{k}}(n_{\mathbf{k}} - 1) + \sum_{\mathbf{k}}\left\langle \hat{a}^\dagger_{\mathbf{k}}\hat{a}^\dagger_{-\mathbf{k}}\hat{a}_{\mathbf{k}}\hat{a}_{-\mathbf{k}}\right\rangle \\ &= N(N-1) + \sum_{\mathbf{k}}\left\langle \hat{a}^\dagger_{\mathbf{k}}\hat{a}^\dagger_{-\mathbf{k}}\hat{a}_{\mathbf{k}}\hat{a}_{-\mathbf{k}}\right\rangle;\end{aligned} \tag{2.90}$$

the Fock term gives

$$\begin{aligned}&\sum_{\mathbf{q}(\neq 0),\mathbf{k}} V_{\mathbf{q}}\left\langle \hat{a}^\dagger_{\mathbf{k}+\mathbf{q}}\hat{a}^\dagger_{\mathbf{k}}\hat{a}_{\mathbf{k}}\hat{a}_{\mathbf{k}+\mathbf{q}}\right\rangle \\ &= \sum_{\mathbf{q}(\neq 0)} V_{\mathbf{q}}\left(\langle \hat{a}^\dagger_{\mathbf{q}}\hat{a}_{\mathbf{q}}\rangle n_0 + \left\langle \hat{a}^\dagger_{-\mathbf{q}}\hat{a}_{-\mathbf{q}}\right\rangle n_0 + \sum_{\mathbf{k}(\neq 0,-\mathbf{q})}\left\langle \hat{a}^\dagger_{\mathbf{k}+\mathbf{q}}\hat{a}_{\mathbf{k}+\mathbf{q}}\hat{a}^\dagger_{\mathbf{k}}\hat{a}_{\mathbf{k}}\right\rangle\right) \\ &= 2n_0\sum_{\mathbf{q}(\neq 0)} V_{\mathbf{q}}n_{\mathbf{q}} + \sum_{\mathbf{k}(\neq 0,-\mathbf{q}),\mathbf{q}} V_{\mathbf{q}}n_{\mathbf{k}+\mathbf{q}}n_{\mathbf{k}},\end{aligned} \tag{2.91}$$

where in the first and second terms of the middle equation, we have assumed $\mathbf{k} = 0$ and $\mathbf{k} = -\mathbf{q}$, respectively. The pairing term combined with the last term in Eq. (2.90) gives

$$\begin{aligned}&\sum_{\mathbf{kq}} V_{\mathbf{q}}\left\langle \hat{a}^\dagger_{\mathbf{k}+\mathbf{q}}\hat{a}^\dagger_{-\mathbf{k}-\mathbf{q}}\hat{a}_{\mathbf{k}}\hat{a}_{-\mathbf{k}}\right\rangle \\ &= \sum_{\mathbf{q}} V_{\mathbf{q}}\left(\left\langle \hat{a}^\dagger_{\mathbf{q}}\hat{a}^\dagger_{-\mathbf{q}}\right\rangle\left\langle \hat{a}_0^2\right\rangle + \left\langle \hat{a}_0^{\dagger 2}\right\rangle\left\langle \hat{a}_{-\mathbf{q}}\hat{a}_{\mathbf{q}}\right\rangle + \sum_{\mathbf{k}(\neq 0,-\mathbf{q})}\left\langle \hat{a}^\dagger_{\mathbf{k}+\mathbf{q}}\hat{a}_{-\mathbf{k}-\mathbf{q}}\hat{a}^\dagger_{\mathbf{k}}\hat{a}_{-\mathbf{k}}\right\rangle\right) \\ &= \sum_{\mathbf{q}} V_{\mathbf{q}}\left[\chi^*_{\mathbf{q}}\phi^2 + \phi^{*2}\chi_{\mathbf{q}} + \sum_{\mathbf{k}}\chi^*_{\mathbf{k}+\mathbf{q}}\chi_{\mathbf{k}}\right].\end{aligned} \tag{2.92}$$

Using the relations

$$\phi \equiv \sqrt{n_0} e^{i\theta_0}, \quad \chi_{\mathbf{k}} = \frac{\alpha_k}{|\alpha_k|}\sqrt{n_{\mathbf{k}}(1+n_{\mathbf{k}})} = \sqrt{n_{\mathbf{k}}(1+n_{\mathbf{k}})}e^{i\theta_{\mathbf{k}}}, \quad (2.93)$$

we obtain

$$\sum_{\mathbf{qk}} V_{\mathbf{q}} \left\langle \hat{a}^\dagger_{\mathbf{k+q}} \hat{a}^\dagger_{-\mathbf{k-q}} \hat{a}_{\mathbf{k}} \hat{a}_{-\mathbf{k}} \right\rangle = 2n_0 \sum_{\mathbf{q}} V_{\mathbf{q}} |\chi_{\mathbf{q}}| \cos(2\theta_0 - \theta_{\mathbf{q}}) + \sum_{\mathbf{kq}} V_{\mathbf{q}} \chi^*_{\mathbf{k+q}} \chi_{\mathbf{k}}.$$

Adding the above results, we obtain

$$\left\langle \hat{H} - \mu \hat{H} \right\rangle = \sum_{\mathbf{k}} (\epsilon_{\mathbf{k}} - \mu) n_{\mathbf{k}} + \frac{V_0}{2} N(N-1) + n_0 \sum_{\mathbf{q} \neq 0} V_{\mathbf{q}} n_{\mathbf{q}}$$
$$+ \frac{1}{2} \sum_{\mathbf{k,q}} V_{\mathbf{q}} n_{\mathbf{k+q}} n_{\mathbf{k}} + n_0 \sum_{\mathbf{q}} V_q |\chi_{\mathbf{q}}| \cos(2\theta_0 - \theta_{\mathbf{q}}) + \frac{1}{2} \sum_{\mathbf{kq}} V_{\mathbf{q}} \chi^*_{\mathbf{k+q}} \chi_{\mathbf{k}}, \quad (2.94)$$

where the second term on the right-hand side describes the Hartree energy, the third and fourth terms describe the Fock exchange energy, and the last two terms arise from pair correlation, also called anomalous correlation, which depends on the phase difference between the condensate and virtual excitations.

For the case of repulsive interaction $V_{\mathbf{q}} > 0$, the states with $\mathbf{k} \neq \mathbf{0}$ are not occupied by a macroscopic number of particles because it costs the exchange interaction $n_0 \sum_{\mathbf{q} \neq 0} V_{\mathbf{q}} n_{\mathbf{q}}$. If $\phi_0 = 0$ and hence $n_0 = 0$, it is not possible for a pair of states $(\mathbf{k}, -\mathbf{k})$ to be macroscopically occupied; this is because $\chi_{\mathbf{k}} \cong n_{\mathbf{k}} e^{i\theta_{\mathbf{k}}}$ if $n_{\mathbf{k}} \gg 1$. Then, the sum of the fourth and last terms on the right-hand side of Eq. (2.94) gives

$$\frac{1}{2} \sum_{\mathbf{k,q}} V_{\mathbf{q}} n_{\mathbf{k+q}} n_{\mathbf{k}} + \frac{1}{2} \sum_{\mathbf{k,q}} V_{\mathbf{q}} \chi^*_{\mathbf{k+q}} \chi_{\mathbf{k}}$$
$$\cong \frac{1}{2} \sum_{\mathbf{k,q}} V_{\mathbf{q}} n_{\mathbf{k+q}} n_{\mathbf{k}} \left[1 + \cos\left(\theta_{\mathbf{k+q}} - \theta_{\mathbf{k}}\right)\right] > 0,$$

which results in a large (extensive) loss of energy.

Let us assume that only the $\mathbf{k} = 0$ state is occupied by a macroscopic number of particles. Then, the dominant term, which depends on the phase, on the right-hand side of Eq. (2.94) is

$$n_0 \sum_{\mathbf{k}} V_{\mathbf{k}} |\chi_{\mathbf{k}}| \cos(2\theta_0 - \theta_{\mathbf{k}}).$$

To minimize this term, the phase $\theta_{\mathbf{k}}$ of the amplitude of pair excitations must be locked to the phase θ_0 of the condensate so that

$$\theta_{\mathbf{k}} = 2\theta_0 + \pi. \quad (2.95)$$

Thus, in the Bose–Einstein condensed system, all the particles (condensate and virtual excitations) are correlated in a phase coherent manner. This is the reason why the superfluid fraction can be equal to unity at zero temperature even though the condensate fraction is not so.

2.4 Bogoliubov Theory of Quasi-One-Dimensional Torus

Major consequences of the Bogoliubov theory can be obtained analytically by considering a quasi-one-dimensional torus geometry as shown in Fig. 2.2, where the circumference and cross-sectional area of the torus are $L \equiv 2\pi R$ and $S = \pi r^2$. It is assumed that $r \ll R$ and $\hbar^2/(mr^2) \gg k_B T$. This system resides in the lowest radial mode and can be considered to be effectively one-dimensional.

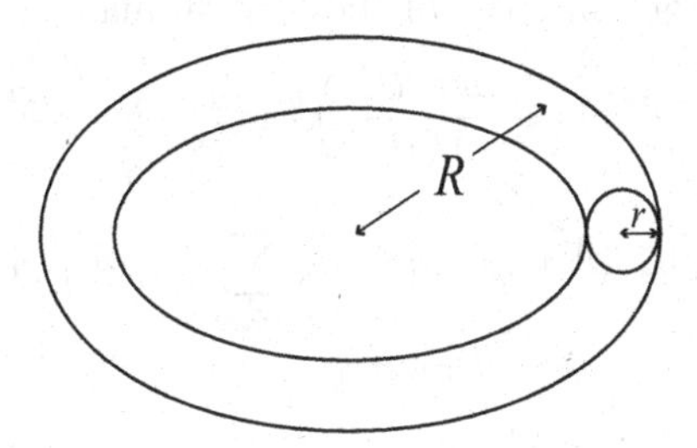

Fig. 2.2 Quasi-one-dimensional torus of circumference $L = 2\pi R$ and cross-sectional area πr^2 in which N identical bosons are enclosed, where $r \ll R$.

A complete set of basis functions is given by

$$\phi_n(x) = \frac{1}{\sqrt{L}} e^{ik_n x}, \tag{2.96}$$

where

$$k_n = \frac{2\pi}{L} n \quad (n = 0, \pm 1, \pm 2, \cdots) \tag{2.97}$$

is the wave vector along the circumference of the torus. The basis functions satisfy the orthonormal relations

$$\int_0^L \phi_n^*(x)\phi_m(x)dx = \delta_{nm}, \tag{2.98}$$

and the completeness relation

$$\begin{aligned}\sum_{n=-\infty}^{\infty} \phi_n(x)\phi_n^*(x') &= \frac{1}{L} \sum_{n=-\infty}^{\infty} e^{i\frac{2\pi}{L}n(x-x')} = \sum_{n=-\infty}^{\infty} \delta(x - x' - Ln) \\ &= \delta(x - x'),\end{aligned} \tag{2.99}$$

where the second equality is derived from the Poisson summation formula, and the lost equality is derived using the condition $0 \le x, x' < L$. The field operators are expressed in terms of the basis functions as

$$\hat{\psi}(x) = \sum_{n=-\infty}^{\infty} \hat{a}_n \phi_n(x), \quad \hat{\psi}^\dagger(x) = \sum_{n=-\infty}^{\infty} \hat{a}_n^\dagger \phi_n(x), \tag{2.100}$$

where $\hat{a}_n^\dagger$ and $\hat{a}_n$ are the creation and annihilation operators of bosons, respectively, with the wavenumber k_n and satisfy the canonical commutation relations of bosons

$$\left[\hat{a}_n, \hat{a}_m^\dagger\right] = \delta_{nm}, \quad [\hat{a}_n, \hat{a}_m] = 0, \quad \left[\hat{a}_n^\dagger, \hat{a}_m^\dagger\right] = 0.$$

It can be shown that the field operators satisfy

$$\left[\psi(x), \psi^\dagger(x')\right] = \delta(x - x'), \; \left[\hat{\psi}(x), \hat{\psi}(x')\right] = 0, \; \left[\hat{\psi}^\dagger(x), \hat{\psi}^\dagger(x')\right] = 0. \tag{2.101}$$

The kinetic and interaction parts of the Hamiltonian are given by

$$\hat{H}_{\rm KE} = \int_0^L dx \; \hat{\psi}^\dagger(x) \left(-\frac{\hbar^2}{2M}\frac{\partial^2}{\partial x^2}\right) \hat{\psi}(x) = \sum_n \epsilon_n \hat{a}_n^\dagger \hat{a}_n, \tag{2.102}$$

$$\hat{H}_{\rm int} = \frac{gL}{2}\int_0^L dx \; \hat{\psi}^{\dagger^2}(x)\hat{\psi}^2(x) = \frac{g}{2}\sum_{lmn} \hat{a}_l^\dagger \hat{a}_m^\dagger \hat{a}_{l+n}\hat{a}_{m-n}, \tag{2.103}$$

respectively, where

$$\epsilon_n \equiv \frac{\hbar^2 k_n^2}{2M} = \frac{\hbar^2 n^2}{2MR^2} = \hbar\omega_{\rm c} n^2, \quad \omega_{\rm c} \equiv \frac{\hbar}{2MR^2}, \tag{2.104}$$

$$g \equiv \frac{4\pi\hbar^2 a}{M}\frac{1}{LS} = \frac{2\hbar^2 a}{MRS}, \tag{2.105}$$

and

$$gN = \frac{4\pi\hbar^2 a}{M} n, \tag{2.106}$$

gives the mean-field interaction energy per particle.

2.4.1 *Case of BEC at rest: stability of BEC*

We first consider a case in which the torus is at rest and N_0 particles undergo BEC in the $k = 0$ state. In the Bogoliubov approximation, we retain terms up to the second order in $\hat{a}_n$ and $\hat{a}_n^\dagger$ $(n \neq 0)$. Then, the interaction Hamiltonian is approximated as

$$\hat{H}_{\rm int} \simeq \frac{g}{2}\left[N^2 + N_0 \sum_{n\neq 0}\left(\hat{a}_n\hat{a}_{-n} + \hat{a}_n^\dagger\hat{a}_{-n}^\dagger + 2\hat{a}_n^\dagger\hat{a}_n\right)\right], \tag{2.107}$$

where $N = N_0 + \sum_{n\neq 0} \hat{a}_n^\dagger \hat{a}_n$. We may also subsititute $N_0 \simeq N$ in the last term of Eq. (2.107) with the same degree of approximation. We thus obtain

$$\hat{H} \simeq \frac{gN^2}{2} + \sum_{n=1}^{\infty} \left[\tilde{\epsilon}_n (\hat{a}_n^\dagger \hat{a}_n + \hat{a}_{-n}^\dagger \hat{a}_{-n}) + gN(\hat{a}_n \hat{a}_{-n} + \hat{a}_n^\dagger \hat{a}_{-n}^\dagger) \right], \quad (2.108)$$

where $\tilde{\epsilon}_n \equiv \epsilon_n + gN$. To diagonalize Eq. (2.108), we use the following Bogoliubov transformations:

$$\hat{a}_n = \hat{b}_n \cosh\theta_n - \hat{b}_{-n}^\dagger \sinh\theta_n, \quad \hat{a}_n^\dagger = \hat{b}_n^\dagger \cosh\theta_n - \hat{b}_{-n} \sinh\theta_n. \quad (2.109)$$

The off-diagonal terms vanish if

$$\sinh 2\theta_n = \frac{gN}{\sqrt{\epsilon_n(\epsilon_n + 2gN)}}, \quad \cosh 2\theta_n = \frac{\tilde{\epsilon}_n}{\sqrt{\epsilon_n(\epsilon_n + 2gN)}}. \quad (2.110)$$

Then, the Hamiltonian is diagonalized as

$$H = E_0 + \sum_{n=1}^{\infty} \hbar\omega_c n \sqrt{n^2 + 2\gamma}\, (\hat{b}_n^\dagger \hat{b}_n + \hat{b}_{-n}^\dagger \hat{b}_{-n}), \quad (2.111)$$

where

$$\gamma \equiv \frac{gN}{\hbar\omega_c} \quad (2.112)$$

and

$$E_0 = \frac{gN^2}{2} + \sum_{n=1}^{\infty} \hbar\omega_c (n\sqrt{n^2 + 2\gamma} - n^2 - \gamma). \quad (2.113)$$

All excitation energies $\hbar\omega_c n\sqrt{n^2 + 2\gamma}$ in Eq. (2.111) must be positive for the system to be stable against excitations. This is guaranteed for the case of repulsive interaction ($\gamma > 0$). For the case of attractive interaction, the condition $1 + 2\gamma > 0$ must be satisfied, thereby yielding the following stability condition:

$$|\gamma| = \frac{|g|N}{\hbar\omega_c} < \frac{1}{2}. \quad (2.114)$$

2.4.2 *Case of rotating BEC: Landau criterion*

We next consider a case in which a condensate of N_m atoms is rotating with the wavenumber k_m, while the torus is at rest. The kinetic part of the Hamiltonian is given by

$$\begin{aligned} \hat{H}_{\mathrm{KE}} &= \epsilon_m N_m + \sum_{n\neq 0} \epsilon_{m+n} \hat{a}_{m+n}^\dagger \hat{a}_{m+n} \\ &= \epsilon_m N + \sum_{n\neq 0} (\epsilon_{m+n} - \epsilon_m)\, \hat{a}_{m+n}^\dagger \hat{a}_{m+n}. \end{aligned}$$

Using the Bogoliubov approximation, the interaction Hamiltonian is given by

$$\hat{H}_{\rm int} \simeq \frac{g}{2}\left[\hat{a}_m^{\dagger 2}\hat{a}_m^2 + \sum_{n\neq 0}\left(\hat{a}_m^{\dagger 2}\hat{a}_{m+n}\hat{a}_{m-n} + \hat{a}_{m+n}^{\dagger}\hat{a}_{m-n}^{\dagger}\hat{a}_m^2 + 4\hat{a}_m^{\dagger}\hat{a}_m\hat{a}_{m+n}^{\dagger}\hat{a}_{m+n}\right)\right]$$

$$\simeq \frac{gN^2}{2} + \frac{gN}{2}\sum_{n\neq 0}\left(\hat{a}_{m+n}\hat{a}_{m-n} + \hat{a}_{m+n}^{\dagger}\hat{a}_{m-n}^{\dagger} + 2\hat{a}_{m+n}^{\dagger}\hat{a}_{m+n}\right),$$

where $N = N_m + \sum_{n\neq 0}\hat{a}_{m+n}^{\dagger}\hat{a}_{m+n}$. Thus,

$$\hat{H} = \epsilon_m N + \frac{gN^2}{2} + \sum_{n=1}^{\infty}\left[\tilde{\epsilon}_{m+n}\hat{a}_{m+n}^{\dagger}\hat{a}_{m+n} + \tilde{\epsilon}_{m-n}\hat{a}_{m-n}^{\dagger}\hat{a}_{m-n} + gN\left(\hat{a}_{m+n}\hat{a}_{m-n} + \hat{a}_{m+n}^{\dagger}\hat{a}_{m-n}^{\dagger}\right)\right], \tag{2.115}$$

where

$$\begin{aligned}\tilde{\epsilon}_{m+n} &\equiv \epsilon_{m+n} - \epsilon_m + gN = \hbar\omega_{\rm c}\left(n^2 + 2mn + \gamma\right),\\ \tilde{\epsilon}_{m-n} &\equiv \epsilon_{m-n} - \epsilon_m + gN = \hbar\omega_{\rm c}\left(n^2 - 2mn + \gamma\right).\end{aligned} \tag{2.116}$$

To diagonalize Eq. (2.115), we use the Bogoliubov transformations

$$\begin{aligned}\hat{a}_{m+n} &= \hat{b}_n\cosh\theta_n - \hat{b}_{-n}^{\dagger}\sinh\theta_n,\\ \hat{a}_{m-n} &= \hat{b}_{-n}\cosh\theta_n - \hat{b}_n^{\dagger}\sinh\theta_n.\end{aligned} \tag{2.117}$$

The off-diagonal terms vanish if

$$\sinh 2\theta_n = \frac{2gN}{\sqrt{\left(\tilde{\epsilon}_{m+n} + \tilde{\epsilon}_{m-n}\right)^2 - (2gN)^2}} = \frac{\gamma}{n\sqrt{n^2+2\gamma}},$$

$$\cosh 2\theta_n = \frac{\tilde{\epsilon}_{m+n} + \tilde{\epsilon}_{m-n}}{\sqrt{\left(\tilde{\epsilon}_{m+n} + \tilde{\epsilon}_{m-n}\right)^2 - (2gN)^2}} = \frac{n^2+\gamma}{n\sqrt{n^2+2\gamma}},$$

where θ_n depends on n, but not on m. We finally obtain

$$\hat{H} = E_m + \sum_{n=1}^{\infty}\hbar\omega_{\rm c}\left[\left(n\sqrt{n^2+2\gamma} + 2mn\right)\hat{b}_n^{\dagger}\hat{b}_n + \left(n\sqrt{n^2+2\gamma} - 2mn\right)\hat{b}_{-n}^{\dagger}\hat{b}_{-n}\right], \tag{2.118}$$

where

$$E_m = \hbar\omega_{\rm c}m^2N + \frac{gN^2}{2} + \sum_{n=1}^{\infty}\hbar\omega_{\rm c}\left(n\sqrt{n^2+2\gamma} - n^2 - \gamma\right). \tag{2.119}$$

When the condensate is at rest (*i.e.*, $m = 0$), Eqs. (2.118) and (2.119) reduce to Eqs. (2.111) and (2.113), respectively.

For a rotating condensate to be stable, all excitation energies must be positive. For the case $m > 0$, the excitation energy is minimal for $n = 1$. It follows from Eq. (2.118) that the stability criterion for a rotating condensate is given by

$$m < \frac{1}{2}\sqrt{1+2\gamma}, \quad i.e., \; v_s < \frac{\hbar}{2MR}\sqrt{1+\frac{2gN}{\hbar\omega_c}}, \tag{2.120}$$

where v_s is the superfluid velocity:

$$v_s \equiv \frac{\hbar k_m}{M} = \frac{\hbar m}{MR}. \tag{2.121}$$

In particular, for the case $gN \gg \hbar\omega_c$, the right-hand side of Eq. (2.120) yields the sound velocity:

$$v_s < \sqrt{\frac{gN}{M}} = \frac{\hbar}{MR}\sqrt{\frac{2aRN}{S}}. \tag{2.122}$$

The condition (2.122) is referred to as the Landau criterion. For the case of attractive interaction ($\gamma < 0$), Eq. (2.120) is not satisfied, even for $m = 1$. Hence, persistent currents cannot exist stably.

For comparison, let us consider the stability of the system within the Hartree–Fock (HF) approximation. The energy of the system, in which N_m atoms are in the $n = m$ state and $N - N_m$ atoms are in the $n = 0$ state, is

$$E(N_m) = \epsilon_m N_m + gN_m(N - N_m) + \frac{gN^2}{2}. \tag{2.123}$$

Suppose that all the atoms are initialy in the m state. Then,

$$E(N) = \epsilon_m N + \frac{gN^2}{2}. \tag{2.124}$$

The energy of the system in which one of the atoms in the m state has decayed into the $n = 0$ state is given by

$$E(N-1) = \epsilon_m(N-1) + g(N-1) + \frac{gN^2}{2} \simeq \epsilon_m(N-1) + gN + \frac{gN^2}{2}.$$

For the persistent current to be stable, the following condition must be satisfied:

$$E(N-1) - E(N) = -\epsilon_m + gN > 0 \to m < \sqrt{\gamma}. \tag{2.125}$$

This result differs from the one given in Eq. (2.120) by a factor of $\sqrt{2}$ for $\gamma \gg 1$, indicating that the HF approximation overestimates the stability of the persistent current as compared to the Bogoliubov approximation.

2.4.3 *Ground state of BEC in rotating torus*

We finally consider the case in which the torus is rotating at angular frequency ω. The Hamiltonian of the system in the rotating frame of reference is given by $\hat{H} - \omega\hat{L}$, where $\hat{L}$ is the total angular momentum of the system given by

$$\begin{aligned}\hat{L} &= \sum_m \hbar m \hat{a}_m^\dagger \hat{a}_m = \hbar m \hat{N}_m + \sum_{n\neq 0} \hbar(m+n)\hat{a}_{m+n}^\dagger \hat{a}_{m+n} \\ &= \hbar m N + \sum_{n\neq 0} \hbar n \hat{a}_{m+n}^\dagger \hat{a}_{m+n}. \end{aligned} \tag{2.126}$$

From Eqs. (2.126) and (2.115), we obtain

$$\begin{aligned}\hat{K} \equiv \hat{H} - \omega\hat{L} &= \hbar\left(\omega_c m^2 - \omega m\right) N + \frac{gN^2}{2} \\ &+ \sum_{n=1}^{\infty}\Big[\left(\tilde{\epsilon}_{m+n} - \hbar\omega n\right)\hat{a}_{m+n}^\dagger \hat{a}_{m+n} + \left(\tilde{\epsilon}_{m-n} + \hbar\omega n\right)\hat{a}_{m-n}^\dagger \hat{a}_{m-n} \\ &+ gN\left(\hat{a}_{m+n}\hat{a}_{m-n} + \hat{a}_{m+n}^\dagger \hat{a}_{m-n}^\dagger\right)\Big]. \end{aligned} \tag{2.127}$$

By diagonalizing $\hat{K}$ in a manner similar to that described in the preceding subsection, we obtain

$$\begin{aligned}\hat{K} = E_m + \sum_{n=1}^{\infty} \hbar\omega_c n &\left\{\left[\sqrt{n^2+2\gamma} + 2\left(m - \frac{\omega}{2\omega_c}\right)\right]\hat{b}_n^\dagger \hat{b}_n \right. \\ &+ \left.\left[\sqrt{n^2+2\gamma} - 2\left(m - \frac{\omega}{2\omega_c}\right)\right]\hat{b}_{-n}^\dagger \hat{b}_{-n}\right\}, \end{aligned} \tag{2.128}$$

where

$$E_m = \hbar\omega_c\left(m^2 - \frac{m\omega}{\omega_c}\right)N + \frac{gN^2}{2} + \sum_{n=1}^{\infty} \hbar\omega_c\left(n\sqrt{n^2+2\gamma} - n^2 - \gamma\right). \tag{2.129}$$

For the system to be stable against excitations, the following condition should be satisfied:

$$\left|m - \frac{\omega}{2\omega_c}\right| < \frac{1}{2}\sqrt{1+2\gamma}. \tag{2.130}$$

This result shows that the stability range is a periodic function of ω with period $2\omega_c$. In general, the properties of a system in a quasi-one-dimensional torus are periodic functions of ω with the same period. Figure 2.3 compares the stability region of an attractive condensate (shaded regions) obtained by the Bogoliubov approximation with that obtained by the HF approximation.

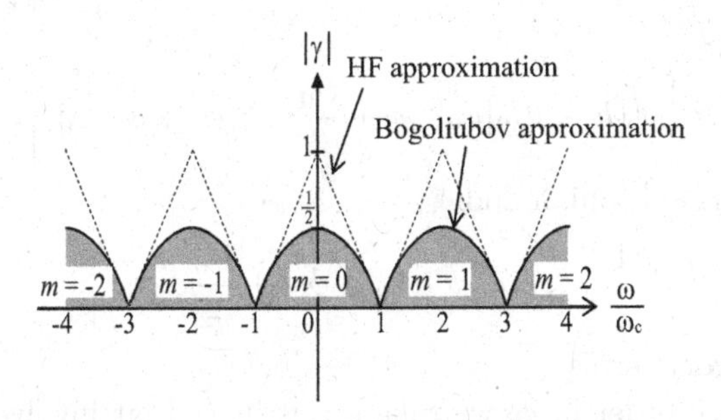

Fig. 2.3 Stability diagram of an attractive condensate under rotation. The dotted (solid) curve show the stability boundary calculated by the HF (Bogoliubov) approximation, where m denotes the value of the quantized circulation in units of h/M

2.5 Bogoliubov–de Gennes (BdG) Theory

Many physical systems are confined in a spatially nonuniform potential $U(\mathbf{r})$. Hence, it is important to generalize the Bogoliubov theory to such cases [de Gennes (1989)].

We start with the second-quantized Hamiltonian

$$\hat{H} = \int d\mathbf{r}\hat{\psi}^\dagger(\mathbf{r})h(\mathbf{r})\hat{\psi}(\mathbf{r}) + \frac{U_0}{2}\int d\mathbf{r}\hat{\psi}^\dagger(\mathbf{r})\hat{\psi}^\dagger(\mathbf{r})\hat{\psi}(\mathbf{r})\hat{\psi}(\mathbf{r}), \quad (2.131)$$

where the field operators satisfy the canonical commutation relations (2.33) and

$$h(\mathbf{r}) \equiv -\frac{\hbar^2\nabla^2}{2M} + U(\mathbf{r}). \quad (2.132)$$

We decompose the field operator into the condensate part $\psi_0(\mathbf{r})$, which we assume to be a c-number, and the noncondensate part $\hat{\phi}(\mathbf{r})$, as follows:

$$\hat{\psi}(\mathbf{r}) = \psi_0(\mathbf{r}) + \hat{\phi}(\mathbf{r}), \quad (2.133)$$

where the condensate and noncondensate modes are assumed to be orthogonal to each other, *i.e.*,

$$\int d\mathbf{r}\psi_0^*(\mathbf{r})\hat{\phi}(\mathbf{r}) = \int d\mathbf{r}\psi_0(\mathbf{r})\hat{\phi}^\dagger(\mathbf{r}) = 0. \quad (2.134)$$

The number operator is then expressed as

$$\hat{N} = \int d\mathbf{r}\left[|\psi_0(\mathbf{r})|^2 + \hat{\phi}^\dagger(\mathbf{r})\hat{\phi}(\mathbf{r})\right]. \quad (2.135)$$

Substituting Eq. (2.133) into Eq. (2.131), we obtain

$$\hat{H} = \int d\mathbf{r}\ \psi_0^*(h + \frac{U_0}{2}|\psi_0|^2)\psi_0 + \int d\mathbf{r}\left[\hat{\phi}^\dagger(h + U_0|\psi_0|^2)\psi_0 + \psi_0^*(h + U_0|\psi_0|^2)\hat{\phi}\right] + \hat{H}_{\mathrm{B}} + \hat{H}', \quad (2.136)$$

where

$$\hat{H}_{\rm B} = \int d\mathbf{r} \left[\hat{\phi}^\dagger (h + 2U_0|\psi_0|^2)\hat{\phi} + \frac{U_0}{2}(\psi_0^{*2}\hat{\phi}^2 + \hat{\phi}^{\dagger 2}\psi_0^2) \right] \tag{2.137}$$

is the Bogoliubov Hamiltonian and

$$\hat{H}' = \int d\mathbf{r} \left[U_0(\psi_0\hat{\phi}^{\dagger 2}\hat{\phi} + \psi_0^*\hat{\phi}^\dagger\hat{\phi}^2) + \frac{U_0}{2}\hat{\phi}^{\dagger 2}\hat{\phi}^2 \right] \tag{2.138}$$

describes higher-order terms.

Up to the zeroth order in $\hat{\phi}$, we take terms in $\hat{H}$ that involve only ψ_0 and ψ_0^*. Assuming $\hat{H} - \mu\hat{N}$ to be stationary with respect to the variations in ψ_0^*, we obtain the Gross–Pitaevskii equation:

$$\left(h + U_0|\psi_0|^2\right)\psi_0 = \mu\psi_0. \tag{2.139}$$

Substituting Eq. (2.139) into Eq. (2.136) and using Eq. (2.134), we find that the terms linear in $\hat{\phi}$ or $\hat{\phi}^\dagger$ vanish; thus, we obtain

$$\hat{H} = \int d\mathbf{r} \left(\mu|\psi_0|^2 - \frac{U_0}{2}|\psi_0|^4 \right) + \hat{H}_{\rm B} + \hat{H}'. \tag{2.140}$$

The BdG theory diagonalizes $\hat{H}_{\rm B}$ by introducing generalized Bogoliubov transformations [Fetter (1972)]:

$$\hat{\phi}(\mathbf{r}) = e^{iS(\mathbf{r})} \sum_j{}' \left[u_j(\mathbf{r})\hat{\alpha}_j - v_j^*(\mathbf{r})\hat{\alpha}_j^\dagger \right], \tag{2.141}$$

where the prime symbol (′) indicates that the sum is to be taken over all noncondensate modes and $\hat{\alpha}_j$ are the Bogoliubov quasiparticle operators that satisfy

$$[\hat{\alpha}_j, \hat{\alpha}_k^\dagger] = \delta_{jk}, \quad [\hat{\alpha}_j, \hat{\alpha}_k] = 0 \quad [\hat{\alpha}_j^\dagger, \hat{\alpha}_k^\dagger] = 0. \tag{2.142}$$

Let $S(\mathbf{r})$ denote the phase of $\psi_0(\mathbf{r})$, *i.e.*, $\psi_0(\mathbf{r}) = |\psi_0(\mathbf{r})|e^{iS(\mathbf{r})}$. Substituting Eq. (2.141) into Eq. (2.137), we obtain

$$\begin{aligned} \hat{H}_{\rm B} = \sum_{jk}{}' \int d\mathbf{r} \Bigg\{ & \hat{\alpha}_j^\dagger\hat{\alpha}_k \left[u_j^* L u_k - \frac{U_0}{2}|\psi_0|^2(u_j^*v_k + u_k v_j^*) \right] \\ & + \hat{\alpha}_j\hat{\alpha}_k^\dagger \left[v_j L v_k^* - \frac{U_0}{2}|\psi_0|^2(u_j v_k^* + u_k^* v_j) \right] \\ & - \hat{\alpha}_j^\dagger\hat{\alpha}_k^\dagger \left[u_j^* L v_k^* - \frac{U_0}{2}|\psi_0|^2(u_j^* u_k^* + v_j^* v_k^*) \right] \\ & - \hat{\alpha}_j\hat{\alpha}_k \left[v_j L u_k - \frac{U_0}{2}|\psi_0|^2(u_j u_k + v_j v_k) \right] \Bigg\}, \end{aligned} \tag{2.143}$$

where

$$L \equiv -\frac{\hbar^2}{2M}\left[\nabla + i\nabla S(\mathbf{r})\right]^2 + U(\mathbf{r}) + 2U_0|\psi_0(\mathbf{r})|^2. \tag{2.144}$$

For the diagonalization of $\hat{H}_{\rm B}$, u_j and v_j should satisfy the following BdG equations:

$$Lu_j - U_0|\psi_0|^2 v_j = E_j u_j, \tag{2.145}$$

$$L^* v_j - U_0|\psi_0|^2 u_j = -E_j v_j, \tag{2.146}$$

where u_j and v_j are normalized so that they satisfy the condition:

$$\int d\mathbf{r}(|u_j|^2 - |v_j|^2) = 1. \tag{2.147}$$

As the BdG equations suggest, u_j and v_j represent the amplitudes of particles and holes, respectively. It can be shown from Eqs. (2.145)–(2.147) that the eigenvalues E_j are real, and u_j and v_j satisfy the following orthonormalization conditions:

$$\int d\mathbf{r}\ (u_j^* u_k - v_j^* v_k) = \delta_{jk}, \tag{2.148}$$

$$\int d\mathbf{r}\ (u_j v_k - u_k v_j) = 0 \qquad \text{if } E_j + E_k \neq 0. \tag{2.149}$$

Thus, the low-energy excitations are described by Bogoliubov quasiparticles that are expressed as linear combinations of particles and holes. Equations (2.145) and (2.146) can be used to simplify Eq. (2.143) as

$$\hat{H}_{\rm B} = \frac{1}{2}\sum_{jk}{}' \int d\mathbf{r}\ \Big[\ (E_j + E_k)(\hat{\alpha}_j^\dagger \hat{\alpha}_k u_j^* u_k - \hat{\alpha}_j \hat{\alpha}_k^\dagger v_j v_k^*) + (E_j - E_k)(\hat{\alpha}_j \hat{\alpha}_k u_k v_j - \hat{\alpha}_j^\dagger \hat{\alpha}_k^\dagger u_j^* v_k^*)\Big]. \tag{2.150}$$

Because of Eq. (2.149), the last term in Eq. (2.150) vanishes for

$$\sum_{jk}{}' \int d\mathbf{r}(E_j - E_k)(\hat{\alpha}_j \hat{\alpha}_k u_k v_j - \hat{\alpha}_j^\dagger \hat{\alpha}_k u_j^* v_k^*)$$

$$= \frac{1}{2}\sum_{jk}{}'(E_j - E_k)\int d\mathbf{r}\left[\hat{\alpha}_j \hat{\alpha}_k (u_k v_j - u_j v_k) - \hat{\alpha}_j^\dagger \hat{\alpha}_k^\dagger (u_j^* v_k^* - u_k^* v_j^*)\right] = 0.$$

Using Eq. (2.148), $\hat{H}_{\rm B}$ is diagonalized as

$$\hat{H}_{\rm B} = \sum_j{}' E_j \hat{\alpha}_j^\dagger \hat{\alpha}_j - \sum_j{}' E_j \int d\mathbf{r} |v_j(\mathbf{r})|^2. \tag{2.151}$$

Following are a few remarks about the BdG equations:

- It can be seen from Eqs. (2.145) and (2.146) that if (u_j, v_j) is a solution of the BdG equations with the eigenvalue E_j with positive normalization, (v_j^*, u_j^*) is a solution with eigenvalue $-E_j$ with negative normalization. For uniform condensates with plane-wave solutions, the positive normalization always leads to positive eigenvalues. However, nonuniform condensates such as one with a vortex can have physical states with positive normalization and negative eigenvalues [Fetter and Svidzinsky (2001)]. If a positive-norm solution has a negative eigenvalue, the system exhibits the Landau instability because it can lower the energy of the system by exciting quasiparticles with negative eigenvalues.

- If the system has a complex eigenvalue, the assumed ground state ψ_0 is not stable and the system exhibits a dynamical instability. The corresponding solutions (u_j, v_j) have zero norm. In fact, from Eqs. (2.145) and (2.146), we can show

$$(E_j - E_j^*) \int d\mathbf{r}(|u_j|^2 - |v_j|^2) = 0. \tag{2.152}$$

Hence, if $E_j \neq E_j^*$, the integral must vanish.

- The substitution of

$$\psi_0(\mathbf{r}) = \sqrt{\frac{N_0}{\Omega}} e^{iS(\mathbf{r})} f(\mathbf{r}) \tag{2.153}$$

in Eq. (2.139) gives

$$Lf(\mathbf{r}) - U_0|\psi_0|^2 f(\mathbf{r}) = 0. \tag{2.154}$$

Comparing this with Eq. (2.145), we find that

$$u_0(\mathbf{r}) = v_0(\mathbf{r}) = f(\mathbf{r})$$

is the solution of the BdG equations with the eigenvalue $E_0 = 0$. Thus, the condensate wave function is the zero-energy eigenfunction of the BdG equations.

- From the canonical commutation relation for $\hat{\psi}$, it follows that the completeness relation

$$\frac{1}{\Omega} f(\mathbf{r}) f(\mathbf{r}') + \sum_j{}' \left[u_j(\mathbf{r}) u_j^*(\mathbf{r}') - v_j^*(\mathbf{r}) v_j(\mathbf{r}') \right]$$
$$= \delta(\mathbf{r} - \mathbf{r}') \tag{2.155}$$

holds, where $f(\mathbf{r})$ satisfies the normalization condition

$$\frac{1}{\Omega}\int d\mathbf{r}\,[f(\mathbf{r})]^2 = 1 \tag{2.156}$$

and

$$\int d\mathbf{r} f(\mathbf{r}) u_j(\mathbf{r}) = \int d\mathbf{r} f(\mathbf{r}) v_j(\mathbf{r}). \tag{2.157}$$

The last equation can be proved using Eqs. (2.145), (2.146), and (2.154).

As an illustrative example, let us consider a case in which the condensate is moving in a uniform system (*i.e.*, $U(\mathbf{r}) = 0$) with velocity

$$\mathbf{v} = \frac{\hbar\mathbf{k}}{M}. \tag{2.158}$$

The condensate wave function is

$$\psi_0(\mathbf{r}) = \sqrt{n_0}\, e^{i\mathbf{kr}}, \tag{2.159}$$

and the chemical potential is obtained from Eq. (2.139) as

$$\mu = n_0 U_0 + \frac{M}{2}\mathbf{v}^2.$$

The BdG equations can be solved using the assumptions:

$$u_{\mathbf{k}}(\mathbf{r}) = \frac{1}{\sqrt{\Omega}} A_{\mathbf{k}} e^{i\mathbf{kr}}, \quad v_{\mathbf{k}} = \frac{1}{\sqrt{\Omega}} B_{\mathbf{k}} e^{i\mathbf{kr}}. \tag{2.160}$$

Substituting Eq. (2.160) in Eqs. (2.145) and (2.146), we obtain coupled equations, which yield two eigenenergies:

$$E_{\mathbf{k}}^{(\pm)} = \hbar\mathbf{k}\cdot\mathbf{v} \pm E_k, \tag{2.161}$$

where

$$E_k = \sqrt{\epsilon_k(\epsilon_k + 2n_0U_0)}. \tag{2.162}$$

The corresponding eigenfunctions are given by

$$u_{\mathbf{k}}^{(+)}(\mathbf{r}) = \frac{1}{\sqrt{\Omega}} u_{\mathbf{k}} e^{i\mathbf{kr}}, \quad v_{\mathbf{k}}^{(+)} = \frac{1}{\sqrt{\Omega}} v_{\mathbf{k}} e^{i\mathbf{kr}}, \tag{2.163}$$

$$u_{\mathbf{k}}^{(-)}(\mathbf{r}) = \frac{i}{\sqrt{\Omega}} v_{\mathbf{k}} e^{i\mathbf{kr}}, \quad v_{\mathbf{k}}^{(-)} = \frac{i}{\sqrt{\Omega}} u_{\mathbf{k}} e^{i\mathbf{kr}}, \tag{2.164}$$

where

$$u_k = \left[\frac{\epsilon_k + n_0U_0}{2E_k} + \frac{1}{2}\right]^{\frac{1}{2}}, \quad v_k = \left[\frac{\epsilon_k - n_0U_0}{2E_k} - \frac{1}{2}\right]^{\frac{1}{2}}. \tag{2.165}$$

2.6 Method of Binary Collision Expansion

Lee and Yang [Lee and Yang (1957)] developed a method of binary collision expansion of the partition function based on the "binary kernel," which describes two-body scattering exactly. Because two-body scattering is the dominant interaction mechanism for dilute gases, this method is a powerful tool for determining their thermodynamic properties. This section provides a brief introduction to this method.

2.6.1 *Equation of state*

In the following discussions, we shall assume $\hbar = 2M = 1$. The Hamiltonian of our system is

$$\hat{H} = \sum_{i=1}^{N} \hat{T}_i + \sum_{i<j} \hat{V}_{ij}, \tag{2.166}$$

where $\hat{T}_i = \hat{\mathbf{p}}_i^2$ is the kinetic energy of the i-th particle and $\hat{V}_{ij} = \hat{V}(|\mathbf{r}_i - \mathbf{r}_j|)$ describes the interaction between the i-th and j-th particles. The partition function Z_N is

$$Z_N \equiv \mathrm{Tr} e^{-\beta \hat{H}} = \sum_m e^{-\beta E_m}, \tag{2.167}$$

where $\beta \equiv 1/k_{\mathrm{B}}T$ and Tr indicaates summation over all symmetrized (or anti-symmetrized) states for Bose (or Fermi) systems. When Z_N is given, the pressure P of the system can be obtained as

$$\beta P = -\frac{\partial}{\partial V} \ln Z_N, \tag{2.168}$$

where V is the volume of the system. The grand partition function Z_{G} can be defined using Z_N as

$$Z_{\mathrm{G}}(z) = \sum_{N=0}^{\infty} z^N Z_N, \tag{2.169}$$

where $Z = e^{\beta\mu}$ is the fugacity. The pressure P and the mass density ρ can be obtained from Z_{G} as

$$\beta P = \frac{1}{V} \ln Z_{\mathrm{G}}(z), \quad \rho = z\frac{d}{dz}\left[\frac{1}{\Omega} \ln Z_{\mathrm{G}}(z)\right]. \tag{2.170}$$

Eliminating z from Eq. (2.170) will yield pressure as a function of the density n. The energy of the ground state E_0 can be obtained from the relation $P = P(\rho)$ at absolute zero as

$$P = \rho^2 \frac{d}{d\rho}\frac{E_0}{N} \to \frac{E_0}{N} = \int_0^{\rho} \frac{P}{\rho^2} d\rho. \tag{2.171}$$

2.6.2 *Cluster expansion of partition function*

Usually, it is not possible to carry out the summation in Eq. (2.167) exactly. We use the systematic method developed by Kahn and Uhlenbeck [Kahn and Uhlenbeck (1938)] to calculate the partition function. Consider a Hamiltonian consisting of two parts:

$$\hat{H} = \hat{H}_0 + \hat{H}_I. \tag{2.172}$$

Let $\{\Psi_n\}$ be a complete set of eigenfunctions of $\hat{H}$. Then, the partition function can be expanded in terms of Ψ_n as

$$\begin{aligned} Z_N &= \mathrm{Tr}\, e^{-\beta(\hat{H}_0+\hat{H}_I)} = \sum_n \int d\mathbf{r}_1 \cdots d\mathbf{r}_N \Psi_n^* e^{-\beta(H_0+H_I)} \Psi_n \\ &= \frac{1}{N!} \int d\mathbf{r}_1 \cdots d\mathbf{r}_N W_N^{\mathrm{S}}(1, \cdots, N; 1, \cdots, N), \end{aligned} \tag{2.173}$$

where

$$W_N^{\mathrm{S}}(1', \cdots, N'; 1, \cdots, N) \equiv N! \sum_n \Psi_n^*(1', \cdots, N') e^{-\beta H} \Psi_n(1, \cdots, N), \tag{2.174}$$

and the superscript S indicates summation over completely (anti-) symmetrized Ψ_n. In Eqs. (2.173) and (2.174), $1, \cdots, N$, and $1', \cdots, N'$ in fact denote $\mathbf{r}_1, \cdots, \mathbf{r}_N$, and $\mathbf{r}'_1, \cdots, \mathbf{r}'_N$, respectively.

We decompose W_N^{S} into linked clusters. From Eq. (2.174), we find that W_N^{S} is a matrix element in which the coordinates of the N-particle system make a transition via the interaction $e^{-\beta H}$ from $1, 2, \cdots, N$ to $1', 2', \cdots, N'$. In general, W_N^{S} can be decomposed into the product of linked clusters as

$$W_N^{\mathrm{S}} = \sum_{\{m_l\}} \sum_D \underbrace{(U_1^{\mathrm{S}} \cdots U_1^{\mathrm{S}})}_{m_1} \underbrace{(U_2^{\mathrm{S}} \cdots U_2^{\mathrm{S}})}_{m_2} \cdots \underbrace{(U_N^{\mathrm{S}})}_{m_N}, \tag{2.175}$$

where m_1, m_2, $\cdots$, m_N denote the numbers of the linked clusters in each group within parentheses and satisfy $\sum_{l=1}^N l m_l = N$. Here, D indicates summation over all combinations of coordinates for a given $\{m_l\}$.

The partition function can be obtained by substituting $\mathbf{r}'_i = \mathbf{r}_i$ in Eq. (2.175) and integrating the resultant expression over $\mathbf{r}_1, \mathbf{r}_2, \cdots, \mathbf{r}_N$. Here, since the integrals of $U_1^{\mathrm{S}}(i;i)$ are all equal, we have

$$Z_N = \frac{1}{N!} \sum_{\{m_l\}} \frac{N!}{m_1! m_2! \cdots m_N!} \prod_{l=1}^N \left(\frac{1}{l!} \int d\mathbf{r}_1 \cdots d\mathbf{r}_l U_l^{\mathrm{S}}(1, \cdots, l; 1, \cdots, l) \right)^{m_l}. \tag{2.176}$$

Introducing the cluster integration

$$b_l^{\mathrm{S}} \equiv \frac{\lambda^3}{V}\frac{1}{l!}\int d\mathbf{r}_1\cdots d\mathbf{r}_l U_l^{\mathrm{S}}(1,\cdots,l;1,\cdots,l), \tag{2.177}$$

where $\lambda = \sqrt{4\pi\beta}$ denotes the thermal de Broglie wavelength, we obtain

$$Z_N = \sum_{\{m_l\}} \delta_{\sum_l l m_l, N} \prod_{l=1}^{N} \frac{1}{m_l!}\left(\frac{\Omega}{\lambda^3} b_l^{\mathrm{S}}\right)^{m_l}. \tag{2.178}$$

The grand partition function can be obtained by substituting Eq. (2.178) in Eq. (2.169):

$$\begin{aligned} Z_{\mathrm{G}}(z) &= \sum_{N=0}^{\infty}\sum_{\{m_l\}} \delta_{\sum_l l m_l, N} \prod_{l=1}^{N} \frac{1}{m_l!}\left(\frac{\Omega}{\lambda^3} z^l b_l^{\mathrm{S}}\right)^{m_l} = \prod_{l=1}^{\infty}\sum_{m_l=0}^{\infty} \frac{1}{m_l!}\left(\frac{\Omega}{\lambda^3} z^l b_l^{\mathrm{S}}\right)^{m_l} \\ &= \exp\left(\frac{\Omega}{\lambda^3}\sum_{l=1}^{\infty} b_l^{\mathrm{S}} z^l\right). \end{aligned} \tag{2.179}$$

Therefore, we obtain the expressions for the pressure and density of the system from Eq. (2.170) as

$$\beta P = \frac{1}{\lambda^3}\sum_{l=1}^{\infty} b_l^{\mathrm{S}} z^l, \quad \rho = \frac{1}{\lambda^3}\sum_{l=1}^{\infty} l b_l^{\mathrm{S}} z^l. \tag{2.180}$$

The equation of state of the system can be obtained from the cluster integration b_l^{S} in Eq. (2.180).

2.6.3 *Ideal Bose and Fermi gases*

We first consider the case of an ideal Bose gas. The grand partition function Z_{G} is given by

$$\begin{aligned} Z_{\mathrm{G}} &= \prod_k \sum_{n_k=0}^{\infty}\left(e^{\beta(\mu-\epsilon_k)}\right)_k^n = \prod_k\left(1-e^{\beta(\mu-\epsilon_k)}\right)^{-1} \\ &= \exp\left(-\sum_k \ln\left(1-e^{\beta(\mu-\epsilon_k)}\right)\right) \\ &= \exp\left(-\frac{V}{4\pi^2}\left(\frac{2m}{\hbar^2}\right)^{\frac{3}{2}}\int_0^{\infty} dE\, E^{\frac{1}{2}}\ln\left(1-e^{\beta(\mu-E)}\right)\right) \\ &= \exp\left(\frac{V}{6\pi^2}\left(\frac{2m}{\hbar^2}\right)^{\frac{3}{2}}\beta\int_0^{\infty} dE\frac{E^{\frac{3}{2}}}{e^{\beta(E-\mu)}-1}\right). \end{aligned}$$

Substituting $e^{\beta\mu} = z$ and $e^{\beta E} = t$, we have

$$Z_{\rm G} = \exp\left(\frac{V}{6\pi^2}\left(\frac{2M}{\beta\hbar^2}\right)^{\frac{3}{2}}\int_1^\infty dt\frac{t^{\frac{3}{2}}}{z^{-1}e^t - 1}\right) = \exp\left(\frac{V}{\lambda^3}g_{\frac{5}{2}}(z)\right), \tag{2.181}$$

where

$$\int_1^\infty dt\frac{t^{\frac{3}{2}}}{z^{-1}e^t - 1} = \Gamma\left(\frac{5}{2}\right), \quad g_{\frac{5}{2}}(z) = \frac{3\sqrt{\pi}}{4}\sum_{l=1}^{\infty}\frac{z^l}{l^{\frac{5}{2}}}. \tag{2.182}$$

Comparing Eq. (2.181) with Eq. (2.179), we find that

$$b_l^{\rm S} = l^{-\frac{5}{2}}. \tag{2.183}$$

For the case of an ideal Fermi gas, we have

$$\begin{aligned}
Z_{\rm G} &= \prod_k \sum_{n_k=0}^{1}\left(e^{\beta(\mu-\epsilon_k)}\right)^{n_k} = \prod_k\left(1 + e^{\beta(\mu-\epsilon_k)}\right) \\
&= \exp\left(\sum_k \ln\left(1 + e^{\beta(\mu-\epsilon_k)}\right)\right) \\
&= \exp\left(\frac{V}{4\pi^2}\left(\frac{2M}{\hbar^2}\right)^{\frac{3}{2}}\int_0^\infty dE\ E^{\frac{1}{2}}\ln\left(1 + e^{\beta(\mu-E)}\right)\right) \\
&= \exp\left(-\frac{V}{6\pi^2}\left(\frac{2M}{\hbar^2}\right)^{\frac{3}{2}}\beta\int_0^\infty dE\frac{E^{\frac{3}{2}}}{e^{\beta(E-\mu)} + 1}\right) \\
&= \exp\left(\frac{V}{\lambda^3}\sum_{l=1}^{\infty}\frac{(-z)^l}{l^{\frac{5}{2}}}\right),
\end{aligned}$$

where

$$\int_0^\infty dt\frac{t^{\frac{3}{2}}}{(-z)^{-1}e^t - 1} = \Gamma\left(\frac{5}{2}\right)\sum_{l=1}^{\infty}\frac{(-z)^l}{l^{\frac{5}{2}}}. \tag{2.184}$$

Comparing Eq. (2.184) with Eq. (2.179), we find that

$$b_l^{\rm S} = (-1)^l l^{-\frac{5}{2}}. \tag{2.185}$$

As shown in Eqs. (2.183) and (2.185), the cluster integration exists up to $l = \infty$ even in the absence of interaction between particles because of the (anti)symmetrization of the wave function.

For comparison, let us consider a classical ideal gas which obeys the Boltzmann statistics. Consider a one-particle partition function given as

$$Z_1 = \sum_k e^{-\beta\epsilon_k} = \frac{V}{4\pi^2}\left(\frac{2M}{\hbar^2}\right)^{\frac{3}{2}}\int_0^\infty dE\ E^{\frac{1}{2}}e^{-\beta E} = \frac{V}{\lambda^3}.$$

Then, the N-particle partition is given by

$$Z_N = \frac{1}{N!} Z_1 = \frac{1}{N!} \left(\frac{V}{\lambda^3} \right)^N .$$

We thus obtain

$$Z_G = \sum_{N=0}^{\infty} z^N Z_N = \exp\left(\frac{\Omega}{\lambda^3} z \right). \tag{2.186}$$

Comparing Eq. (2.186) with Eq. (2.179), we find that

$$b_l^{cl} = \delta_{l,1} \tag{2.187}$$

That is, in a classical ideal gas, linked clusters among particles ($l \geq 2$) do not exist. This is because of the absence of interaction between particles. However, in quantum systems, the entanglement among many particles exists even in the absence of interactions because of the (anti)symmetrization of the wave function.

2.6.4 *Matsubara formula*

The grand partition function can be calculated directly using operators.

$$\begin{aligned} Z_G &= \mathrm{Tr}\left(e^{-\beta(\hat{H}-\mu\hat{N})} \right) \\ &= \mathrm{Tr}\left(e^{-\beta(\hat{H}_0-\mu\hat{N})} \right) \frac{\mathrm{Tr}\left(e^{-\beta(\hat{H}_0-\mu\hat{N})} e^{\beta(\hat{H}_0-\mu\hat{N})} e^{-\beta(\hat{H}-\mu\hat{N})} \right)}{\mathrm{Tr}\left(e^{-\beta(\hat{H}_0-\mu\hat{N})} \right)} \\ &= Z_G^0 \langle \hat{S}(\beta) \rangle_0, \end{aligned} \tag{2.188}$$

where

$$Z_G^0 \equiv \mathrm{Tr}\left(e^{-\beta(\hat{H}_0-\mu\hat{N})} \right),$$

$$\hat{S}(\beta) \equiv e^{\beta(\hat{H}_0-\mu\hat{N})} e^{-\beta(\hat{H}-\mu\hat{N})},$$

and

$$\langle \cdots \rangle_0 \equiv \frac{\mathrm{Tr}\left(e^{-\beta(\hat{H}_0-\mu\hat{N})} e^{\beta(\hat{H}_0-\mu\hat{N})} e^{-\beta(\hat{H}-\mu\hat{N})} \right)}{\mathrm{Tr}\left(e^{-\beta(\hat{H}_0-\mu\hat{N})} \right)}. \tag{2.189}$$

The derivative of $\hat{S}(\beta)$ is

$$\begin{aligned} \frac{d}{d\beta} \hat{S}(\beta) &= e^{\beta(\hat{H}_0-\mu\hat{N})} (\hat{H}_0 - \hat{H}) e^{-\beta(\hat{H}-\mu\hat{N})} \\ &= -e^{\beta(\hat{H}_0-\mu\hat{N})} \hat{H}_I e^{-\beta(\hat{H}_0-\mu\hat{N})} \hat{S}(\beta) \\ &= -\hat{H}_I(\beta) \hat{S}(\beta), \end{aligned} \tag{2.190}$$

where $\hat{H}_\mathrm{I} = \hat{H} - \hat{H}_0$ and $\hat{H}_\mathrm{I}(\beta) = e^{\beta(\hat{H}_0-\mu\hat{N})}\hat{H}_\mathrm{I}e^{-\beta(\hat{H}_0-\mu\hat{N})}$. Integrating Eq. (2.190), we get

$$\begin{aligned}\hat{S}(\beta) &= 1 - \int_0^\beta d\tau \hat{H}_\mathrm{I}(\tau)\hat{S}(\tau) \\ &= 1 - \int_0^\beta d\tau \hat{H}_\mathrm{I}(\tau) + \int_0^\beta d\tau_1 d\tau_2 \hat{H}_\mathrm{I}(\tau_1)\hat{H}_\mathrm{I}(\tau_2) - \cdots \\ &= T_\tau \exp\left(-\int_0^\beta d\tau \hat{H}_\mathrm{I}(\tau)\right), \end{aligned} \tag{2.191}$$

where T_τ is the time-ordered operator and

$$\int_0^\beta d\tau_1 d\tau_2 \hat{H}_\mathrm{I}(\tau_1)\hat{H}_\mathrm{I}(\tau_2) = \frac{1}{2!}T_\tau\left(\int_0^\beta d\tau \hat{H}_\mathrm{I}(\tau)\right)^2, \text{ etc.} \tag{2.192}$$

Taking the average of Eq. (2.191), we have

$$\begin{aligned}\langle\hat{S}(\beta)\rangle_0 = 1 + \sum_{n=1}^\infty \frac{(-1)^n}{n!}\int_0^\beta d\tau_1 \int_0^\beta d\tau_2 \cdots \int_0^\beta d\tau_n \\ \times \langle T_\tau\{\hat{H}_\mathrm{I}(\tau_1)\hat{H}_\mathrm{I}(\tau_2)\cdots\hat{H}_\mathrm{I}(\tau_n)\}\rangle_0. \end{aligned} \tag{2.193}$$

Here, second or higher-order terms in the perturbation series include disconnected diagrams as well as connected diagrams. In detail,

$$\begin{aligned}&\frac{(-1)^n}{n!}\int_0 \beta d\tau_1 \cdots \int_0 \beta d\tau_n \langle T_\tau\{\hat{H}_\mathrm{I}(\tau_1)\cdots\hat{H}_\mathrm{I}(\tau_n)\}\rangle_0 \\ &= \frac{(-1)^n}{n!}\sum_{\{m_l\}}\delta_{\sum_l l m_l, n}\frac{n!}{m_1!m_2!\cdots m_n!}\left[\int_0^\beta d\tau \langle\hat{H}_\mathrm{I}(\tau)\rangle_0^{\mathrm{con}}\right]^{m_1} \\ &\quad \times \left[\frac{1}{2!}\int_0^\beta d\tau_1 \int_0^\beta d\tau_2 \langle T_\tau\{\hat{H}_\mathrm{I}(\tau_1)\hat{H}_\mathrm{I}(\tau_2)\}\rangle_0^{\mathrm{con}}\right]^{m_2} \cdots \\ &= \sum_{\{m_l\}}\delta_{\sum_l l m_l, n}\prod_{l=1}^n \frac{1}{m_l!}C_l^{ml}, \end{aligned} \tag{2.194}$$

where

$$C_1 = -\int_0^\beta d\tau \langle\hat{H}_\mathrm{I}(\tau)\rangle_0^{\mathrm{con}}, \quad C_2 = \frac{1}{2!}\int_0^\beta d\tau_1 \int_0^\beta d\tau_2 \langle T_\tau\{\hat{H}_\mathrm{I}(\tau_1)\hat{H}_\mathrm{I}(\tau_2)\}\rangle_0^{\mathrm{con}}$$

and

$$C_l \equiv \frac{(-1)^l}{l!}\int_0^\beta d\tau_1 \cdots d\tau_l \langle T_\tau\{\hat{H}_\mathrm{I}(\tau_1)\hat{H}_\mathrm{I}(\tau_2)\cdots\hat{H}_\mathrm{I}(\tau_l)\}\rangle_0^{\mathrm{con}}.$$

Substituting Eq. (2.194) in Eq. (2.193), we have

$$\langle \hat{S}(\beta) \rangle = 1 + \sum_{n=1}^{\infty} \sum_{\{m_l\}} \delta_{\sum_l l m_l, n} \prod_{l=1}^{n} \frac{1}{m l!} C_l^{ml} = \prod_{l=1}^{\infty} \sum_{ml=0} \frac{C_l^{ml}}{ml!}$$
$$= e^{\sum_{l=1}^{\infty} C_l}. \tag{2.195}$$

Here, we have

$$\sum_{l=1}^{\infty} C_l = \left\langle T_\tau \exp\left(-\beta \int_0 d\tau \hat{H}_{\mathrm{I}}(\tau)\right)\right\rangle_0^{\mathrm{con}} - 1 = \langle \hat{S}(\beta) \rangle_0^{\mathrm{con}} - 1. \tag{2.196}$$

We thus obtain

$$\langle \hat{S}(\beta) \rangle_0 = \exp\left(\langle \hat{S}(\beta) \rangle_0^{\mathrm{con}} - 1\right). \tag{2.197}$$

Substituting Eq. (2.197) in Eq. (2.188) gives

$$Z_{\mathrm{G}} = Z_{\mathrm{G}}^0 \exp\left(\langle \hat{S}(\beta) \rangle_0^{\mathrm{con}} - 1\right),$$
$$\hat{S}(\beta) = T_\tau \exp\left(-\int_0^\beta d\tau \hat{H}_{\mathrm{I}}(\tau)\right). \tag{2.198}$$

Therefore, we obtain

$$\Omega = -\beta^{-1} \ln Z_{\mathrm{G}} = \Omega_0 - \beta^{-1} \left(\langle \hat{S}(\beta) \rangle_0^{\mathrm{con}} - 1\right). \tag{2.199}$$

Equation (2.199) gives the Matsubara formula. The thermodynamic potential can be described using two-body temperature Green's functions.

$$\hat{S}(\beta, g) \equiv T_\tau \exp\left(-g \int_0^\beta d\tau \hat{H}_{\mathrm{I}}(\tau)\right) \tag{2.200}$$

Using Eq. (2.200), we obtain

$$\Omega(g) \stackrel{(20)}{=} \Omega_0 - \beta^{-1} \ln \langle S(\beta, g) \rangle_0. \tag{2.201}$$

Thus,

$$\frac{\partial \Omega(g)}{\partial g} = -\beta^{-1} \frac{\langle \frac{\partial}{\partial g} S(\beta, g) \rangle_0}{\langle S(\beta, g) \rangle_0}$$
$$= \beta^{-1} \int_0^\beta d\tau \frac{\langle T_\tau \{\hat{H}_{\mathrm{I}}(\tau) S(\beta, g)\} \rangle_0}{\langle S(\beta, g) \rangle_0}, \tag{2.202}$$

where Eq. (2.188) was used for obtaining the last equality. Integrating both sides of Eq. (2.202) over g from 0 to 1 and substituting $\Omega(g = 0) = \Omega_0$ and $\Omega(g = 1) = \Omega$, we obtain

$$\Omega = \Omega_0 + \beta^{-1} \int_0^1 dg \int_0^\beta dt \frac{\langle T_\tau \{\hat{H}_{\mathrm{I}}(\tau) \hat{S}(\beta, g)\} \rangle_0}{\langle \hat{S}(\beta, g) \rangle_0}. \tag{2.203}$$

Assuming two-body interactions,

$$\hat{H}_{\rm I}(\tau) = \frac{1}{2}\int d\mathbf{r}_1 d\mathbf{r}_2 \hat{\Psi}^\dagger(\mathbf{r}_1\tau)\hat{\Psi}^\dagger(\mathbf{r}_2\tau)V(\mathbf{r}_1,\mathbf{r}_2)\hat{\Psi}(\mathbf{r}_2\tau)\hat{\Psi}(\mathbf{r}_1\tau), \quad (2.204)$$

Eq. (2.203) gives

$$\Omega = \Omega_0 + \frac{1}{2\beta}\int_0^1 dg \int_0^\beta d\tau \int d\mathbf{r}_1 d\mathbf{r}_2 V(\mathbf{r}_1,\mathbf{r}_2) \times \frac{\langle T_\tau\{\Psi^\dagger(\mathbf{r}_1\tau)\Psi^\dagger(\mathbf{r}_2\tau)\Psi(\mathbf{r}_2\tau)\Psi(\mathbf{r}_1\tau)S(\beta,g)\}\rangle_0}{\langle S(\beta,g)\rangle_0}. \quad (2.205)$$

Chapter 3

Trapped Systems

3.1 Ideal Bose Gas in a Harmonic Potential

Let us consider a system of N noninteracting particles with mass M confined in a harmonic potential. The Hamiltonian of the system is given by

$$H = \sum_{i=1}^{N} \frac{1}{2M} \left(p_{xi}^2 + p_{yi}^2 + p_{zi}^2 \right) + \sum_{i=1}^{N} \frac{M}{2} \left(\omega_x^2 x_i^2 + \omega_y^2 y_i^2 + \omega_z^2 z_i^2 \right). \quad (3.1)$$

It follows from the canonical commutation relations

$$[x_i, p_{xj}] = [y_i, p_{yj}] = [z_j, p_{zj}] = i\hbar\delta_{ij} \quad (3.2)$$

that the energy spectrum of the system is given by

$$E_{n_x n_y n_z} = \hbar(\omega_x n_x + \omega_y n_y + \omega_z n_z) + E_0, \quad (3.3)$$

where $n_{x,y,z} = 0, 1, 2, \cdots$, and E_0 is the zero-point energy given by

$$E_0 = \frac{\hbar}{2}(\omega_x + \omega_y + \omega_z). \quad (3.4)$$

At thermal equilibrium, the bosons distribute over the energy levels (3.3) according to the Bose–Einstein distribution

$$N_{n_x n_y n_z} = \frac{1}{\exp[\beta(E_{n_x n_y n_z} - \mu)] - 1},$$

where $\beta \equiv 1/(k_B T)$ and μ is the chemical potential that is determined such that the average total number of particles is equal to N:

$$N = \sum_{n_x, n_y, n_z = 0}^{\infty} \frac{1}{\exp\left[\beta(E_{n_x n_y n_z} - \mu)\right] - 1}. \quad (3.5)$$

When $k_B T$ is much higher than the discrete energy-level spacings, the sum in Eq. (3.5) can be replaced with the integral

$$\sum_{n_x n_y n_z} \cdots \to N_0 + \int_0^\infty dE D(E) \cdots , \tag{3.6}$$

where $D(E)$ is the density of states and

$$N_0 = \frac{1}{\exp[\beta(E_0 - \mu)] - 1} \tag{3.7}$$

is the number of bosons in the lowest-energy state. Since this number can be macroscopic as μ approaches E_{000} below the BE transition temperature, we keep it separate from the integral in Eq. (3.6).

In Eq. (3.6), $D(E)\hbar\omega$ denotes the number of lattice points (n_x, n_y, n_z) in the energy range of $E - \hbar\omega < \hbar(\omega_x n_x + \omega_y n_y + \omega_z n_z) \leq E$. For simplicity, we assume that the parabolic potential is isotropic, $\omega_x = \omega_y = \omega_z = \omega$, and let n be the maximum integer that does not exceed $E/\hbar\omega$. Then, $D(E)\hbar\omega$ gives the number of lattice points $(n_x,\ n_y,\ n_z)$ that satisfy

$$n_x + n_y + n_z = n \quad (n_{x,y,z} = 0, 1, 2, \cdots).$$

A combinatorial calculation gives

$$D(E)\hbar\omega = {}_{n+2}C_2 = \frac{(n+2)(n+1)}{2} \simeq \frac{1}{2}(\frac{E}{\hbar\omega})^2 + \frac{3}{2}\frac{E}{\hbar\omega} + 1. \tag{3.8}$$

The coefficient 3/2 on the right-hand side is modified for an anisotropic potential. Let us write this coefficient for a general case as γ and ignore 1 in Eq. (3.8) by assuming $E \gg \hbar\bar{\omega}$; then, the density of states can be described as follows.

$$D(E) \simeq \frac{1}{2}\frac{E^2}{(\hbar\bar{\omega})^3} + \gamma\frac{E}{(\hbar\bar{\omega})^2}, \tag{3.9}$$

where $\bar{\omega} \equiv (\omega_x \omega_y \omega_z)^{\frac{1}{3}}$. A numerical study suggests that γ is well approximated by [Grossmann and Holthaus (1995)]

$$\gamma \simeq \frac{\omega_x + \omega_y + \omega_z}{2\bar{\omega}}. \tag{3.10}$$

Replacing the discrete sum in Eq. (3.5) by the integral using Eqs. (3.6) and (3.9), we obtain

$$\begin{aligned} N &= N_0 + \frac{1}{2(\hbar\bar{\omega})^3}\int_0^\infty \frac{E^2 dE}{e^{\beta(E+E_0-\mu)} - 1} + \frac{\gamma}{(\hbar\bar{\omega})^2}\int_0^\infty \frac{E dE}{e^{\beta(E+E_0-\mu)} - 1} \\ &= N_0 + \left(\frac{k_B T}{\hbar\bar{\omega}}\right)^3 g_3(z) + \gamma\left(\frac{k_B T}{\hbar\bar{\omega}}\right)^2 g_2(z), \end{aligned} \tag{3.11}$$

where $z \equiv e^{\beta(\mu - E_0)}$ and $g_n(z)$ is the Bose–Einstein distribution (or polylogarithmic) function defined by

$$g_n(z) \equiv \frac{1}{\Gamma(n)} \int_0^\infty \frac{x^{n-1}}{z^{-1}e^x - 1} dx = \sum_{l=1}^\infty \frac{z^l}{l^n}. \tag{3.12}$$

As shown below, z lies in the range $0 < z < 1$ in the classical regime and approaches 1 as the transition temperature is neared. At $z = 1$, $g_n(z)$ becomes the Riemann zeta function $\zeta(n)$:

$$g_n(1) = \sum_{l=1}^\infty \frac{1}{l^n} = \zeta(n), \tag{3.13}$$

where

$$\zeta(2) = \frac{\pi^2}{6}, \quad \zeta\left(\frac{3}{2}\right) = 2.612, \text{ and } \zeta(3) = 1.202. \tag{3.14}$$

3.1.1 *Transition temperature*

The number of bosons that occupy the lowest-energy level is given from Eq. (3.7) and $z \equiv e^{\beta(\mu - E_0)}$ by

$$N_0 = \frac{z}{1 - z}. \tag{3.15}$$

Solving Eq. (3.15) for z gives

$$z = \frac{N_0}{N_0 + 1}. \tag{3.16}$$

At T_c, we may substitute $z = 1$ in Eq. (3.11) because $N_0 \gg 1$, and ignore N_0 in Eq. (3.11) because $N \gg N_0$. We then obtain

$$N \simeq \left(\frac{k_B T_c}{\hbar\bar{\omega}}\right)^3 \zeta(3) + \gamma \left(\frac{k_B T_c}{\hbar\bar{\omega}}\right)^2 \zeta(2).$$

Solving this for T_c gives

$$T_c \simeq \frac{\hbar\bar{\omega}}{k_B} \left(\frac{N}{\zeta(3)}\right)^{1/3} \left[1 - \frac{\gamma}{3} \frac{\zeta(2)}{\zeta(3)^{2/3}} N^{-\frac{1}{3}}\right]. \tag{3.17}$$

We thus find that when $N \gg 1$, the transition temperature T_0 is given by

$$T_0 = \frac{\hbar\bar{\omega}}{k_B} \left(\frac{N}{\zeta(3)}\right)^{\frac{1}{3}} \simeq 0.94 \frac{\hbar\bar{\omega}}{k_B} N^{\frac{1}{3}} \simeq 4.5 \left(\frac{\bar{\omega}/2\pi}{100\text{Hz}}\right) N^{\frac{1}{3}} \text{ nK}. \tag{3.18}$$

For the case of finite N, the deviation of the transition temperature from T_0 is found from Eq. (3.17) to be

$$\frac{T_c - T_0}{T_0} \simeq -\frac{\gamma}{3}\frac{\zeta(2)}{\zeta(3)^{2/3}}N^{-\frac{1}{3}} \simeq -0.243\frac{\omega_x + \omega_y + \omega_z}{(\omega_x\omega_y\omega_z)^{1/3}}N^{-\frac{1}{3}}. \tag{3.19}$$

In dilute-gas BEC systems, it is possible to make a detailed comparison between theories and experiments [Ensher, *et al.* (1996); Giorgini, *et al.* (1996)]. In the case of a uniform system, the transition temperature rises due to a repulsive interaction by an amount proportional to $an^{1/3}$ [Baym, *et al.* (1999, 2001); Yukalov (2004)]. This is because the repulsive interaction suppresses density fluctuations in favor of BEC. In the case of a harmonically trapped system, the transition temperature decreases because the repulsive interaction reduces the peak density. A semiclassical estimate gives [Giorgini, *et al.* (1996)]

$$\frac{\delta T_c}{T_0} \simeq -1.326\frac{a}{d_0}N^{\frac{1}{6}}, \tag{3.20}$$

where $d_0 \equiv \sqrt{\hbar/(M\omega)}$ is the width of the ground-state wave function of a particle in a harmonic trap. The experiment on a harmonically trapped ^{87}Rb condensate deviates significantly from the ideal gas formula (3.17) and agrees well with Eq. (3.20) [Gerbier, *et al.* (2004)]. Critical fluctuations and diverging behavior of the correlation length are expected near the second-order phase transition. The correlation length was measured just above the BEC transition temperature and found to diverge as $(T - T_c)^{-\nu}$ with $\nu = 0.67 \pm 0.13$ [Donner, *et al.* (2007)] in agreement with the critical exponent of the 3D XY model.

Equation (3.17) suggests that the thermodynamic limit in the trapped system is obtained by taking the following limits:

$$N \to \infty, \quad \bar{\omega} \to 0, \quad N\bar{\omega}^3 = \text{constant}. \tag{3.21}$$

3.1.2 *Condensate fraction*

Let us examine the temperature dependence of the condensate fraction, *i.e.*, the ratio of the number of condensate particles to the total particle number for $T < T_c$. Since $N_0 \gg 1$ for $T < T_c$, we may substitute $z = 1$ in Eq. (3.11) to obtain

$$\frac{N_0}{N} = 1 - \left(\frac{k_B T}{\hbar\bar{\omega}}\right)^3 \frac{\zeta(3)}{N} - \gamma\left(\frac{k_B T}{\hbar\bar{\omega}}\right)^2 \frac{\zeta(2)}{N}. \tag{3.22}$$

Substituting Eq. (3.18) into Eq. (3.22), we obtain

$$\frac{N_0}{N} = 1 - \left(\frac{T}{T_0}\right)^3 - \gamma\frac{\zeta(2)}{\zeta(3)^{2/3}}\left(\frac{T}{T_0}\right)^2 N^{-\frac{1}{3}}, \tag{3.23}$$

where $\zeta(2)/\zeta(3)^{2/3} \simeq 1.46$. Comparing this result with Eq. (1.23), we find that the trapped condensate grows more rapidly than the uniform condensate as a function of temperature.

3.1.3 *Chemical potential*

Eliminating $\hbar\bar{\omega}$ in Eq. (3.11) in favor of T_0 using Eq. (3.18), we obtain

$$N = N_0 + \left(\frac{T}{T_0}\right)^3 \frac{N}{\zeta(3)} g_3(z) + \gamma\left(\frac{T}{T_0}\right)^2 \left(\frac{N}{\zeta(3)}\right)^{\frac{2}{3}} g_2(z). \tag{3.24}$$

When $N \gg 1$, we can ignore the last term. Furthermore, because $N_0 \approx 0$ for $T > T_c$, $z = N_0/(N_0+1) \ll 1$ and hence $g_3(z) \approx z = e^{\beta(\mu - E_0)}$. Thus,

$$N \simeq \left(\frac{T}{T_0}\right)^3 \frac{N}{\zeta(3)} e^{\beta(\mu - E_0)},$$

and hence,

$$\mu \simeq E_0 - 3k_B T \ln\frac{T}{T_0}. \tag{3.25}$$

The chemical potential increases as the temperature is lowered and it becomes equal to the zero-point energy E_0 of the system at the transition temperature.

3.1.4 *Specific heat*

In the limit of $N \to \infty$, the specific heat exhibits a discontinuous jump at $T = T_0$. To show this, let us consider the internal energy of the system:

$$E = \int_{E_0}^{\infty} \frac{\epsilon}{e^{\beta(\epsilon-\mu)} - 1}\rho(\epsilon)d\epsilon. \tag{3.26}$$

Since E changes continuously as a function of temperature, the jump in the specific heat, ΔC, should arise from the temperature dependence of μ:

$$\Delta C \equiv C(T_0 + 0) - C(T_0 - 0) = \left(\frac{\partial E}{\partial \mu}\right)_T \frac{\partial \mu}{\partial T}\Big|_{T_0-0}^{T_0+0}, \tag{3.27}$$

where T_0+0 (T_0-0) represents a temperature that is infinitesimally larger (smaller) than T_0, and

$$\begin{aligned}\left(\frac{\partial E}{\partial \mu}\right)_T &= \int_{E_0}^{\infty} \epsilon\rho(\epsilon)\frac{\partial}{\partial \mu}\frac{1}{e^{\beta(\epsilon-\mu)}-1}d\epsilon \\ &= \int_{E_0}^{\infty} \frac{1}{e^{\beta(\epsilon-\mu)}-1}\frac{\partial(\epsilon\rho)}{\partial\epsilon}d\epsilon. \end{aligned} \tag{3.28}$$

In three dimensions and for a large-N limit, $\rho(\epsilon)=\epsilon^2/2(\hbar\bar{\omega})^3$, and therefore,

$$\left(\frac{\partial E}{\partial \mu}\right)_T = 3N. \tag{3.29}$$

Thus,

$$\Delta C = 3N\frac{\partial\mu}{\partial T}\bigg|_{T_0-0}^{T_0+0}. \tag{3.30}$$

Since μ is constant and equal to E_0 below T_0, $\partial\mu/\partial T=0$ at $T=T_0-0$. On the other hand, $\partial\mu/\partial T$ at $T=T_0+0$ is obtained by differentiating Eq. (3.24) with respect to T. Noting that $N_0=0$ for $T>T_0$ and that the last term in Eq. (3.24) is negligible for $N\gg 1$, we have

$$\begin{aligned} 0 &= 3T^2 g_3(z) + T^3 g_3'(z)\frac{\partial z}{\partial T} \\ &= 3T^2 g_3(z) + T^3 g_2(z)\left[-\frac{\mu-E_0}{k_B T^2}+\beta\frac{\partial\mu}{\partial T}\right], \end{aligned}$$

where $g_3'(z)=g_2(z)/z$ is used. Thus, at $T=T_0+0$, where $z=1$ and $\mu=E_0$, we obtain

$$\frac{\partial\mu}{\partial T} = -3k_B\frac{g_3(1)}{g_2(1)} = -3k_B\frac{\zeta(3)}{\zeta(2)}. \tag{3.31}$$

Substituting this into Eq. (3.30) gives

$$\frac{\Delta C^{(N=\infty)}}{Nk_B} = 9\frac{\zeta(3)}{\zeta(2)} \simeq 6.577. \tag{3.32}$$

Thus, the BEC phase transition of a harmonically trapped ideal Bose gas is of the second order. This is in sharp contrast with the uniform system of noninteracting bosons, where the constant-volume specific heat is continuous at $T=T_0$. When N is finite, the specific heat derived from Eq. (3.26) exhibits a continuous change. However, when N is sufficiently large ($\gtrsim 10,000$), the specific heat exhibits an almost discontinuous jump [Grossmann and Holthaus (1995)].

3.2 BEC in One- and Two-Dimensional Parabolic Potentials

3.2.1 *Density of states*

Let us count the number of combinations (n_x, n_y, n_z) that satisfy

$$E - dE < \hbar(\omega_x n_x + \omega_y n_y + \omega_z n_z) \leq E. \tag{3.33}$$

A *one-dimensional* (1D) *system* is defined as the system that satisfies

$$\hbar\omega_z \ll k_B T \ll \hbar\omega_x, \hbar\omega_y \tag{3.34}$$

so that $n_x = n_y = 0$. Condition (3.33) then reduces to

$$E - dE < \hbar\omega_z n_z \leq E.$$

Since there exists only one state per energy interval $dE = \hbar\omega_z$, the density of states is given by

$$D(E) = \frac{1}{\hbar\omega_z} \quad \text{(1D)}. \tag{3.35}$$

A *two-dimensional* (2D) *system* is defined as the system that satisfies

$$\hbar\omega_x, \hbar\omega_y \ll k_B T \ll \hbar\omega_z \tag{3.36}$$

so that $n_z = 0$. To simplify matters, let us consider an isotropic case in which $\omega_x = \omega_y \equiv \omega$. Then, we have

$$E - dE < \hbar\omega(n_x + n_y) \leq E.$$

Since there exist $E/(\hbar\omega)$ states per energy interval $dE = \hbar\omega$, the density of states is given by

$$D(E) = \frac{E}{(\hbar\omega)^2} \quad \text{(2D)}. \tag{3.37}$$

For comparison, in the uniform system, $D(E)$ is constant in 2D and $D(E) \propto E^{-\frac{1}{2}}$ in 1D.

3.2.2 *Transition temperature*

The transition temperature is defined as the temperature at which the chemical potential becomes equal to the zero-point energy of the system. For the 1D case, we have

$$N = \sum_{n_z} \frac{1}{e^{\hbar\omega_z n_z / k_B T_0} - 1} = \int_{\frac{\hbar\omega}{2}}^{\infty} \frac{D(E)}{e^{E/k_B T_0} - 1} dE$$
$$= -\frac{k_B T_0}{\hbar\omega} \ln \left| \frac{e^{\frac{\hbar\omega}{2k_B T_0}} - 1}{e^{\frac{\hbar\omega}{2k_B T_0}}} \right| \simeq \frac{k_B T_0}{\hbar\omega} \ln \frac{2k_B T_0}{\hbar\omega}.$$

$$N \simeq \frac{k_B T_0}{\hbar\omega} \ln \frac{2k_B T_0}{\hbar\omega} \quad \text{(1D)}. \tag{3.38}$$

The transition temperature can be determined by numerically solving this equation. According to the Hohenberg–Mermin–Wagner theorem [see Sec. 9.2], spatially uniform infinite 1D systems do not undergo BEC even at zero temperature. An important distinction in the harmonically trapped 1D system is the constancy of the density of states that makes the condensation of bosons in the lowest-energy state easier than that in the uniform case in which the density of states diverges at $E = 0$.

Similarly, in two and three dimensions, we obtain

$$N \simeq \left(\frac{k_B T_0}{\hbar\omega}\right)^2 \zeta(2) \quad \text{(2D)}, \tag{3.39}$$

$$N \simeq \left(\frac{k_B T_0}{\hbar\omega}\right)^3 \zeta(3) \quad \text{(3D)}. \tag{3.40}$$

Solving these for T_0, we obtain

$$T_0 = \hbar\omega \left(\frac{N}{\zeta(2)}\right)^{\frac{1}{2}} \quad \text{(2D)}, \tag{3.41}$$

$$T_0 = \hbar\omega \left(\frac{N}{\zeta(3)}\right)^{\frac{1}{3}} \quad \text{(3D)}. \tag{3.42}$$

3.2.3 *Condensate fraction*

In general, we have

$$N = N_0 + \int_{E_0}^{\infty} \frac{D(E)}{e^{\beta E} - 1} dE. \tag{3.43}$$

For the 1D case, we substitute Eq. (3.35) into Eq. (3.43), obtaining

$$N = N_0 + \frac{1}{\hbar\omega} \int_{E_0}^{\infty} \frac{dE}{e^{\beta E} - 1} = N_0 - \frac{k_B T}{\hbar\omega} \ln(1 - e^{-\beta E_0}).$$

Assuming $\beta E_0 \ll 1$, we obtain

$$\frac{N_0}{N} \simeq 1 - \frac{k_B T}{\hbar\omega} \frac{1}{N} \ln \frac{\hbar\omega}{2k_B T}. \tag{3.44}$$

Substituting Eq. (3.38) in Eq. (3.44), we obtain

$$\frac{N_0}{N} \simeq 1 - \frac{T}{T_0} \frac{\ln \frac{\hbar\omega}{2k_B T}}{\ln \frac{\hbar\omega}{2k_B T_0}} \quad \text{(1D)}. \tag{3.45}$$

Table 3.1 Summary of density of states $D(E)$, critical temperature T_0, and condensate fraction N_0/N of harmonically trapped Bose–Einstein condensates, where $W(x)$ is a positive solution to $We^W = x$ and it is approximated as $W(x) \simeq \ln x - \ln\ln x + \ln\ln x/\ln x + \cdots$ for large x.

	3D	2D	1D	uniform
$D(E)$	$\frac{E^2}{2(\hbar\omega)^3}$	$\frac{E}{(\hbar\omega)^2}$	$\frac{1}{\hbar\omega}$	$\frac{\sqrt{2M^3E}}{2\pi^2\hbar^3}$
T_0	$\frac{\hbar\omega}{k_B}\left(\frac{N}{\zeta(3)}\right)^{\frac{1}{3}}$	$\frac{\hbar\omega}{k_B}\left(\frac{N}{\zeta(2)}\right)^{\frac{1}{2}}$	$\frac{\hbar\omega N}{k_B W(2N)}$	$\frac{2\pi}{[\zeta(\frac{3}{2})]^{\frac{2}{3}}}\frac{\hbar^2}{k_B M}n^{\frac{2}{3}}$
$\frac{N_0}{N}$	$1-\left(\frac{T}{T_0}\right)^3$	$1-\left(\frac{T}{T_0}\right)^2$	$1-\frac{T\ln\frac{\hbar\omega}{2k_BT}}{T_0\ln\frac{\hbar\omega}{2k_BT_0}}$	$1-\left(\frac{T}{T_0}\right)^{\frac{3}{2}}$

For the 2D case, we substitute Eq. (3.37) into Eq. (3.43), obtaining

$$N \simeq N_0 + \left(\frac{k_B T}{\hbar\omega}\right)^2 \zeta(2).$$

Combining this with Eq. (3.39), we obtain

$$\frac{N_0}{N} \simeq 1 - \left(\frac{T}{T_0}\right)^2.$$

We can reproduce the above results by taking the direct sum of the discrete levels [de Groot, *et al.* (1950); Ketterle and van Druten (1996)].

The results obtained in this section are summarized in Table. 3.1. A comprehensive survey of the condensates in harmonic traps is given by Mullin [Mullin (1997)].

3.3 Semiclassical Distribution Function

When the thermal energy of the system k_BT is greater than the quantum-mechanical zero-point energy, we may use the semiclassical distribution function

$$f_{\mathbf{p}}(\mathbf{r}) = \frac{1}{e^{\beta[\epsilon_{\mathbf{p}}(\mathbf{r})-\mu]} - 1}, \tag{3.46}$$

where

$$\epsilon_{\mathbf{p}}(\mathbf{r}) = \frac{\mathbf{p}^2}{2m} + V(\mathbf{r}). \tag{3.47}$$

The function $f_{\mathbf{p}}(\mathbf{r})$ gives the number of particles in the phase space of $\mathbf{r} \sim \mathbf{r} + d\mathbf{r}$ and $\mathbf{p} \sim \mathbf{p} + d\mathbf{p}$:

$$f_{\mathbf{p}}(\mathbf{r}) \frac{d\mathbf{p}d\mathbf{r}}{(2\pi\hbar)^3}. \tag{3.48}$$

The integral of $f_{\mathbf{p}}(\mathbf{r})$ over momentum gives the semiclassical density distribution of the particle-number density:

$$n(\mathbf{r}) = \int \frac{d\mathbf{p}}{(2\pi\hbar)^3} \frac{1}{e^{\beta[\epsilon_{\mathbf{p}}(\mathbf{r}) - \mu]} - 1}. \tag{3.49}$$

Assuming $z(\mathbf{r}) = e^{\beta(\mu - V(\mathbf{r}))}$ and changing the variable of integration to $x = \mathbf{p}^2/(2mk_{\rm B}T)$, Eq. (3.49) becomes

$$n(\mathbf{r}) = \frac{2}{\sqrt{\pi}\lambda_{\rm th}^3} \int_0^\infty dx \frac{x^{1/2}}{z^{-i}e^x - 1} = \frac{1}{\lambda_{\rm th}^3} \sum_{n=1}^\infty \frac{z^n}{n^{\frac{3}{2}}} = \frac{1}{\lambda_{\rm th}^3} g_{\frac{3}{2}}(z), \tag{3.50}$$

where $g_{\frac{3}{2}}(z)$ is defined in Eq. (3.12). If we keep only the $n = 1$ term in the sum, we obtain the Maxwell–Boltzmann distribution

$$n^{MB}(\mathbf{r}) = \frac{Z}{\lambda_{\rm th}^3} e^{\beta(\mu - V(\mathbf{r}))},$$

where Z is the normalization constant. The semiclassical distribution fails over a length scale shorter than $\lambda_{\rm th}$.

The integral of $n(\mathbf{r})$ over $\mathbf{r}$ gives the total number of particles. In the case of the harmonic potential $V(\mathbf{r}) = m\omega^2\mathbf{r}^2/2$, we have

$$N = \int d\mathbf{r}\, n(\mathbf{r}) = \left(\frac{k_{\rm B}T}{\hbar\omega}\right)^3 \sum_{n=1}^\infty \frac{(e^{\beta\mu})}{n^3} = \left(\frac{k_{\rm B}T}{\hbar\omega}\right)^3 g_3(e^{\beta\mu}). \tag{3.51}$$

The chemical potential μ is determined so as to satisfy this equation. In particular, at $T = T_{\rm c}$, we have $\mu = 0$, and $g_3(1) = \zeta(3) = 1.202$. Equation (3.51) thus gives the transition temperature that is correct up to the leading order in N [*cf.* (3.17)]:

$$T_0 = \frac{\hbar\omega}{\zeta(3)^{1/3}k_{\rm B}} N^{\frac{1}{3}}. \tag{3.52}$$

For the harmonic potential, the zero-point energy is $3\hbar\omega/2$; thus, the semiclassical distribution function (3.46) is applicable if

$$k_{\rm B}T \gg \hbar\omega. \tag{3.53}$$

Comparing this with Eq. (3.52), we find that the temperature range over which Eq. (3.46) is valid is given by

$$T \gg T_0 N^{-\frac{1}{3}}. \tag{3.54}$$

This condition is consistent with the fact that when N is finite, the transition temperature (3.52) has a correction of the order of $N^{-1/3}$ [*cf.* (3.17)].

3.4 Gross–Pitaevskii Equation

We consider a system of N identical bosons with mass M that are confined in a one-body potential $V(\mathbf{r})$ and that interact via the delta-function potential. The Hamiltonian of the system is

$$H = \sum_{i=1}^{N} \frac{\mathbf{p}_i^2}{2M} + \sum_{i=1}^{N} V(\mathbf{r}_i) + \frac{U_0}{2} \sum_{i \neq j} \delta(\mathbf{r}_i - \mathbf{r}_j), \tag{3.55}$$

where $\mathbf{p}_i = -i\hbar\nabla_i$. Let $\Psi(\mathbf{r}_1, \mathbf{r}_2, \cdots, \mathbf{r}_N)$ be an eigenstate of the Hamiltonian (3.55) that is normalized to unity and that satisfies the Schrödinger equation

$$H\Psi(\mathbf{r}_1, \mathbf{r}_2, \cdots, \mathbf{r}_N) = E\Psi(\mathbf{r}_1, \mathbf{r}_2, \cdots, \mathbf{r}_N). \tag{3.56}$$

Multiplying $\Psi^*(\mathbf{r}_1, \mathbf{r}_2, \cdots, \mathbf{r}_N)$ from the left and integrating over the coordinates gives

$$E = \int \psi^* H \psi \, d\mathbf{r}_1 \cdots d\mathbf{r}_N. \tag{3.57}$$

Now, assume that the system undergoes BEC and that a majority of the bosons share the same single-particle state $\psi_1(\mathbf{r})$. Then, the many-body wave function may be approximated by the product of $\psi_1(\mathbf{r}_i)$:

$$\Psi(\mathbf{r}_1, \mathbf{r}_2, \cdots, \mathbf{r}_N) = \prod_{i=1}^{N} \psi_1(\mathbf{r}_i), \tag{3.58}$$

where ψ_1 is normalized to unity. Substituting Eq. (3.58) in Eq. (3.57) yields

$$E = N \int d\mathbf{r} \, \psi_1^*(\mathbf{r}) \left[-\frac{\hbar^2}{2M}\nabla^2 + V(\mathbf{r}) + \frac{U_0}{2}(N-1)|\psi_1(\mathbf{r})|^2 \right] \psi_1(\mathbf{r}). \tag{3.59}$$

Since the wave function of the condensate is typically normalized to N, we substitute

$$\psi(\mathbf{r}) = \sqrt{N}\psi_1(\mathbf{r}), \tag{3.60}$$

and then, Eq. (3.59) becomes

$$E[\psi] = \int d\mathbf{r}\psi^*(\mathbf{r}) \left[-\frac{\hbar^2}{2M}\nabla^2 + V(\mathbf{r}) + \frac{U_0}{2}|\psi(\mathbf{r})|^2 \right] \psi(\mathbf{r}), \tag{3.61}$$

where we replace $(1-1/N)U_0$ with U_0 in the last term because $N \gg 1$. The last term in Eq. (3.61) is the Hartree energy that describes the mean-field interaction between bosons. We refer to Eq. (3.61) as the Gross–Pitaevskii (GP) energy functional. The condensate wave function ψ is determined by requiring that the GP energy functional (3.61) be extremal subject to the normalization condition $\int |\psi|^2 d\mathbf{r} = N$. Introducing the Lagrange multiplier associated with this normalization condition, we have

$$\begin{aligned}&\frac{\delta}{\delta\psi^*(\mathbf{r})}\left(E[\psi] - \mu \int d\mathbf{r}|\psi|^2\right)\\&= \left[-\frac{\hbar^2}{2M}\nabla^2 + V(\mathbf{r}) + U_0|\psi(\mathbf{r})|^2 - \mu\right]\psi(\mathbf{r}) = 0,\end{aligned} \tag{3.62}$$

where ψ and ψ^* must be considered to be independent in functional differentiation because ψ is a complex field that has two degrees of freedom (*i.e.*, real and imaginary parts). We thus find that the condensate wave function obeys a nonlinear Schrödinger equation known as the GP equation

$$\left[-\frac{\hbar^2}{2M}\nabla^2 + V(\mathbf{r}) + U_0|\psi(\mathbf{r})|^2\right]\psi(\mathbf{r}) = \mu\psi(\mathbf{r}), \tag{3.63}$$

where μ plays the role of the chemical potential of the system. The dynamics of the condensate is governed by the time-dependent GP equation

$$i\hbar\frac{\partial}{\partial t}\psi(\mathbf{r},t) = \left[-\frac{\hbar^2}{2M}\nabla^2 + V(\mathbf{r}) + U_0|\psi(\mathbf{r},t)|^2\right]\psi(\mathbf{r},t). \tag{3.64}$$

The consistency between Eqs. (3.63) and (3.64) is guaranteed by the fact that the time dependence of the condensate wave function is governed by the chemical potential [see Eq. (1.67)].

The crucial assumption in the derivation of the GP equation (3.63) is the mean-field approximation (3.58) for the many-body wave function. In dilute BEC systems in which the condensate fraction is close to unity, this approximation is expected to be valid to a high degree, and the deviations from the mean field are well accounted for by the Bogoliubov theory of weakly interacting bosons. It can also be shown [Lieb, *et al.* (2000)] that the Gross–Pitaevskii energy functional gives the exact ground-state energy and particle density of a dilute Bose gas with a repulsive interaction in the limit of $N \to \infty$ and $a \to 0$ with Na fixed. For a dilute Bose gas with a repulsive interaction, the Gross–Pitaevskii equation is shown to be asymptotically exact in the limit of $N \to \infty$ and $a \to 0$ with Na/d_0 kept constant [Lieb, *et al.* (2000)].

3.5 Thomas–Fermi Approximation

Consider the case of a repulsive interaction which expands the condensate and suppresses density fluctuations. As a consequence, the kinetic energy becomes less important as the number of bosons, N, increases. In fact, as shown in Eq. (3.77), the ratio of the kinetic energy to the potential energy scales as $N^{-\frac{4}{5}}$, and therefore, the kinetic energy becomes negligible when N is very large. This regime is referred to as the Thomas-Fermi (TF) limit in which the energy of the system consists only of the one-body potential $V(\mathbf{r})$ and the mean-field interaction (Hartree) energy. Then, it follows from Eq. (3.63) that the TF wave function is given by

$$\psi_{\mathrm{TF}}(\mathbf{r}) = \sqrt{\frac{1}{U_0}(\mu - V(\mathbf{r}))\,\theta(\mu - V(\mathbf{r}))}, \tag{3.65}$$

where $\theta(x)$ is the unit step function and the TF density is

$$n_{\mathrm{TF}}(\mathbf{r}) = \frac{1}{U_0}(\mu - V(\mathbf{r}))\,\theta(\mu - V(\mathbf{r})). \tag{3.66}$$

For the case of an isotropic harmonic potential

$$V(\mathbf{r}) = \frac{M\omega^2}{2}\mathbf{r}^2, \tag{3.67}$$

Eq. (3.65) gives

$$\psi_{\mathrm{TF}}(\mathbf{r}) = \sqrt{\frac{M\omega^2}{2U_0}(R^2 - \mathbf{r}^2)\,\theta(R - |\mathbf{r}|)}, \tag{3.68}$$

where $R \equiv \sqrt{2\mu/M\omega^2}$ is a characteristic radius of the condensate. Substituting $U_0 = 4\pi\hbar^2 a/M$ in Eq. (3.68) gives

$$\psi_{\mathrm{TF}}(\mathbf{r}) = \sqrt{\frac{R^2 - \mathbf{r}^2}{8\pi a d_0^4}\,\theta(R - |\mathbf{r}|)}. \tag{3.69}$$

The chemical potential μ is determined from the normalization condition

$$\int \psi_{\mathrm{TF}}^2(\mathbf{r})d\mathbf{r} = N. \tag{3.70}$$

Substituting Eq. (3.69) in Eq. (3.70) gives

$$R = \left(\frac{15Na}{d_0}\right)^{\frac{1}{5}} d_0. \tag{3.71}$$

Hence,

$$\mu = \frac{M\omega^2}{2}R^2 = \frac{\hbar\omega}{2}\left(\frac{15Na}{d_0}\right)^{\frac{2}{5}} = \frac{\hbar\omega}{2}\left(\frac{R}{d_0}\right)^2. \tag{3.72}$$

It follows from the relation $\mu = \partial E/\partial N$ that the total energy is given by

$$E = \frac{5\hbar\omega}{14}\left(\frac{15Na}{d_0}\right)^{\frac{2}{5}} N = \frac{5}{7}\mu N. \tag{3.73}$$

In the TF approximation, the total energy consists of the potential energy $E_{\rm pot}$ and the interaction energy $E_{\rm int}$. The virial theorem then gives $2E_{\rm pot} = 3E_{\rm int}$. Thus,

$$E_{\rm pot} = \frac{3}{7}\mu N, \quad E_{\rm int} = \frac{2}{7}\mu N, \tag{3.74}$$

where $E_{\rm int}$ can be found from the release-energy measurement[1] [Ensher, *et al.* (1996)].

The peak density of the condensate is given by

$$n(0) = \psi_{\rm TF}^2(0) = \frac{R^2}{8\pi a d_0^4} = \frac{5}{2}\frac{N}{4\pi R^3/3}. \tag{3.75}$$

It follows from Eqs. (3.72) and (3.75) that

$$\mu = \frac{\hbar\omega}{2}\left(\frac{R}{d_0}\right)^2 = 4\pi\hbar\omega a d_0^2 n(0) = U_0 n(0). \tag{3.76}$$

The measurement of the release energy per particle, $E_{\rm int}/N = 2\mu/7 = 2U_0 n(0)/7$, thus provides direct determination of the peak density and the chemical potential of the condensate.

When the size of the condensate is R, the kinetic energy per particle is of the order of $\hbar^2/2MR^2$. The ratio of the kinetic energy to the potential energy is estimated to be

$$\frac{\text{kinetic energy}}{\text{potential energy}} \sim \frac{\hbar^2/(2MR^2)}{M\omega^2R^2/2} = \left(\frac{d_0}{R}\right)^4 = \left(\frac{15Na}{d_0}\right)^{-\frac{4}{5}}. \tag{3.77}$$

Thus, the kinetic energy is negligible if

$$15N\frac{a}{d_0} \gg 1. \tag{3.78}$$

Because the interaction causes virtual excitations of bosons out of the condensate, the number of condensate bosons is actually less than the total number of bosons, N, in the system. The ratio of the number of virtually excited bosons to N is called the fraction of the condensate depletion, and it is roughly given by [see Eq. (2.79)]

$$\sqrt{n(0)a^3} = \sqrt{\frac{15}{8\pi}\left(\frac{a}{R}\right)^3 N}. \tag{3.79}$$

[1] The release-energy measurement is performed by switching off the trapping potential and measuring the total energy of the gas; therefore, the released energy is the sum of the kinetic energy and the interaction energy. In the TF limit, the kinetic energy is negligible, and hence, the release-energy measurement gives $E_{\rm int}$.

Because the GP theory ignores the depletion of the condensate, this fraction must be much smaller than unity for both the GP theory and the Bogoliubov theory to be valid. This gives the condition

$$N \ll \left(\frac{R}{a}\right)^3. \tag{3.80}$$

For typical parameters $R \sim 1$ μm and $a \sim 10$ Å, it gives $N \ll 10^9$, which has been satisfied for most experiments performed thus far. However, it is possible that condition (3.80) does not hold when a is very large [Papp, *et al.* (2008); Pollack, *et al.* (2009)]. In this case, neither the GP theory nor the Bogoliubov theory is valid.

When the system is locally perturbed by external fields, container walls, or by topological defects such as vortices, the system tries to restore its unperturbed density. A characteristic length scale ξ over which the local density restores its equilibrium value is called the healing length or correlation length; it is determined by the balance between the zero-point kinetic energy $\hbar^2/2M\xi^2$ and the mean-field interaction energy $4\pi\hbar^2 an/M$:

$$\xi = \frac{1}{\sqrt{8\pi an}}. \tag{3.81}$$

In the case of an isotropic harmonic trap, the healing length at the density peak can be obtained from Eqs. (3.75) and (3.81) as

$$\xi = \frac{1}{\sqrt{8\pi an(0)}} = \frac{d_0^2}{R}. \tag{3.82}$$

Because $R \gg d_0$ in the TF regime, the healing length is much shorter than the size of the condensate R. However, it is much longer than the mean interparticle distance because

$$n^{\frac{1}{3}}\xi = \frac{1}{\sqrt{8\pi a n^{\frac{1}{3}}}} \gg 1. \tag{3.83}$$

Thus, in the TF regime, we have a clear separation of various length scales: $R \gg d_0 \gg \xi \gg n^{-1/3}$.

It follows from Eqs. (3.72) and (3.76) that

$$\frac{\hbar\omega}{U_0 n(0)} = \frac{\hbar\omega}{\mu} = 2\left(\frac{d_0}{R}\right)^2 = 2\left(\frac{15Na}{d_0}\right)^{-\frac{2}{5}}. \tag{3.84}$$

When $15Na/d \gg 1$, the mean-field interaction energy per particle, $U_0 n(0)$, is much larger than the single-particle energy level spacing $\hbar\omega$, implying that single-particle energy levels are substantially blurred by the interaction. When calculating the properties of BEC, we therefore need not care about the discreteness of the harmonic potential. It is also worthwhile to note that the usual definition of BEC as $n_0/N = O(1)$, where n_0 is the number of bosons in the lowest single-particle state, cannot be used. Instead, we must resort to a more general definition of BEC formulated by Penrose and Onsager [Penrose and Onsager (1956)], which we restate here [see Sec. 1.3 for details]. Irrespective of the presence of interactions, we can always define the single-particle reduced density operator ρ_1. Let the largest eigenvalue of the reduced density operator be $n_{\max}$. Then, we may say that BEC exists if $n_{\max}/N = O(1)$. If there is one and only one eigenvalue that is of the order of N, the matrix element $\langle \mathbf{r}|\rho_1|\mathbf{r}'\rangle$ is shown to have an asymptotic form

$$\langle \mathbf{r}|\rho_1|\mathbf{r}'\rangle \to \psi(\mathbf{r})\psi^*(\mathbf{r}') \quad \text{for} \quad |\mathbf{r}-\mathbf{r}'| \to \infty. \tag{3.85}$$

Conversely, when condition (3.85) holds, it can be shown that $\psi(\mathbf{r})$ is a very good approximation of the eigenfunction of ρ_1 corresponding to the largest eigenvalue $n_{\max}$. In this sense, we call $\psi(\mathbf{r})$ the condensate wave function and $n_{\max}/N$, the condensate fraction. As given in Eq. (3.58), the GP wave function is well approximated by $\psi(\mathbf{r})$ in Eq. (3.85) if the condensate fraction is close to unity.

3.6 Collective Modes in the Thomas–Fermi Regime

In the Thomas–Fermi regime, the frequencies of various collective modes can be calculated analytically. Substituting

$$\psi(\mathbf{r},t) = \sqrt{n(\mathbf{r},t)}\, e^{i\phi(\mathbf{r},t)} \tag{3.86}$$

into Eq. (3.64) and separating the real and imaginary parts, we obtain

$$\frac{\partial n}{\partial t} = -\boldsymbol{\nabla}(n\mathbf{v}), \tag{3.87}$$

$$M\frac{\partial \mathbf{v}}{\partial t} = -\boldsymbol{\nabla}\left(\mu + \frac{M}{2}\mathbf{v}^2\right), \tag{3.88}$$

where

$$\mathbf{v} = \frac{\hbar}{M}\nabla\phi \tag{3.89}$$

is the superfluid velocity and

$$\mu = V + U_0 n - \frac{\hbar^2}{2M\sqrt{n}}\boldsymbol{\nabla}^2\sqrt{n} \tag{3.90}$$

is the chemical potential. The last term in (3.69) is called the quantum pressure term and it may be ignored in the Thomas–Fermi regime. We linearize n and μ around their equilibrium values:

$$n(\mathbf{r},t) = n_{\mathrm{TF}}(\mathbf{r}) + \delta n(\mathbf{r},t), \tag{3.91}$$

$$\mu(\mathbf{r},t) = \mu + \delta\mu(\mathbf{r},t). \tag{3.92}$$

Substituting Eqs. (3.91) and (3.92) into Eqs. (3.87) and (3.88) and ignoring the higher-order terms such as $\delta n\mathbf{v}$ and $\mathbf{v}^2$, we obtain

$$\frac{\partial \delta n}{\partial t} = -\boldsymbol{\nabla}(n_{\mathrm{TF}}\mathbf{v}), \tag{3.93}$$

$$M\frac{\partial \mathbf{v}}{\partial t} = -U_0\boldsymbol{\nabla}\delta n, \tag{3.94}$$

where $\delta\mu = U_0\delta n$ is used. Taking the time derivative of both sides of Eq. (3.93) and substituting Eq. (3.94), we obtain

$$\frac{\partial^2 \delta n}{\partial t^2} = \frac{U_0}{M}\boldsymbol{\nabla}(n_{\mathrm{TF}}\boldsymbol{\nabla}\delta n). \tag{3.95}$$

Assuming the time dependence of the collective mode to be $\delta n \propto e^{-i\Omega t}$, we obtain

$$\Omega^2\delta n(\mathbf{r}) = -\frac{U_0}{M}\boldsymbol{\nabla}\big(n_{\mathrm{TF}}(\mathbf{r})\boldsymbol{\nabla}\delta n(\mathbf{r})\big). \tag{3.96}$$

By solving this equation for a given potential $V(\mathbf{r})$, we obtain the frequency Ω of the collective mode.

3.6.1 *Isotropic harmonic potential*

For an isotropic harmonic potential (3.67) and the TF density (3.66), Eq. (3.96) reduces to

$$\frac{\Omega^2}{\omega^2}\delta n = r\frac{\partial}{\partial r}\delta n - \frac{1}{2}(R^2 - r^2)\left[\frac{1}{r}\frac{\partial^2}{\partial r^2}(r\delta n) - \frac{\ell^2}{r^2}\delta n\right], \tag{3.97}$$

where we used the polar-coordinate representation of the Laplacian

$$\nabla^2 = \frac{1}{r}\frac{\partial^2}{\partial r^2}r - \frac{\ell^2}{r^2} \tag{3.98}$$

with

$$\ell^2 \equiv -\frac{1}{\sin\theta}\frac{\partial}{\partial\theta}\left(\sin\theta\frac{\partial}{\partial\theta}\right) - \frac{1}{\sin^2\theta}\frac{\partial^2}{\partial\phi^2}. \tag{3.99}$$

As a solution to Eq. (3.97), we assume

$$\delta n(\mathbf{r}) = f(r)r^{\ell}Y_{\ell m}(\theta,\phi), \tag{3.100}$$

where $Y_{\ell m}$ is the spherical harmonic function that is an eigenfunction of $\boldsymbol{\ell}^2$ with eigenvalue $\ell(\ell+1)$:

$$\boldsymbol{\ell}^2 Y_{\ell m} = \ell(\ell+1) Y_{\ell m}. \tag{3.101}$$

Substituting Eq. (3.100) into Eq. (3.97), we obtain

$$x(x-1)\frac{d^2 f}{dx^2} + \left(\frac{2\ell+3}{2} - \frac{2\ell+5}{2}x\right)\frac{df}{dx} + \frac{\left(\frac{\Omega}{\omega}\right)^2 - \ell}{2} f = 0, \tag{3.102}$$

where $x \equiv r^2/R^2$. This equation takes the same form as the differential equation of the hypergeometric function ${}_2F_1(a,b,c;x)$:

$$x(x-1)\frac{d^2\,{}_2F_1}{dx^2} + [c - (a+b+1)x]\frac{d_2F_1}{dx} - ab_2F_1 = 0, \tag{3.103}$$

where

$$\begin{aligned} {}_2F_1(a,b,c;x) &= 1 + \frac{ab}{c}\frac{x}{1!} + \frac{a(a+1)b(b+1)}{c(c+1)}\frac{x^2}{2!} + \cdots \\ &\equiv \sum_{n=0}^{\infty} \frac{(a)_n (b)_n}{(c)_n}\frac{x^n}{n!}. \end{aligned} \tag{3.104}$$

Comparing Eq. (3.102) with (3.103), we find that

$$c = \frac{2\ell+3}{2}, \quad a+b = \frac{2\ell+3}{2}, \quad \text{and} \quad ab = \frac{\left(\frac{\Omega}{\omega}\right)^2 - \ell}{2}. \tag{3.105}$$

For the series in Eq. (3.104) to terminate at finite n, a or b must be zero or a negative integer. Let $a = -n$ $(n = 0, 1, 2, \cdots)$. Then, $b = n + \ell + 3/2$ and [Stringari (1996)]

$$\Omega = \omega\sqrt{2n^2 + 2n\ell + 3n + \ell} \quad (n = 0, 1, 2, \cdots) \tag{3.106}$$

and

$$f(r) = {}_2F_1\left(-n,\ n+\ell+\frac{3}{2},\ \ell+\frac{3}{2};\ \frac{r^2}{R^2}\right). \tag{3.107}$$

Equation (3.106) is to be compared with the ideal-gas result $\Omega = \omega(2n+\ell)$.

The mode with $n \neq 0$ and $\ell = 0$ oscillates along the radial direction and it is called the monopole mode or the breathing mode with the frequency given by

$$\Omega = \omega\sqrt{2n^2 + 3n}. \tag{3.108}$$

The mode with $\ell \neq 0$ and $n = 0$ has the frequency

$$\Omega = \sqrt{\ell}\,\omega \tag{3.109}$$

and the density oscillation

$$\delta n \propto r^\ell Y_{\ell m}(\theta, \phi) e^{-i\Omega t}. \tag{3.110}$$

The amplitude of the oscillation increases for larger r and peaks near the surface of the condensate. This mode is therefore called the surface mode.

3.6.2 *Axisymmetric trap*

For an axisymmetric potential,

$$V(\rho, z) = \frac{M\omega^2}{2}(\rho^2 + \lambda^2 z^2), \tag{3.111}$$

where $\rho \equiv \sqrt{x^2 + y^2}$ and λ is the trap aspect ratio. Then, the TF density is

$$n_{\mathrm{TF}} = \frac{M\omega^2}{2U_0}(R^2 - \rho^2 - \lambda^2 z^2)\theta(R^2 - \rho^2 - \lambda^2 z^2), \tag{3.112}$$

and Eq. (3.96) reduces to

$$\frac{\Omega^2}{\omega^2}\delta n = \left(\rho\frac{\partial}{\partial\rho} + \lambda^2 z\frac{\partial}{\partial z}\right)\delta n - \frac{1}{2}(R^2 - \rho^2 - \lambda^2 z^2)\nabla^2\delta n. \tag{3.113}$$

We note that the Laplacian is expressed in cylindrical coordinates as

$$\nabla^2 = \frac{\partial^2}{\partial\rho^2} + \frac{1}{\rho}\frac{\partial}{\partial\rho} + \frac{1}{\rho^2}\frac{\partial^2}{\partial\phi^2} + \frac{\partial^2}{\partial z^2}. \tag{3.114}$$

Then, by inspection, we find that a solution to Eq. (3.113) is given by the surface mode

$$\delta n \propto \rho e^{\pm i\ell\phi} e^{-i\Omega t} \tag{3.115}$$

which satisfies $\nabla^2\delta n = 0$ and gives

$$\Omega = \sqrt{\ell}\,\omega. \tag{3.116}$$

For an axisymmetric trap, the magnetic quantum number m is conserved, but ℓ is, in general, not conserved. Thus, the quadrupolar mode with $\ell = 2$ and $m = 0$ is coupled with the breathing mode with $n = 1$ and $\ell = m = 0$. Since the density oscillations of these modes are given by $\delta n \propto r^2 Y_{20} \propto 2z^2 - x^2 - y^2$ and $\delta n \propto x^2 + y^2 + z^2$, respectively, it is reasonable to assume that the density oscillation of the coupled mode is given by $\delta n = a\rho^2 + bz^2 + c$. Substituting this into Eq. (3.113), we find that

$$\Omega = \omega\left(2 + \frac{3}{2}\lambda^2 \pm \sqrt{4 - 4\lambda^2 + \frac{9}{4}\lambda^4}\right)^{\frac{1}{2}}. \tag{3.117}$$

3.6.3 *Scissors mode*

The irrotational property of a superfluid gives rise to a unique response of the system, known as a scissors mode, against a sudden rotation of the trapping potential [Guéry-Odelin and Stringari (1999); Maragò, *et al.* (2000)]. We consider a general harmonic potential

$$V(\mathbf{r}) = \frac{M}{2}(\omega_x^2 x^2 + \omega_y^2 y^2 + \omega_z^2 z^2) \tag{3.118}$$

and assume density fluctuations of the form

$$\delta n = \alpha xy e^{-i\Omega t}, \tag{3.119}$$

where α is a constant. Substituting Eq. (3.119) into Eq. (3.95), we obtain

$$\begin{aligned}\Omega^2 \delta n &= -\frac{\alpha U_0}{M}\left(\frac{\partial n_{\mathrm{TF}}}{\partial x} y + \frac{\partial n_{\mathrm{TF}}}{\partial y}\right) e^{-i\Omega t} \\ &= \alpha(\omega_x^2 + \omega_y^2) xy e^{-i\Omega t}.\end{aligned}$$

Thus, the frequency of the scissors mode is given by

$$\Omega = \sqrt{\omega_x^2 + \omega_y^2}. \tag{3.120}$$

The velocity field of this mode can be found from Eq. (3.94) as

$$\mathbf{v} = -i\frac{U_0 \alpha}{M\Omega}(y, x, 0), \tag{3.121}$$

which satisfies the irrotationality condition rot$\mathbf{v}$ = 0. In contrast, if the condensate rotated like a rigid body, the velocity field would be proportional to $(-y, x, 0)$. Instead of a uniform rotation, the scissors mode describes an oscillation of the condensate axes in a manner similar to periodic opening and closing of a pair of scissors, as shown in Fig. 3.1. .

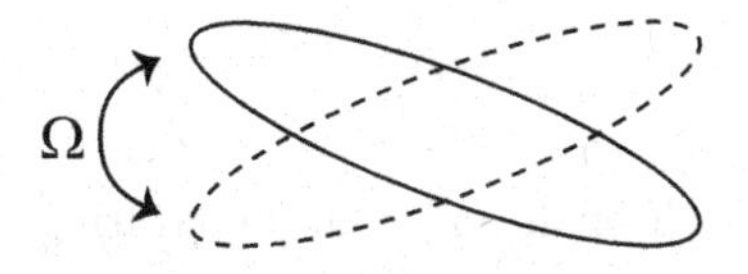

Fig. 3.1 Scissors mode. The axis of the condensate oscillates around an equilibrium position in a manner similar to periodic opening and closing of a pair of scissors.

3.7 Variational Method

Variational methods based on the minimal action principle have successfully been applied to many phenomena, especially time-dependent ones such as collective modes. The time-dependent GP equation can be derived from the minimal action principle with the action given by

$$S = \int dt \int d\mathbf{r} \left[i\hbar\psi^* \frac{\partial \psi}{\partial t} + \frac{\hbar^2}{2M}\psi^*\nabla^2\psi - \frac{M\omega^2}{2}(x^2+y^2+\lambda^2 z^2)\psi^*\psi \right.$$
$$\left. - \frac{2\pi\hbar^2 a}{M}(\psi^*\psi)^2 \right], \tag{3.122}$$

where we assume an axially symmetric harmonic potential and λ is an asymmetry parameter. In fact, the functional derivative of S with respect to ψ^* gives

$$\frac{\delta S}{\delta \psi^*} = i\hbar\frac{\partial \psi}{\partial t} + \frac{\hbar^2}{2M}\nabla^2\psi - \frac{M\omega^2}{2}(x^2+y^2+\lambda^2 z^2)\psi - \frac{4\pi\hbar^2 a}{M}\psi^*\psi^2 = 0,$$

that is

$$i\hbar\frac{\partial \psi}{\partial t} = \left[-\frac{\hbar^2}{2M}\nabla^2 + \frac{M\omega^2}{2}(x^2+y^2+\lambda^2 z^2) + \frac{4\pi\hbar^2 a}{M}|\psi|^2 \right]\psi, \tag{3.123}$$

which is nothing but the time-dependent GP equation.

Let us measure the length, time, and ψ in units of $d_0 \equiv \sqrt{\hbar/M\omega}$, ω^{-1}, and $\sqrt{N/d_0^3}$, respectively. This is equivalent to the substitutions

$$t \to \omega^{-1}t, \quad \mathbf{r} \to d_0\mathbf{r}, \quad \text{and} \quad \psi \to \sqrt{\frac{N}{d^3}}\psi. \tag{3.124}$$

Then, the action (3.122) becomes

$$S = N\hbar \int dt \int d\mathbf{r} \left[i\psi^* \frac{\partial \psi}{\partial t} + \frac{1}{2}\psi^*\nabla^2\psi - \frac{1}{2}(x^2+y^2+\lambda^2 z^2)\psi^*\psi \right.$$
$$\left. - \frac{g}{2}(\psi^*\psi)^2 \right], \tag{3.125}$$

where ψ is normalized to unity $\int |\psi|^2 d\mathbf{r} = 1$ and

$$g \equiv \frac{4\pi N a}{d_0} \tag{3.126}$$

is the dimensionless strength of interaction. Separating ψ into amplitude A and phase ϕ as

$$\psi(\mathbf{r},t) = A(\mathbf{r},t)e^{i\phi(\mathbf{r},t)}, \tag{3.127}$$

and substituting this into Eq. (3.125), we obtain

$$S = N\hbar \int dt \int d\mathbf{r} \left[\frac{i}{2}\frac{\partial A^2}{\partial t} + \frac{i}{2}\nabla(A\nabla\phi) - A^2\frac{\partial\phi}{\partial t} + \frac{1}{2}A\nabla^2 A - \frac{1}{2}A^2(\nabla\phi)^2 \right.$$
$$\left. -\frac{1}{2}(x^2+y^2+\lambda^2 z^2)A^2 - \frac{g}{2}A^4 \right]. \tag{3.128}$$

Here, the first two terms on the right-hand side play no role when the variational principle is applied because the first term can be integrated with respect to time, while the second term can be transformed via Stoke's theorem into a surface integral.

The requirement that the action be stationary with respect to small variations in phase ϕ yields

$$\frac{\partial A^2}{\partial t} + \nabla(A^2\mathbf{v}_\mathrm{s}) = 0, \tag{3.129}$$

where

$$\mathbf{v}_\mathrm{s} = \nabla\phi \tag{3.130}$$

is the superfluid velocity, and hence, $A^2\mathbf{v}_s$ describes the mass current. Equation (3.129) is the equation of continuity that guarantees the conservation of the particle number. The requirement that the action be stationary with respect to small variations in amplitude A leads to

$$-\frac{\partial\phi}{\partial t} = \frac{1}{2}\mathbf{v}_\mathrm{s}^2 + \frac{1}{2}(x^2+y^2+\lambda^2 z^2) + gA^2 - \frac{1}{2A}\nabla^2 A. \tag{3.131}$$

This is a quantum version of the Hamilton–Jacobi equation, where the last term is referred to as the quantum pressure term or quantum potential. The solution of Eqs. (3.129) and (3.131) is equivalent to the solution of the GP equation

$$i\frac{\partial\psi}{\partial t} = \left(-\frac{\nabla^2}{2} + \frac{1}{2}(x^2+y^2+\lambda^2 z^2) + g|\psi|^2 \right)\psi. \tag{3.132}$$

3.7.1 *Gaussian variational wave function*

Any variational wave function other than the exact one cannnot satisfy Eqs. (3.129) and (3.131) simultaneously. Instead, we minimize the action S in Eq. (3.125) within the functional subspace of a given variational wave function. Let us consider a Gaussian vaiational wave function whose amplitude is of the form

$$A(\mathbf{r},t) = \sqrt{\frac{1}{\pi^{\frac{3}{2}} d_x(t) d_y(t) d_z(t)}} \exp\left[-\frac{x^2}{2d_x^2(t)} - \frac{y^2}{2d_y^2(t)} - \frac{z^2}{2d_z^2(t)} \right], \tag{3.133}$$

where $d_x(t)$, $d_y(t)$, and $d_z(t)$ are variational parameters that are made time-dependent to account for shape oscillations of the condensate in a harmonic trap. The preexponential factor in Eq. (3.133) is chosen to satisfy the normalization condition $\int A^2 d\mathbf{r} = 1$. The phase of the wave function is determined to satisfy the equation of continuity (3.129). Substituting Eq. (3.133) in Eq. (3.129) gives

$$f(x, d_x) + f(y, d_y) + f(z, d_z) = 0, \tag{3.134}$$

where

$$\begin{aligned} f(x, d_x) &= -\frac{\dot{d}_x}{d_x} + \frac{2\dot{d}_x}{d_x^3}x^2 - \frac{2x}{d_x^2}\frac{\partial \phi}{\partial x} + \frac{\partial^2 \phi}{\partial x^2} \\ &= -\frac{\dot{d}_x}{d_x} + \frac{2\dot{d}_x}{d_x^3}x^2 + e^{\frac{x^2}{d_x^2}}\frac{\partial}{\partial x}\left(e^{-\frac{x^2}{d_x^2}}\frac{\partial \phi_x}{\partial x}\right). \end{aligned} \tag{3.135}$$

Here, the overdot denotes differentiation with respect to t. Since x, y, and z are independent, we may substitute $f(\alpha, d_\alpha) = a_\alpha$ for $\alpha = x, y, z$ where a_α's are constants that satisfy $a_x + a_y + a_z = 0$. Integrating $f(x, d_x) = a_x$ with respect to x gives

$$\phi_x = \frac{\dot{d}_x}{2d_x}x^2 + a_x\int_0^x dx'\int_0^{x'} dx'' e^{\frac{x'^2 - x''^2}{d_x^2}} + b_x\int_0^x dx' e^{\frac{x'^2}{d_x^2}} + C_x, \tag{3.136}$$

where b_x and C_x are integration constants. Because the global phase is arbitrary, we can substitute $C_x = 0$. By symmetry, we assume that the superfluid velocity at the origin is zero; thus

$$b_x = \left.\frac{\partial \phi_x}{\partial x}\right|_{x=0}.$$

The mass current density is

$$\begin{aligned} j_x &\propto \frac{1}{i}(\psi^*\partial_x\psi - \psi\partial_x\psi^*) \propto A^2\frac{\partial \phi}{\partial x} \\ &\propto e^{-\frac{x^2}{d_x^2} - \frac{y^2}{d_y^2} - \frac{z^2}{d_z^2}}\left[\frac{\dot{d}_x}{d_x}x + a_x e^{\frac{x^2}{d_x^2}}\int_0^x dx' e^{-\frac{x'^2}{d_x^2}}\right]. \end{aligned}$$

By requiring that j_α should vanish at $x \to \pm\infty$, we have $a_x = 0$. Thus, $f(x, d_x) = (\dot{d}_x/2d_x)x^2$ and

$$\phi(\mathbf{r}, t) = \frac{\dot{d}_x}{2d_x}x^2 + \frac{\dot{d}_y}{2d_y}y^2 + \frac{\dot{d}_z}{2d_z}z^2. \tag{3.137}$$

Combining Eqs. (3.133) and (3.137) yields [Pérez-García, *et al.* (1996, 1997)]

$$\psi(\mathbf{r},t) = \sqrt{\frac{1}{\pi^{3/2}d_x d_y d_z}} \exp\left[-\frac{x^2}{2d_x^2}(1-i\dot{d}_x d_x) - \frac{y^2}{2d_y^2}(1-i\dot{d}_y d_y) - \frac{z^2}{2d_z^2}(1-i\dot{d}_z d_z)\right]; \tag{3.138}$$

this gives the Gaussian variational wave function subject to the conservation of particle number and to the requirement that the mass current vanishes at the origin and at infinity.

The variational parameters d_α are determined so as to make the action extremal. Substituting Eq. (3.138) in Eq. (3.125) gives

$$S = \frac{N\hbar}{4}\int dt \left[\sum_{\alpha=x,y,z}\left(-d_\alpha \ddot{d}_\alpha - d_\alpha^{-2} - \lambda_\alpha^2 d_\alpha^2\right) - \frac{\gamma}{d_x d_y d_z}\right], \tag{3.139}$$

where $\lambda_\alpha = 1$ for $\alpha = x, y$ and $\lambda_\alpha = \lambda$ for $\alpha = z$, and

$$\gamma \equiv \frac{g}{\sqrt{2\pi^3}} = \frac{4N}{\sqrt{2\pi}}\frac{a}{d_0}. \tag{3.140}$$

Taking the functional derivative of S with respect to d_α gives the equation of motion for d_α:

$$\ddot{d}_\alpha = -\lambda^2 d_\alpha + d_\alpha^{-3} + \frac{\gamma}{2d_x d_y d_z d_\alpha} \equiv -\frac{\partial V^{\text{eff}}}{\partial d_\alpha} \quad (\alpha = x, y, a), \tag{3.141}$$

where

$$V^{\text{eff}} = \frac{1}{2}\sum_{\alpha=x,y,z}(d_\alpha^2 + d_\alpha^{-2}) + \frac{\gamma}{2d_x d_y d_z} \tag{3.142}$$

is an effective potential for d_α.

3.7.2 *Collective modes*

The equations of motion (3.141) for variational parameters d_α's can be used to find the frequencies of the collective modes of a condenstate such that the widths of the condensate, which are characterized by d_α, oscillate in time. We consider a trap potential that is symmetric with respect to the z-axis, and let $\bar{d}_r$ and $\bar{d}_z$ be the equilibrium values of d_α along the radial and axial directions, respectively. From Eq. (3.141), we see that they obey

$$\bar{d}_r = \bar{d}_r^{-3} + \frac{\gamma}{2\bar{d}_r^3 \bar{d}_z}, \tag{3.143}$$

$$\lambda^2 \bar{d}_z = \bar{d}_z^{-3} + \frac{\gamma}{2\bar{d}_r^2 \bar{d}_z^2}. \tag{3.144}$$

Expanding Eq. (3.141) up to linear terms from these equilibrium values leads to

$$\Delta\ddot{d}_x = -\left(1 + 3\bar{d}_r^{-4} + \frac{\gamma}{\bar{d}_r^4\bar{d}_z}\right)\Delta d_x - \frac{\gamma}{2\bar{d}_r^4\bar{d}_z}\Delta d_y - \frac{\gamma}{2\bar{d}_r^3\bar{d}_z^2}\Delta d_z, \quad (3.145)$$

$$\Delta\ddot{d}_y = -\frac{\gamma}{2\bar{d}_r^4\bar{d}_z}\Delta d_x - \left(1 + 3\bar{d}_r^{-4} + \frac{\gamma}{\bar{d}_r^4\bar{d}_z}\right)\Delta d_y - \frac{\gamma}{2\bar{d}_r^3\bar{d}_z^2}\Delta d_z, \quad (3.146)$$

$$\Delta\ddot{d}_z = -\frac{\gamma}{2\bar{d}_r^3\bar{d}_z^2}\Delta d_x - \frac{\gamma}{2\bar{d}_r^3\bar{d}_z^2}\Delta d_y - \left(\lambda^2 + 3\bar{d}_z^{-4} + \frac{\gamma}{\bar{d}_r^2\bar{d}_z^3}\right)\Delta d_z, \quad (3.147)$$

where $\Delta d_\alpha \equiv d_\alpha - \bar{d}_\alpha$. Substituting

$$a \equiv 1 + 3\bar{d}_r^{-4} + \frac{\gamma}{\bar{d}_r^4\bar{d}_z} = 3 + \bar{d}_r^{-4}, \quad (3.148)$$

$$b \equiv \lambda^2 + 3\bar{d}_z^{-4} + \frac{\gamma}{\bar{d}_r^2\bar{d}_z^3} = 3\lambda^2 + \bar{d}_z^{-4}, \quad (3.149)$$

$$\alpha \equiv \frac{\gamma}{2\bar{d}_r^4\bar{d}_z}, \quad \beta \equiv \frac{\gamma}{2\bar{d}_r^3\bar{d}_z^2}, \quad (3.150)$$

the eigenvalue equation becomes

$$\begin{vmatrix} \omega^2 - a & -\alpha & -\beta \\ -\alpha & \omega^2 - a & -\beta \\ -\beta & -\beta & \omega^2 - b \end{vmatrix}$$
$$= (\omega^2 - a + \alpha)[\omega^4 - (a + b + \alpha)\omega^2 + ab + \alpha b - 2\beta^2] = 0.$$

Hence,

$$\omega = \begin{cases} 2\sqrt{1 - \frac{\gamma}{4\bar{d}_r^4\bar{d}_z}}, \\ \sqrt{2 + 2\lambda^2 - \frac{\gamma}{4\bar{d}_r^2\bar{d}_z^3} \pm \sqrt{4\left(1 - \lambda^2 + \frac{\gamma}{8\bar{d}_r^2\bar{d}_z^3}\right)^2 + 2\left(\frac{\gamma}{2\bar{d}_r^3\bar{d}_z^2}\right)^2}}. \end{cases} \quad (3.151)$$

For a given strength of interaction γ and asymmetry parameter λ, we can solve Eqs. (3.143) and (3.144) for $\bar{d}_r$ and $\bar{d}_z$. Substituting these in Eq. (3.151) give the frequencies of the collective modes.

In the TF regime, the kinetic-energy terms $\bar{d}_r^{-3}$ and $\bar{d}_z^{-3}$ in Eqs. (3.143) and (3.144) can be neglected, so that we have

$$\frac{\gamma}{2\bar{d}_r^4\bar{d}_z} = 1, \quad \frac{\gamma}{2\bar{d}_r^2\bar{d}_z^3} = \lambda^2, \quad \frac{\gamma}{2\bar{d}_r^3\bar{d}_x^2} = \lambda.$$

Substituting these in Eq. (3.151) gives

$$\omega_{\mathrm{Q}} = \sqrt{2}, \quad (3.152)$$

$$\omega_{\mathrm{M-Q}} = \sqrt{2 + \frac{3}{2}\lambda^2 \pm \sqrt{4 - 4\lambda^2 + \frac{9}{4}\lambda^4}}, \quad (3.153)$$

where $\omega_{\rm Q}$ is the frequency of the quadrupole mode and $\omega_{\rm M-Q}$ is that of the combined monopole and quadrupole modes. For the case of an isotropic potential ($\lambda = 1$), Eqs. (3.152) and (3.153) reduce to $\omega = \sqrt{2}$ (doubly degenerate) and $\sqrt{5}$. For the case of a prolonged potential ($\lambda \ll 1$), Eqs. (3.152) and (3.153) give $\omega = \sqrt{2}$, 2, and $\sqrt{5/2}\lambda$. For the case of an oblate trap ($\lambda \gg 1$), we have $\omega = \sqrt{2}$, $\sqrt{10/3}$, and $\sqrt{3}\lambda$.

3.8 Attractive Bose–Einstein Condensate

Attractive bosons are believed not to undergo BEC in a uniform system because the system would collapse into a high-density state. However, Bose–Einstein condensates with attractive interactions have been realized in systems of confined atomic gases [Bradley, *et al.* (1995, 1997)], and remarkable collapsing dynamics have been observed [Gerton, *et al.* (2000); Donley, *et al.* (2001)]. We begin by offering a quantative argument of why attractive condensates can exist in a confined system.

A confined atomic gas has three characteristic energies. The first is the zero-point kinetic energy that is proportional to d^{-2}, where d is the size of the condensate. The second is the energy of a harmonic potential that is proportional to d^2. The third is the interaction energy that is proportional to the density of particles and is therefore proportional to N_0/d^3, where N_0 is the number of condensate bosons. Because the interction is attractive, the interaction energy is negative. The total energy E is the sum of these three energies,

$$E(d) = N_0(Ad^{-2} + Bd^2 - CN_0d^{-3}), \tag{3.154}$$

where A, B, and C are positive constants. It follows that the total energy has a metastable minimum at $d = d_{\rm m}$, where $E'(d_{\rm m}) = 0$, if N_0 is below a critical value $N_{\rm c}$ that is determined from $E'(d_{\rm c}) = E''(d_{\rm c}) = 0$ by

$$N_{\rm c} = \frac{8A}{15C}\left(\frac{A}{5B}\right)^{\frac{1}{4}}. \tag{3.155}$$

The condensate is believed to be formed at this local energy minimum. However, because of the interaction energy $-CN_0d^{-3}$, E tends to minus infinity as d tends to zero. This implies that the condensate is not the true ground state but it is in a metastable one. The energy barrier separating these two states arises from the term Ad^{-2} in Eq. (3.154). Thus, the metastability of BEC is ensured by the zero-point kinetic energy that counterbalances the attractive interaction. If N_0 exceeds $N_{\rm c}$, however, the energy barrier vanishes and no condensate can exist. Since the zero-point energy arises from the confinement of the condensate due to a trapping potential, we may say that the BEC with attractive interaction is a mesoscopic phenomenon and it does not exist in the thermodynamic limit.

The above qualitative argument suggests that for the metastable Bose–Einstein condensate to exist, the zero-point energy $\sim \hbar\omega$ must exceed the mean-field interaction energy per particle, *i.e.*, $\hbar\omega > N_0 U_0/V$. The critical number of condensate bosons $N_{\rm c}$ can therefore be estimated from the condition $\hbar\omega \sim N_{\rm c}|U_0|/V$. Since the volume of the condensate is roughly given by $V \sim 4\pi d_0^3$, $N_{\rm c}$ is estimated to be

$$N_{\rm c} \sim \frac{d_0}{|a|}. \tag{3.156}$$

A more quantitative argument below shows that $N_{\rm c}$ is indeed proportional to $d_0/|a|$, and the constant of proportionality is of the order of unity. One might conclude from (3.156) that $N_{\rm c}$ can be made infinite by increasing d_0. However, this is not the case, since the density of particles $n \sim N_{\rm c}/d_0^3 \sim 1/|a|d_0^2$ decreases with increasing d_0. Since n must be larger than λ_{dB}^{-3} for the system to undergo BEC, d_0 cannot be larger than $\sim \sqrt{\lambda_{\rm th}^3/|a|}$, and thus, we have the following fundamental upper limit for $N_{\rm c}$:

$$N_{\rm c} \leq \left(\frac{\lambda_{\rm th}}{|a|}\right)^{\frac{3}{2}}. \tag{3.157}$$

3.8.1 *Collective modes*

We consider the case of an isotropic potential. Then, the equation of motion for the variational parameter d_α is obtained from Eq. (3.141) with $\lambda = 1$ as

$$\ddot{d}_\alpha = -d_\alpha + d_\alpha^{-3} + \frac{\gamma}{2d_x d_y d_z d_\alpha} = -\frac{\partial V^{\rm eff}}{\partial d_\alpha} \quad (\alpha = x, y, z), \tag{3.158}$$

where

$$V^{\rm eff} = \frac{1}{2}\sum_{\alpha=x,y,z}(d_\alpha^2 + d_\alpha^{-2}) + \frac{\gamma}{2d_x d_y d_z}, \tag{3.159}$$

$$\gamma = \frac{g}{\sqrt{2\pi^3}} = \frac{4N}{\sqrt{2\pi}}\frac{a}{d_0}. \tag{3.160}$$

The equilibrium value $\bar{d}$ is the same for all α and it is determined from

$$\bar{d}^{\,5} - \bar{d} - \frac{\gamma}{2} = 0. \tag{3.161}$$

Linearizing Eq. (3.158) with respect to $\Delta d_\alpha \equiv d_\alpha - \bar{d}$ yields

$$\Delta \ddot{d}_\alpha = \sum_\beta [\bar{d}^{-4} - 1 - 2(\bar{d}^{-4} + 1)\delta_{\alpha\beta}]\Delta d_\beta. \tag{3.162}$$

Substituting $\Delta d_\alpha(t) = \Delta d_\alpha(0)e^{-i\omega t}$ leads to the eigenvalue equation

$$\begin{aligned} &\det(\omega^2\delta_{\alpha\beta} - [\bar{d}^{-4} - 1 - 2(\bar{d}^{-4} + 1)\delta_{\alpha\beta}]) \\ &= (\omega^2 + \bar{d}^{-4} - 5)(\omega^2 - 2\bar{d}^{-4} - 2)^2 = 0, \end{aligned} \tag{3.163}$$

and hence we obtain [Stringari (1996)]

$$\omega_{\mathrm{M}} = \sqrt{5 - \bar{d}^{-4}}, \tag{3.164}$$

$$\omega_{\mathrm{Q}} = \sqrt{2(1 + \bar{d}^{-4})}, \tag{3.165}$$

where ω_{M} is the frequency of the monopole mode and ω_{Q} is that of the doubly degenerate quadrupole modes. For a given γ, Eq. (3.161) can be solved for $\bar{d}$, which then gives the frequencies of the monopole and quadrupole modes via Eqs. (3.164) and (3.165), respectively. As the strength of the attractive interaction increases, the equilibrium width $\bar{d}$ of the condensate decreases. When $\bar{d}$ becomes smaller than $r_{\mathrm{c}} = 5^{-\frac{1}{4}}$, the frequency of the monopole mode becomes pure imaginary, which signals the onset of a dynamical instability of an attractive condensate, *i.e.*, of the collapse of the condensate. The critical strength of interaction is determined from Eq. (3.161) with $\bar{d} = r_{\mathrm{c}}$ as [Baym and Pethick (1996)]

$$\gamma_{\mathrm{c}} = -\frac{8}{5^{5/4}} \simeq -1.070. \tag{3.166}$$

The corresponding critical number of condensate bosons is given by

$$N_{\mathrm{c}} \simeq \frac{2\sqrt{2\pi}}{5^{5/4}}\frac{d_0}{|a|} \simeq 0.6705\frac{d_0}{|a|}. \tag{3.167}$$

This number is somewhat larger than the more precise value of

$$N_{\mathrm{c}} \simeq 0.575\frac{d_0}{|a|} \tag{3.168}$$

that is obtained from the numerical integration of the GP equation [Dodd, *et al.* (1996)]. The difference arises from the fact that the variational method overestimates the stable phase. To understand the nature of the collective modes, let us substitute $\Delta d_\alpha(t) = \Delta d_\alpha(0)e^{-i\omega_{Q}t}$ in Eq. (3.162). Then, we have

$$\Delta d_x + \Delta d_y + \Delta d_z = 0. \tag{3.169}$$

Thus, the quadrupole oscillation occurs in such a manner that the volume of the condensate is conserved. For the monopole mode, we substitute $\Delta d_\alpha(t) = \Delta d_\alpha(0)e^{-i\omega_{\rm M}t}$ in Eq. (3.162), obtaining

$$\Delta d_x = \Delta d_y = \Delta d_z.$$

This implies that the condensate alternately expands and contracts in an isotropic manner. We can thus substitute $d_\alpha(t) = r(t)$ for all α in Eq. (3.139), obtaining

$$S_{\rm M} = \frac{N\hbar}{4}\int dt(3\dot{r}^2 - 3r^{-2} - 3r^2 - \gamma r^{-3}), \tag{3.170}$$

where the constant term that results from the partial integration is dropped. The first term on the right-hand side may be interpreted as arising from the collective kinetic motion of the condensate. When transformed back to the original units, it becomes

$$\frac{N\hbar}{4}\int dt 3\dot{r}^2 \to \frac{4\hbar}{4}\int d(\omega t) 3\left[\frac{d}{d(\omega t)}\left(\frac{R}{d_0}\right)\right]^2 = \int dt \frac{1}{2}M^*\dot{R}^2, \tag{3.171}$$

where

$$M^* = \frac{3}{2}Nm \tag{3.172}$$

is the effective mass of the condensate for the monopole motion [Ueda and Leggett (1998)].

To understand the behavior of the monopole mode near the critical point, let us assume that N is close to but below $N_{\rm c}$, and substitute $\gamma = \gamma_{\rm c} + \delta\gamma$ and $r = r_{\rm c} + \delta r$, where $\gamma_{\rm c}$ is given in Eq. (3.166) and $r_{\rm c} = 5^{-\frac{1}{4}}$ is obtained from Eq. (3.164) with the condition $\omega_{\rm M} = 0$. Substituting these in Eqs. (3.161), (3.164), and (3.165) yield

$$\frac{\delta r}{r_{\rm c}} = \sqrt{\frac{2}{5}\left(1 - \frac{N}{N_{\rm c}}\right)}, \tag{3.173}$$

$$\omega_{\rm M} = \sqrt[4]{160\left(1 - \frac{N}{N_{\rm c}}\right)}, \tag{3.174}$$

$$\omega_{\rm Q} = \sqrt{12 - 8\sqrt{10\left(1 - \frac{N}{N_{\rm c}}\right)}}. \tag{3.175}$$

We thus find that near the critical point, the frequency of the monopole mode vanishes according to the one-fourth power of $1 - N/N_{\rm c}$.

3.8.2 *Collapsing dynamics of an attractive condensate*

Let us consider a situation in which the strength of attractive interaction $|\gamma|$ slightly exceeds its critical value $|\gamma_c|$. Then, $\bar{d}$ becomes smaller than $5^{-\frac{1}{4}}$ so that the frequency of the monopole mode (3.164) becomes pure imaginary. The radius of the condensate evolves with time as $r(t) \propto e^{-|\omega_M|t}$, so that the condensate implodes upon itself. The Lagrangian of the system is given from Eq. (3.170) as

$$L = \frac{N\hbar}{4}[3\dot{r}^2 - f(r)], \tag{3.176}$$

where

$$f(r) = 3r^{-2} + 3r^2 - |\gamma|r^{-3}. \tag{3.177}$$

The energy E of the system, which is a conserved quantity, is given by

$$E = \dot{r}\frac{\partial L}{\partial \dot{r}} - L = \frac{N\hbar}{4}[3\dot{r}^2 + f(r)] \equiv \frac{N\hbar}{4}f(r_c). \tag{3.178}$$

Hence,

$$\frac{dr}{dt} = -\sqrt{\frac{1}{3}[f(r_c) - f(r)]}. \tag{3.179}$$

We substitute $\delta\gamma \equiv |\gamma| - |\gamma_c|$ and expand $f(r_c) - f(r)$ up to the third power in $x \equiv r_c - r$. Then,

$$\frac{dx}{dt} = \sqrt{5\delta\gamma x + 10r_c^{-1}\delta\gamma x^2 + \frac{20}{3}r_c^{-1}x^3}. \tag{3.180}$$

In an initial stage of the collapse with $x \ll 1$, it is sufficient to keep up to the x^2 term in Eq. (3.180), giving

$$x(t) = \frac{r_c}{4}\left[\cosh(\sqrt{10r_c^{-1}\delta\gamma}\, t) - 1\right] \simeq \frac{5}{4}\gamma_c\left(\frac{N}{N_c} - 1\right)t^2. \tag{3.181}$$

When x becomes of the order of 1, the last term in Eq. (3.180) becomes dominant because other terms include a small factor $\delta\gamma$. Hence,

$$x(t) = \frac{3 \cdot 5^{-\frac{5}{4}}}{(t_0 - t)^2}. \tag{3.182}$$

The fact that $x(t)$ diverges at a finite time implies that the collapse occurs in a finite time. Here, t_0 is determined by an initial slow dynamics governed by the first term in Eq. (3.180). Thus, taking the first and third terms in Eq. (3.180), we obtain

$$\frac{dx}{dt} = \sqrt{ax^3 + bx}, \tag{3.183}$$

where $a = 20.5^{1/4}/3$ and $b = 5\delta\gamma = (8/5^{1/4})(N/N_c - 1)$. Integrating this gives

$$t = \frac{1}{\sqrt[4]{ab}} \int_0^{\sinh^{-1}\sqrt{\frac{a}{b}}x} \sinh^{-1} y \, dy. \tag{3.184}$$

When N is close to N_c, $a/b \gg 1$. Thus, the collapse time is estimated to be

$$t_{\text{collapse}} \simeq \frac{1}{\sqrt[4]{ab}} \int_0^{\infty} \sinh^{-\frac{1}{2}} y dy = \frac{1}{\sqrt[4]{ab}} \frac{\Gamma(\frac{1}{4})^2}{2\sqrt{\pi}}$$
$$\simeq 1.37 \left(\frac{N}{N_c} - 1\right)^{-\frac{1}{4}}. \tag{3.185}$$

At the final stage of collapse, the implosion is accelerated by higher-order terms that are neglected in Eq. (3.180). The time $t_{\text{implosion}}$ required for the system to implode is therefore shorter than t_{collapse} by a constant amount Δt:

$$t_{\text{implosion}} = t_{\text{collapse}} - \Delta t,$$

where Δt is numerically evaluated to be 1.83 [Saito and Ueda (2001)].

Chapter 4

Linear Response and Sum Rules

4.1 Linear Response Theory

In this section, we present basic tools to investigate the excitation spectrum of a many-body system. All results obtained in this section are applicable to both bosons and fermions.

4.1.1 *Linear response of density fluctuations*

The excitation spectrum of quasiparticles can be probed through the interaction of a test particle with the system of interest. Let $U(\mathbf{r},t)$ be a time-dependent external potential that couples to the system at position $\mathbf{r}$. In second-quantized language, the corresponding Hamiltonian is given by

$$\hat{H}_{\mathrm{ext}}(t) = \int d\mathbf{r}\, U(\mathbf{r},t)\hat{\psi}^{\dagger}(\mathbf{r})\hat{\psi}(\mathbf{r}), \tag{4.1}$$

where $\hat{\psi}(\mathbf{r})$ is the field operator of the system. Substituting Fourier transforms

$$\hat{\psi}(\mathbf{r}) = \frac{1}{\sqrt{V}}\sum_{\mathbf{k}} \hat{a}_{\mathbf{k}} e^{i\mathbf{kr}}, \tag{4.2}$$

$$U(\mathbf{r},t) = \sum_{\mathbf{k}}\int \frac{d\omega}{2\pi} U(\mathbf{k},\omega) e^{i\mathbf{kr}} e^{-i\omega t}, \tag{4.3}$$

into Eq. (4.1) gives

$$\hat{H}_{\mathrm{ext}}(t) = \sum_{\mathbf{k}}\int \frac{d\omega}{2\pi} U(\mathbf{k},\omega)\hat{\rho}_{-\mathbf{k}} e^{-i\omega t}. \tag{4.4}$$

Here,

$$\hat{\rho}_{\mathbf{k}} \equiv \int d\mathbf{r}\, \hat{n}(\mathbf{r}) e^{-i\mathbf{kr}} = \sum_{\mathbf{p}} \hat{a}^{\dagger}_{\mathbf{p}} \hat{a}_{\mathbf{p}+\mathbf{k}} = \hat{\rho}^{\dagger}_{-\mathbf{k}} \tag{4.5}$$

is the Fourier transform of the number-density operator $\hat{n}(\mathbf{r}) \equiv \hat{\psi}^\dagger(\mathbf{r})\hat{\psi}(\mathbf{r})$.

Let us consider a situation in which the external potential has a single wave vector $\mathbf{k}$ and single frequency ω. Then,

$$\hat{H}_{\text{ext}}(t) = U(\mathbf{k},\omega)\hat{\rho}_{-\mathbf{k}}e^{-i\omega t} + U^*(\mathbf{k},\omega)\hat{\rho}^\dagger_{-\mathbf{k}}e^{i\omega t}. \tag{4.6}$$

The state of the system evolves with time according to the Schrödinger equation

$$i\hbar\frac{\partial}{\partial t}|\psi(t)\rangle = \left(\hat{H} + \hat{H}_{\text{ext}}(t)e^{\epsilon t}\right)|\psi(t)\rangle, \tag{4.7}$$

where $\hat{H}$ is the Hamiltonian of the system and ϵ, an infinitesimal positive number. The factor $e^{\epsilon t}$ is introduced to ensure that the external potential is adiabatically switched off in the remote past. The initial condition is assumed to be

$$|\psi(-\infty)\rangle = |0\rangle, \tag{4.8}$$

where $|0\rangle$ is the ground state of $\hat{H}$.

In linear response theory, we solve Eq. (4.7) up to first order in $U(\mathbf{k},\omega)$. We expand the state vector in terms of a complete set of eigenstates $\{|n\rangle\}$ of $\hat{H}$:

$$|\psi(t)\rangle = \sum_n c_n(t)e^{-\frac{i}{\hbar}E_n t}|n\rangle, \tag{4.9}$$

where

$$\hat{H}|n\rangle = E_n|n\rangle \quad (n = 0, 1, 2, \cdots). \tag{4.10}$$

The initial condition (4.8) is satisfied if the following condition is met:

$$c_n(-\infty) = \delta_{n0}. \tag{4.11}$$

Substituting Eq. (4.9) in Eq. (4.7), we obtain

$$\begin{aligned}\dot{c}_m(t) &= -\frac{i}{\hbar}\sum_n c_n(t)e^{(i\omega_{mn}+\epsilon)t}\langle m|\hat{H}_{\text{ext}}(t)|n\rangle \\ &= -\frac{i}{\hbar}\sum_n c_n(t)\left[U(\mathbf{k},\omega)\langle m|\hat{\rho}_{-\mathbf{k}}|n\rangle e^{i(\omega_{mn}-\omega-i\epsilon)t}\right. \\ &\quad \left. + U^*(\mathbf{k},\omega)\langle m|\hat{\rho}^\dagger_{-\mathbf{k}}|n\rangle e^{i(\omega_{mn}+\omega-i\epsilon)t}\right],\end{aligned} \tag{4.12}$$

where $\omega_{mn} \equiv (E_m - E_n)/\hbar$ and Eq. (4.6) is substituted in the last equation.

Integrating Eq. (4.12) with respect to t from $-\infty$ to t up to the first order in U gives

$$c_m(t) = \delta_{m0} + (1-\delta_{m0})\left[\frac{U(\mathbf{k},\omega)\langle m|\hat{\rho}_{-\mathbf{k}}|0\rangle}{\hbar(\omega-\omega_{m0}+i\epsilon)}e^{i(\omega_{m0}-\omega-i\epsilon)t} - \frac{U^*(\mathbf{k},\omega)\langle m|\hat{\rho}^\dagger_{-\mathbf{k}}|0\rangle}{\hbar(\omega+\omega_{m0}-i\epsilon)}e^{i(\omega_{m0}+\omega-i\epsilon)t}\right]. \tag{4.13}$$

The change in density due to $\hat{H}_{\text{ext}}$ is given as

$$\delta\langle\hat{\rho}_{\mathbf{k}}(t)\rangle \equiv \langle\psi(t)|\hat{\rho}_{\mathbf{k}}|\psi(t)\rangle - \langle 0|\hat{\rho}_{\mathbf{k}}|0\rangle. \tag{4.14}$$

Substituting Eq. (4.9) for $|\psi(t)\rangle$ gives

$$\delta\langle(\hat{\rho}_{\mathbf{k}}(t)\rangle = \sum_n{}' \left(c_n(t)e^{-i\omega_{n0}t}\langle 0|\hat{\rho}_{\mathbf{k}}|n\rangle + c_n^*(t)e^{i\omega_{n0}t}\langle n|\hat{\rho}_{\mathbf{k}}|0\rangle\right), \tag{4.15}$$

where $\sum_n'$ denotes the summation over n except $n=0$. Substituting Eq. (4.13) in Eq. (4.15) and simplifying the result using[1]

$$\langle n|\hat{\rho}_{\mathbf{k}}|0\rangle\langle n|\hat{\rho}_{-\mathbf{k}}|0\rangle = \langle 0|\hat{\rho}_{\mathbf{k}}|n\rangle\langle n|\hat{\rho}^\dagger_{-\mathbf{k}}|0\rangle = 0 \tag{4.16}$$

leads to

$$\delta\langle\hat{\rho}_{\mathbf{k}}(t)\rangle = \frac{1}{\hbar}U(\mathbf{k},\omega)e^{-(i\omega-\epsilon)t}\sum_n{}'\left(\frac{|\langle 0|\hat{\rho}_{\mathbf{k}}|n\rangle|^2}{\omega-\omega_{n0}+i\epsilon} - \frac{|\langle n|\hat{\rho}_{\mathbf{k}}|0\rangle|^2}{\omega+\omega_{n0}+i\epsilon}\right). \tag{4.17}$$

We assume that the system possesses the space-inversion symmetry ($\hat{\rho}_{\mathbf{k}} = \hat{\rho}_{-\mathbf{k}}$), so that

$$\langle n|\hat{\rho}_{\mathbf{k}}|0\rangle = \langle n|\hat{\rho}_{-\mathbf{k}}|0\rangle = \langle n|\hat{\rho}^\dagger_{\mathbf{k}}|0\rangle = \langle 0|\hat{\rho}_{\mathbf{k}}|n\rangle^*. \tag{4.18}$$

Then, Eq. (4.17) reduces to

$$\delta\langle\hat{\rho}_{\mathbf{k}}(t)\rangle = \frac{1}{\hbar}U(\mathbf{k},\omega)e^{-(i\omega-\epsilon)t}\sum_n{}'|\langle n|\hat{\rho}^\dagger_{\mathbf{k}}|0\rangle|^2\frac{2\omega_{n0}}{(\omega+i\epsilon)^2-\omega_{n0}^2}. \tag{4.19}$$

Applying the Fourier transform to this equation gives

$$\delta\langle\hat{\rho}(\mathbf{k},\omega)\rangle = U(\mathbf{k},\omega)D^{\text{ret}}(\mathbf{k},\omega), \tag{4.20}$$

where

$$D^{\text{ret}}(\mathbf{k},\omega) = \frac{1}{\hbar}\sum_n{}'|\langle n|\hat{\rho}^\dagger_{\mathbf{k}}|0\rangle|^2\frac{2\omega_{n0}}{(\omega+i\epsilon)^2-\omega_{n0}^2}. \tag{4.21}$$

Equation (4.20) gives the linear response of density against the external perturbation $U(\mathbf{k},\omega)$, and the ratio

$$\chi_\rho(\mathbf{k},\omega) \equiv \frac{\delta\langle\hat{\rho}(\mathbf{k},\omega)\rangle}{U(\mathbf{k},\omega)} = D^{\text{ret}}(\mathbf{k},\omega) \tag{4.22}$$

defines the linear susceptibility of density fluctuations.

---

[1]Since $\hat{\rho}_{\mathbf{k}} = \sum_{\mathbf{p}}\hat{a}^\dagger_{\mathbf{p}}\hat{a}_{\mathbf{p}+\mathbf{k}}$ annihilates the net momentum $\mathbf{k}$ from the system, $\langle n|\hat{\rho}_{\mathbf{k}}|0\rangle \neq 0$ only if the total momentum of $|n\rangle$ is $-\mathbf{k}$. Thus, $\langle n|\hat{\rho}_{\mathbf{k}}|0\rangle\langle n|\hat{\rho}_{-\mathbf{k}}|0\rangle = 0$. Similarly, we obtain the second equation using Eq. (4.5).

4.1.2 *Retarded response function*

The function $D^{\rm ret}(\mathbf{k},\omega)$ defined in Eq. (4.21) is referred to as the retarded response function or the retarded Green's function of the density. Taking the imaginary part of Eq. (4.21) and using

$$\frac{1}{x+i\epsilon} = P\left(\frac{1}{x}\right) - i\pi\delta(x), \tag{4.23}$$

where $P\left(\frac{1}{x}\right)$ denotes the principal value of $\frac{1}{x}$, we obtain the dynamic structure factor

$$S(\mathbf{k},\omega) \equiv -\frac{\hbar}{\pi}{\rm Im}D^{\rm ret}(\mathbf{k},\omega) = \sum_n{}' |\langle n|\hat{\rho}^\dagger_{\mathbf{k}}|0\rangle|^2[\delta(\omega-\omega_{n0}) - \delta(\omega+\omega_{n0})], \tag{4.24}$$

which gives the excitation spectrum of density, in which the perturbation transfers energy $\hbar\omega$ and momentum $\hbar\mathbf{k}$ to the system.

The inverse Fourier transform of $D^{\rm ret}(\mathbf{k},\omega)$ is given by

$$\begin{aligned} D^{\rm ret}(\mathbf{k},t) &= \int_{-\infty}^{\infty}\frac{d\omega}{2\pi}D^{\rm ret}(\mathbf{k},\omega)e^{-i\omega t} \\ &= \frac{1}{\hbar}\sum_n{}' |\langle n|\hat{\rho}^\dagger_{\mathbf{k}}|0\rangle|^2 \int_{-\infty}^{\infty}\frac{d\omega}{2\pi}\left(\frac{1}{\omega+i\epsilon-\omega_{n0}} - \frac{1}{\omega+i\epsilon+\omega_{n0}}\right)e^{-i\omega t}. \end{aligned} \tag{4.25}$$

Because of the factor $e^{-i\omega t}$, the integration contour in Eq. (4.25) must be taken in the lower (or upper) half of the complex ω-plane if $t > 0$ (or if $t < 0$). Since the poles $\omega = \pm\omega_{n0} - i\epsilon$ lie in the lower-half plane, the integral is nonzero only for $t > 0$. Hence,

$$D^{\rm ret}(\mathbf{k},t) = -\frac{i}{\hbar}\theta(t)\sum_n{}' |\langle n|\hat{\rho}^\dagger_{\mathbf{k}}|0\rangle|^2\left(e^{-i\omega_{n0}t} - e^{i\omega_{n0}t}\right), \tag{4.26}$$

where $\theta(t)$ is the unit step function. Comparing this with Eq. (4.24), we find that

$$D^{\rm ret}(\mathbf{k},t) = -\frac{i}{\hbar}\theta(t)\int_{-\infty}^{\infty} d\omega S(\mathbf{k},\omega)e^{-i\omega t}. \tag{4.27}$$

Applying the Fourier transform to this equation with respect to time gives

$$D^{\rm ret}(\mathbf{k},\omega) = \frac{1}{\hbar}\int_{-\infty}^{\infty}\frac{S(\mathbf{k},\omega')}{\omega+i\epsilon-\omega'}d\omega'. \tag{4.28}$$

We may use Eq. (4.18) and the completeness relation to eliminate the sum over n in Eq. (4.26); in fact,

$$\sum_n{}' |\langle n|\hat{\rho}_{\mathbf{k}}^\dagger|0\rangle|^2 e^{-i\omega_{n0}t} = \sum_n \langle 0|e^{\frac{i}{\hbar}\hat{H}t}\hat{\rho}_{\mathbf{k}}e^{-\frac{i}{\hbar}\hat{H}t}|n\rangle\langle n|\hat{\rho}_{\mathbf{k}}^\dagger|0\rangle = \langle 0|\hat{\rho}_{\mathbf{k}}(t)\hat{\rho}_{\mathbf{k}}^\dagger|0\rangle,$$

$$\begin{aligned}\sum_n{}' |\langle n|\hat{\rho}_{\mathbf{k}}^\dagger|0\rangle|^2 e^{i\omega_{n0}t} &= \sum_n |\langle n|\hat{\rho}_{\mathbf{k}}|0\rangle|^2 e^{i\omega_{n0}t} \\ &= \sum_n \langle 0|\hat{\rho}_{\mathbf{k}}^\dagger|n\rangle\langle n|e^{\frac{i}{\hbar}\hat{H}t}\hat{\rho}_{\mathbf{k}}e^{-\frac{i}{\hbar}\hat{H}t}|0\rangle = \langle 0|\hat{\rho}_{\mathbf{k}}^\dagger\hat{\rho}_{\mathbf{k}}(t)|0\rangle,\end{aligned}$$

where $\sum_n{}'$ is replaced by $\sum_n$ because $\langle 0|\hat{\rho}_{\mathbf{k}}^\dagger|0\rangle = 0$. We thus find that

$$D^{\text{ret}}(\mathbf{k},t) = -\frac{i}{\hbar}\theta(t)\Big\langle 0\Big|\left[\hat{\rho}_{\mathbf{k}}(t),\hat{\rho}_{\mathbf{k}}^\dagger(0)\right]\Big|0\Big\rangle, \tag{4.29}$$

where

$$\hat{\rho}_{\mathbf{k}}(t) \equiv e^{\frac{i}{\hbar}\hat{H}t}\hat{\rho}_{\mathbf{k}}e^{-\frac{i}{\hbar}\hat{H}t}. \tag{4.30}$$

In a special case in which $S(\mathbf{k},\omega)$ has a single peak of the form

$$S(\mathbf{k},\omega) = S(\mathbf{k})\delta(\omega-\omega_{\mathbf{k}}), \tag{4.31}$$

we obtain

$$D^{\text{ret}}(\mathbf{k},t) = -\frac{i}{\hbar}\theta(t)S(\mathbf{k})e^{-i\omega_{\mathbf{k}}t} \tag{4.32}$$

from Eq. (4.27) and Eq. (4.28). Applying the Fourier transform to this equation with respect to time gives

$$D^{\text{ret}}(\mathbf{k},\omega) = \frac{S(\mathbf{k})}{\hbar(\omega-\omega_{\mathbf{k}}+i\epsilon)}. \tag{4.33}$$

Equation (4.33) implies that the pole of the retarded Green's function of the density gives the frequency of the collective mode.

4.2 Sum Rules

When the system is translationally invariant in space, the excitation spectrum satisfies some exact relations known as sum rules.

4.2.1 *Longitudinal f-sum rule*

When the system possesses space-translation invariance, the single-particle Hamiltonian can be diagonalized with respect to the wave vector $\mathbf{k}$:

$$\hat{H}_0 = \sum_{\mathbf{k}} \epsilon_{\mathbf{k}} \hat{a}^\dagger_{\mathbf{k}} \hat{a}_{\mathbf{k}}, \quad \epsilon_{\mathbf{k}} = \frac{\hbar^2 \mathbf{k}^2}{2m}. \tag{4.34}$$

The interaction Hamiltonian is expressed in the second-quantized form as

$$\hat{V} = \frac{1}{2} \int d\mathbf{r} \int d\mathbf{r}' \hat{\psi}^\dagger(\mathbf{r}) \hat{\psi}^\dagger(\mathbf{r}') V(\mathbf{r}-\mathbf{r}') \hat{\psi}(\mathbf{r}') \hat{\psi}(\mathbf{r}). \tag{4.35}$$

Substituting Eq. (4.2) and

$$V(\mathbf{r}) = \sum_{\mathbf{k}} V_{\mathbf{k}} e^{i\mathbf{k}\mathbf{r}} \tag{4.36}$$

into Eq. (4.30), we obtain

$$\hat{V} = \frac{1}{2} \sum_{\mathbf{p},\mathbf{q},\mathbf{k}} V_{\mathbf{k}} \hat{a}^\dagger_{\mathbf{p}} \hat{a}^\dagger_{\mathbf{q}} \hat{a}_{\mathbf{q}+\mathbf{k}} \hat{a}_{\mathbf{p}-\mathbf{k}} = \frac{1}{2} \sum_{\mathbf{k}} V_{\mathbf{k}} \left(\hat{\rho}_{-\mathbf{k}} \hat{\rho}_{\mathbf{k}} - \hat{N} \right), \tag{4.37}$$

where $\hat{\rho}_{\mathbf{k}}$ is given in Eq. (4.5) and

$$\hat{N} = \sum_{\mathbf{p}} \hat{a}^\dagger_{\mathbf{p}} \hat{a}_{\mathbf{p}} \tag{4.38}$$

is the total number operator. The total Hamiltonian is given by

$$\hat{H} = \hat{H}_0 + \hat{V}. \tag{4.39}$$

A straightforward calculation gives

$$\left[\hat{\rho}_{-\mathbf{k}}, \left[\hat{\rho}_{\mathbf{k}}, \hat{H}\right]\right] = -\sum_{\mathbf{p}} (\epsilon_{\mathbf{p}+\mathbf{k}} + \epsilon_{\mathbf{p}-\mathbf{k}} - 2\epsilon_{\mathbf{p}}) \hat{a}^\dagger_{\mathbf{p}} \hat{a}_{\mathbf{p}}. \tag{4.40}$$

Since $\epsilon_{\mathbf{k}} = \hbar^2 \mathbf{k}^2 / 2m$,

$$\epsilon_{\mathbf{p}+\mathbf{k}} + \epsilon_{\mathbf{p}-\mathbf{k}} - 2\epsilon_{\mathbf{p}} = \frac{\hbar^2 \mathbf{k}^2}{m} = 2\epsilon_{\mathbf{k}},$$

and thus,

$$\left[\hat{\rho}_{-\mathbf{k}}, [\hat{\rho}_{\mathbf{k}}, \hat{H}]\right] = -2\epsilon_{\mathbf{k}} \hat{N}. \tag{4.41}$$

Taking the expectation value of the left-hand side of Eq. (4.41) over $|0\rangle$ and inserting the completeness relation, we have

$$\left\langle 0 \right| \left[\hat{\rho}_{-\mathbf{k}}, \left[\hat{\rho}_{\mathbf{k}}, \hat{H}\right]\right] \left| 0 \right\rangle = -\sum_n \hbar\omega_{n0} \left(|\langle n|\hat{\rho}_{\mathbf{k}}|0\rangle|^2 + |\langle n|\hat{\rho}_{-\mathbf{k}}|0\rangle|^2 \right), \tag{4.42}$$

where $\hbar\omega_{n0} = E_n - E_0$. Substituting Eq. (4.41) in Eq. (4.42), we obtain

$$\sum_n \hbar\omega_{n0}\left(|\langle n|\hat{\rho}_{\mathbf{k}}|0\rangle|^2 + |\langle n|\hat{\rho}_{-\mathbf{k}}|0\rangle|^2\right) = 2\epsilon_{\mathbf{k}}N. \tag{4.43}$$

Equation (4.43) is referred to as the longitudinal f-sum rule. When the system possesses space-inversion symmetry, the relation $|\langle n|\hat{\rho}_{-\mathbf{k}}|0\rangle| = |\langle n|\hat{\rho}_{\mathbf{k}}|0\rangle|$ holds, and thus Eq. (4.43) reduces to

$$\sum_n f_{n0} = N, \tag{4.44}$$

where

$$f_{n0} \equiv \frac{\hbar\omega_{n0}}{\epsilon_{\mathbf{k}}}|\langle n|\hat{\rho}_{\mathbf{k}}|0\rangle|^2 \tag{4.45}$$

is called the oscillator strength. As Eq. (4.44) suggests, the f-sum rule reflects the conservation of the particle number. Equation (4.44) may be regarded as a generalization of the Thomas–Reiche–Kuhn sum rule: for a single particle,

$$\sum_n (E_0 - E_n)|\langle n|\hat{x}|0\rangle|^2 = \frac{\hbar^2}{2M}, \tag{4.46}$$

where $\hat{H}|n\rangle = E_n|n\rangle$.

In terms of the dynamic structure factor $S(\mathbf{k},\omega)$ in Eq. (4.24), the longitudinal f-sum rule is expressed as

$$\int_0^\infty d\omega\hbar\omega S(\mathbf{k},\omega) = \epsilon_{\mathbf{k}}N. \tag{4.47}$$

For the special case of $S(\mathbf{k},\omega) = NS(\mathbf{k})\delta(\omega - \omega_{\mathbf{k}})$, Eq. (4.47) gives the energy of an elementary excitation as [Bijl (1940); Feynman (1954)]

$$\hbar\omega_{\mathbf{k}} = \frac{\hbar^2\mathbf{k}^2}{2mS(\mathbf{k})}, \tag{4.48}$$

where the static structure factor $S(\mathbf{k})$ determines the dispersion relation, *i.e.*, the relation between $\omega_{\mathbf{k}}$ and $\mathbf{k}$.

An extension to finite temperature is straightforward. Multiplying both sides of Eq. (4.41) by

$$\hat{\rho} = \frac{e^{-\beta\hat{H}}}{Z}, \tag{4.49}$$

where $Z = \mathrm{Tr}\, e^{-\beta\hat{H}}$, we obtain Eq. (4.47) with

$$S(\mathbf{k},\omega) = \frac{1}{Z}\sum_{m,n} e^{-\beta E_m}(|\langle m|\hat{\rho}_{\mathbf{k}}^\dagger|n\rangle|^2 + |\langle m|\hat{\rho}_{\mathbf{k}}|n\rangle|^2)\delta(\omega - \omega_{nm}), \tag{4.50}$$

where $\omega_{nm} \equiv (E_n - E_m)/\hbar$. In the presense of space-inversion symmetry, Eq. (4.50) reduces to

$$S(\mathbf{k},\omega) = \frac{2}{Z}\sum_{m,n} e^{-\beta E_m}|\langle m|\hat{\rho}_{\mathbf{k}}|n\rangle|^2\delta(\omega-\omega_{nm}). \tag{4.51}$$

Under the same assumption, we obtain the detailed balance of the dynamic structure factor:

$$S(\mathbf{k},-\omega) = e^{-\beta\hbar\omega}S(\mathbf{k},\omega). \tag{4.52}$$

4.2.2 *Compressibility sum rule*

The compressibility κ measures the degree of volume reduction against an increase in pressure at a fixed number of particles.

$$\kappa = -\frac{1}{V}\left(\frac{\partial V}{\partial P}\right)_N. \tag{4.53}$$

The pressure P is defined as the derivative of energy with respect to volume,

$$P = -\left(\frac{\partial E}{\partial V}\right)_N. \tag{4.54}$$

Substituting this in Eq. (4.53) gives

$$\kappa^{-1} = V\left(\frac{\partial^2 E}{\partial V^2}\right)_N. \tag{4.55}$$

Noting that V is related to the particle density n through $V = N/n$, we have

$$\frac{\partial}{\partial V} = \frac{\partial n}{\partial V}\frac{\partial}{\partial n} = -\frac{n^2}{N}\frac{\partial}{\partial n}. \tag{4.56}$$

Substituting $E = N\epsilon_g$, where ϵ_g is the ground-state energy per particle, we obtain

$$\kappa^{-1} = n\frac{d}{dn}\left(n^2\frac{d\epsilon_g}{dn}\right) = n^2\frac{d^2}{dn^2}(n\epsilon_g). \tag{4.57}$$

On the other hand, the chemical potential μ is given by

$$\mu = \left(\frac{\partial E}{\partial N}\right)_V = \left(\frac{\partial(E/V)}{\partial(N/V)}\right)_V = \frac{d}{dn}(n\epsilon_g). \tag{4.58}$$

Comparing Eqs. (4.57) and (4.58), we obtain

$$\kappa^{-1} = n^2\frac{d\mu}{dn}. \tag{4.59}$$

A microscopic expression of the compressibility can be found from Eqs. (4.22) and (4.28):

$$\frac{\delta\langle\hat{\rho}(\mathbf{k},\omega)\rangle}{U(\mathbf{k},\omega)} = D^{\mathrm{ret}}(\mathbf{k},\omega) = \frac{1}{\hbar}\int_{-\infty}^{\infty}\frac{S(\mathbf{k},\omega')}{\omega+i\epsilon-\omega'}d\omega'. \tag{4.60}$$

Suppose that we first take the limit $\mathbf{k}\to\mathbf{0}$ and then take the limit $\omega\to 0$. Then, the denominator on the left-hand side of Eq. (4.60) provides a uniform scalar potential $U(\mathbf{k}=0,\omega=0)$ which may be interpreted as a minus shift in the chemical potential. On the other hand, the numerator gives the concomitant change in the number of particles. Thus, we obtain

$$-\left(\frac{\partial N}{\partial\mu}\right)_V = \lim_{\omega\to 0}\lim_{\mathbf{k}\to 0} D^{\mathrm{ret}}(\mathbf{k},\omega) = -\frac{1}{\hbar}\int_{-\infty}^{\infty}\frac{S(\mathbf{k}=\mathbf{0},\omega)}{\omega}d\omega. \tag{4.61}$$

It follows from Eqs. (4.59) and (4.61) that

$$\int_{-\infty}^{\infty}\frac{S(\mathbf{k}=\mathbf{0},\omega)}{\hbar\omega}d\omega = \kappa n^2 V. \tag{4.62}$$

This relation is known as the compressibility sum rule.

The compressibility gives the isothermal and adiabatic sound velocity c. In fact, defining the mass density as $\rho\equiv mn$, we have

$$c = \sqrt{\frac{\partial P}{\partial\rho}} = \sqrt{\frac{1}{m}\frac{dP}{dn}}. \tag{4.63}$$

Since

$$P = -\left(\frac{\partial E}{\partial V}\right)_N = n^2\frac{d\epsilon_g}{dn}, \tag{4.64}$$

from Eq. (4.57), we find that

$$\frac{dP}{dn} = \frac{d}{dn}\left(n^2\frac{d\epsilon_g}{dn}\right) = \frac{1}{n\kappa}. \tag{4.65}$$

Thus,

$$c = \frac{1}{\sqrt{mn\kappa}}. \tag{4.66}$$

We may use this relation to obtain another expression for the compressibility sum rule.

$$\int_{-\infty}^{\infty}\frac{S(\mathbf{k}=\mathbf{0},\omega)}{\hbar\omega}d\omega = \frac{N}{mc^2}. \tag{4.67}$$

The combination of the f-sum rule and the compressibility sum rule gives an upper bound for the static structure factor. The Schwartz inequality gives

$$S(\mathbf{k}) \equiv \int_{-\infty}^{\infty} S(\mathbf{k},\omega)d\omega \leq \sqrt{\int_{-\infty}^{\infty} \hbar\omega S(\mathbf{k},\omega)d\omega \int_{-\infty}^{\infty} \frac{S(\mathbf{k},\omega)}{\hbar\omega}d\omega}. \quad (4.68)$$

Substituting Eq. (4.47), we have (note that the range of integration is doubled here)

$$S(\mathbf{k}) \leq \sqrt{2\epsilon_k N \int_{-\infty}^{\infty} \frac{S(\mathbf{k},\omega)}{\hbar\omega}d\omega}. \quad (4.69)$$

Taking the limit of $k \to 0$ and using Eq. (4.67), we obtain

$$S(\mathbf{k}) \leq \frac{\hbar k}{mc}N \quad (k \to 0). \quad (4.70)$$

This inequality implies that the density fluctuations of the system with a finite compressibility become negligible in the long-wavelength limit.

4.2.3 *Zero energy gap theorem*

The zero energy gap theorem holds for translationally invariant systems with positive compressibility.

Theorem. If a system is translationally invariant in space, the excitation spectrum has zero energy gap in the long-wavelength limit as long as the compressibility is positive.

Proof. Let us assume that the excitation spectrum has an energy gap Δ in the limit $\mathbf{k} \to 0$. Then, $S(\mathbf{k}=0,\omega) = 0$ for $\hbar\omega < \Delta$, and therefore, the compressibility sum rule (4.62) gives

$$\frac{1}{2}\kappa n^2 V = \int_{\Delta/\hbar}^{\infty} \frac{S(\mathbf{k}=\mathbf{0},\omega)}{\hbar\omega}d\omega \leq \frac{1}{\Delta}\int_{\Delta/\hbar}^{\infty} S(\mathbf{k}=\mathbf{0},\omega)d\omega. \quad (4.71)$$

On the other hand, the f-sum rule (4.47) leads to

$$\frac{\hbar^2\mathbf{k}^2}{2m}N = \int_{\Delta/\hbar}^{\infty} d\omega\ \hbar\omega\ S(\mathbf{k},\omega) \geq \Delta \int_{\Delta/\hbar}^{\infty} S(\mathbf{k},\omega)d\omega. \quad (4.72)$$

Combining Eq. (4.71) and Eq. (4.72), we find

$$\lim_{\mathbf{k}\to 0} \frac{\hbar^2\mathbf{k}^2}{2m}N \geq \frac{\Delta^2}{2}\kappa n^2 V. \quad (4.73)$$

This inequality implies that as long as $\kappa > 0$, Δ must vanish in the long-wavelength limit.

As a special application of this theorem, we find that a spatially uniform Bose system with repulsive interaction is gapless in the long-wavelength limit.

4.2.4 *Josephson sum rule*

The condensate density is defined in terms of the eigenfunction of the single-particle density matrix corresponding to a macroscopic (*i.e.*, extensive) eigenvalue and it is a thermodynamic quantity. The superfluid density, on the other hand, is defined in terms of linear response theory and it is a transport quantity. These two quantities therefore belong to different notions despite their apparent similarity. However, Josephson reported an interesting relation between them that is referred to as the Josephson sum rule [Josephson (1966)].

We consider a situation in which a superfluid is flowing with velocity $\mathbf{v}_s$ through a long container that is at rest with respect to the laboratory frame. Then, the mass current density operator $\hat{\mathbf{j}}$ is given by

$$\hat{\mathbf{j}}(\mathbf{r}) = \frac{1}{2i}\left[\hat{\psi}^\dagger(\mathbf{r})\nabla\hat{\psi}(\mathbf{r}) - (\nabla\hat{\psi}^\dagger(\mathbf{r}))\hat{\psi}(\mathbf{r})\right]. \tag{4.74}$$

The quantum-statistical average of $\hat{\mathbf{j}}$ defines the superfluid mass density ρ_s through the relation

$$\langle\hat{\mathbf{j}}\rangle = \rho_s \mathbf{v}_s. \tag{4.75}$$

The condensate density $|\psi_0|^2$ is defined in terms of the eigenfunction ψ_0 corresponding to a macroscopic eigenvalue of the single-particle density matrix. Because ψ_0 is complex, we may decompose it into the amplitude and the phase

$$\psi_0(\mathbf{r}) = A(\mathbf{r})e^{i\phi(\mathbf{r})}. \tag{4.76}$$

When the amplitude A may be considered as a constant, a variation in the wave function is related to a change in the phase through

$$\delta\psi_0 = i\psi_0\delta\phi. \tag{4.77}$$

Since the spatial variation in ϕ is related to the superfluid velocity $\mathbf{v}_s$ through

$$\mathbf{v}_s = \frac{\hbar}{m}\nabla\phi, \tag{4.78}$$

one may expect to find a relationship between ρ_s and $|\psi_0|^2$ by examining the responses $\langle\mathbf{j}\rangle$ and ψ_0 to a common external perturbation.

A variation in ψ_0 is caused by a Hamiltonian that includes a term conjugate to $\hat{\psi}$. Here, we consider the response of the system to the following Hamiltonian:

$$\hat{H}_{\text{ext}}(t) = \int d\mathbf{r}\left(\xi(\mathbf{k},\omega)e^{i(\mathbf{kr}-\omega t)}\hat{\psi}^\dagger(\mathbf{r}) + \xi^*(\mathbf{k},\omega)e^{-i(\mathbf{kr}-\omega t)}\hat{\psi}(\mathbf{r})\right). \tag{4.79}$$

The state evolution due to $\hat{H}_{\rm ext}(t)$ can be found by following a procedure similar to the one in Sec. 4.1.1. We expand the state vector in tems of the eigenstates $\{|n\rangle\}$ of $\hat{H}$ as in Eq. (4.9), where the expansion coefficients can be calculated up to the first order in $\hat{H}_{\rm ext}$ as

$$c_n(t) = \delta_{n0} + (1-\delta_{n0}) \int d\mathbf{r} \left[\frac{\xi(\mathbf{k},\omega)\langle n|\hat{\psi}^\dagger(\mathbf{r})|0\rangle}{\hbar(\omega-\omega_{n0}+i\epsilon)} e^{i\mathbf{k}\mathbf{r}} e^{i(\omega_{n0}-\omega-i\epsilon)t} - \frac{\xi^*(\mathbf{k},\omega)\langle n|\hat{\psi}(\mathbf{r})|0\rangle}{\hbar(\omega+\omega_{n0}-i\epsilon)} e^{-i\mathbf{k}\mathbf{r}} e^{i(\omega_{n0}+\omega-i\epsilon)t} \right]. \tag{4.80}$$

The response of $\hat{\psi}$ is given by

$$\begin{aligned} \delta\langle\hat{\psi}(\mathbf{r},t)\rangle &\equiv \langle\psi(t)|\hat{\psi}(\mathbf{r})|\psi(t)\rangle - \langle 0|\hat{\psi}(\mathbf{r})|0\rangle \\ &= \sum_n{}' \left(c_n(t) e^{-i\omega_{n0}t} \langle 0|\hat{\psi}(\mathbf{r})|n\rangle + c_n^*(t) e^{i\omega_{n0}t} \langle n|\hat{\psi}(\mathbf{r})|0\rangle \right). \end{aligned} \tag{4.81}$$

We assume that each state $|n\rangle$ has a fixed number of particles, so that

$$\langle 0|\hat{\psi}(\mathbf{r})|n\rangle\langle n|\hat{\psi}(\mathbf{r}')|0\rangle = \langle n|\hat{\psi}(\mathbf{r})|0\rangle\langle n|\hat{\psi}^\dagger(\mathbf{r}')|0\rangle = 0. \tag{4.82}$$

Substituting $c_n(t)$ in Eq. (4.80) into Eq. (4.81) and using Eq. (4.82), we obtain

$$\begin{aligned} \delta\langle\hat{\psi}(\mathbf{r},t)\rangle &= \frac{1}{\hbar}\xi(\mathbf{k},\omega)e^{-i\omega t+\epsilon t} \sum_n{}' \int d\mathbf{r}' \left(\frac{\langle 0|\hat{\psi}(\mathbf{r})|n\rangle\langle n|\hat{\psi}^\dagger(\mathbf{r}')|0\rangle}{\omega-\omega_{n0}+i\epsilon} - \frac{\langle n|\hat{\psi}(\mathbf{r}')|0\rangle^*\langle n|\hat{\psi}(\mathbf{r})|0\rangle}{\omega+\omega_{n0}+i\epsilon} \right) e^{i\mathbf{k}\mathbf{r}'} \\ &= -\frac{i}{\hbar}\xi(\mathbf{k},\omega)e^{-i\omega t+\epsilon t} \sum_n{}' \int d\mathbf{r}' \int_0^\infty dt' \left(\langle 0|\hat{\psi}(\mathbf{r},t')|n\rangle\langle n|\hat{\psi}^\dagger(\mathbf{r}',0)|0\rangle \right. \\ &\quad \left. - \langle 0|\hat{\psi}^\dagger(\mathbf{r}',0)|n\rangle\langle n|\hat{\psi}(\mathbf{r},t)|0\rangle \right) e^{i(\omega+i\epsilon)t'} e^{i\mathbf{k}\mathbf{r}'}, \end{aligned} \tag{4.83}$$

where

$$\hat{\psi}(\mathbf{r},t) \equiv e^{\frac{i}{\hbar}\hat{H}t}\hat{\psi}(\mathbf{r})e^{-\frac{i}{\hbar}\hat{H}t}. \tag{4.84}$$

Since $\langle 0|\hat{\psi}(\mathbf{r},t)|0\rangle = 0$, we may replace the restricted sum $\sum_n{}'$ in Eq. (4.83) with the unrestricted one $\sum_n$. Then, it follows from the completeness relation

$$\sum_n |n\rangle\langle n| = \hat{1} \tag{4.85}$$

that Eq. (4.83) reduces to

$$\delta\langle\hat{\psi}(\mathbf{r},t)\rangle = \xi(\mathbf{k},\omega)e^{-i\omega t+\epsilon t}\int d\mathbf{r}'\int_0^\infty dt' G^{\mathrm{ret}}(\mathbf{r},t';\mathbf{r}',0)e^{i\mathbf{k}\mathbf{r}'}e^{i(\omega+i\epsilon)t'}, \quad (4.86)$$

where we introduced the single-particle retarded Green's function

$$G^{\mathrm{ret}}(\mathbf{r},t;\mathbf{r}',t') \equiv -\frac{i}{\hbar}\theta(t-t')\Big\langle 0\Big|\left[\hat{\psi}(\mathbf{r},t),\hat{\psi}^\dagger(\mathbf{r}',t')\right]\Big|0\Big\rangle. \quad (4.87)$$

For convenience in later discussions, let us introduce the spectral density function $A(\mathbf{k},\omega)$:

$$A(\mathbf{k},\omega) \equiv i\hbar\int d(\mathbf{r}-\mathbf{r}')e^{-i\mathbf{k}(\mathbf{r}-\mathbf{r}')}\int_{-\infty}^{\infty} d(t-t')e^{i\omega(t-t')}G^{\mathrm{ret}}(\mathbf{r},t;\mathbf{r}',t'), \quad (4.88)$$

$$G^{\mathrm{ret}}(\mathbf{r},t;\mathbf{r}',t') = -\frac{i}{\hbar}\int\frac{d\mathbf{k}}{(2\pi)^3}e^{i\mathbf{k}(\mathbf{r}-\mathbf{r}')}\int_{-\infty}^{\infty}\frac{d\omega}{2\pi}e^{-i\omega(t-t')}A(\mathbf{k},\omega), \quad (4.89)$$

where $A(\mathbf{k},\omega)$ satisfies

$$\int_{-\infty}^{\infty}\frac{d\omega}{2\pi}A(\mathbf{k},\omega) = 1. \quad (4.90)$$

Substituting Eq. (4.89) in Eq. (4.86) gives

$$\delta\langle\hat{\psi}(\mathbf{r},t)\rangle = \frac{1}{\hbar}\xi(\mathbf{k},\omega)e^{i(\mathbf{k}\mathbf{r}-\omega t)+\epsilon t}\int_{-\infty}^{\infty}\frac{d\omega'}{2\pi}\frac{A(\mathbf{k},\omega')}{\omega+i\epsilon-\omega'}. \quad (4.91)$$

Taking the limit of $\omega \to 0$ and $\epsilon \to 0$, we obtain

$$\delta\langle\hat{\psi}(\mathbf{r})\rangle = -\frac{1}{\hbar}\xi(\mathbf{k},0)e^{i\mathbf{k}\mathbf{r}}\int_{-\infty}^{\infty}\frac{d\omega}{2\pi}\frac{A(\mathbf{k},\omega)}{\omega}. \quad (4.92)$$

In a similar manner, the response of the mass current density is given by

$$\delta\langle\hat{\mathbf{j}}(\mathbf{r},t)\rangle = -\frac{i}{\hbar}\xi(\mathbf{k},\omega)e^{-i\omega t+\epsilon t}\int d\mathbf{r}'\int_0^\infty dt' \times\Big\langle 0\Big|\left[\hat{\mathbf{j}}(\mathbf{r},t'),\hat{\psi}^\dagger(\mathbf{r}',0)\right]\Big|0\Big\rangle e^{i\mathbf{k}\mathbf{r}'}e^{i(\omega+i\epsilon)t'}. \quad (4.93)$$

We introduce another spectral density function $B(\mathbf{k},\omega)$ of the correlation function

$$\Big\langle 0\Big|\left[\hat{\mathbf{j}}(\mathbf{r},t'),\hat{\psi}^\dagger(\mathbf{r},0)\right]\Big|0\Big\rangle = \int\frac{d\mathbf{k}'}{(2\pi)^3}\int\frac{d\omega'}{2\pi}B(\mathbf{k}',\omega')e^{i\mathbf{k}'(\mathbf{r}-\mathbf{r}')-i\omega't'}, \quad (4.94)$$

$$B(\mathbf{k},\omega) = \int d\mathbf{r}\int dt\Big\langle\left[\hat{\mathbf{j}}(\mathbf{r},t),\hat{\psi}^\dagger(\mathbf{r}',t')\right]\Big\rangle e^{-i\mathbf{k}(\mathbf{r}-\mathbf{r}')+i\omega(t-t')}. \quad (4.95)$$

Substituting this in Eq. (4.93) gives

$$\delta\langle\hat{\mathbf{j}}(\mathbf{r},t)\rangle = \frac{1}{\hbar}\xi(\mathbf{k},\omega)e^{i(\mathbf{kr}-\omega t)+\epsilon t}\int_{-\infty}^{\infty}\frac{d\omega'}{2\pi}\frac{B(\mathbf{k},\omega')}{\omega+i\epsilon-\omega'}. \tag{4.96}$$

Taking the limit of $\omega \to 0$ and $\epsilon \to 0$, we obtain

$$\delta\langle\hat{\mathbf{j}}(\mathbf{r})\rangle = -\frac{1}{\hbar}\xi(\mathbf{k},0)e^{i\mathbf{kr}}\int_{-\infty}^{\infty}\frac{d\omega}{2\pi}\frac{B(\mathbf{k},\omega)}{\omega}. \tag{4.97}$$

The right-hand side is proportional to $\delta\langle\hat{\psi}(\mathbf{r})\rangle$, since the Fourier transformation of the continuity equation

$$\nabla\hat{\mathbf{j}}(\mathbf{r},t) + m\frac{\partial\hat{\rho}(\mathbf{r},t)}{\partial t} = 0 \tag{4.98}$$

gives

$$i\mathbf{k}\cdot\hat{\mathbf{j}}(\mathbf{k},t) + m\frac{\partial\hat{\rho}(\mathbf{k},t)}{\partial t} = 0. \tag{4.99}$$

Assuming that $\hat{\mathbf{j}}(\mathbf{k},t)$ is proportional to $\mathbf{k}$, we obtain

$$\hat{\mathbf{j}}(\mathbf{k},t) = \frac{im\mathbf{k}}{\hbar^2}\frac{\partial\hat{\rho}(\mathbf{k},t)}{\partial t}. \tag{4.100}$$

Hence,

$$\begin{aligned}\hat{\mathbf{j}}(\mathbf{r},t) &= \int\frac{d\mathbf{k}}{(2\pi)^3}e^{i\mathbf{kr}}\frac{im\mathbf{k}}{k^2}\frac{\partial\hat{\rho}(\mathbf{k},t)}{\partial t}\\ &= im\frac{\partial}{\partial t}\int d\mathbf{r}'\int\frac{d\mathbf{k}}{(2\pi)^3}\frac{\mathbf{k}}{k^2}e^{i\mathbf{k}(\mathbf{r}-\mathbf{r}')}\hat{\rho}(\mathbf{r}',t).\end{aligned} \tag{4.101}$$

Substituting Eq. (4.101) in Eq. (4.95), we have

$$\begin{aligned}B(\mathbf{k},\omega) &= im\int d\mathbf{r}\int dt\; e^{-i\mathbf{k}(\mathbf{r}-\mathbf{r}')+i\omega(t-t')}\frac{\partial}{\partial t}\int d\mathbf{r}''\\ &\quad\times\int\frac{d\mathbf{k}'}{(2\pi)^3}\frac{\mathbf{k}'}{k'^2}e^{i\mathbf{k}'(\mathbf{r}-\mathbf{r}'')}\left\langle\left[\hat{\rho}(\mathbf{r}'',t),\hat{\psi}^\dagger(\mathbf{r}',t')\right]\right\rangle\\ &= \frac{m\omega\mathbf{k}}{k^2}\int dt\; e^{i\omega(t-t')}\int d\mathbf{r}''e^{-i\mathbf{k}(\mathbf{r}''-\mathbf{r}')}\left\langle\left[\hat{\rho}(\mathbf{r}'',t),\hat{\psi}^\dagger(\mathbf{r}',t')\right]\right\rangle.\end{aligned} \tag{4.102}$$

Substituting this in Eq. (4.97), we obtain

$$\delta\langle\hat{\mathbf{j}}(\mathbf{r})\rangle = -\frac{m\mathbf{k}}{\hbar k^2}\xi(\mathbf{k},0)e^{i\mathbf{kr}}\int d\mathbf{r}\; e^{-i\mathbf{k}(\mathbf{r}-\mathbf{r}')}\left\langle\left[\hat{\rho}(\mathbf{r},t),\hat{\psi}^\dagger(\mathbf{r}',t)\right]\right\rangle. \tag{4.103}$$

Using

$$\begin{aligned}\left\langle\left[\hat{\rho}(\mathbf{r},t),\hat{\psi}^\dagger(\mathbf{r}',t)\right]\right\rangle &= \left\langle\left[\hat{\rho}(\mathbf{r}),\hat{\psi}^\dagger(\mathbf{r}')\right]\right\rangle = \left\langle\hat{\psi}^\dagger(\mathbf{r})\left[\hat{\psi}(\mathbf{r}),\hat{\psi}^\dagger(\mathbf{r}')\right]\right\rangle\\ &= \langle\hat{\psi}^\dagger(\mathbf{r})\rangle\delta(\mathbf{r}-\mathbf{r}'),\end{aligned} \tag{4.104}$$

we obtain

$$\delta\langle\hat{\mathbf{j}}(\mathbf{r})\rangle = -\frac{m\mathbf{k}}{\hbar k^2}\xi(\mathbf{k},0)e^{i\mathbf{kr}}\langle\hat{\psi}^\dagger(\mathbf{r})\rangle. \tag{4.105}$$

Comparing Eqs. (4.92) and (4.105), we find that the desired relation between $\delta\langle\hat{\mathbf{j}}\rangle$ and $\delta\langle\hat{\psi}\rangle$ is given by

$$\delta\langle\hat{\mathbf{j}}(\mathbf{r})\rangle = m\frac{\mathbf{k}}{k^2}\left[\int_{-\infty}^{\infty}\frac{d\omega}{2\pi}\frac{A(\mathbf{k},\omega)}{\omega}\right]^{-1}\langle\hat{\psi}^\dagger(\mathbf{r})\rangle\delta\langle\hat{\psi}(\mathbf{r})\rangle, \tag{4.106}$$

where we substitute $\psi_0(\mathbf{r}) = \langle\hat{\psi}(\mathbf{r})\rangle$ and use Eq. (4.77) to obtain

$$\delta\langle\hat{\psi}(\mathbf{r})\rangle = \delta\psi_0(\mathbf{r}) = i\psi_0(\mathbf{r})\delta\phi(\mathbf{r}). \tag{4.107}$$

Then, Eq. (4.106) may be rewritten as

$$\begin{aligned}\delta\langle\hat{\mathbf{j}}(\mathbf{r})\rangle &= im\frac{\mathbf{k}}{k^2}\left[\int_{-\infty}^{\infty}\frac{d\omega}{2\pi}\frac{A(\mathbf{k},\omega)}{\omega}\right]^{-1}|\psi_0(\mathbf{r})|^2\delta\phi(\mathbf{r}).\\ &= \frac{m}{k^2}\left[\int_{-\infty}^{\infty}\frac{d\omega}{2\pi}\frac{A(\mathbf{k},\omega)}{\omega}\right]^{-1}|\psi_0(\mathbf{r})|^2\nabla\delta\phi(\mathbf{r}),\end{aligned} \tag{4.108}$$

since $\delta\phi \propto e^{i\mathbf{kr}}$. From Eqs. (4.75) and (4.78), on the other hand, we have

$$\delta\langle\mathbf{j}(\mathbf{r})\rangle = \frac{\hbar}{m}\rho_s\nabla\delta\phi(\mathbf{r}). \tag{4.109}$$

Equating Eqs. (4.108) and (4.109), we finally obtain the relation among the superfluid density ρ_s, condensate density $|\psi_0|^2$, and spectral density function $A(\mathbf{k},\omega)$ as

$$\int_{-\infty}^{\infty}\frac{d\omega}{2\pi}\frac{A(\mathbf{k},\omega)}{\omega} = \frac{m^2|\psi_0|^2}{\hbar k^2\rho_s}. \tag{4.110}$$

This relation may be interpreted as a sum rule obeyed by the single-particle spectral density function $A(\mathbf{k},\omega)$, and is referred to as the Josephson sum rule.

Another sum rule obeyed by $A(\mathbf{k},\omega)$ is

$$\int_{-\infty}^{\infty}\frac{d\omega}{2\pi}A(\mathbf{k},\omega) = 1, \tag{4.111}$$

which can be shown directly from Eq. (4.88).

4.3 Sum-Rule Approach to Collective Modes

We investigate the collective mode of a system described by Hamiltonian $\hat{H}$. Let $\{|n\rangle\}$ and $\{E_n\}$ be a complete set of exact eigenstates and that of the corresponding eigenvalues:

$$\hat{H}|n\rangle = E_n|n\rangle,$$

where we assume that $E_0 \le E_1 \le E_2 \le \cdots$. In general, a system will exhibit various types of collective modes characterized by symmetries and excitation energies. Let $\hat{F}$ be an excitation operator of the system. When $\hat{F}$ acts on the ground state $|0\rangle$, various states $|F_1\rangle, |F_2\rangle, \cdots$ can, in general, be excited, where $|F_i\rangle$ belongs to the set $\{|n\rangle\}$ and satisfies

$$\hat{H}|F_i\rangle = E_{F_i}|F_i\rangle \quad (i = 1, 2, 3, \cdots) \tag{4.112}$$

with $E_{F_1} \le E_{F_2} \le \cdots$. The following theorem is useful for finding an upper bound of a collective mode.

Theorem. An upper bound $\hbar\omega^{\rm upper}$ to the minimum excitation energy $\hbar\omega^{\rm min} \equiv E_{F_1} - E_0$ of the states excited by $\hat{F}$ is given by

$$\hbar\omega^{\rm upper} = \sqrt{\frac{m_3}{m_1}}, \tag{4.113}$$

where E_0 is the ground state energy and

$$m_p \equiv \sum_i |\langle F_i|\hat{F}|0\rangle|^2 (E_{F_i} - E_0)^p \tag{4.114}$$

is the p-th energy-weighted moment of the excitation.

Proof. A straightforward calculation shows that

$$\frac{(\hbar\omega^{\rm upper})^2 - (\hbar\omega^{\rm min})^2}{(F_{F_1} - E_0)^3} = \frac{\sum_i |\langle F_i|\hat{F}|0\rangle|^2 \left[\left(\frac{E_{F_i}-E_0}{E_{F_1}-E_0}\right)^3 - \frac{E_{F_i}-E_0}{E_{F_1}-E_0}\right]}{\sum_i |\langle F_i|\hat{F}|0\rangle|^2 (E_{F_i} - E_0)}.$$

Since $(E_{F_i} - E_0)/(E_{F_i} - E_0) \ge 1$, we have $\hbar\omega^{\rm min} \le \hbar\omega^{\rm upper}$.

When $\hat{F}$ is Hermitian, m_1 and m_3 can be rewritten as

$$m_1 = \frac{1}{2}\left\langle 0\right| \left[\hat{F}^\dagger, \left[\hat{H}, \hat{F}\right]\right] \left|0\right\rangle, \tag{4.115}$$

$$m_3 = \frac{1}{2}\left\langle 0\right| \left[\left[\hat{F}^\dagger, \hat{H}\right], \left[\hat{H}, \left[\hat{H}, \hat{F}\right]\right]\right] \left|0\right\rangle. \tag{4.116}$$

Equations (4.115) and (4.116) can be shown by inserting the completeness relation $\sum_n |n\rangle\langle n| = 1$ and noting that only states $\{|F_i\rangle\}$ are connected to the ground state $|0\rangle$ via $\hat{F}$. In fact, calculating the right-hand sides of Eqs. (4.115) and (4.116), we have

$$\frac{1}{2}\left\langle 0\right|\left[\hat{F}^\dagger,\left[\hat{H},\hat{F}\right]\right]\left|0\right\rangle = \frac{1}{2}\sum_i\left[|\langle F_i|\hat{F}|0\rangle|^2+|\langle F_i|\hat{F}^\dagger|0\rangle|^2\right](E_{F_i}-E_0), \tag{4.117}$$

$$\frac{1}{2}\left\langle 0\right|\left[\left[\hat{F}^\dagger,\hat{H}\right],\left[\hat{H},\left[\hat{H},\hat{F}\right]\right]\right]\left|0\right\rangle = \frac{1}{2}\sum_i\left[|\langle F_i|\hat{F}|0\rangle|^2+|\langle F_i|\hat{F}^\dagger|0\rangle|^2\right](E_{F_i}-E_0)^3. \tag{4.118}$$

When $\hat{F}$ is Hermitian ($\hat{F}=\hat{F}^\dagger$), Eqs. (4.117) and (4.118) respectively give m_1 and m_3, as defined in Eq. (4.114). When $\hat{F}$ is not Hermitian, as in Eq. (4.140), only one among $\langle F_i|\hat{F}|0\rangle$ and $\langle F_i|\hat{F}_i^\dagger|0\rangle$ can be nonzero. In this case, Eqs. (4.117) and (4.118) give $m_1/2$ and $m_3/2$, respectively. In forming the ratio, the factor of 1/2 is canceled out, and thus, Eq. (4.113) still holds.

The advantage of the sum-rule approach is that no information concerning the excited states is required to find the excitation energies. In particular, given a correct excitation operator $\hat{F}$ and the exact ground state, $\hbar\omega^{\text{upper}}$ gives the exact excitation energy.

4.3.1 *Excitation operators*

Consider an excitation operator

$$\hat{F}(n,l,m)=\sum_{i=1}^N r_i^{2n+l}Y_{lm}(\theta_i,\phi_i) \tag{4.119}$$

that excites a state characterized by radial quantum number n, angular-momentum quantum number l, and magnetic quantum number m, where N is the number of atoms, and r_i, θ_i, and ϕ_i are polar coordinates of the i-th atom, that is,

$$x_i=r_i\sin\theta_i\cos\phi_i,\ y_i=r_i\sin\theta_i\sin\phi_i,\ z_i=r_i\cos\theta_i. \tag{4.120}$$

The spherical harmonic function $Y_{lm}(\theta_i,\phi_i)$ is given by

$$Y_{lm}(\theta,\phi)=e^{im\phi}P_l^{-m}(\cos\theta)=(-1)^m\frac{(l-m)!}{(l+m)!}e^{im\phi}P_l^m(\cos\theta), \tag{4.121}$$

where P_l^m is the associated Laguerre polynomial defined as

$$P_l^m(x)=\frac{(1-x^2)^{\frac{m}{2}}}{2^l l!}\frac{d^{l+m}}{dx^{l+m}}(x^2-1)^l. \tag{4.122}$$

The excitation with $n = 1$ and $l = m = 0$ is called the monopole mode. In this case, $Y_{00} = 1$ and the corresponding excitation operator is given from Eq. (4.119) as

$$\hat{F} = \sum_i r_i^2 = \sum_i (x_i^2 + y_i^2 + z_i^2). \tag{4.123}$$

The excitations with $n = 0$ and $l = 1$ are called dipole modes that are classified into three types depending on the value of m. In this case, the corresponding excitation operator is given from Eq. (4.119) as

$$\hat{F} = \sum_i r_i Y_{lm}(\theta_i, \phi_i) = \begin{cases} \sum_i (x_i + iy_i) & \text{for } m = 1, \\ \sum_i z_i & \text{for } m = 0, \\ \sum_i (x_i - iy_i) & \text{for } m = -1. \end{cases} \tag{4.124}$$

The excitations with $n = 0$ and $l = 2$ are called quadrupole modes that are classified into five types depending on the value of m. The corresponding excitation operators are given by

$$\hat{F} = \sum_i r_i^2 Y_{2m}(\theta_i, \phi_i) = \begin{cases} \sum_i (x_i \pm iy_i)^2 & \text{for } m = \pm 2, \\ \sum_i (x_i \pm iy_i) z_i & \text{for } m = \pm 1, \\ \sum_i (x_i^2 + y_i^2 - 2z_i^2) & \text{for } m = 0. \end{cases} \tag{4.125}$$

4.3.2 *Virial theorem*

In the following discussions, we shall restrict ourselves to a situation in which N particles are confined in a harmonic potential and undergo contact interactions described by a delta function. The Hamiltonian of our system is then given by

$$\hat{H} = \sum_i \frac{\mathbf{p}_i^2}{2m} + \sum_i \frac{m}{2}(\omega_x^2 x_i^2 + \omega_y^2 y_i^2 + \omega_z^2 z_i^2) + \frac{U_0}{2} \sum_{i \neq j} \delta(\mathbf{r}_i - \mathbf{r}_j). \tag{4.126}$$

We assume that the state of our system is stationary. Then, the expectation value of $\sum_i x_i p_{ix}$ is constant in time:

$$\frac{d}{dt}\left\langle \sum_i x_i p_{ix} \right\rangle = 0, \tag{4.127}$$

where p_{ix} is the x-component of the momentum of the i-th particle. On the other hand, Heisenberg's equation of motion gives

$$\frac{d}{dt}\sum_i x_i p_{ix} = \frac{i}{\hbar}\left[H, \sum_i x_i p_{ix}\right]. \tag{4.128}$$

Substituting Eq. (4.126) in Eq. (4.127) gives

$$\frac{d}{dt}\sum_i x_i p_{ix} = 2\sum_i \frac{p_{ix}^2}{2m} - 2\sum_i \frac{m\omega_x^2}{2}x_i^2 - U_0 \sum_{i\neq j} x_i \frac{\partial\delta(\mathbf{r}_i - \mathbf{r}_j)}{\partial x_i}. \tag{4.129}$$

Hence, we have

$$2\langle T_x\rangle - 2\langle U_x\rangle - U_0\Big\langle \sum_{i\neq j} x_i \frac{\partial\delta(\mathbf{r}_i - \mathbf{r}_j)}{\partial x_i}\Big\rangle = 0, \tag{4.130}$$

where T_x and U_x are the x-component of the kinetic energy and the potential energy, respectively. The last term in Eq. (4.130) may be rewritten as

$$\begin{aligned}\Big\langle \sum_{i\neq j} x_i \frac{\partial\delta(\mathbf{r}_i - \mathbf{r}_j)}{\partial x_i}\Big\rangle &= \int d\mathbf{r}_i d\mathbf{r}_j x_i \frac{\partial\delta(\mathbf{r}_i - \mathbf{r}_j)}{\partial x_i}\psi^2(\mathbf{r}_i)\psi^2(\mathbf{r}_j) \\ &= -\int d\mathbf{r}_i \frac{\partial\left(x_i\psi^2(\mathbf{r}_i)\right)}{\partial x_i}\psi^2(\mathbf{r}_i) \\ &= -\int d\mathbf{r}_i\left[\psi^4(\mathbf{r}_i) + \frac{x_i}{2}\frac{\partial\psi^4(\mathbf{r}_i)}{\partial x_i}\right] = -\frac{1}{2}\int d\mathbf{r}_i\psi^4(\mathbf{r}_i) \\ &= -\frac{1}{2}\Big\langle\sum_{i\neq j}\delta(\mathbf{r}_i - \mathbf{r}_j)\Big\rangle.\end{aligned} \tag{4.131}$$

We thus obtain

$$2\langle T_x\rangle - 2\langle U_x\rangle + \langle V\rangle = 0, \tag{4.132}$$

where

$$V = \frac{U_0}{2}\sum_{i\neq j}\delta(\mathbf{r}_i - \mathbf{r}_j). \tag{4.133}$$

We can obtain equations similar to Eq. (4.132) for the y- and z-components. Summing up the x-, y-, z-components, we obtain

$$2\langle T\rangle - 2\langle U\rangle + 3\langle V\rangle = 0. \tag{4.134}$$

The relations (4.132) and (4.134) are known as the virial theorem.

4.3.3 *Kohn theorem*

The collective-mode frequency of the dipole mode ($n = 0$, $l = 1$) is independent of interactions and equal to the frequencies of the trapping potential. This fact is known as the Kohn theorem. To show this, we consider some cases of the axisymmetric harmonic potential

$$U = \sum_i \frac{m}{2}\left[\omega_\perp^2\left(x_i^2 + y_i^2\right) + \omega_z^2 z_i^2\right]. \tag{4.135}$$

4.3.3.1 *Case of* $m = 0$

The excitation operator with $m = 0$ is given by

$$\hat{F} = \sum_i z_i. \tag{4.136}$$

Straightforward calculations give

$$\left[\hat{F}^\dagger, \left[\hat{H}, \hat{F}\right]\right] = \frac{\hbar^2}{m} N, \tag{4.137}$$

$$\left[\left[\hat{F}^\dagger, \hat{H}\right], \left[\hat{H}, \left[\hat{H}, \hat{F}\right]\right]\right] = \frac{\hbar^4 \omega_z^2}{m} N. \tag{4.138}$$

We note that the right-hand sides of Eqs. (4.137) and (4.138) are constants, and independent of the state of the system. Substituting Eqs. (4.137) and (4.138) in Eq. (4.113) gives

$$\hbar\omega^{\text{upper}} = \hbar\omega_z. \tag{4.139}$$

4.3.3.2 *Case of* $m = \pm 1$

The excitation operators with $m = \pm 1$ are given by

$$\hat{F} = \sum_i (x_i \pm i y_i). \tag{4.140}$$

Straightforward calculations give

$$\left[\hat{F}^\dagger, \left[\hat{H}, \hat{F}\right]\right] = \frac{2\hbar^2}{m} N, \tag{4.141}$$

$$\left[\left[\hat{F}^\dagger, \hat{H}\right], \left[\hat{H}, \left[\hat{H}, \hat{F}\right]\right]\right] = \frac{2\hbar^4 \omega_\perp^2}{m} N. \tag{4.142}$$

The right-hand sides of Eqs. (4.141) and (4.142) are again independent of the state of the system. Substituting Eqs. (4.141) and (4.142) in Eq. (4.113) gives

$$\hbar\omega^{\text{upper}} = \hbar\omega_\perp. \tag{4.143}$$

Equations (4.139) and (4.143), in fact, give the exact frequencies of the dipole modes.

4.3.4 *Isotropic trap*

When the trap is isotropic, the Hamiltonian is given by

$$\hat{H} = \sum_i \frac{\mathbf{p}_i^2}{2m} + \sum_i \frac{m\omega^2}{2} \mathbf{r}_i^2 + \frac{U_0}{2} \sum_{i \neq j} \delta(\mathbf{r}_i - \mathbf{r}_j). \tag{4.144}$$

We discuss two important collective modes.

4.3.4.1 *Monopole mode*

The excitation operator of the monopole (or breathing) mode with $n = 1$ and $l = 0$ is given by

$$\hat{F} = \sum_i \mathbf{r}_i^2 = \sum_i \left(x_i^2 + y_i^2 + z_i^2\right). \tag{4.145}$$

Calculating the commutation relations for m_1 gives

$$\left[\hat{F}^\dagger, \left[\hat{H}, \hat{F}\right]\right] = \frac{4\hbar^2}{m} \sum_i \mathbf{r}_i^2. \tag{4.146}$$

Hence,

$$m_1 = \frac{4\hbar^2}{m^2\omega^2} \langle U \rangle. \tag{4.147}$$

Calculating the commutation relations for m_3 is slightly complicated. We first note that

$$\left[\hat{F}^\dagger, \hat{H}\right] = \frac{2i\hbar}{m} \sum_i \left(\mathbf{p}_i \mathbf{r}_i + \frac{3}{2} i\hbar\right), \tag{4.148}$$

$$\left[\hat{H}, \left[\hat{H}, \hat{F}\right]\right] = -\frac{4\hbar^2}{m} \Big(\sum_i \frac{\mathbf{p}_i^2}{2m} - \sum_i \frac{m\omega^2}{2} \mathbf{r}_i^2 - \frac{U_0}{2} \sum_{i \neq j} \mathbf{r}_i \frac{\partial \delta(\mathbf{r}_i - \mathbf{r}_j)}{\partial \mathbf{r}_i} \Big). \tag{4.149}$$

Hence,

$$\left[\left[\hat{F}^\dagger, \hat{H}\right], \left[\hat{H}, \left[\hat{H}, \hat{F}\right]\right]\right] = \frac{16\hbar^4}{m^2} \Bigg\{ \sum_i \frac{\mathbf{p}_i^2}{2m} + \sum_i \frac{m\omega^2}{2} \mathbf{r}_i^2$$
$$+ \frac{U_0}{4} \sum_{i \neq j} \left(\mathbf{r}_i \cdot \nabla_i + \mathbf{r}_j \cdot \nabla_j\right) \mathbf{r}_i \cdot \nabla_i \delta\left(\mathbf{r}_i - \mathbf{r}_j\right) \Bigg\}. \tag{4.150}$$

The expectation value of the last term can be evaluated as

$$\begin{aligned} A &= \langle \sum_{i \neq j} \left(\mathbf{r}_i \cdot \nabla_i + \mathbf{r}_j \cdot \nabla_j\right) \mathbf{r}_i \cdot \nabla_i \delta\left(\mathbf{r}_i - \mathbf{r}_j\right) \rangle \\ &= \int d\mathbf{r} d\mathbf{r}' \psi^2(\mathbf{r}) \psi^2(\mathbf{r}') \left(x_i \partial_i + x'_i \partial'_i\right) x_j \partial_j \delta(\mathbf{r} - \mathbf{r}'), \end{aligned}$$

where $x_1 = x$, $x_2 = y$, $x_3 = z$, etc. Integration by parts gives

$$
\begin{aligned}
A = &- \int d\mathbf{r}d\mathbf{r}' \left\{\psi^2(\mathbf{r}') \left[\partial i \left(x_i\psi^2(\mathbf{r})\right)\right] x_j\partial_j\delta(\mathbf{r}-\mathbf{r}')\right. \\
&\left. +\psi^2(\mathbf{r}) \left[\partial'_i \left(x'_i\psi^2(\mathbf{r}')\right)\right] x_j\partial_j\delta(\mathbf{r}-\mathbf{r}')\right\} \\
= &\int d\mathbf{r} \left\{\psi^2\partial_j x_j \left[\partial_i \left(x_i\psi^2\right)\right] + \left[\partial_j \left(x_j\psi^2\right)\right] \left[\partial_i \left(x_i\psi^2\right)\right]\right\} \\
= &\int d\mathbf{r} \left\{-\left(\partial_j\psi^2\right) x_j \left[\partial_i \left(x_i\psi^2\right)\right] + \left[\left(\partial_j x_j\right)\psi^2 + x_j\left(\partial_j\psi^2\right)\right] \left[\partial_i \left(x_i\psi^2\right)\right]\right\} \\
= &\,3\int d\mathbf{r}\psi^2\partial_i\left(x_i\psi^2\right) = 3\int d\mathbf{r}\left[\psi^4\left(\partial_i x_i\right) + x_i\psi^2\left(\partial_i\psi^2\right)\right\} \\
= &\,3\int d\mathbf{r}\left(3\psi^4 + \frac{1}{2}x_i\partial_i\psi^4\right) = \frac{9}{2}\int d\mathbf{r}\psi^4 = \frac{9}{2}\left\langle\sum_{i\neq j}\delta(\mathbf{r}_i-\mathbf{r}_j)\right\rangle \\
= &\,\frac{9}{U_0}\langle V\rangle.
\end{aligned}
$$

Hence, we have

$$m_3 = \frac{8\hbar^2}{m^2}\langle T + U + \frac{9}{4}V\rangle. \tag{4.151}$$

We may use the virial theorem (4.134) to eliminate $\langle V\rangle$ in Eq. (4.151), thus obtaining

$$m_3 = \frac{4\hbar^2}{m^2}\langle 5U - T\rangle. \tag{4.152}$$

Substituting Eqs. (4.146) and (4.152) in Eq. (4.113) gives

$$\hbar\omega^{\mathrm{upper}} = \hbar\omega\sqrt{5 - \frac{\langle T\rangle}{\langle U\rangle}}. \tag{4.153}$$

In the absence of interactions, the vitial theorem gives $\langle T\rangle = \langle U\rangle$, and hence, we have $\hbar\omega^{\mathrm{upper}} = 2\hbar\omega$. In the Thomas–Fermi limit, where $\langle T\rangle = 0$, we have $\hbar\omega^{\mathrm{upper}} = \sqrt{5}\hbar\omega$.

4.3.4.2 *Quadrupole mode*

The excitations with $n = 0$ and $l = 2$ are called quadrupole modes. In the case of an isotropic trap, the excitation frequency is independent of m. It is therefore sufficient to consider the case of $m = 0$, where the excitation operator is given by

$$F = \sum_i \left(x_i^2 + y_i^2 - 2z_i^2\right). \tag{4.154}$$

Straightforward calculations give

$$m_1 = \frac{8\hbar^2}{m^2\omega^2}\langle U\rangle, \tag{4.155}$$

$$m_3 = \frac{16\hbar^4}{m^2}\langle T + U\rangle. \tag{4.156}$$

Hence, we obtain

$$\hbar\omega^{\text{upper}} = \hbar\omega\sqrt{2\left(1+\frac{\langle T\rangle}{\langle U\rangle}\right)}. \tag{4.157}$$

In the absence of interactions, $\hbar\omega^{\text{upper}} = \sqrt{3}\hbar\omega$, while in the Thomas–Fermi limit, $\hbar\omega^{\text{upper}} = \sqrt{2}\omega$.

4.3.5 *Axisymmetric trap*

When the trap is axisymmetric, the Hamiltonian is given by

$$\hat{H} = \sum_i \frac{\mathbf{p}_i^2}{2m} + \sum_i \frac{m}{2}\left[\omega_\perp^2\left(x_i^2+y_i^2\right)+\omega_z^2 z_i^2\right] + \frac{U_0}{2}\sum_{i\neq j}\delta\left(\mathbf{r}_i-\mathbf{r}_j\right). \tag{4.158}$$

In this case, the frequency of the quadrupole mode depends on the value of m. When $n=0$, $l=2$, and $m=\pm 2$, the excitation operators are given by

$$\hat{F} = \sum_i \left(x_i \pm i y_i\right)^2. \tag{4.159}$$

In this case,

$$m_1 = \frac{4\hbar^2}{m}\left\langle \sum_i \left(x_i^2+y_i^2\right)\right\rangle \equiv \frac{16\hbar^2}{m^2\omega^2}\left\langle U_\perp\right\rangle, \tag{4.160}$$

$$m_3 = \frac{16\hbar^4}{m^2}\left\langle \sum_i \frac{p_{ix}^2+p_{iy}^2}{2m} + \sum_i \frac{m\omega^2}{2}\left(x_i^2+y_i^2\right)\right\rangle \equiv \frac{32\hbar^2}{m^2}\langle T_\perp + U_\perp\rangle. \tag{4.161}$$

Hence, we have

$$\hbar\omega^{\text{upper}} = \hbar\omega_\perp\sqrt{2\left(1+\frac{\langle T_\perp\rangle}{\langle U_\perp\rangle}\right)}. \tag{4.162}$$

The mode with $m=0$ is called the radial breathing mode. When $\omega_\perp \neq \omega_z$, the angular momentum is no longer a good quantum number, and the mode with $n=1$, $l=m=0$ $\left(\hat{F}=\sum_i r_i^2\right)$ couples with the mode with $n=0$,

$l = 2$, $m = 0$ $\left(\hat{F} = \sum_i (x_i^2 + y_i^2 - 2z_i^2)\right)$. The excitation operator for the coupled mode is, in general, given by

$$\hat{F} = \sum_i \left(x_i^2 + y_i^2 - \alpha z_i^2\right), \tag{4.163}$$

where α is a variational parameter to be determined later. Calculations of the commutation relations give

$$m_1 = \frac{4\hbar^2}{m^2\omega_\perp^2}\left[2\langle U_\perp\rangle + \frac{\alpha^2}{\lambda^2}\langle U_z\rangle\right], \tag{4.164}$$

$$m_3 = \frac{8\hbar^4}{m^2}\left[2(\langle T_\perp\rangle + \langle U_\perp\rangle) + \alpha^2(\langle T_\perp\rangle + \langle U_z\rangle) + \left(1 - \frac{\alpha}{2}\right)^2\langle V\rangle\right], \tag{4.165}$$

where $\lambda \equiv \omega_z/\omega_\perp$. Hence, we have

$$\hbar\omega^{\text{upper}} = \sqrt{2}\hbar\omega_\perp\left[\frac{2(\langle T_\perp\rangle + \langle U_\perp\rangle) + \alpha^2(\langle T_z\rangle + \langle U_z\rangle) + \left(1 - \frac{\alpha}{2}\right)^2\langle V\rangle}{2\langle U_\perp\rangle + \frac{\alpha^2}{\lambda^2}\langle U_z\rangle}\right]^{\frac{1}{2}}. \tag{4.166}$$

In the Thomas–Fermi limit, we have $\langle T_\perp\rangle = \langle T_z\rangle = 0$. It can also be shown that $\langle U_\perp\rangle = \langle U_z\rangle$. On the other hand, according to the virial theorem, we obtain $\langle V\rangle = \frac{2}{3}\langle U\rangle = 2\langle U_\perp\rangle$. Hence, we have

$$\hbar\omega^{\text{upper}} = \hbar\omega_z\sqrt{\frac{3\alpha^2 - 4\alpha + 8}{\alpha^2 + 2\lambda^2}}. \tag{4.167}$$

Minimizing this with respect to α gives

$$\omega^{\text{upper}} = \omega_\perp\left[2 + \frac{3}{2}\lambda^2 \pm \sqrt{\frac{9}{4}\lambda^4 - 4\lambda^2 + 4}\right]^{\frac{1}{2}}. \tag{4.168}$$

Chapter 5

Statistical Mechanics of Superfluid Systems in a Moving Frame

5.1 Transformation to Moving Frames

Suppose that a system of particles is contained in an infinitely long cylinder that is moving with velocity $\mathbf{u}$ in the laboratory frame. When the state of the system is normal, the system will be dragged by the cylinder via scattering of particles with microscopic rugosities of the wall, and eventually move, on average, at the velocity of the wall. However, when the system is a superfluid, a part of the system, which is called the superfluid component, can move faster or slower than the wall, whereas the remainder of the system, which is called the normal component, moves at the same velocity as the wall and is in thermal equilibrium with it.

From the viewpoint of many-body physics, the normal component is constituted of quasiparticle excitations that are at thermal equilibrium with the moving wall. To find the correct Hamiltonian in the moving frame of reference, we invoke the principle of least action that is applicable independently of the selected frame of reference. According to this principle, the equation of motion is determined so as to minimize the action of the system

$$S = \int L dt, \tag{5.1}$$

where L is the Lagrangian of the system. It follows that the *form* of Lagrange's equation of motion

$$\frac{d}{dt}\frac{\partial L}{\partial \dot{\mathbf{r}}_i} = \frac{\partial L}{\partial \mathbf{r}_i} \tag{5.2}$$

is invariant with respect to the choice of the reference frame. Let

$$H = \sum_i \frac{\mathbf{p}_i^2}{2m} + \frac{1}{2}\sum_{i \neq j} v\left(\mathbf{r}_i - \mathbf{r}_j\right) \tag{5.3}$$

be the Hamiltonian of the system of particles in the laboratory frame. The corresponding Lagrangian is given by

$$L = \sum_i \frac{m}{2}\dot{\mathbf{r}}_i^2 - \frac{1}{2}\sum_{i\neq j} v\left(\mathbf{r}_i - \mathbf{r}_j\right). \tag{5.4}$$

Now, consider a frame of reference that is moving with velocity $\mathbf{u}$ against the laboratory frame. Then, the coordinates of the particles in both frames are related by

$$\mathbf{r}_i(t) = \mathbf{r}'_i(t) + \mathbf{u}t, \tag{5.5}$$

where the prime refers to the moving frame. Substituting this into Eq. (5.4) gives the Lagrangian of the system in the moving frame:

$$L' = \sum_i \frac{m}{2}\left(\dot{\mathbf{r}}'_i + \mathbf{u}\right)^2 - \frac{1}{2}\sum_{i\neq j} v\left(\mathbf{r}'_i - \mathbf{r}'_j\right). \tag{5.6}$$

The canonical momentum in the moving frame is given by

$$\mathbf{p}'_i = \frac{\partial L'}{\partial \dot{\mathbf{r}}'_i} = m\left(\dot{\mathbf{r}}' + \mathbf{u}\right) = m\dot{\mathbf{r}}_i = \mathbf{p}_i. \tag{5.7}$$

Thus, the canonical momentum is invariant under the Galilean transformation. The Hamiltonian in the moving frame is given by

$$H' = \sum_i \dot{\mathbf{r}}'_i \frac{\partial L}{\partial \dot{\mathbf{r}}'_i} - L' = \sum_i \frac{\mathbf{p}'^2_i}{2m} + \frac{1}{2}\sum_{i\neq j} v\left(\mathbf{r}'_i - \mathbf{r}'_j\right) - \sum_i \mathbf{p}'_i \mathbf{u}. \tag{5.8}$$

From Eqs. (5.5), (5.7), and (5.8), we find that the Hamiltonians in the two frames are related to each other as

$$H' = H - \sum_i \mathbf{p}_i \mathbf{u}. \tag{5.9}$$

Now, consider a situation in which particles are confined in a torus that is rotating with angular velocity Ω about its symmetry axis. Then, the wall at position $\mathbf{r}_i$ is moving with velocity $\mathbf{u} = \boldsymbol{\Omega} \times \mathbf{r}_i$, and therefore, the last term in Eq. (5.9) becomes

$$\sum_i \mathbf{p}_i \cdot (\boldsymbol{\Omega} \times \mathbf{r}_i) = \sum_i \boldsymbol{\Omega} \cdot (\mathbf{r}_i \times \mathbf{p}_i) = \boldsymbol{\Omega} \cdot \mathbf{L},$$

where $\mathbf{L} = \sum_i \mathbf{r}_i \times \mathbf{p}_i$ is the total angular momentum of the system. Thus, the Hamiltonian H' in the rotating frame of reference is related to the Hamiltonian H in the laboratory frame by

$$H' = H - \boldsymbol{\Omega} \cdot \mathbf{L}. \tag{5.10}$$

This relation also holds true when the system is contained in a cylinder.

In both of the above cases, the equilibrium statistical mechanics is defined with respect to the frame of reference that is at rest with the wall; otherwise, the system will be dragged by the wall and driven out of equilibrium. The equilibrium density operator of the system is therefore given by

$$\rho = \frac{e^{-\beta H'}}{\mathrm{Tr}\, e^{-\beta H'}}. \tag{5.11}$$

5.2 Elementary Excitations of a Superfluid

First, consider a situation in which a superfluid is at rest with the container wall. Let E_0 be the internal energy of the ground state and $\epsilon_{\mathbf{p}}$, the energy of an elementary excitation with momentum $\mathbf{p}$. Then, the total energy E and total momentum $\mathbf{P}$ of the system are given by

$$E = E_0 + \sum_{\mathbf{p}} \epsilon_{\mathbf{p}} n_{\mathbf{p}}, \tag{5.12}$$

$$\mathbf{P} = \sum_{\mathbf{p}} \mathbf{p} n_{\mathbf{p}}, \tag{5.13}$$

where $n_{\mathbf{p}}$ is the number of elementary excitations with momentum $\mathbf{p}$, and it is given by

$$n_{\mathbf{p}} = \frac{1}{e^{\beta\epsilon_{\mathbf{p}}} - 1}.$$

Next, we consider a situation in which the superfluid is moving with velocity $\mathbf{v}_{\mathrm{s}}$ with respect to the wall of the container. The energy $\epsilon_{\mathbf{p}}$ and momentum $\mathbf{p}$ of an elementary excitation is defined in the frame of reference in which the superfluid is at rest, because elementary excitations are defined as excitations from the ground state of the system, which, in the present case, is the superfluid. It follows from Eqs. (5.7) and (5.9) that the energy $\epsilon'_{\mathbf{p}}$ and momentum $\mathbf{p}'$ of the elementary excitation in the frame of reference of the wall (*i.e.*, the laboratory frame) are given by (note that the wall is moving with velocity $-\mathbf{v}_{\mathrm{s}}$ against the superfluid)

$$\epsilon'_{\mathbf{p}} = \epsilon_{\mathbf{p}} + \mathbf{p}\cdot\mathbf{v}_{\mathrm{s}}, \quad \mathbf{p}' = \mathbf{p}. \tag{5.14}$$

The equilibrium distribution function of the elementary excitation is thus given by

$$n_{\mathbf{p}} = \frac{1}{e^{\beta(\epsilon_{\mathbf{p}}+\mathbf{p}\cdot\mathbf{v}_{\mathrm{s}})} - 1}. \tag{5.15}$$

The energy and momentum of the entire system as measured from the laboratory frame are given by

$$E' = E_0 + \frac{M}{2}\mathbf{v}_s^2 + \sum_{\mathbf{p}} (\epsilon_{\mathbf{p}} + \mathbf{p}\mathbf{v}_s)\, n_{\mathbf{p}}. \tag{5.16}$$

$$\mathbf{P}' = M\mathbf{v}_s + \mathbf{P} = M\mathbf{v}_s + \sum_{\mathbf{p}} \mathbf{p} n_{\mathbf{p}}. \tag{5.17}$$

In deriving these formulae, we assume that the dispersion relation $\epsilon_{\mathbf{p}}$ is independent of $\mathbf{v}_s$.

Finally, we consider a situation in which an external perturbation acts on the system, and therefore, the elementary excitations drift with velocity $\mathbf{v}_n$ with respect to the wall. If we may assume that the excitations drift as a whole while maintaining thermal equilibrium, we should consider the dispersion relation $\epsilon''_{\mathbf{p}}$ in the frame of reference moving with this thermal cloud:

$$\epsilon''_{\mathbf{p}} = \epsilon_{\mathbf{p}} + \mathbf{p} \cdot (\mathbf{v}_s - \mathbf{v}_n), \tag{5.18}$$

where $\mathbf{v}_s$ may or may not be affected by the perturbation. The distribution function of the elementary excitation is thus given by

$$n_{\mathbf{p}} = \frac{1}{e^{\beta[\epsilon_{\mathbf{p}} + \mathbf{p}\cdot(\mathbf{v}_s - \mathbf{v}_n)]} - 1}. \tag{5.19}$$

The average mass current density of the system is given by

$$\langle \mathbf{j} \rangle = \frac{1}{V}\sum_{\mathbf{p}} \mathbf{p} n_{\mathbf{p}} = \int \frac{d\mathbf{p}}{(2\pi\hbar)^3} \frac{\mathbf{p}}{e^{\beta[\epsilon_{\mathbf{p}} + \mathbf{p}\cdot(\mathbf{v}_s - \mathbf{v}_n)]} - 1}. \tag{5.20}$$

Assuming that $\mathbf{v}_s - \mathbf{v}_n$ is small, we may expand the integrand to the first order in $\mathbf{v}_s - \mathbf{v}_n$, thus obtaining

$$\begin{aligned} \langle \mathbf{j} \rangle &\simeq \int \frac{d\mathbf{p}}{(2\pi\hbar)^3} \mathbf{p}\,[\mathbf{p}\cdot(\mathbf{v}_s - \mathbf{v}_n)] \frac{\partial}{\partial \epsilon_{\mathbf{p}}} \frac{1}{e^{\beta\epsilon_{\mathbf{p}}} - 1} \\ &= \frac{1}{3}(\mathbf{v}_s - \mathbf{v}_n) \int \frac{d\mathbf{p}}{(2\pi\hbar)^3} p^2 \frac{\partial}{\partial \epsilon_{\mathbf{p}}} \frac{1}{e^{\beta\epsilon_{\mathbf{p}}} - 1}. \end{aligned} \tag{5.21}$$

When $\mathbf{v}_s = 0$, we may define the normal fluid density ρ_n through the relation

$$\langle \mathbf{j} \rangle = \rho_n \mathbf{v}_n, \tag{5.22}$$

where

$$\rho_n = -\frac{1}{3} \int \frac{d\mathbf{p}}{(2\pi\hbar)^3} p^2 \frac{\partial}{\partial \epsilon_{\mathbf{p}}} \frac{1}{e^{\beta\epsilon_{\mathbf{p}}} - 1}. \tag{5.23}$$

The superfluid density ρ_s is defined as

$$\rho_s \equiv \rho - \rho_n, \tag{5.24}$$

where ρ is the total mass density.

At sufficiently low temperatures, the normal mass current is carried by phonons, where $\epsilon_{\mathbf{p}} = cp$ with c being the sound velocity. Equation (5.23) then becomes

$$\rho_n = -\frac{1}{6\pi^2\hbar^3 c}\int_0^\infty dp\, p^4 \frac{\partial}{\partial p}\frac{1}{e^{\beta cp}-1} = \frac{2(k_B T)^4}{3\pi^2\hbar^3 c^5}\int_0^\infty dx \frac{x^3}{e^x - 1}.$$

The value of the integral is $\pi^2/15$, thus giving

$$\rho_n = \frac{2(k_B T)^4}{45\hbar^3 c^5}. \tag{5.25}$$

5.3 Landau Criterion

Landau [Landau (1941, 1947)] argued that a superfluid can flow without dissipation if the velocity of the flow is below a certain critical value v_{crit}. The superfluid dissipates energy via excitations of quasiparticles whose dispersion relation is given by Eq. (5.14). Here, $\epsilon'_{\mathbf{p}} = \epsilon_{\mathbf{p}} + \mathbf{p}\cdot\mathbf{v}_s$ is minimized when $\mathbf{p}$ is antiparallel to $\mathbf{v}_s$. If $(\epsilon'_{\mathbf{p}})_{\min} = \epsilon_{\mathbf{p}} - pv_s$ is negative, the system can lower its energy by exciting quasiparticles. The critical velocity is thus given by

$$v_{\mathrm{crit}} = \min_{\mathbf{p}}\left(\frac{\epsilon_{\mathbf{p}}}{p}\right). \tag{5.26}$$

If $v_s > v_{\mathrm{crit}}$, the superflow decays by emitting quasiparticles in the direction of $\theta = \arccos^{-1}(-\epsilon_{\mathbf{p}}/(pv_s))$. Since v_s exceeds the phase velocity $\epsilon_{\mathbf{p}}/p$ of the excitation, this emission may be interpreted as Cherenkov radiation.

If $v_s < v_{\mathrm{crit}}$, on the other hand, the superflow cannot decay and should persist permanently. In fact, however, the experimentally observed critical velocity is usually much smaller than v_{crit}. Contrary to widespread belief, the Landau criterion cannot be applied to the stability of a superflow, because microscopic rugosities of the wall, which violate the Galilean invariance and may cause energy and momentum dissipation of the system, are not at all reflected in the Hamiltonian (5.3). The criterion is more suitably applied to situations where an impurity is moving through a superfluid at rest. Then, as long as the velocity of the impurity is below v_{crit}, it can propagate through the superfluid without causing dissipation. The Landau criterion has been confirmed in experiments using ions in superfluid ^{4}He [Ellis and McClintock (1985)] or an optical spoon in a gaseous Bose–Einstein condensate [Raman, *et al.* (1999)].

5.4 Correlation Functions at Thermal Equilibrium

The expectation value of an arbitrary operator $\hat{\mathcal{O}}$ over the grand canonical ensemble is defined as

$$\langle\hat{\mathcal{O}}\rangle \equiv Z^{-1}\mathrm{Tr}\left[e^{-\beta(\hat{H}-\mu\hat{N})}\hat{\mathcal{O}}\right], \quad Z \equiv \mathrm{Tr}\left[e^{-\beta(\hat{H}-\mu\hat{N})}\right], \tag{5.27}$$

where $\hat{N}$ is the number operator and μ is the chemical potential. The time evolution of the operator in the Heisenberg representation is defined as

$$\hat{\mathcal{O}}(t) \equiv e^{\frac{i}{\hbar}\hat{H}t}\hat{\mathcal{O}}e^{-\frac{i}{\hbar}\hat{H}t}. \tag{5.28}$$

Because $\hat{H}$ and $\hat{N}$ commute, $\langle\hat{\mathcal{O}}(t)\rangle$ is independent of time:

$$\langle\hat{\mathcal{O}}(t)\rangle = \langle\hat{\mathcal{O}}\rangle. \tag{5.29}$$

For two arbitrary operators $\hat{A}$ and $\hat{B}$, the following relation holds.

$$\langle\hat{A}(t)\hat{B}(t')\rangle = \langle\hat{B}(t')e^{\beta\mu\hat{N}}\hat{A}(t+i\beta\hbar)e^{-\beta\mu\hat{N}}\rangle. \tag{5.30}$$

In particular, when $\hat{A}$ commutes with $\hat{N}$, we have

$$\langle\hat{A}(t)\hat{B}(t')\rangle = \langle\hat{B}(t')\hat{A}(t+i\beta\hbar)\rangle. \tag{5.31}$$

To prove Eq. (5.30), we substitute $e^{\beta(\hat{H}-\mu\hat{N})}e^{-\beta(\hat{H}-\mu\hat{N})} = \hat{1}$ in the left-hand side:

$$\begin{aligned}\langle\hat{A}(t)\hat{B}(t')\rangle &= Z^{-1}\mathrm{Tr}\left[e^{-\beta(\hat{H}-\mu\hat{N})}\hat{A}(t)\hat{B}(t')\right]\\ &= Z^{-1}\mathrm{Tr}\left[e^{-\beta(\hat{H}-\mu\hat{N})}\hat{A}(t)e^{\beta(\hat{H}-\mu\hat{N})}e^{-\beta(\hat{H}-\mu\hat{N})}\hat{B}(t')\right].\end{aligned}$$

Using the cyclic property of the trace, *i.e.*, $\mathrm{Tr}[\hat{A}\hat{B}\hat{C}] = \mathrm{Tr}[\hat{C}\hat{A}\hat{B}]$, we obtain Eq. (5.30):

$$\begin{aligned}\langle\hat{A}(t)\hat{B}(t')\rangle &= Z^{-1}\mathrm{Tr}\left[e^{-\beta(\hat{H}-\mu\hat{N})}\hat{B}(t')e^{-\beta(\hat{H}-\mu\hat{N})}\hat{A}(t)e^{\beta(\hat{H}-\mu\hat{N})}\right]\\ &= Z^{-1}\mathrm{Tr}\left[e^{-\beta(\hat{H}-\mu\hat{N})}\hat{B}(t')e^{\beta\mu\hat{N}}e^{\frac{i}{\hbar}\hat{H}(i\beta\hbar)}\hat{A}(t)e^{-\frac{i}{\hbar}\hat{H}(i\beta\hbar)}e^{-\beta\mu\hat{N}}\right]\\ &= \langle\hat{B}(t')e^{\beta\mu\hat{N}}\hat{A}(t+i\beta\hbar)e^{-\beta\mu\hat{N}}\rangle.\end{aligned}$$

At thermal equilibrium, $\langle\hat{A}(t)\hat{B}(t')\rangle$ depends on t and t' only through their difference $t-t'$, and neither $\langle\hat{A}(t)\rangle$ nor $\langle\hat{B}(t')\rangle$ depend on time. It is then convenient to introduce the Fourier transforms of the correlation functions:

$$C_{AB}(\omega) \equiv \int_{-\infty}^{\infty}(\langle\hat{A}(t)\hat{B}(t')\rangle - \langle\hat{A}\rangle\langle\hat{B}\rangle)e^{i\omega(t-t')}dt, \tag{5.32}$$

$$C_{BA}(\omega) \equiv \int_{-\infty}^{\infty}(\langle\hat{B}(t')\hat{A}(t)\rangle - \langle\hat{A}\rangle\langle\hat{B}\rangle)e^{i\omega(t-t')}dt. \tag{5.33}$$

Substituting Eq. (5.31) in Eq. (5.32) yields

$$C_{AB}(\omega) = \int_{-\infty}^{\infty} (\langle \hat{B}(t')\hat{A}(t+i\beta\hbar)\rangle - \langle \hat{A}\rangle\langle \hat{B}\rangle) e^{i\omega(t-t')} dt$$
$$= \int_{-\infty}^{\infty} (\langle \hat{B}(t')\hat{A}(t)\rangle - \langle \hat{A}\rangle\langle \hat{B}\rangle) e^{i\omega(t-i\beta\hbar-t')} dt.$$

Hence,

$$C_{AB}(\omega) = e^{\beta\hbar\omega} C_{BA}(\omega). \tag{5.34}$$

It follows that

$$\gamma_{AB}(\omega) \equiv \int_{-\infty}^{\infty} \left\langle \left[\hat{A}(t), \hat{B}(t')\right] \right\rangle e^{i\omega(t-t')} dt$$
$$= C_{AB}(\omega) - C_{BA}(\omega) = (1 - e^{-\beta\hbar\omega}) C_{AB}(\omega). \tag{5.35}$$

Thus, we obtain

$$C_{AB}(\omega) = \frac{\gamma_{AB}(\omega)}{1 - e^{-\beta\hbar\omega}}. \tag{5.36}$$

We may use this relation to express the correlation function of $\hat{A}$ and $\hat{B}$ in terms of γ_{AB}:

$$\langle \hat{A}(t)\hat{B}(t')\rangle - \langle \hat{A}\rangle\langle \hat{B}\rangle = \int_{-\infty}^{\infty} \frac{d\omega}{2\pi} C_{AB}(\omega) e^{-i\omega(t-t')}$$
$$= \int_{-\infty}^{\infty} \frac{d\omega}{2\pi} \frac{\gamma_{AB}(\omega)}{1 - e^{-\beta\hbar\omega}} e^{-i\omega(t-t')}. \tag{5.37}$$

Integrating this equation with respect to $t - t'$ from 0 to $-i\beta\hbar$ gives

$$\int_0^{-i\beta\hbar} d(t-t') \left[\langle \hat{A}(t)\hat{B}(t')\rangle - \langle \hat{A}\rangle\langle \hat{B}\rangle\right] = -i \int_{-\infty}^{\infty} \frac{d\omega}{2\pi} \frac{\gamma_{AB}(\omega)}{\omega}. \tag{5.38}$$

In a special case in which

$$\int_0^{-i\beta\hbar} d(t-t') \langle \hat{A}(t)\hat{B}(t')\rangle = -i\beta\hbar \langle \hat{A}(0)\hat{B}(0)\rangle, \tag{5.39}$$

Eq. (5.38) reduces to

$$\langle \hat{A}\hat{B}\rangle - \langle \hat{A}\rangle\langle \hat{B}\rangle = \frac{1}{\beta\hbar} \int_{-\infty}^{\infty} \frac{d\omega}{2\pi} \frac{\gamma_{AB}(\omega)}{\omega}. \tag{5.40}$$

5.5 Normal Fluid Density

Suppose that a superfluid is flowing through a cylinder with velocity $\mathbf{v}_s$. At zero temperature, the superfluid mass density ρ_s is equal to the total mass density ρ [see Sec. 1.8] and the entire system flows without dissipation. At finite temperatures, the normal component will be scattered by microscopic roughness of the wall before eventually coming to rest with the wall and achieving thermal equilibrium with it. Let us find the normal mass density ρ_n by considering the problem in the frame of reference in which the superfluid is at rest. Then, the wall moves with velocity $\mathbf{u} = -\mathbf{v}_s$, and only the normal component, which is dragged by the wall, contributes to the mass current density

$$\hat{\mathbf{j}}(\mathbf{r}) = \frac{\hbar}{2i}\left[\hat{\psi}^\dagger(\mathbf{r})\nabla\hat{\psi}(\mathbf{r}) - \left(\nabla\hat{\psi}^\dagger(\mathbf{r})\right)\hat{\psi}(\mathbf{r})\right]. \tag{5.41}$$

The expectation value of $\hat{\mathbf{j}}$ is given by

$$\langle\hat{\mathbf{j}}(\mathbf{r})\rangle_\mathbf{u} = \frac{\mathrm{Tr}\left[e^{-\beta(\hat{H}-\hat{\mathbf{P}}\cdot\mathbf{u}-\mu\hat{N})}\hat{\mathbf{j}}(\mathbf{r})\right]}{\mathrm{Tr}\left[e^{-\beta(\hat{H}-\hat{\mathbf{P}}\cdot\mathbf{u}-\mu\hat{N})}\right]}, \tag{5.42}$$

where $\hat{\mathbf{P}}$ is the total momentum operator

$$\hat{\mathbf{P}} = \int d\mathbf{r}\,\hat{\mathbf{j}}(\mathbf{r}). \tag{5.43}$$

We expand Eq. (5.42) up to the first order in $\mathbf{u}$. Since $\hat{\mathbf{P}}$ commutes with $\hat{H}$ and $\hat{N}$, we obtain

$$\begin{aligned}\langle\hat{\mathbf{j}}(\mathbf{r})\rangle_\mathbf{u} &\simeq \frac{\mathrm{Tr}\left[(1+\beta\hat{\mathbf{P}}\cdot\mathbf{u})e^{-\beta(\hat{H}-\mu\hat{N})}\hat{\mathbf{j}}(\mathbf{r})\right]}{\mathrm{Tr}\left[e^{-\beta(\hat{H}-\mu\hat{N})}(1+\beta\hat{\mathbf{P}}\cdot\mathbf{u})\right]} = \frac{\langle\hat{\mathbf{j}}(\mathbf{r})\rangle + \beta\sum_i\langle\hat{\mathbf{j}}(\mathbf{r})\hat{P}_j\rangle u_j}{1+\beta\sum_j\langle\hat{P}_j\rangle u_j} \\ &\simeq \langle\hat{\mathbf{j}}(\mathbf{r})\rangle + \beta\sum_{j=x,y,z}\int d\mathbf{r}'\left(\langle\hat{\mathbf{j}}(\mathbf{r})\hat{j}_j(\mathbf{r}')\rangle - \langle\hat{\mathbf{j}}(\mathbf{r})\rangle\langle\hat{j}_j(\mathbf{r}')\rangle\right)u_j.\end{aligned} \tag{5.44}$$

The difference $\langle\hat{\mathbf{j}}(\mathbf{r})\rangle_\mathbf{u} - \langle\hat{\mathbf{j}}(\mathbf{r})\rangle$ gives the mass current of the normal component, and we define the normal mass density tensor ρ_{ij}^n through the relation

$$\langle\hat{j}_i(\mathbf{r})\rangle_\mathbf{u} - \langle\hat{j}_i(\mathbf{r})\rangle = \sum_{j=x,y,z}\rho_{ij}^n(\mathbf{r})u_j, \tag{5.45}$$

where

$$\rho_{ij}^n(\mathbf{r}) \equiv \beta\int d\mathbf{r}'\left(\langle\hat{j}_i(\mathbf{r})\hat{j}_j(\mathbf{r}')\rangle - \langle\hat{j}_i(\mathbf{r})\rangle\langle\hat{j}_j(\mathbf{r}')\rangle\right). \tag{5.46}$$

In accordance with Eq. (5.35), we introduce the spectral density of the current-current correlation function

$$\gamma_{ij}(\mathbf{r},\mathbf{r}';\omega) = \int_{-\infty}^{\infty} \left\langle \left[\hat{j}_i(\mathbf{r},t), \hat{j}_j(\mathbf{r}',t')\right] \right\rangle e^{i\omega(t-t')} dt \quad (i,j = x,y,z). \tag{5.47}$$

Then, it follows from Eq. (5.40) and (5.46) that

$$\rho^{\rm n}_{ij}(\mathbf{r}) = \frac{1}{\hbar}\int d\mathbf{r}' \int_{-\infty}^{\infty} \frac{d\omega}{2\pi}\frac{\gamma_{ij}(\mathbf{r},\mathbf{r}';\omega)}{\omega}. \tag{5.48}$$

When the system is translationally invariant in space, $\gamma_{ij}(\mathbf{r},\mathbf{r}';\omega)$ depends only on $\mathbf{r}-\mathbf{r}'$, and we may introduce its Fourier transform

$$\gamma_{ij}(\mathbf{r},\mathbf{r}';\omega) = \int \frac{d\mathbf{k}}{(2\pi)^3} e^{i\mathbf{k}(\mathbf{r}-\mathbf{r}')}\gamma_{ij}(\mathbf{k},\omega). \tag{5.49}$$

Substituting Eq. (5.49) in Eq. (5.48) gives

$$\rho^{\rm n}_{ij} = \frac{1}{\hbar}\int_{-\infty}^{\infty} \frac{d\omega}{2\pi}\int d\mathbf{k}\ \delta(\mathbf{k})\frac{\gamma_{ij}(\mathbf{k},\omega)}{\omega}. \tag{5.50}$$

The right-hand side no longer depends on $\mathbf{r}$ because of the assumption of translation invariance in space. However, care must be taken when we perform integration over $\mathbf{k}$, for the value of the response function $\gamma_{ij}(\mathbf{k},\omega)$ at $\mathbf{k}=\mathbf{0}$ depends on how the limit $\mathbf{k}\to\mathbf{0}$ is taken. To understand this, let us first recall that γ_{ij} is a tensor of rank two, and there are only two such tensors available, namely δ_{ij} and k_ik_j. We may therefore decompose γ_{ij} into a linear combination of these,

$$\gamma_{ij}(\mathbf{k},\omega) = \frac{k_ik_j}{k^2}\gamma^{\rm L}(\mathbf{k},\omega) + \left(\delta_{ij} - \frac{k_ik_j}{k^2}\right)\gamma^{\rm T}(\mathbf{k},\omega), \tag{5.51}$$

where $\gamma^{\rm L}$ and $\gamma^{\rm T}$ describe the "longitudinal" and "transverse" components, respectively, of γ_{ij} in the following sense. We may construct a vector from γ_{ij} by contracting it with k_j, that is, $\sum_j \gamma_{ij}k_j$ is a vector with respect to the index i. Counting out the same procedure on the right-hand side of Eq. (5.51) gives $k_i\gamma^{\rm L}$, which is a vector parallel to k_i, *i.e.*, in the longitudinal direction. The last term in Eq. (5.51) therefore represents the transverse component. Because γ_{ij} is symmetric with respect to i and j, the same conclusion follows for index j. It can readily be checked that $\gamma^{\rm L}$ and $\gamma^{\rm T}$ are given by

$$\gamma^{\rm L}(\mathbf{k},\omega) = \sum_{i,j}\frac{k_ik_j}{k^2}\gamma_{ij}(\mathbf{k},\omega), \tag{5.52}$$

$$\gamma^{\rm T}(\mathbf{k},\omega) = \frac{1}{2}\sum_{i,j}\left(\delta_{ij} - \frac{k_ik_j}{k^2}\right)\gamma_{ij}(\mathbf{k},\omega). \tag{5.53}$$

Here, from Eqs. (5.47) and (5.49), we have

$$\gamma_{ij}(\mathbf{k},\omega) = \int_{-\infty}^{\infty} dt\, e^{i\omega(t-t')} \int d\mathbf{r}\, e^{-i\mathbf{k}(\mathbf{r}-\mathbf{r}')} \left\langle \left[\hat{\jmath}_i(\mathbf{r},t), \hat{\jmath}_j(\mathbf{r}',t')\right]\right\rangle. \tag{5.54}$$

Substituting this into Eqs. (5.52) and (5.53), we obtain

$$\gamma^{\mathrm{L}}(\mathbf{k},\omega) = \int_{-\infty}^{\infty} dt\, e^{i\omega(t-t')} \int d\mathbf{r}\, e^{-i\mathbf{k}(\mathbf{r}-\mathbf{r}')} \left\langle \left[\hat{\jmath}^{\mathrm{L}}(\mathbf{r},t), \hat{\jmath}^{\mathrm{L}}(\mathbf{r}',t')\right]\right\rangle, \tag{5.55}$$

$$\gamma^{\mathrm{T}}(\mathbf{k},\omega) = \int_{-\infty}^{\infty} dt e^{i\omega(t-t')} \int d\mathbf{r} e^{-i\mathbf{k}(\mathbf{r}-\mathbf{r}')} \Big\{ \sum_i \left[\hat{\jmath}_i(\mathbf{r},t), \hat{\jmath}_i(\mathbf{r}',t')\right] - \left\langle \left[\hat{\jmath}^{\mathrm{L}}(\mathbf{r},t), \hat{\jmath}^{\mathrm{L}}(\mathbf{r}',t')\right]\right\rangle \Big\}, \tag{5.56}$$

where

$$\hat{\jmath}^{\mathrm{L}}(\mathbf{r},t) \equiv \sum_i \frac{k_i}{k} \hat{\jmath}_i(\mathbf{r},t). \tag{5.57}$$

To be specific, let us consider a situation in which the system is moving along the z-direction. Assuming that the normal component is isotropic, *i.e.*, $\rho^{\mathrm{n}}_{ij} \propto \delta_{ij}$, Eq. (5.50) reduces to

$$\begin{aligned} \rho^{\mathrm{n}}_{zz} &= \frac{1}{\hbar}\int_{-\infty}^{\infty} \frac{d\omega}{2\pi} \int d\mathbf{k}\, \delta(\mathbf{k}) \frac{\gamma_{zz}(\mathbf{k},\omega)}{\omega} \\ &= \frac{1}{\hbar}\int_{-\infty}^{\infty} \frac{d\omega}{2\pi}\frac{1}{\omega} \int d\mathbf{k}\, \delta(\mathbf{k}) \left[\frac{k_z^2}{k^2}\gamma^{\mathrm{L}}(\mathbf{k},\omega) + \left(1-\frac{k_z^2}{k^2}\right)\gamma^{\mathrm{T}}(\mathbf{k},\omega)\right]. \end{aligned} \tag{5.58}$$

Here, the value of k_z^2/k^2 at $\mathbf{k}=\mathbf{0}$ depends on how the limit $\mathbf{k}\to\mathbf{0}$ is taken; in fact,

$$\lim_{k_x,k_y\to 0}\lim_{k_z\to 0}\frac{k_z^2}{k^2} = 0, \quad \lim_{k_z\to 0}\lim_{k_x,k_y\to 0}\frac{k_z^2}{k^2} = 1. \tag{5.59}$$

It follows that

$$\lim_{k_x,k_y\to 0}\lim_{k_z\to 0}\frac{1}{\hbar}\int_{-\infty}^{\infty}\frac{d\omega}{2\pi}\frac{\gamma_{zz}(\mathbf{k},\omega)}{\omega} = \lim_{\mathbf{k}\to\mathbf{0}}\frac{1}{\hbar}\int_{-\infty}^{\infty}\frac{d\omega}{2\pi}\frac{\gamma^{\mathrm{T}}(\mathbf{k},\omega)}{\omega}, \tag{5.60}$$

$$\lim_{k_z\to 0}\lim_{k_x,k_y\to 0}\frac{1}{\hbar}\int_{-\infty}^{\infty}\frac{d\omega}{2\pi}\frac{\gamma_{zz}(\mathbf{k},\omega)}{\omega} = \lim_{\mathbf{k}\to\mathbf{0}}\frac{1}{\hbar}\int_{-\infty}^{\infty}\frac{d\omega}{2\pi}\frac{\gamma^{\mathrm{L}}(\mathbf{k},\omega)}{\omega}. \tag{5.61}$$

Thus, the order of the limiting procedures affects the final value. In Eq. (5.60), the limit $k_z \to 0$ is first taken with k_y and k_z kept constant. This implies that the system is uniform in the direction of flow and only the transverse components contribute to the density in response to the motion of the transverse walls. This part of the density, after the limits $k_x, k_y \to 0$

are taken subsequently, is referred to as the normal fluid density and it is denoted as $\rho_{\rm n}$:

$$\rho_{\rm n} = \lim_{k_x,k_y\to 0}\lim_{k_z\to 0}\int_{-\infty}^{\infty}\frac{d\omega}{2\pi}\frac{\gamma_{zz}(\mathbf{k},\omega)}{\hbar\omega} = \lim_{\mathbf{k}\to\mathbf{0}}\int_{-\infty}^{\infty}\frac{d\omega}{2\pi}\frac{\gamma^{\rm T}(\mathbf{k},\omega)}{\hbar\omega}. \tag{5.62}$$

In Eq. (5.61), on the other hand, the limits $k_x, k_y \to 0$ are first taken with k_z kept constant. In this case, the response of the current subject to the periodic boundary condition along the z-direction is probed. Clearly, the total density ρ should respond in this case, leading to

$$\rho = \lim_{k_z\to 0}\lim_{k_x,k_y\to 0}\int_{-\infty}^{\infty}\frac{d\omega}{2\pi}\frac{\gamma_{zz}(\mathbf{k},\omega)}{\hbar\omega} = \lim_{\mathbf{k}\to\mathbf{0}}\int_{-\infty}^{\infty}\frac{d\omega}{2\pi}\frac{\gamma^{\rm L}(\mathbf{k},\omega)}{\hbar\omega}. \tag{5.63}$$

When the system exhibits superfluidity, $\rho_{\rm n} < \rho$ since the superfluid fraction does not respond to the motion of the transverse walls. When the system is normal, $\rho_{\rm n} = \rho$.

Finally, we show explicitly that the right-hand side of Eq. (5.63) gives the total density. To show this, we calculate the commutation relation between $\hat{\rho}(\mathbf{r},t) = \hat{\psi}^\dagger(\mathbf{r},t)\hat{\psi}(\mathbf{r},t)$ and $\hat{\mathbf{j}}(\mathbf{r}',t)$:

$$\left[\hat{\rho}(\mathbf{r},t),\hat{\mathbf{j}}(\mathbf{r}',t)\right] = i\hbar\hat{\psi}^\dagger(\mathbf{r},t)\hat{\psi}(\mathbf{r},t)\nabla\delta(\mathbf{r}-\mathbf{r}') - \frac{i\hbar}{2}\delta(\mathbf{r}-\mathbf{r}')\nabla[\hat{\psi}^\dagger(\mathbf{r},t)\hat{\psi}(\mathbf{r},t)]. \tag{5.64}$$

Taking the expectation value of Eq. (5.64) and denoting $\langle\hat{\psi}^\dagger(\mathbf{r},t)\hat{\psi}(\mathbf{r},t)\rangle \equiv \rho$, we obtain

$$\left\langle\left[\hat{\rho}(\mathbf{r},t),\hat{\mathbf{j}}(\mathbf{r}',t)\right]\right\rangle = i\hbar\rho\nabla'\delta(\mathbf{r}-\mathbf{r}') - \frac{i\hbar}{2}\delta(\mathbf{r}-\mathbf{r}')\nabla'\rho. \tag{5.65}$$

We assume that the particle density is constant, so that the last term in Eq. (5.65) vanishes. We next use Eq. (5.52) to rewrite the right-hand side of Eq. (5.63) as

$$\begin{aligned}
\int_{-\infty}^{\infty}\frac{d\omega}{2\pi}\frac{\gamma^{\rm L}(\mathbf{k},\omega)}{\hbar\omega} &= \int_{-\infty}^{\infty}\frac{d\omega}{2\pi}\frac{1}{\hbar\omega}\sum_{i,j}\frac{k_ik_j}{k^2}\gamma_{ij}(\mathbf{k},\omega)\\
&= \int_{-\infty}^{\infty}\frac{d\omega}{2\pi}\frac{1}{\hbar\omega}\sum_{i,j}\frac{k_ik_j}{k^2}\langle[\hat{j}_i,\hat{j}_j]\rangle(\mathbf{k},\omega)\\
&= \frac{1}{\hbar}\int_{-\infty}^{\infty}\frac{d\omega}{2\pi}\sum_j\frac{k_j}{k^2}\langle[\hat{\rho},\hat{j}_j]\rangle(\mathbf{k},\omega),
\end{aligned} \tag{5.66}$$

where the Fourier-transformed form of the continuity equation, $\sum_i k_i\hat{j}_i(\mathbf{k},\omega) = \omega\hat{\rho}(\mathbf{k},\omega)$, is used in deriving the last equality. Fourier-transforming $\langle[\hat{\rho},\hat{j}_j]\rangle(\mathbf{k},\omega)$ in Eq. (5.66) back to the real space and time, we obtain

$$\int_{-\infty}^{\infty}\frac{d\omega}{2\pi}\frac{\gamma^{\rm L}(\mathbf{k},\omega)}{\hbar\omega} = \frac{1}{\hbar k^2}\mathbf{k}\cdot\int d(\mathbf{r}-\mathbf{r}')e^{-i\mathbf{k}(\mathbf{r}-\mathbf{r}')}\langle[\hat{\rho}(\mathbf{r},t),\mathbf{j}(\mathbf{r}',t)]\rangle. \tag{5.67}$$

Substituting Eq. (5.64) in Eq. (5.67), we obtain Eq. (5.63)

5.6 Low-Lying Excitations of a Superfluid

When a system exhibits superfluidity, ρ_n is, in general, smaller than ρ. This result follows from the fact that the low-lying excitations of a superfluid are longitudinal ones that do not contribute to ρ_n but only to ρ. To show this explicitly, we substitute $\hat{A} = \hat{j}_z(\mathbf{r}, t)$ and $\hat{B} = \hat{j}_z(\mathbf{r}', t')$ in Eq. (5.35):

$$\gamma_{zz}(\mathbf{r}, \mathbf{r}'; \omega) = \int_{-\infty}^{\infty} \langle \hat{j}_z(\mathbf{r}, t)\hat{j}_z(\mathbf{r}', t')\rangle e^{i\omega(t-t')} dt. \tag{5.68}$$

Substituting

$$\hat{j}_z(\mathbf{r}, t) = \int \frac{d\mathbf{k}}{(2\pi)^3} \hat{j}_z(\mathbf{k}, t) e^{i\mathbf{kr}} \tag{5.69}$$

in Eq. (5.68), we obtain

$$\gamma_{zz}(\mathbf{r}, \mathbf{r}'; \omega) = \int_{-\infty}^{\infty} dt\, e^{i\omega(t-t')} \int \frac{d\mathbf{k}}{(2\pi)^3} e^{i\mathbf{k}(\mathbf{r}-\mathbf{r}')} \langle \hat{j}_z(\mathbf{k}, t)\hat{j}_z(-\mathbf{k}, t')\rangle, \tag{5.70}$$

where we used $\langle \hat{j}_z(\mathbf{k}, t)\hat{j}_z(\mathbf{k}', t')\rangle = (2\pi)^3 \delta(\mathbf{k}+\mathbf{k}')\langle \hat{j}_z(\mathbf{k}, t)\hat{j}_z(-\mathbf{k}, t')\rangle$, which holds true when the system has translational invariance in space. Performing the Fourier transformation of Eq. (5.70), we obtain

$$\gamma_{zz}(\mathbf{k}, \omega) = \int_{-\infty}^{\infty} dt\, e^{i\omega(t-t')} \langle \hat{j}_z(\mathbf{k}, t)\hat{j}_z(-\mathbf{k}, t')\rangle. \tag{5.71}$$

Substituting the completeness relation $\sum_n |n\rangle\langle n| = \hat{1}$, where $|n\rangle$ is the eigenstate of $\hat{H}$ with energy E_n, we obtain

$$\gamma_{zz}(\mathbf{k}, \omega) = 2\pi\hbar \sum_n |\langle n|\hat{j}_z(-\mathbf{k})|0\rangle|^2 \delta(\hbar\omega - E_n + E_0), \tag{5.72}$$

and hence,

$$\int_{-\infty}^{\infty} \frac{d\omega}{2\pi} \frac{\gamma_{zz}(\mathbf{k}, \omega)}{\hbar\omega} = \sum_n \frac{|\langle n|\hat{j}_z(-\mathbf{k})|0\rangle|^2}{E_n - E_0}. \tag{5.73}$$

In the limit $\mathbf{k} \to \mathbf{0}$, $\hat{j}_z(-\mathbf{k})$ becomes the z-component $\hat{P}_z$ of the total momentum operator:

$$\lim_{\mathbf{k}\to\mathbf{0}} \hat{j}_z(-\mathbf{k}) = \int d\mathbf{r}\, \hat{j}_z(\mathbf{r}) = \hat{P}_z. \tag{5.74}$$

Since $\hat{P}_z|0\rangle = 0$, only those states whose energies approach E_0 as $\mathbf{k} \to \mathbf{0}$ can contribute to the right-hand side of Eq. (5.73). In a superfluid, the relevant states are those of the phonons that are longitudinal and parallel to $\mathbf{k}$. Thus,

$$\lim_{\mathbf{k}\to\mathbf{0}} \int_{-\infty}^{\infty} \frac{d\omega}{2\pi} \frac{\gamma_{zz}(\mathbf{k}, \omega)}{\hbar\omega} = \begin{cases} \rho_n = 0 & \text{if } \mathbf{k} \perp z \\ \rho & \text{if } \mathbf{k} \parallel z. \end{cases} \tag{5.75}$$

5.7 Examples

It is instructive to apply the formalism developed in the preceding sections to some typical examples. Let $\hat{a}_{\mathbf{k}}$ be the annihilation operator of bosons with wave vector $\mathbf{k}$ and dipersion relation $\omega = \omega(\mathbf{k})$. The field operator $\hat{\psi}$ can be expanded in terms of $\hat{a}_{\mathbf{k}}$ as

$$\hat{\psi}(\mathbf{r},t) = \frac{1}{\sqrt{V}} \sum_{\mathbf{k}} \hat{a}_{\mathbf{k}} e^{i(\mathbf{kr}-\omega(\mathbf{k})t)}, \tag{5.76}$$

and the mass current density operator is given as

$$\begin{aligned} \hat{\mathbf{j}}(\mathbf{r},t) &= \frac{i\hbar}{2}\left[(\nabla\hat{\psi}^\dagger(\mathbf{r},t))\hat{\psi}(\mathbf{r},t) - \hat{\psi}^\dagger(\mathbf{r},t)\nabla\hat{\psi}(\mathbf{r},t)\right] \\ &= \frac{\hbar}{2V}\sum_{\mathbf{k},\mathbf{k}'}(\mathbf{k}+\mathbf{k}')\, \hat{a}^\dagger_{\mathbf{k}}\hat{a}_{\mathbf{k}'}\, e^{-i(\mathbf{k}-\mathbf{k}')\mathbf{r}+i(\omega(\mathbf{k})-\omega(\mathbf{k}'))t}. \end{aligned} \tag{5.77}$$

The current-current correlation function is given by

$$\begin{aligned} \chi_{\mathbf{r},\beta}(\mathbf{r},t) =&\ \langle [j_\alpha(\mathbf{r},t), j_\beta(\mathbf{r},t)]\rangle \\ =&\ \left(\frac{\hbar}{2V}\right)^2 \sum_{\mathbf{k}_1,\cdots,\mathbf{k}_4} (k_{1\alpha}+k_{2\alpha})(k_{3\beta}+k_{4\beta}) \\ &\times \left\langle \left[\hat{a}^\dagger_{\mathbf{k}_1}\hat{a}_{\mathbf{k}_2}, \hat{a}^\dagger_{\mathbf{k}_3}\hat{a}_{\mathbf{k}_4}\right]\right\rangle e^{-i(\mathbf{k}_1-\mathbf{k}_2)\mathbf{r}+i(\omega_1-\omega_2)t}, \end{aligned} \tag{5.78}$$

where $\alpha, \beta = x, y, z$ and $\omega_i \equiv \omega(\mathbf{k}_i)$. A straightforward calculation gives

$$\left[\hat{a}^\dagger_{\mathbf{k}_1}\hat{a}_{\mathbf{k}_2}, \hat{a}^\dagger_{\mathbf{k}_3}\hat{a}_{\mathbf{k}_4}\right] = \hat{a}^\dagger_{\mathbf{k}_1}\hat{a}_{\mathbf{k}_4}\delta_{\mathbf{k}_2\mathbf{k}_3} - \hat{a}^\dagger_{\mathbf{k}_3}\hat{a}_{\mathbf{k}_2}\delta_{\mathbf{k}_1\mathbf{k}_4}, \tag{5.79}$$

and therefore, Eq. (5.78) reduces to

$$\begin{aligned} \chi_{\alpha\beta}(\mathbf{r},t) =&\ \left(\frac{\hbar}{2V}\right)^2 \sum_{\mathbf{k}_1,\mathbf{k}_2,\mathbf{k}_3} (k_{1\alpha}+k_{2\alpha}) e^{-i(\mathbf{k}_1-\mathbf{k}_2)\mathbf{r}+i(\omega_1-\omega_2)t} \\ &\times \left[(k_{2\beta}+k_{3\beta})\langle\hat{a}^\dagger_{\mathbf{k}_1}\hat{a}_{\mathbf{k}_3}\rangle - (k_{1\beta}+k_{3\beta})\langle\hat{a}^\dagger_{\mathbf{k}_3}\hat{a}_{\mathbf{k}_2}\rangle\right]. \end{aligned} \tag{5.80}$$

5.7.1 *Ideal Bose gas*

Let us first consider an ideal Bose gas at thermal equilibrium. Then, we use Eq. (1.34) to simplify Eq. (5.80) as

$$\begin{aligned} \chi_{\alpha\beta}(\mathbf{r},t) =&\ \left(\frac{\hbar}{2V}\right)^2 \sum_{\mathbf{k},\mathbf{k}_2} (k_{1\alpha}+k_{2\alpha})(k_{1\beta}+k_{2\beta}) e^{-i(\mathbf{k}_1-\mathbf{k}_2)\mathbf{r}+i(\omega_1-\omega_2)t} \\ &\times (n(\omega_1) - n(\omega_2)), \end{aligned} \tag{5.81}$$

where $\omega_1 = \hbar \mathbf{k}_i^2/2M$ and

$$n(\omega_i) = \frac{1}{e^{\beta(\hbar\omega_i - \mu)} - 1}. \tag{5.82}$$

The Fourier transform of Eq. (5.81) gives

$$\chi_{\alpha\beta}(\mathbf{k},\omega) = \frac{\hbar^2}{16\pi^2} \int d\mathbf{k}_1 (2k_{1\alpha} + k_\alpha)(2k_{1\beta} + k_\beta)(n(\omega_1) - n(\omega_2)) \times \delta\left(\omega - \frac{\hbar}{2M}(\mathbf{k}^2 + 2\mathbf{k}\cdot\mathbf{k}_1)\right). \tag{5.83}$$

Here,

$$n(\omega_1) - n(\omega_2) \simeq \frac{\partial n(\omega_1)}{\partial \omega_1}(\omega_1 - \omega_2) = \frac{\beta\hbar^2}{2M}(\mathbf{k}^2 + 2\mathbf{k}\cdot\mathbf{k}_1)\frac{e^{\beta(\hbar\omega_1-\mu)}}{(e^{\beta(\hbar\omega_1-\mu)}-1)^2}.$$

Thus,

$$\chi_{\alpha\beta}(\mathbf{k},\omega) = \frac{\beta\hbar^4}{32\pi^2 M} \int d\mathbf{k}_1 (2k_{1\alpha} + k_\alpha)(2k_{1\beta} + k_\beta)(\mathbf{k}^2 + 2\mathbf{k}\cdot\mathbf{k}_1) \times \delta\left(\omega - \frac{\hbar}{2M}(\mathbf{k}^2 + 2\mathbf{k}\cdot\mathbf{k}_1)\right)\frac{e^{\beta(\hbar\omega_1-\mu)}}{(e^{\beta(\hbar\omega_1-\mu)}-1)^2}. \tag{5.84}$$

Substituting this in Eq. (5.62) yields

$$\rho_n = \lim_{k_x,k_y\to 0}\lim_{k_z\to 0}\int_{-\infty}^{\infty}\frac{d\omega}{2\pi}\frac{\chi_{zz}(\mathbf{k},\omega)}{\hbar\omega} = \frac{\beta\hbar^2}{6\pi^2}\int_0^\infty dk_1 k_1^4 \frac{e^{\beta(\hbar\omega_1-\mu)}}{(e^{\beta(\hbar\omega_1-\mu)}-1)^2}.$$

Performing partial integration leads to

$$\begin{aligned}\rho_n &= \frac{2M}{\sqrt{\pi}\lambda_{\text{th}}^3}\int_0^\infty dx \frac{\sqrt{\pi}}{e^{x-\beta\mu}-1} = \frac{2M}{\sqrt{\pi}\lambda_{\text{th}}^3}\sum_{n=1}^{\infty} e^{\beta\mu n}\int_0^\infty dx\sqrt{x}e^{-nx} \\ &= \frac{M}{\lambda_{\text{th}}^3}\sum_{n=1}^{\infty}\frac{e^{\beta\mu n}}{n^{\frac{3}{2}}},\end{aligned} \tag{5.85}$$

where $\lambda_{\text{th}} \equiv h/\sqrt{2\pi M k_B T}$ is the thermal de Broglie length defined in Eq. (1.24).

Equation (5.85) may be rewritten using the Bose–Einstein function defined in Eq. (3.12) as

$$\rho_n = \frac{M}{\lambda_{\text{th}}^3} g_{\frac{3}{2}}(e^{\beta\mu}). \tag{5.86}$$

In particular, at the transition temperature T_0, $\mu = 0$ and $g_{\frac{3}{2}}(1) = \zeta(3/2) \simeq 2.612$. We thus reproduce the well-known formula for the phase-space density of an ideal Bose gas at the transition temperature:

$$(\rho_n/M)\lambda_{\text{th}}^3 = 2.612 \quad \text{at } T_0. \tag{5.87}$$

5.7.2 *Weakly interacting Bose gas*

In the presence of weak interaction between bosons, the elementary excitations are Bogoliubov quasiparticles whose creation and annihilation operators $\hat{b}^\dagger_{\mathbf{k}}, \hat{b}_{\mathbf{k}}$ are related to $\hat{a}^\dagger_{\mathbf{k}}$, $\hat{a}_{\mathbf{k}}$ by the Bogoliubov transformation:

$$\hat{a}_{\mathbf{k}} = \gamma_{\mathbf{k}}(\hat{b}_{\mathbf{k}} - \alpha_{\mathbf{k}}\hat{b}^\dagger_{-\mathbf{k}}), \quad \hat{a}^\dagger_{\mathbf{k}} = \gamma^\dagger_{\mathbf{k}}(\hat{b}^\dagger_{\mathbf{k}} - \alpha_{\mathbf{k}}\hat{b}_{-\mathbf{k}}), \tag{5.88}$$

where $\gamma_{\mathbf{k}} = (1-\alpha^2_{\mathbf{k}})^{-\frac{1}{2}}$ and $\alpha_{\mathbf{k}}$ is given in Eq. (2.44). At thermal equilibrium, we have

$$\langle \hat{b}^\dagger_{\mathbf{k}}\hat{b}_{\mathbf{k}'}\rangle = n_{\mathbf{k}}\delta_{\mathbf{k},\mathbf{k}'}, \quad \langle \hat{b}_{\mathbf{k}}\hat{b}_{\mathbf{k}'}\rangle = \langle \hat{b}^\dagger_{\mathbf{k}}\hat{b}^\dagger_{\mathbf{k}'}\rangle = 0, \tag{5.89}$$

where

$$n_{\mathbf{k}} = \frac{1}{e^{\beta(E_{\mathbf{k}}-\mu)}-1}, \tag{5.90}$$

and $E_{\mathbf{k}}$ is the Bogoliubov spectrum given in Eq. (2.50):

$$E_{\mathbf{k}} = \frac{\hbar^2 k}{2M}\sqrt{k^2+16\pi a n}. \tag{5.91}$$

Then, we obtain

$$\langle \hat{a}^\dagger_{\mathbf{k}}\hat{a}_{\mathbf{k}'}\rangle = \gamma^2_{\mathbf{k}}\left[(1+\alpha^2_{\mathbf{k}})n_{\mathbf{k}} + \alpha^2_{\mathbf{k}}\right]\delta_{\mathbf{k},\mathbf{k}'} \equiv \tilde{n}_{\mathbf{k}}\delta_{\mathbf{k},\mathbf{k}'}. \tag{5.92}$$

Substituting this in Eq. (5.80) gives

$$\begin{aligned}\chi_{\alpha\beta}(\mathbf{k},\omega) = \frac{\hbar^2}{16\pi^2}\int d\mathbf{k}_1(2k_{1\alpha}+k_\alpha)(2k_{1\beta}+k_\beta)\\ \times\delta(\omega+\omega(\mathbf{k}_1)-\omega(\mathbf{k}+\mathbf{k}_1))(\tilde{n}_{\mathbf{k}_1}-\tilde{n}_{\mathbf{k}+\mathbf{k}_1}).\end{aligned} \tag{5.93}$$

Here,

$$\tilde{n}_{\mathbf{k}_1} - \tilde{n}_{\mathbf{k}+\mathbf{k}_1} \simeq \frac{\partial \tilde{n}_{\mathbf{k}_1}}{\partial \epsilon_{\mathbf{k}_1}}(\epsilon_{\mathbf{k}_1}-\epsilon_{\mathbf{k}+\mathbf{k}_1}) = \hbar(\omega(\mathbf{k}_1)-\omega(\mathbf{k}+\mathbf{k}_1))\frac{\partial \tilde{n}_{\mathbf{k}_1}}{\partial \epsilon_{\mathbf{k}_1}}.$$

Thus,

$$\begin{aligned}\chi_{\alpha\beta}(\mathbf{k},\omega) = -\frac{\hbar^3\omega}{16\pi^2}\int d\mathbf{k}_{\mathrm{T}}(2k_{1\alpha}+k_\alpha)(2k_{1\beta}+k_\beta)\frac{\partial \tilde{n}_{\mathbf{k}_1}}{\partial \epsilon_{\mathbf{k}_1}}\\ \times\delta(\omega+\omega(\mathbf{k}_1)-\omega(\mathbf{k}+\mathbf{k}_1)).\end{aligned} \tag{5.94}$$

Substituting this in Eq. (5.62) yields

$$\begin{aligned}\rho_{\mathrm{n}} &= \lim_{k_x,k_y\to 0}\lim_{k_z\to 0}\frac{\hbar^2}{32\pi^3}\int d\mathbf{k}_1(2k_{1z}+k_z)^2\left(-\frac{\partial \tilde{n}_{\mathbf{k}_1}}{\partial \epsilon_{\mathbf{k}_1}}\right)\\ &= \frac{M}{2\pi^2}\int_0^\infty dk k^2\frac{(1+\alpha^2_{\mathbf{k}})n_{\mathbf{k}}+\alpha^2_{\mathbf{k}}}{1-\alpha^2_{\mathbf{k}}}.\end{aligned} \tag{5.95}$$

We may use Eq. (2.44) to change the variable of integration from k to $\alpha_{\mathbf{k}}$:

$$\rho_{\mathrm{n}} = \frac{M}{4\pi^2}(4\pi a n)^{\frac{3}{2}}\int_0^1 d\alpha_{\mathbf{k}}\frac{1-\alpha_{\mathbf{k}}}{\alpha^{3/2}_{\mathbf{k}}}[(1+\alpha^2_{\mathbf{k}})n_{\mathbf{k}}+\alpha^2_{\mathbf{k}}]. \tag{5.96}$$

This formula gives the temperature dependence of ρ_{n}. In particular, at $T=0$, where $n_{\mathbf{k}}=0$, we obtain the fraction of the quantum depletion

$$\frac{\rho_{\mathrm{n}}}{M} = \frac{8}{3}\sqrt{\frac{na^3}{\pi}}, \tag{5.97}$$

which is in agreement with Eq. (2.78).

Chapter 6

Spinor Bose–Einstein Condensate

6.1 Internal Degrees of Freedom

Most Bose–Einstein condensates of atomic gases have internal degrees of freedom originating from the spin. Alkali atoms have an electronic spin of 1/2 and a nuclear spin of 3/2 (^{7}Li, ^{23}Na, ^{39}K, ^{41}K, ^{87}Rb), 5/2 (^{85}Rb), or 7/2 (^{123}Cs), whereas ^{52}Cr has an electronic spin of 3 and no nuclear spin.

In alkali atoms, the electronic spin $\mathbf{s}$ and the nuclear spin $\mathbf{i}$ are coupled by the hyperfine interaction $V_{\rm hf} = A\,\mathbf{s}\cdot\mathbf{i}$ to form the hyperfine spin $\mathbf{f} = \mathbf{s} + \mathbf{i}$. The hyperfine energy splitting is of the order of 0.1 K and it is much larger than the typical transition temperature of BEC $\sim 1\ \mu$K. Thus, the hyperfine spin is a good quantum number. The values of f are 1 and 2 for $i = 3/2$, 2 and 3 for $i = 5/2$, and 3 and 4 for $i = 7/2$. For most alkali atoms, the smaller-f state lies lower in energy. This implies that the higher-f state has a shorter lifetime because atoms decay into the lower-f state. An important exception is ^{40}K, for which the $f = 9/2$ state is lower in energy than the $f = 7/2$ state because of the sign change in the g-factor.

When a Bose–Einstein condensate is trapped in a magnetic potential, the spin aligns along the direction of a local magnetic field, and the internal degrees of freedom are virtually frozen. The condensate is therefore described with a scalar order parameter. When it is trapped in an optical potential, the internal degrees of freedom are liberated because the optical potential exerts the same force on an atom irrespective of which magnetic sublevel it is in. Such a condensate is called a spinor Bose–Einstein condensate; the order parameter of this condensate has $2f + 1$ components, where f is the hyperfine spin for alkali atoms and the electronic spin for chromium. The values of s, i, and f of the atomic species that have undergone BEC or Fermi degeneracy are listed in Table 6.1.

Table 6.1 List of electronic spin s, nuclear spin i, and their composition $f = |s \pm i|$ for the atomic species that have undergone BEC or Fermi degeneracy.

s	i	f	atomic species
0	0	0	^{40}Ca, ^{84}Sr, ^{170}Yb, ^{174}Yb, ^{176}Yb
0	$\frac{1}{2}$	$\frac{1}{2}$	^{171}Yb
0	$\frac{5}{2}$	$\frac{5}{2}$	^{173}Yb
$\frac{1}{2}$	$\frac{1}{2}$	0, 1	^{1}H
$\frac{1}{2}$	1	$\frac{1}{2}, \frac{3}{2}$	^{6}Li
$\frac{1}{2}$	$\frac{3}{2}$	1, 2	^{7}Li, ^{23}Na, ^{39}K, ^{41}K, ^{87}Rb
$\frac{1}{2}$	$\frac{5}{2}$	2, 3	^{85}Rb
$\frac{1}{2}$	$\frac{7}{2}$	3, 4	^{133}Cs
1	0	1	^{4}He* (metastable)
$\frac{1}{2}$	4	$\frac{7}{2}, \frac{9}{2}$	^{40}K
3	0	3	^{52}Cr

The many-body wave function of identical particles of spin f acquires a phase factor of $(-1)^{2f}$ under the exchange of any pair of particles. By the same exchange, the spin and orbital parts of the wave function change by a factor of $(-1)^{F+2f}$ and $(-1)^{L}$, respectively, where F and L are the total spin and the relative orbital angular momentum of the pair, respectively. For the sake of consistency, we have $(-1)^{2f} = (-1)^{F+2f} \times (-1)^{L}$, and therefore, $F + L$ must be even. This rule applies to both bosons and fermions. For example, when identical bosons undergo an s-wave interaction ($L = 0$), the total spin F of two interacting bosons must be even. Consequently, when $f = 1$, F must be 0 or 2; when $f = 2$, F must be 0, 2, or 4.

6.2 General Hamiltonian of Spinor Condensates

We consider a system of identical bosons of spin f, and let $\hat{\psi}_m(\mathbf{r})$ ($m = f, f-1, \cdots, -f$) be the corresponding field operator that annihilates a boson having a magnetic quantum number m at position $\mathbf{r}$. The field operators are assumed to satisfy the following canonical commutation relations:

$$[\hat{\psi}_m(\mathbf{r}), \hat{\psi}_n^\dagger(\mathbf{r}')] = \delta_{nm}\delta(\mathbf{r}-\mathbf{r}'),$$
$$[\hat{\psi}_n(\mathbf{r}), \hat{\psi}_m(\mathbf{r}')] = 0, \quad [\hat{\psi}_m^\dagger(\mathbf{r}), \hat{\psi}_n^\dagger(\mathbf{r}')] = 0. \tag{6.1}$$

We consider a situation in which bosons interact via an s-wave channel. Then, the symmetry consideration described above implies that the total spin F of any two interacting bosons with spin f must be $0, 2, \cdots, 2f$. The interaction Hamiltonian $\hat{V}$ is therefore classified according to F:

$$\hat{V} = \sum_{F=0,2,\cdots,2f} \hat{V}^{(F)}. \tag{6.2}$$

To construct $\hat{V}^{(F)}$, we consider the operator $\hat{A}_{FM}(\mathbf{r},\mathbf{r}')$ that annihilates a pair of bosons with total spin F and total magnetic quantum number M at positions $\mathbf{r}$ and $\mathbf{r}'$. The pair annihilation operator is expressed in terms of the field operator as

$$\hat{A}_{FM}(\mathbf{r},\mathbf{r}') = \sum_{m_1,m_2=-f}^{f} \langle F, M | f, m_1; f, m_2 \rangle \hat{\psi}_{m_1}(\mathbf{r})\hat{\psi}_{m_2}(\mathbf{r}'), \tag{6.3}$$

where $\langle F, M | f, m_1; f, m_2 \rangle$ is the Clebsch–Gordan coefficient. Because the Hamiltonian must be scalar, *i.e.*, it must be invariant under rotation, $\hat{V}^{(F)}$ takes the following form:

$$\hat{V}^{(F)} = \frac{1}{2}\int d\mathbf{r} \int d\mathbf{r}' v^{(F)}(\mathbf{r}-\mathbf{r}') \sum_{M=-F}^{F} \hat{A}_{FM}^\dagger(\mathbf{r},\mathbf{r}')\hat{A}_{FM}(\mathbf{r},\mathbf{r}'), \tag{6.4}$$

where $v^{(F)}(\mathbf{r},\mathbf{r}')$ describes the interaction potential for the total spin F channel. Using the completeness relation

$$\sum_F \sum_{M=-F}^{F} |F, M\rangle\langle F, M| = \hat{1}, \tag{6.5}$$

where $\hat{1}$ is the identity operator, we can show that

$$\sum_F \sum_{M=-F}^{F} \hat{A}_{FM}^\dagger(\mathbf{r},\mathbf{r}')\hat{A}_{FM}(\mathbf{r},\mathbf{r}') =: \hat{n}(\mathbf{r})\hat{n}(\mathbf{r}') :. \tag{6.6}$$

Here, $\hat{n}(\mathbf{r})$ is the density operator of the particle number

$$\hat{n}(\mathbf{r}) \equiv \sum_{m=-f}^{f} \hat{\psi}_m^\dagger(\mathbf{r})\hat{\psi}_m(\mathbf{r}), \tag{6.7}$$

and the symbol : : denotes normal ordering that dictates that the annihilation operators are placed to the right of the creation operators. It follows from Eq. (6.6) that if $v^{(F)}$ is independent of F (*i.e.*, $v^{(F)}(\mathbf{r}) = v(\mathbf{r})$), the interaction Hamiltonian (6.2) reduces to the Hartree interaction

$$\hat{V} = \frac{1}{2}\int d\mathbf{r} \int d\mathbf{r}' v(\mathbf{r}-\mathbf{r}') : \hat{n}(\mathbf{r})\hat{n}(\mathbf{r}') : . \tag{6.8}$$

In a gaseous Bose–Einstein condensate, the range of the interaction potential is negligible as compared to the average interatomic distance, and the detailed profile of the interaction potential is irrelevant. We may therefore approximate $v^{(F)}(\mathbf{r})$ as

$$v^{(F)}(\mathbf{r}) = g_F \delta(\mathbf{r}), \tag{6.9}$$

where g_F characterizes the strength of the interaction between two particles having a total spin of F and it is related to the corresponding s-wave scattering length a_F as

$$g_F = \frac{4\pi\hbar^2}{M} a_F. \tag{6.10}$$

The interaction Hamiltonian then becomes

$$\hat{V}^{(F)} = \frac{g_F}{2}\int d\mathbf{r} \sum_{M=-F}^{F} \hat{A}^{\dagger}_{FM}(\mathbf{r})\hat{A}_{FM}(\mathbf{r}), \tag{6.11}$$

where

$$\begin{aligned}\hat{A}_{FM}(\mathbf{r}) &\equiv \hat{A}_{FM}(\mathbf{r},\mathbf{r}) \\ &= \sum_{m_1,m_2=-f}^{f} \langle F, M|f, m_1; f, m_2\rangle \hat{\psi}_{m_1}(\mathbf{r})\hat{\psi}_{m_2}(\mathbf{r}).\end{aligned} \tag{6.12}$$

When $F = 0$, the Clebsch–Gordan coefficient takes the following form:

$$\langle 0, 0|f, m_1; f, m_2\rangle = \delta_{m_1+m_2,0}\frac{(-1)^{f-m_1}}{\sqrt{2f+1}}. \tag{6.13}$$

Substituting this into Eq. (6.3), we obtain the spin-singlet pair operator

$$\hat{A}_{00}(\mathbf{r},\mathbf{r}') = \frac{1}{\sqrt{2f+1}}\sum_{m=-f}^{f}(-1)^{f-m}\hat{\psi}_m(\mathbf{r})\hat{\psi}_{-m}(\mathbf{r}'). \tag{6.14}$$

It is instructive to show that $\hat{A}_{FM}(\mathbf{r})$ vanishes for odd F. For example, when $F = 1$ and $f = 1$, the Clebsch–Gordan coefficients are given by

$$\begin{aligned}\langle 1, M|1, m_1; 1, m_2\rangle &= \frac{(-1)^{1-m_1}}{\sqrt{2}}\delta_{m_1+m_2,M} \\ &\times [\delta_{M,1}(\delta_{M_1,1}+\delta_{m_1,0}) + \delta_{M,0}\, m_1 - \delta_{M,-1}(\delta_{m_1,0}+\delta_{m_1,-1})].\end{aligned} \tag{6.15}$$

Substituting this into Eq. (6.12), we find that $\hat{A}_{1M}(\mathbf{r}) = 0$. Similarly, we can show that $\hat{A}_{FM}(\mathbf{r}) = 0$ for odd F.

Finally, let us derive the interaction Hamiltonians for the cases in which $f = 1$ and $f = 2$. For $f = 1$, the total spin F of any pair of bosons must be 0 or 2, and the interaction Hamiltonian (6.11) reduces to

$$\hat{V}^{(0)} = \frac{g_0}{2} \int d\mathbf{r} \hat{A}_{00}^{\dagger}(\mathbf{r}) \hat{A}_{00}(\mathbf{r}) \tag{6.16}$$

$$\begin{aligned} \hat{V}^{(2)} &= \frac{g_2}{2} \int d\mathbf{r} \sum_{M=-2}^{2} \hat{A}_{2M}^{\dagger}(\mathbf{r}) \hat{A}_{2M}(\mathbf{r}) \\ &= \frac{g_2}{2} \int d\mathbf{r} \left[: \hat{n}^2(\mathbf{r}) : - \hat{A}_{00}^{\dagger}(\mathbf{r}) \hat{A}_{00}(\mathbf{r}) \right], \end{aligned} \tag{6.17}$$

where Eq. (6.6) and $\hat{A}_{1M}(\mathbf{r}) = 0$ are used to obtain the last equality. Adding Eqs. (6.16) and (6.17), we obtain

$$\hat{V} = \int d\mathbf{r} \left[\frac{g_2}{2} : \hat{n}^2(\mathbf{r}) : + \frac{g_0 - g_2}{2} \hat{A}_{00}^{\dagger}(\mathbf{r}) \hat{A}_{00}(\mathbf{r}) \right]. \tag{6.18}$$

For $f = 2$, F must be 0, 2, or 4, and the interaction Hamiltonian (6.11) for $F = 4$ is expressed as

$$\begin{aligned} \hat{V}^{(4)} &= \frac{g_4}{2} \int d\mathbf{r} \sum_{M=-4}^{4} \hat{A}_{4M}^{\dagger}(\mathbf{r}) \hat{A}_{4M}(\mathbf{r}) \\ &= \frac{g_4}{2} \left[: \hat{n}^2(\mathbf{r}) : - \hat{A}_{00}^{\dagger}(\mathbf{r}) \hat{A}_{00}(\mathbf{r}) - \sum_{M=-2}^{2} \hat{A}_{2M}^{\dagger}(\mathbf{r}) \hat{A}_{2M}(\mathbf{r}) \right], \end{aligned} \tag{6.19}$$

where Eq. (6.6) and $\hat{A}_{1M} = \hat{A}_{3M} = 0$ are used to obtain the last equality. The interaction Hamiltonian for $f = 2$ is thus given by

$$\begin{aligned} \hat{V} &= \hat{V}^{(0)} + \hat{V}^{(2)} + \hat{V}^{(4)} \\ &= \int d\mathbf{r} \left[\frac{g_4}{2} : \hat{n}^2(\mathbf{r}) : + \frac{g_0 - g_4}{2} \hat{A}_{00}^{\dagger}(\mathbf{r}) \hat{A}_{00}(\mathbf{r}) \right. \\ &\quad \left. + \frac{g_2 - g_4}{2} \sum_{M=-2}^{2} \hat{A}_{2M}^{\dagger}(\mathbf{r}) \hat{A}_{2M}(\mathbf{r}) \right]. \end{aligned} \tag{6.20}$$

In addition to $\hat{V}$, the total Hamiltonian $\hat{H}$ involves the kinetic term $\hat{H}_{\mathrm{KE}}$, the one-body potential term $\hat{H}_{\mathrm{PE}}$, and linear and quadratic Zeeman terms $\hat{H}_{\rm Z}$:

$$\hat{H}_{\rm KE} = \int d\mathbf{r} \sum_{m=-f}^{f} \hat{\psi}_m^\dagger \left(-\frac{\hbar^2 \nabla^2}{2M} \right) \hat{\psi}_m, \tag{6.21}$$

$$\hat{H}_{\rm PE} = \int d\mathbf{r} \sum_{m=-f}^{f} U(\mathbf{r}) \hat{\psi}_m^\dagger \hat{\psi}_m, \tag{6.22}$$

$$\hat{H}_{\rm Z} = \int d\mathbf{r} \sum_{m,n=-f}^{f} \hat{\psi}_m^\dagger \left[g\, \mu_{\rm B} (\mathbf{B} \cdot \mathbf{f})_{mn} + q' (\mathbf{B} \cdot \mathbf{f})_{mn}^2 \right] \hat{\psi}_n, \tag{6.23}$$

where g is the Landé g-factor, $\mu_{\rm B} = e\hbar/2m \simeq 9.27 \times 10^{-24}$J/T is the Bohr magneton, and $\mathbf{f} = (f_x, f_y, f_z)$ is the vector of spin-f matrices. The last term in Eq. (6.23) describes the quadratic Zeeman effect that arises from the hyperfine interaction between the electronic and the nuclear magnetic moments. A second-order perturbation gives $q' = (g\mu_{\rm B})^2/\Delta E_{\rm hf}$, where $\Delta E_{\rm hf}$ is the hyperfine energy splitting. We define

$$p \equiv |g|\mu_{\rm B} B, \quad q \equiv \frac{(g\mu_{\rm B} B)^2}{\Delta E_{\rm hf}}. \tag{6.24}$$

For the electronic ground state of ^{87}Rb, $g = -1/2$ and $\Delta E_{\rm hf}/\hbar \simeq 6.8$ GHz for $f = 1$, whereas $g = 1/2$ and $\Delta E_{\rm hf}/h \simeq -6.8$ GHz for $f = 2$, where

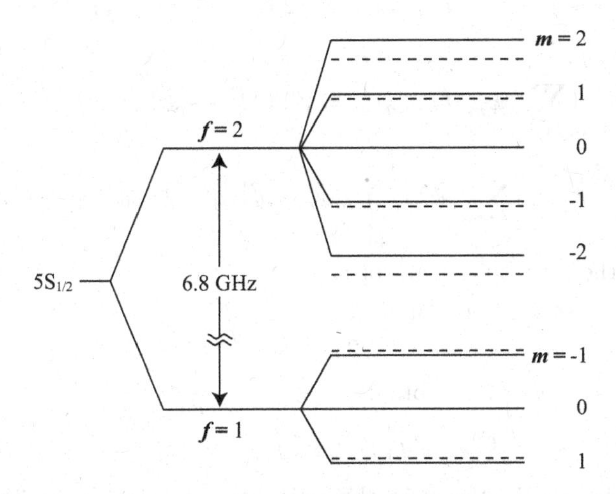

Fig. 6.1 Zeeman sublevels of ^{87}Rb. The rightmost solid lines indicate energy splitting due to the linear Zeeman effect, and the dashed lines indicate further shifts due to the quadratic Zeeman effect. The energy spacings of the Zeeman sublevels are magnified for clarity.

m is the magnetic quantum number. Since the g-factor has opposite signs for $f = 1$ and $f = 2$, the Zeeman sublevels are inverted, as shown in Fig. 6.1. Because of the opposite signs of q, the quadratic Zeeman effect raises (lowers) the energy of the $m \neq 0$ states for $f = 1$ ($f = 2$), as indicated by dashed lines. Note that the quadratic Zeeman effect is absent for ^{52}Cr BEC because of the absence of the hyperfine interaction.

6.3 Spin-1 BEC

The spin-1 matrices are given by

$$f_x = \frac{1}{\sqrt{2}}\begin{pmatrix} 0 & 1 & 0 \\ 1 & 0 & 1 \\ 0 & 1 & 0 \end{pmatrix}, \quad f_y = \frac{i}{\sqrt{2}}\begin{pmatrix} 0 & -1 & 0 \\ 1 & 0 & -1 \\ 0 & 1 & 0 \end{pmatrix}, \quad f_z = \begin{pmatrix} 1 & 0 & 0 \\ 0 & 0 & 0 \\ 0 & 0 & -1 \end{pmatrix}, \tag{6.25}$$

and we define the raising (f_+) and lowering (f_-) operators:

$$f_+ \equiv f_x + if_y = \begin{pmatrix} 0 & \sqrt{2} & 0 \\ 0 & 0 & \sqrt{2} \\ 0 & 0 & 0 \end{pmatrix}, \quad f_- \equiv f_x - if_y = \begin{pmatrix} 0 & 0 & 0 \\ \sqrt{2} & 0 & 0 \\ 0 & \sqrt{2} & 0 \end{pmatrix}. \tag{6.26}$$

In the second-quantized form, the corresponding operators are given by

$$\hat{f}_+ = \sum_{m,n=-1}^{1} \hat{\psi}_m^\dagger (f_+)_{mn} \hat{\psi}_n = \sqrt{2}\,(\hat{\psi}_1^\dagger \hat{\psi}_0 + \hat{\psi}_0^\dagger \hat{\psi}_{-1}), \tag{6.27}$$

$$\hat{f}_- = \sum_{m,n=-1}^{1} \hat{\psi}_m^\dagger (f_-)_{mn} \hat{\psi}_n = \sqrt{2}\,(\hat{\psi}_0^\dagger \hat{\psi}_1 + \hat{\psi}_{-1}^\dagger \hat{\psi}_0), \tag{6.28}$$

$$\hat{f}_z = \sum_{m,n=-1}^{1} \hat{\psi}_m^\dagger (f_z)_{mn} \hat{\psi}_n = \hat{\psi}_1^\dagger \hat{\psi}_1 - \hat{\psi}_{-1}^\dagger \hat{\psi}_{-1}. \tag{6.29}$$

The square of the total spin operator is

$$\hat{\mathbf{f}}^2 = \hat{f_x}^2 + \hat{f_y}^2 + \hat{f_z}^2 = \frac{1}{2}(\hat{f}_+\hat{f}_- + \hat{f}_-\hat{f}_+) + \hat{f}_z^2. \tag{6.30}$$

Since $:\hat{f}_+\hat{f}_-: \,=\, :\hat{f}_-\hat{f}_+:$, we obtain

$$:\hat{\mathbf{f}}^2: \,=\, :\hat{f}_+\hat{f}_-: + :\hat{f_z}^2: . \tag{6.31}$$

Substituting Eqs. (6.27)–(6.29) in the right-hand side of Eq. (6.31) and comparing the result with Eq. (6.14), we find that

$$\hat{A}_{00}^\dagger \hat{A}_{00} = \frac{1}{3}\,(:\hat{n}^2: - :\hat{\mathbf{f}}^2:). \tag{6.32}$$

Substituting this in Eq. (6.18), we obtain

$$\hat{V} = \int d\mathbf{r} \left[\frac{g_0 + 2g_2}{6} : \hat{n}^2 : + \frac{g_2 - g_0}{6} : \hat{\mathbf{f}}^2 : \right]. \tag{6.33}$$

Combining Eqs. (6.21)–(6.23) with Eq. (6.33), we obtain the total Hamiltonian for the spin-1 Bose–Einstein condensate:

$$\hat{H} = \int d\mathbf{r} \left\{ \sum_{m=-1}^{1} \hat{\psi}_m^\dagger \left[-\frac{\hbar^2 \nabla^2}{2M} + U(\mathbf{r}) - pm + qm^2 \right] \hat{\psi}_m + \frac{1}{2} \left(c_0 : \hat{n}^2 : + c_1 : \hat{\mathbf{f}}^2 : \right) \right\}, \tag{6.34}$$

where we assume that the magnetic field is applied along the z-direction, that $g < 0$ for the $f = 1$ condensate, and

$$c_0 \equiv \frac{g_0 + 2g_2}{3} = \frac{4\pi\hbar^2}{3M}(a_0 + 2a_2),$$
$$c_1 \equiv \frac{g_2 - g_0}{3} = \frac{4\pi\hbar^2}{3M}(a_2 - a_0). \tag{6.35}$$

The interaction Hamiltonian (6.33) can also be derived using the projection-operator method. Let P_F be the projection operator that projects an atom pair into a total spin F state. For identical spin-1 bosons, we have

$$P_0 + P_2 = 1. \tag{6.36}$$

On the other hand, the composition law of the spin operator gives

$$\mathbf{f}^2 = (\mathbf{f}_1 + \mathbf{f}_2)^2 = \mathbf{f}_1^2 + \mathbf{f}_2^2 + 2\mathbf{f}_1 \cdot \mathbf{f}_2 = \begin{cases} 0 & (F = 0); \\ 6 & (F = 2). \end{cases} \tag{6.37}$$

Since $\mathbf{f}_1^2 = \mathbf{f}_2^2 = 2$ for spin-1 bosons, we find that

$$\mathbf{f}_1 \cdot \mathbf{f}_2 = -2P_0 + P_2. \tag{6.38}$$

Solving Eqs. (6.36) and (6.38), we find that

$$P_0 = \frac{1}{3}(1 - \mathbf{f}_1 \cdot \mathbf{f}_2), \quad P_2 = \frac{1}{3}(2 + \mathbf{f}_1 \cdot \mathbf{f}_2). \tag{6.39}$$

In terms of P_0 and P_2, the interaction Hamiltonian is constructed as

$$V = \frac{g_0}{2} P_0 + \frac{g_2}{2} P_2 = \frac{g_0 + 2g_2}{6} + \frac{g_2 - g_0}{6} \mathbf{f}_1 \cdot \mathbf{f}_2. \tag{6.40}$$

Writing this in the second-quantized form, we reproduce Eq. (6.33).

6.3.1 *Mean-field theory of a spin-1 BEC*

The mean-field theory is obtained by replacing the field operator $\hat{\psi}_m$ in Eq. (6.34) with a c-number counterpart ψ_m. This can be justified in a number-conserving manner as follows. We expand $\hat{\psi}_m$ in terms of a complete orthonormal set of basis functions $\{\varphi_{mi}(\mathbf{r})\}$:

$$\hat{\psi}_m(\mathbf{r}) = \sum_i \hat{a}_{mi}\varphi_{mi}(\mathbf{r}) \quad (m = 1, 0, -1), \tag{6.41}$$

$$\int d\mathbf{r}\, \varphi^*_{mi}(\mathbf{r})\, \varphi_{mj}(\mathbf{r}) = \delta_{ij}, \tag{6.42}$$

where $\hat{a}_{mi}$ is the annihilation operator of a boson with magnetic quantum number m and spatial wave function $\varphi_{mi}(\mathbf{r})$, and it is assumed to obey the canonical commutation relations:

$$[\hat{a}_{mi}, \hat{a}^\dagger_{nj}] = \delta_{mn}\delta_{ij},\ [\hat{a}_{mi}, \hat{a}_{nj}] = 0,\ [\hat{a}^\dagger_{mi}, \hat{a}^\dagger_{nj}] = 0. \tag{6.43}$$

It can be shown that $\hat{\psi}_m$ obeys the field commutation relations (6.1), provided that φ_{mi}'s satisfy the completeness relation

$$\sum_i \varphi_{mi}(\mathbf{r})\varphi^*_{mi}(\mathbf{r}') = \delta(\mathbf{r} - \mathbf{r}'). \tag{6.44}$$

Now, the mean-field theory assumes that all bosons share the same single-particle state, say φ_{m0}. Then, the state vector is given by

$$|\boldsymbol{\xi}\rangle = \frac{1}{\sqrt{N!}} \left(\sum_{m=-1}^{1} \xi_m \hat{a}^\dagger_{m0} \right)^N |\mathrm{vac}\rangle, \tag{6.45}$$

where

$$\sum_{m=-1}^{1} |\xi_m|^2 = 1. \tag{6.46}$$

It is straightforward to show that

$$\langle \hat{\psi}_m(\mathbf{r}) \rangle = \langle \hat{\psi}^\dagger_m(\mathbf{r}) \rangle = 0, \tag{6.47}$$

$$\langle \hat{\psi}^\dagger_m(\mathbf{r}) \hat{\psi}_n(\mathbf{r}') \rangle = \psi^*_m(\mathbf{r})\psi_n(\mathbf{r}'), \tag{6.48}$$

$$\langle \hat{\psi}^\dagger_m(\mathbf{r}) \hat{\psi}^\dagger_n(\mathbf{r}') \hat{\psi}_k(\mathbf{r}'') \hat{\psi}_\ell(\mathbf{r}''') \rangle = \psi^*_m(\mathbf{r})\psi^*_n(\mathbf{r}')\psi_k(\mathbf{r}'')\psi_\ell(\mathbf{r}'''), \tag{6.49}$$

where $\langle \cdots \rangle \equiv \langle \boldsymbol{\xi} | \cdots | \boldsymbol{\xi} \rangle$ and

$$\psi_m(\mathbf{r}) \equiv \sqrt{N}\xi_m \varphi_{m0}(\mathbf{r}). \tag{6.50}$$

The expectation value of the Hamiltonian (6.34) with respect to the state in Eq. (6.45) is therefore given by

$$E[\psi] \equiv \langle \hat{H} \rangle = \int d\mathbf{r} \Bigg\{ \sum_{m=-1}^{1} \psi_m^* \left[-\frac{\hbar^2 \nabla^2}{2M} + U(\mathbf{r}) - pm + qm^2 \right] \psi_m + \frac{c_0}{2} n^2 + \frac{c_1}{2} \langle \hat{\mathbf{f}} \rangle^2 \Bigg\}, \tag{6.51}$$

where $\langle \hat{\mathbf{f}} \rangle = (\langle \hat{f}_x \rangle, \langle \hat{f}_y \rangle, \langle \hat{f}_z \rangle)$, and

$$n = \sum_{m=-1}^{1} |\psi_m|^2, \tag{6.52}$$

$$\langle \hat{f}_\alpha \rangle = \sum_{m,n=-1}^{1} \psi_m^* (f_\alpha)_{mn} \psi_n \quad (\alpha = x, y, z) \tag{6.53}$$

denote the particle density and the spin density, respectively.

From the energy functional (6.51), we find that c_0 must be nonnegative; otherwise, the system would collapse. In addition, when the Zeeman terms are negligible ($p = q = 0$), the ground state is ferromagnetic ($|\langle \hat{\mathbf{f}} \rangle| = n$) when $c_1 < 0$ and polar ($|\langle \hat{\mathbf{f}} \rangle| = 0$) when $c_1 > 0$. Here, the order parameter of the polar is given by $\sqrt{n}(0, 1, 0)$ or $\sqrt{n/2}(1, 0, 1)$; these two states are transformed into each other by spatial rotations, and they are degenerate in the absence of an external magnetic field.

The dynamics of the mean field is governed by

$$i\hbar \frac{\partial \psi_m(\mathbf{r})}{\partial t} = \frac{\delta E}{\delta \psi_m^*(\mathbf{r})}. \tag{6.54}$$

Substituting Eq. (6.51) into the right-hand side gives

$$i\hbar \frac{\partial \psi_m}{\partial t} = \left[-\frac{\hbar^2 \nabla^2}{2M} + U(\mathbf{r}) - pm + qm^2 \right] \psi_m + c_0 n \psi_m + c_1 \sum_{n=-1}^{1} \langle \hat{\mathbf{f}} \rangle \cdot \mathbf{f}_{mn} \psi_n, \tag{6.55}$$

where $\langle \hat{\mathbf{f}} \rangle \cdot \mathbf{f}_{mn} = \sum_{\alpha=x,y,z} \langle \hat{f}_\alpha \rangle (f_\alpha)_{mn}$. This is the multicomponent Gross–Pitaevskii equation. In a stationary state, we substitute $\psi_m(\mathbf{r}, t) = \psi_m(\mathbf{r})\, e^{i\mu t/\hbar}$ in Eq. (6.55) to obtain

$$\left[-\frac{\hbar^2 \nabla^2}{2M} + U(\mathbf{r}) - pm + qm^2 \right] \psi_m + c_0 n \psi_m + c_1 \sum_{n=-1}^{1} \langle \mathbf{f} \rangle \cdot \mathbf{f}_{mn} \psi_n = \mu \psi_m, \tag{6.56}$$

when μ is the chemical potential, and the components of the spin vector $\langle \hat{\mathbf{f}} \rangle$ are given by

$$\langle \hat{f}_x \rangle = \frac{1}{\sqrt{2}} \left[\psi_1^* \psi_0 + \psi_0^* (\psi_1 + \psi_{-1}) + \psi_{-1}^* \psi_0 \right], \tag{6.57}$$

$$\langle \hat{f}_y \rangle = \frac{i}{\sqrt{2}} \left[-\psi_1^* \psi_0 + \psi_0^* (\psi_1 - \psi_{-1}) + \psi_{-1}^* \psi_0 \right], \tag{6.58}$$

$$\langle \hat{f}_z \rangle = |\psi_1|^2 - |\psi_{-1}|^2. \tag{6.59}$$

Writing down the $m = 1, 0, -1$ components of Eq. (6.56) explicitly, we obtain

$$\left[-\frac{\hbar^2 \nabla^2}{2M} + U(\mathbf{r}) - p + q + c_0 n + c_1 \langle \hat{f}_z \rangle - \mu \right] \psi_1 + \frac{c_1}{\sqrt{2}} \langle \hat{f}_- \rangle \psi_0 = 0, \tag{6.60}$$

$$\frac{c_1}{\sqrt{2}} \langle \hat{f}_+ \rangle \psi_1 + \left[-\frac{\hbar^2 \nabla^2}{2M} + U(\mathbf{r}) + c_0 n - \mu \right] \psi_0 + \frac{c_1}{\sqrt{2}} \langle \hat{f}_- \rangle \psi_{-1} = 0, \tag{6.61}$$

$$\frac{c_1}{\sqrt{2}} \langle \hat{f}_+ \rangle \psi_0 + \left[-\frac{\hbar^2 \nabla^2}{2M} + U(\mathbf{r}) + p + q + c_0 n - c_1 \langle \hat{f}_z \rangle - \mu \right] \psi_{-1} = 0, \tag{6.62}$$

where $\langle \hat{f}_\pm \rangle \equiv \langle \hat{f}_x \rangle \pm i \langle \hat{f}_y \rangle$. By solving these equations, we can investigate the mean-field properties of the spin-1 BEC.

Here, we investigate the ground-state phase diagram of a uniform system [Stenger, *et al.* (1998)] and the Bogoliubov spectrum in the presence of the quadratic Zeeman effect [Murata, *et al.* (2007)]. The ground state of a uniform system is obtained if we set the kinetic energy and $U(\mathbf{r})$ to be zero. We can eliminate $c_0 n$ from Eqs. (6.60)–(6.62) by substituting $\tilde{\mu} \equiv \mu - c_0 n$. We may choose $\langle \hat{f}_y \rangle = 0$ by rotating the system about the z-axis. Furthermore, since the overall phase is arbitrary, we may assume ψ_0 to be real without loss of generality. Then, all $\psi_{1,0,-1}$ can be taken to be, and Eqs. (6.60)–(6.62) reduce to

$$(-p + q + c_1 \langle \hat{f}_z \rangle - \tilde{\mu}) \psi_1 + c_1 (\psi_1 + \psi_{-1}) \psi_0^2 = 0, \tag{6.63}$$

$$[\tilde{\mu} - c_1 (\psi_1 + \psi_{-1})^2] \psi_0 = 0, \tag{6.64}$$

$$c_1 (\psi_1 + \psi_{-1}) \psi_0^2 + [p + q - c_1 \langle \hat{f}_z \rangle - \mu] \psi_{-1} = 0. \tag{6.65}$$

From Eq. (6.64), we have either $\psi_0 = 0$ or $\tilde{\mu} = c_1(\psi_1 + \psi_{-1})^2$. In the former case, the ground state is $\psi_1 = \sqrt{n}$ and $\psi_{-1} = 0$ for $p > 0$ and $\psi_1 = 0$ and $\psi_{-1} = \sqrt{n}$ for $p < 0$. In the latter case, we solve Eqs. (6.63) and (6.65) together with the normalization condition $\sum_{m=-1}^{1} |\psi_m|^2 = n$ to obtain

$$\psi_1^2 = \left(\frac{q+p}{q-p} \right)^2 \psi_{-1}^2 = \left(\frac{p+q}{2q} \right)^2 \frac{-p^2 + q^2 + 2c_1 nq}{2c_1 q},$$

$$\psi_0^2 = \frac{q^2 - p^2}{2q^2} \frac{-p^2 - q^2 + 2c_1 nq}{2c_1 q}. \tag{6.66}$$

Here, we consider the case of $c_1 < 0$. The case of $c_1 > 0$ is discussed in [Stenger, *et al.* (1998); Ueda and Kawaguchi (2010)]. If the conditions

$$p^2 - q^2 + 2|c_1|nq > 0 \quad \text{and} \quad q^2 > p^2 \tag{6.67}$$

are satisfied, all three components $\psi_\pm, \psi_0$ are nonzero and the magnetization tilts against the direction of the external magnetic field. The polar angle of the tilted magnetization is determined by the interaction and the strength of the magnetic field, but the azimuthal angle is spontaneously chosen in each realization of the system. Since the axial symmetry about the magnetic field is broken, this phase is called the broken-axisymmetry phase [Murata, *et al.* (2007)]. When $p^2 - q^2 + 2|c_1|nq < 0$, $\psi_{\pm 1}^2$ in Eq. (6.66) become negative, and therefore, the solution is $\psi_{\pm 1} = 0$ and $\psi_0 = \sqrt{n}$. This phase is referred to as the polar phase. When $p^2 > q^2$, the ground state is $\psi_1 = \sqrt{n}$, $\psi_0 = \psi_{-1} = 0$ for $p > 0$ and $\psi_1 = \psi_0 = 0$, $\psi_{-1} = \sqrt{n}$ for $p < 0$. These two phases are thus ferromagnetic. The phase diagram of a spin-1 Bose–Einstein condensate is shown in Fig. 6.2.

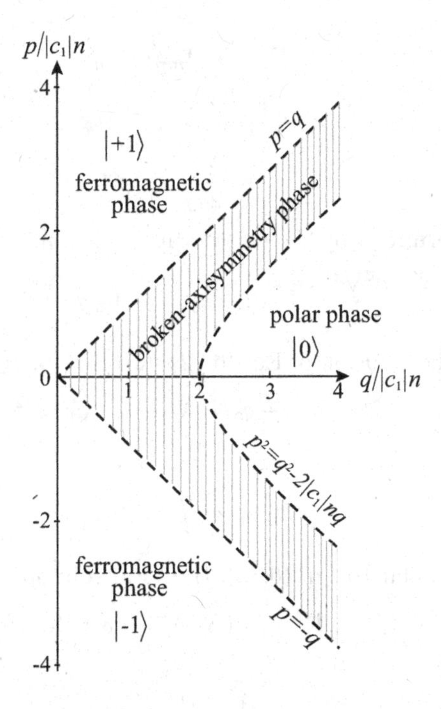

Fig. 6.2 Phase diagram of a spin-1 Bose–Einstein condensate with $c_1 < 0$.

6.3.2 *Many-body states in single-mode approximation*

Under normal circumstances, $|c_0| \gg |c_1|$ holds. It is therefore reasonable to first determine the spin-independent part of the order parameter $\phi(\mathbf{r})$ by setting $c_1 = 0$ in Eq. (6.34), and then focus on the many-body nature of spin states by assuming that all atoms share the same spatial order parameter $\phi(\mathbf{r})$. In this single-mode approximation [Law, *et al.* (1998); Koashi and Ueda (2000); Ho and Yip (2000)], we set

$$\hat{\psi}_m(\mathbf{r}) = \hat{a}_m \phi(\mathbf{r}), \tag{6.68}$$

where $\int |\phi|^2 d\mathbf{r} = 1$ and $\hat{a}_m$ is the annihilation operator of the spatial mode $\phi(\mathbf{r})$ with magnetic quantum number m. For simplicity, we consider the case of zero external magnetic field (*i.e.*, $p = q = 0$). Substituting Eq. (6.68) in Eq. (6.34) and replacing $\sum_m \hat{a}_m^\dagger \hat{a}_m$ and $\sum_{m,n} \hat{a}_m^\dagger \hat{a}_n^\dagger \hat{a}_n \hat{a}_m$ with $\hat{N}$ and $\hat{N}(\hat{N}-1)$, respectively, we obtain

$$\hat{H} = \hat{N} \int d\mathbf{r}\, \phi^* \left(-\frac{\hbar^2 \nabla^2}{2M} + U(\mathbf{r}) + \frac{c_0}{2}(\hat{N}-1)|\phi|^2 \right) \phi + c_1' :\hat{\mathbf{F}}^2:, \tag{6.69}$$

where

$$\hat{\mathbf{F}} = \sum_{m,n} \mathbf{f}_{mn} a_m^\dagger a_n \tag{6.70}$$

and

$$c_1' \equiv \frac{c_1}{2} \int d\mathbf{r} |\phi|^4. \tag{6.71}$$

Here, ϕ is determined by the Gross–Pitaevskii equation:

$$\left(-\frac{\hbar^2 \nabla^2}{2M} + U(\mathbf{r}) + c_0(N-1)|\phi|^2 \right) \phi = \mu\phi, \tag{6.72}$$

Substituting the solution of Eq. (6.72) into Eq. (6.69), we obtain

$$\hat{H} = \mu \hat{N} - c_0' \hat{N}(\hat{N}-1) + c_1' :\hat{\mathbf{F}}^2:, \tag{6.73}$$

where

$$c_0' \equiv \frac{c_0}{2} \int d\mathbf{r} |\phi|^4. \tag{6.74}$$

In a manner similar to Eq. (6.32), the total spin operator $\hat{\mathbf{F}}$ satisfies

$$:\hat{\mathbf{F}}^2: = \hat{N}(\hat{N}-1) - 3\, \hat{S}^\dagger \hat{S}, \tag{6.75}$$

where

$$\hat{S} \equiv \frac{1}{\sqrt{3}} (\hat{a}_0^2 - 2\hat{a}_1 \hat{a}_{-1}) \tag{6.76}$$

is the spin-singlet pair operator. We may use Eq. (6.75) to rewrite Eq. (6.73) as

$$\hat{H} = \mu\hat{N} - c_0'\hat{N}(\hat{N}-1) + c_1'\left[\hat{N}(\hat{N}-1) - 3\hat{S}^\dagger\hat{S}\right]. \tag{6.77}$$

Since $\hat{N}$ and $\hat{S}^\dagger\hat{S}$ commute, the eigenvalue problem for $\hat{H}$ reduces to finding the simultaneous eigenstates of $\hat{N}$ and $\hat{S}^\dagger\hat{S}$.

Let us introduce the vacuum state $|\text{vac}\rangle$ such that $\hat{a}_m|\text{vac}\rangle = 0$ for $m = 0, \pm 1$, and therefore, $\hat{S}|\text{vac}\rangle = 0$. Using the commutation relation

$$\left[\hat{S}, (\hat{S}^\dagger)^k\right] = \frac{2}{3}k\,(2\hat{N} + 2k + 1)(\hat{S}^\dagger)^{k-1}, \tag{6.78}$$

we obtain

$$\hat{S}^\dagger\hat{S}(\hat{S}^\dagger)^k|\text{vac}\rangle = \frac{2}{3}k\,(2\hat{N} - 2k + 1)(\hat{S}^\dagger)^k|\text{vac}\rangle. \tag{6.79}$$

Therefore, by defining

$$|k\rangle \equiv \frac{(S^\dagger)^k|\text{vac}\rangle}{\sqrt{\langle\text{vac}|\hat{S}^k(\hat{S}^\dagger)^k|\text{vac}\rangle}}, \tag{6.80}$$

we obtain

$$\hat{S}^\dagger\hat{S}|k\rangle = \frac{2}{3}k\,(2N - 2k + 1)|k\rangle. \tag{6.81}$$

Now, let us consider a state $|k, \mathbf{n}\rangle$ in which $2k$ atoms form spin-singlet pairs and the spins of the remaining $N - 2k$ atoms reside in the polar state along the $\mathbf{n}$ direction ($|\mathbf{n}| = 1$). Defining

$$\begin{aligned}\hat{a}_{\mathbf{n}} &\equiv n_x\hat{a}_x + n_y\hat{a}_y + n_z\hat{a}_z \\ &= -\frac{n_x + in_y}{\sqrt{2}}\hat{a}_1 + n_z\hat{a}_0 + \frac{n_x - in_y}{\sqrt{2}}\,\hat{a}_{-1},\end{aligned} \tag{6.82}$$

where $\hat{a}_x \equiv -(\hat{a}_1 - \hat{a}_{-1})/\sqrt{2}$, $\hat{a}_y \equiv -i(\hat{a}_1 + \hat{a}_{-1})/\sqrt{2})$, and $\hat{a}_z \equiv \hat{a}_0$, the desired eigenstate and eigenenergy are given by

$$|k, \mathbf{n}\rangle = Z_{\mathbf{n}}^{-\frac{1}{2}}(\hat{a}_{\mathbf{n}}^\dagger)^{N-2k}(\hat{S}^\dagger)^k|\text{vac}\rangle, \tag{6.83}$$

$$E_k = \mu N - c_0'N(N-1) + c_1'\,[N(N-1) - 2k(2N-2k+1)], \tag{6.84}$$

where $Z_{\mathbf{n}}$ is the normalization constant. We note that E_k does not depend on $\mathbf{n}$ because all magnetic sublevels are degenerate in the absence of an external magnetic field. We may alternatively characterize the eigenstate in terms of the total spin F, its projection F_z, and the number of spin-singlet pairs, k, as

$$|k, F, F_z\rangle = Z^{-\frac{1}{2}}(\hat{S}^\dagger)^k(\hat{F}_-)^{F-F_z}(\hat{a}_1^\dagger)^F|\text{vac}\rangle, \tag{6.85}$$

where $F = N - 2k$ and

$$\hat{F}_- \equiv \sum_{m,n} (f_-)_{mn}\, \hat{a}_m^\dagger \hat{a}_n = \sqrt{2}(\hat{a}_0^\dagger \hat{a}_1 + \hat{a}_{-1}^\dagger \hat{a}_0) \tag{6.86}$$

is the spin lowering operator that decreases F_z by one. The energy eigenvalue for the state $|k, F, F_z\rangle$ is again given by Eq. (6.84) since in the absence of an external magnetic field, the energy eigenspectrum is degenerate with respect to F_z.

We now consider a situation in which magnetization points along the z-direction such that $F_z = F$ in Eq. (6.85), and consider how many atoms reside in the $m = 0$ state. For clarity of notation, we introduce a new basis state $|n_1, n_0, n_{-1}\rangle$ in which the magnetic sublevels $m = 1, 0, -1$ are occupied by n_1, n_0, n_{-1} atoms. Then, $|\text{vac}\rangle = |0, 0, 0\rangle$ and

$$\langle \hat{a}_0^\dagger \hat{a}_0 \rangle = \frac{\langle \text{vac}|\hat{a}_1^F \hat{S}^k \hat{a}_0^\dagger \hat{a}_0 (\hat{S}^\dagger)^k (\hat{a}_1^\dagger)^F |\text{vac}\rangle}{\langle \text{vac}|\hat{a}_1^F \hat{S}^k (\hat{S}^\dagger)^k (\hat{a}_1^\dagger)^F |\text{vac}\rangle}. \tag{6.87}$$

Substituting $\hat{a}_0^\dagger \hat{a}_0 = \hat{a}_0 \hat{a}_0^\dagger - 1$ into the right-hand side and using the fact that $\hat{a}_0^\dagger$ commutes with $\hat{S}^\dagger$, we obtain

$$\langle \hat{a}_0^\dagger \hat{a}_0 \rangle = \frac{\langle F, 1, 0|\hat{S}^k (\hat{S}^\dagger)^k |F, 1, 0\rangle}{\langle F, 0, 0|\hat{S}^k (\hat{S}^\dagger)^k |F, 0, 0\rangle} - 1. \tag{6.88}$$

From Eq. (6.78), we have

$$\hat{S}(\hat{S}^\dagger)^k = \frac{2}{3} k\,(2\hat{N} + 2k + 1)(\hat{S}^\dagger)^{k-1} + (\hat{S}^\dagger)^k \hat{S}. \tag{6.89}$$

Substituting this in Eq. (6.88) and noting that $\hat{S}|F, 1, 0\rangle = \hat{S}|F, 0, 0\rangle = 0$ and $\hat{N}|F, 1, 0\rangle = (F + 1)|F, 1, 0\rangle$, we obtain

$$\begin{aligned}
\langle \hat{a}_0^\dagger \hat{a}_0 \rangle &= \frac{(2F + 2k + 3)\langle F, 1, 0|\hat{S}^{k-1} (\hat{S}^\dagger)^{k-1} |F, 1, 0\rangle}{(2F + 2k + 1)\langle F, 0, 0|\hat{S}^{k-1} (\hat{S}^\dagger)^{k-1} |F, 0, 0\rangle} - 1 \\
&= \frac{(2F + 2k + 3)(2F + 2k + 1) \cdots \cdot (2F + 5)}{(2F + 2k + 1)(2F + 2k - 1) \cdots \cdot (2F + 3)} - 1 \\
&= \frac{2k}{2F + 3} = \frac{N - F}{2F + 3}.
\end{aligned}$$

Since $\langle \hat{a}_1^\dagger \hat{a}_1 \rangle - \langle \hat{a}_{-1}^\dagger \hat{a}_{-1} \rangle = F$ and $\sum_{m=-1}^{1} \langle \hat{a}_m^\dagger \hat{a}_m \rangle = N$, we finally obtain

$$\bar{n}_1 = \langle \hat{a}_1^\dagger \hat{a}_1 \rangle = \frac{NF + F^2 + N + 2F}{2F + 3}, \tag{6.90}$$

$$\bar{n}_0 = \langle \hat{a}_0^\dagger \hat{a}_0 \rangle = \frac{N - F}{2F + 3}, \tag{6.91}$$

$$\bar{n}_{-1} = \langle \hat{a}_{-1}^\dagger \hat{a}_{-1} \rangle = \frac{(N - F)(F + 1)}{2F + 3}. \tag{6.92}$$

Two comments are in order here. When all atoms form spin-singlet pairs, *i.e.*, $F = 0$, Eqs. (6.90)–(6.92) give

$$\bar{n}_1 = \bar{n}_0 = \bar{n}_{-1} = \frac{N}{3}. \tag{6.93}$$

That is, the Bose–Einstein condensate is fragmented in the sense described in Sec. 1.9. This is in sharp contrast with the mean-field result that predicts that for $c_1 > 0$, the ground state is polar ($\bar{n}_0 = N$) or ($\bar{n}_1 = \bar{n}_{-1} = N/2$). This discrepancy originates from the mean-field assumption that there exists one and only one BEC. However, such an assumption does not hold when the system possesses a certain exact symmetry, which in the present case, is the rotational symmetry.

A second point to note is the fact that $\bar{n}_0$ in Eq. (6.91) decreases rapidly with increasing F. In fact, when $F = O(\sqrt{N})$, $\bar{n}_0 = O(\sqrt{N})$. This implies that although mean-field results break down at zero magnetic field, the validity of mean-field theory quickly recovers as the magnetization increases, and that the rotational symmetry is broken. This gives yet another example of why the fragmented BEC is fragile against symmetry-breaking perturbations.

Finally, we note a relationship between the many-body spin-singlet state and the corresponding mean-field state, which is a polar state. A mean-field state is one in which all particles occupy the same single-particle state:

$$\frac{(\hat{a}^\dagger_{\mathbf{n}})^N}{\sqrt{N!}} |\text{vac}\rangle, \tag{6.94}$$

where $\hat{a}_{\mathbf{n}}$ is defined in Eq. (6.82). The state (6.94) describes a polar state along the quantization axis $\mathbf{n}$. The familiar polar state in which all particles occupy the $m = 0$ state is obtained when the quantization axis is taken along the z-axis, *i.e.*, $n_x = n_y = 0$ and $n_z = 1$. Expressing $\mathbf{n}$ in polar coordinates, *i.e.*, $n_x = \sin\theta\cos\phi$, $n_y = \sin\theta\sin\phi$, and $n_z = \cos\theta$, we find from Eq. (6.82) that

$$|\theta,\phi\rangle = \frac{1}{\sqrt{N!}}\left(-\frac{\hat{a}^\dagger_1}{\sqrt{2}}\sin\theta e^{i\phi} + \hat{a}^\dagger_0\cos\theta + \frac{\hat{a}^\dagger_{-1}}{\sqrt{2}}\sin\theta e^{-i\phi}\right)^N |\text{vac}\rangle. \tag{6.95}$$

The symmetry-restored state $|\text{sym}\rangle$ is obtained if we average $|\theta,\ \phi\rangle$ over all solid angles $d\Omega = \sin\theta d\theta d\phi$:

$$\begin{aligned} |\text{sym}\rangle &= \frac{1}{4\pi}\int_0^\pi \sin\theta\, d\theta \int_0^{2\pi} d\phi\, |\theta,\phi\rangle \\ &= \begin{cases} 0 & \text{if } N \text{ is odd;} \\ \frac{1}{(N+1)\sqrt{N!}}(a_0^{\dagger 2} - 2a_1^\dagger a_{-1}^\dagger)^{\frac{N}{2}}|\text{vac}\rangle & \text{if } N \text{ is even.} \end{cases} \end{aligned} \tag{6.96}$$

This result shows that the spin-singlet state is the superposition state of a mean-field state $|\theta,\ \phi\rangle$ over all directions.

6.3.3 *Superflow, spin texture, and Berry phase*

Let us first consider the case of a ferromagnetic BEC. When the magnetization points along the z-direction, the normalized order parameter is given by $(1,0,0)^T$, where T denotes transpose. A general order parameter is given by rotating this to an arbitrary direction. A general rotation matrix is characterized by Eulerian angles α, β, γ :

$$U(\alpha,\beta,\gamma) = e^{-if_z\alpha}e^{-if_y\beta}e^{-if_z\gamma}$$
$$= \begin{bmatrix} e^{-i(\alpha+\gamma)}\cos^2\frac{\beta}{2} & -\frac{e^{-i\alpha}}{\sqrt{2}}\sin\beta & e^{-i(\alpha-\gamma)}\sin^2\frac{\beta}{2} \\ \frac{e^{-i\gamma}}{\sqrt{2}}\sin\beta & \cos\beta & -\frac{e^{i\gamma}}{\sqrt{2}}\sin\beta \\ e^{i(\alpha-\gamma)}\sin^2\frac{\beta}{2} & \frac{e^{i\alpha}}{\sqrt{2}}\sin\beta & e^{i(\alpha+\gamma)}\cos^2\frac{\beta}{2} \end{bmatrix}. \quad (6.97)$$

A general order parameter of a ferromagnetic BEC is therefore given by

$$\begin{pmatrix} \psi_1 \\ \psi_0 \\ \psi_{-1} \end{pmatrix} = \sqrt{n}\, e^{i\theta} U(\alpha,\beta,\gamma) \begin{pmatrix} 1 \\ 0 \\ 0 \end{pmatrix} = \sqrt{n}\, e^{i(\theta-\gamma)} \begin{bmatrix} e^{-i\alpha}\cos^2\frac{\beta}{2} \\ \frac{1}{\sqrt{2}}\sin\beta \\ e^{i\alpha}\sin^2\frac{\beta}{2} \end{bmatrix}, \quad (6.98)$$

where n is the total density of particles and θ, the overall phase. Generally, all variables $n, \theta, \alpha, \beta, \gamma$, depend on space and time, and accordingly, the order parameter varies over space and time. The spatial variation of the spinor order parameter at each instant of time is called spin texture, and it gives rise to several interesting phenomena.

First, consider the superflow $\mathbf{j}$ that is determined so as to satisfy the equation of continuity

$$\frac{\partial n}{\partial t} + \boldsymbol{\nabla}\cdot\mathbf{j} = 0. \quad (6.99)$$

Substituting $n = \sum_m |\psi_m|^2$ in Eq. (6.99) and making use of Eq. (6.55), we obtain

$$\mathbf{j} = \frac{\hbar}{2Mi}\sum_m [\psi_m^*\boldsymbol{\nabla}\psi_m - (\boldsymbol{\nabla}\psi_m^*)\psi_m]. \quad (6.100)$$

Substituting Eq. (6.98) into this yields

$$\mathbf{j}^{\mathrm{F}} = \frac{\hbar n}{M}[\boldsymbol{\nabla}(\theta-\gamma) - \cos\beta\,\boldsymbol{\nabla}\alpha] \equiv n\,\mathbf{v}_{\mathrm{s}}^{\mathrm{F}}. \quad (6.101)$$

A crucial observation here is that θ and γ appear in Eqs. (6.98) and (6.101) as a linear combination. This implies that if you rotate the spin by $\delta\gamma$, the effect can be undone by the gauge transformation $\theta \to \theta + \delta\gamma$. Due to this continuous spin-gauge symmetry, the spin texture gives rise to a supercurrent.

In the case of a scalar BEC, the superfluid velocity is given by $\mathbf{v}_s = (\hbar/M)\nabla\theta$, and therefore, it is irrotational, *i.e.*, rot $\mathbf{v}_s = 0$. However, this is not the case in the ferromagnetic phase. In fact, using the formula

$$\mathrm{rot}(f\mathbf{a}) = \nabla f \times \mathbf{a} + f\,\mathrm{rot}\,\mathbf{a}, \tag{6.102}$$

we find that the superfluid velocity is not, in general, irrotational:

$$\mathrm{rot}\,\mathbf{v}_s^F = \frac{\hbar}{M}\sin\beta\,\nabla\beta \times \nabla\alpha. \tag{6.103}$$

This implies that the superfluid is not a potential flow, and therefore, the circulation is not quantized. To see this, we rewrite Eq. (6.101) as

$$\mathbf{v}_s^F - \frac{\hbar}{M}(1-\cos\beta)\nabla\alpha = \frac{\hbar}{M}\nabla(\theta-\gamma-\alpha)$$

and perform a line integral of both sides along a closed contour C. Due to the single-valuedness of the order parameter, the integral of $\nabla(\theta-\gamma-\alpha)$ should give an integral multiple of 2π. Hence, we obtain

$$\oint_C \mathbf{v}_s^F d\mathbf{r} - \frac{\hbar}{M}\oint_C (1-\cos\beta)\nabla\alpha d\mathbf{r} = \frac{\hbar}{M} \times \text{ integer}. \tag{6.104}$$

The term $\oint_C (1-\cos\beta)\nabla\alpha d\mathbf{r}$ gives the Berry phase enclosed by the contour C. We thus conclude that in the ferromagnetic phase, the circulation alone is not quantized, but the difference between the circulation and the contribution due to the Berry phase is quantized.

Suppose that the system is confined in a cylinder of radius R and we describe the system in the cylindrical coordinates (r,ϕ,z). Substituting $\theta-\gamma=\alpha=\phi$ in Eq. (6.98), we obtain

$$\Psi = \sqrt{n(r)}\begin{bmatrix} \cos^2\frac{\beta(r)}{2} \\ \frac{e^{i\phi}}{\sqrt{2}}\sin\beta(r) \\ e^{2i\phi}\sin^2\frac{\beta(r)}{2} \end{bmatrix}, \tag{6.105}$$

and the corresponding spin vector as

$$\mathbf{f}(\mathbf{r}) = (\sin\beta(r)\cos\phi, \sin\beta(r)\sin\phi, \cos\beta(r)). \tag{6.106}$$

If $\beta(r)$ satisfies the boundary conditions $\beta(0) = 0$ and $\beta(R) = \pi/2$, the spin points in the positive z-direction at the symmetry axis and flares out toward the $x-y$ plane at $r = R$. Such a spin texture is called the Mermin-Ho vortex which is nonsingular along the symmetry axis and has a unit circulation at the periphery as can be seen from Eq. (6.105) with $\beta = \pi/2$.

On the other hand, if $\beta(r)$ satisfies the boundary conditions $\beta(0) = 0$ and $\beta(R) = \pi$, the spin texture flares out toward the negative z-direction at $r = R$. Such a spin texture is called the Anderson-Toulouse vortex which is again nonsingular along the symmetry axis and has the circulation of two at $r = R$ as can be seen from Eq. (6.105) with $\beta = \pi$. Such a nonsingular vortex is allowed because the order parameter manifold of a ferromagnetic BEC is $SO(3)$ as discussed in Sec. 12.3.2.2. We also note that the stabilities of Mermin-Ho and Anderson Toulouse vortices are ensured not topologically but by the boundary conditions.

Next, we consider the polar phase. In this case, a general order parameter is given by rotating $(0, 1, 0)^T$ to an arbitrary direction.

$$\begin{pmatrix} \psi_1 \\ \psi_0 \\ \psi_{-1} \end{pmatrix} = \sqrt{n}\, e^{i\theta} U(\alpha, \beta, \gamma) \begin{pmatrix} 0 \\ 1 \\ 0 \end{pmatrix} = \sqrt{n}\, e^{i\theta} \begin{pmatrix} -\frac{e^{-i\alpha}}{\sqrt{2}} \sin\beta \\ \cos\beta \\ \frac{e^{i\alpha}}{\sqrt{2}} \sin\beta \end{pmatrix}. \tag{6.107}$$

This order parameter does not depend on γ and there is no spin-gauge symmetry. Substituting Eq. (6.107) into Eq. (6.100) gives the superfluid velocity

$$\mathbf{v}_{\mathrm{s}}^{\mathrm{P}} = \frac{\hbar}{M} \nabla\theta \tag{6.108}$$

which is irrotational ($\mathrm{rot}\mathbf{v}_{\mathrm{s}}^{\mathrm{P}} = 0$), and the circulation is quantized:

$$\oint \mathbf{v}_{\mathrm{s}}^{\mathrm{P}} d\mathbf{r} = \frac{h}{M} \times \text{ integer}. \tag{6.109}$$

6.4 Spin-2 BEC

The spin-2 matrices are given by

$$f_x = \begin{bmatrix} 0 & 1 & 0 & 0 & 0 \\ 1 & 0 & \frac{\sqrt{6}}{2} & 0 & 0 \\ 0 & \frac{\sqrt{6}}{2} & 0 & \frac{\sqrt{6}}{2} & 0 \\ 0 & 0 & \frac{\sqrt{6}}{2} & 0 & 1 \\ 0 & 0 & 0 & 1 & 0 \end{bmatrix}, \quad f_y = i \begin{bmatrix} 0 & -1 & 0 & 0 & 0 \\ 1 & 0 & -\frac{\sqrt{6}}{2} & 0 & 0 \\ 0 & \frac{\sqrt{6}}{2} & 0 & -\frac{\sqrt{6}}{2} & 0 \\ 0 & 0 & \frac{\sqrt{6}}{2} & 0 & -1 \\ 0 & 0 & 0 & 1 & 0 \end{bmatrix},$$

$$f_z = \begin{bmatrix} 2 & 0 & 0 & 0 & 0 \\ 0 & 1 & 0 & 0 & 0 \\ 0 & 0 & 0 & 0 & 0 \\ 0 & 0 & 0 & -1 & 0 \\ 0 & 0 & 0 & 0 & -2 \end{bmatrix}, \tag{6.110}$$

and we define the raising (f_+) and lowering (f_-) operators:

$$f_+ \equiv f_x + i f_y = \begin{bmatrix} 0 & 2 & 0 & 0 & 0 \\ 0 & 0 & \sqrt{6} & 0 & 0 \\ 0 & 0 & 0 & \sqrt{6} & 0 \\ 0 & 0 & 0 & 0 & 2 \\ 0 & 0 & 0 & 0 & 0 \end{bmatrix},$$

$$f_- \equiv f_x - i f_y = \begin{bmatrix} 0 & 0 & 0 & 0 & 0 \\ 2 & 0 & 0 & 0 & 0 \\ 0 & \sqrt{6} & 0 & 0 & 0 \\ 0 & 0 & \sqrt{6} & 0 & 0 \\ 0 & 0 & 0 & 2 & 0 \end{bmatrix}. \tag{6.111}$$

In the second-quantized form, the spin operators are given by

$$\begin{aligned} \hat{f}_+ = \hat{f}_-^\dagger &= \sum_{m,n=-2}^{2} \hat{\psi}_m^\dagger (f_+)_{mn} \hat{\psi}_n \\ &= 2\,(\hat{\psi}_2^\dagger \hat{\psi}_1 + \hat{\psi}_{-1}^\dagger \hat{\psi}_{-2}) + \sqrt{6}\,(\hat{\psi}_1^\dagger \hat{\psi}_0 + \hat{\psi}_0^\dagger \hat{\psi}_{-1}), \end{aligned} \tag{6.112}$$

$$\begin{aligned} \hat{f}_z &= \sum_{m,n=-2}^{2} \hat{\psi}_m^\dagger (f_z)_{mn} \hat{\psi}_n \\ &= 2\,(\hat{\psi}_2^\dagger \hat{\psi}_2 - \hat{\psi}_{-2}^\dagger \hat{\psi}_{-2}) + \hat{\psi}_1^\dagger \hat{\psi}_1 - \hat{\psi}_{-1}^\dagger \hat{\psi}_{-1}, \end{aligned} \tag{6.113}$$

and the total spin operator can be obtained by substituting Eqs. (6.112) and (6.113) in Eq. (6.30) or (6.31). The interaction Hamiltonian (6.20) for the spin-2 BEC involves a term $\sum_{M=-2}^{2} \hat{A}_{2M}^\dagger \hat{A}_{2M}$ that can be eliminated using the operator identity:

$$\sum_{M=-2}^{2} \hat{A}_{2M}^\dagger \hat{A}_{2M} = \frac{1}{7}\,[4 : \hat{n}^2 : - : \hat{\mathbf{f}}^2 : -10\, \hat{A}_{00}^\dagger \hat{A}_{00}]. \tag{6.114}$$

This can be shown by expressing both sides in terms of field operators [see also Eq. (6.120)]. Substituting this in Eq. (6.20), we obtain

$$\hat{V} = \frac{1}{2} \int d\mathbf{r}\, [c_0 : \hat{n}^2 : + c_1 : \hat{\mathbf{f}}^2 : + c_2\, \hat{A}_{00}^\dagger \hat{A}_{00}] \tag{6.115}$$

where

$$c_0 \equiv \frac{4g_2 + 3g_4}{7}, \quad c_1 \equiv \frac{g_4 - g_2}{7}, \quad c_2 \equiv \frac{7g_0 - 10g_2 + 3g_4}{7}. \tag{6.116}$$

The relation (6.114) and the interaction Hamiltonian (6.115) can also be derived using the projection-operator method. For spin-2 identical bosons undergoing an s-wave interaction, the completeness relation is

$$P_0 + P_2 + P_4 = 1. \tag{6.117}$$

On the other hand, the composition law of spin operators gives

$$\mathbf{f}^2 = (\mathbf{f}_1 + \mathbf{f}_2)^2 = \mathbf{f}_1^2 + \mathbf{f}_2^2 + 2\mathbf{f}_1 \cdot \mathbf{f}_2 = \begin{cases} 0 \ \ (F = 0); \\ 6 \ \ (F = 2); \\ 20 \ (F = 4), \end{cases} \tag{6.118}$$

where F is the total spin. Since $\mathbf{f}_1^2 = \mathbf{f}_2^2 = 6$, we obtain

$$\mathbf{f}_1 \cdot \mathbf{f}_2 = -6P_0 - 3P_2 + 4P_4. \tag{6.119}$$

Solving Eqs. (6.117) and (6.119) for P_2 and P_4, we obtain

$$P_2 = \frac{1}{7}(4 - \mathbf{f}_1 \cdot \mathbf{f}_2 - 10P_0), \tag{6.120}$$

$$P_4 = \frac{1}{7}(3 + \mathbf{f}_1 \cdot \mathbf{f}_2 + 3P_0). \tag{6.121}$$

Thus,

$$\begin{aligned} V &= \frac{g_0}{2}P_0 + \frac{g_2}{2}P_2 + \frac{g_4}{2}P_4 \\ &= \frac{4g_2 + 3g_4}{14} + \frac{g_4 - g_2}{14}\mathbf{f}_1 \cdot \mathbf{f}_2 + \frac{7g_0 - 10g_2 + 3g_4}{14}P_0. \end{aligned} \tag{6.122}$$

In the second-quantized form, Eqs. (6.120) and (6.122) give Eq. (6.114) and (6.115), respectively.

Mean-field theory assumes that all particles share the same single-particle state. As in Eq. (6.45), we take as a mean-field state

$$|\xi\rangle = \frac{1}{\sqrt{N!}}\left(\sum_{m=-2}^{2} \xi_m \hat{a}_{m0}^\dagger\right)^N |\text{vac}\rangle. \tag{6.123}$$

The expectation value of the Hamiltonian with respect to $|\xi\rangle$ can be calculated in a manner similar to the spin-1 case:

$$\begin{aligned} E[\psi] = \int d\mathbf{r} \Bigg\{ &\sum_{m=-2}^{2} \psi_m^* \left[-\frac{\hbar^2\nabla^2}{2M} + U(\mathbf{r}) - pm + qm^2\right] \psi_m \\ &+ \frac{c_0}{2}n^2 + \frac{c_1}{2}\langle \mathbf{f}\rangle^2 + \frac{c_2}{2}|A|^2 \Bigg\}, \end{aligned} \tag{6.124}$$

where

$$A \equiv \frac{1}{\sqrt{5}}\left(2\psi_2\psi_{-2} - 2\psi_1\psi_{-1} + \psi_0^2\right) \tag{6.125}$$

is the amplitude of the spin-singlet pair that is a new term in the energy functional as to the spin-1 case [see Eq. (6.51)].

The dynamics of the system is governed by

$$i\hbar\,\frac{\partial\psi_m}{\partial t}=\frac{\delta E}{\delta\psi_m^*}. \tag{6.126}$$

Substituting Eq. (6.124) in Eq. (6.126), we obtain

$$i\hbar\,\frac{\partial\psi_{\pm2}}{\partial t}=\left(-\frac{\hbar^2\nabla^2}{2M}+U\mp2p+4q+c_0n\pm2c_1\langle f_z\rangle\right)\psi_{\pm2}$$
$$+c_1\langle f_\mp\rangle\psi_{\pm1}+\frac{c_2}{\sqrt{5}}\,A\psi_{\mp2}^*, \tag{6.127}$$

$$i\hbar\,\frac{\partial\psi_{\pm1}}{\partial t}=\left(-\frac{\hbar^2\nabla^2}{2M}+U\mp p+q+c_0n\pm c_1\langle f_z\rangle\right)\psi_{\pm1}$$
$$+c_1\left(\frac{\sqrt{6}}{2}\,\langle f_\mp\rangle\psi_0+\langle f_\pm\rangle\psi_{\pm2}\right)-\frac{c_2}{\sqrt{5}}\,A\psi_{\mp1}^*, \tag{6.128}$$

$$i\hbar\,\frac{\partial\psi_0}{\partial t}=\left(-\frac{\hbar^2\nabla^2}{2M}+U+c_0n\right)\psi_0+\frac{\sqrt{6}}{2}\,c_1\left(\langle f_+\rangle\psi_1+\langle f_-\rangle\psi_{-1}\right)$$
$$+\frac{c_2}{\sqrt{5}}\,A\psi_0^*, \tag{6.129}$$

where

$$\langle f_+\rangle=\langle f_-\rangle^*=2(\psi_2^*\psi_1+\psi_{-1}^*\psi_{-2})+\sqrt{6}\,(\psi_1^*\psi_0+\psi_0^*\psi_{-1}), \tag{6.130}$$

$$\langle f_z\rangle=2(|\psi_2|^2-|\psi_{-2}|^2)+|\psi_1|^2-|\psi_{-1}|^2. \tag{6.131}$$

The stationary solution can be obtained by solving Eqs. (6.127)–(6.129) with $\psi_m(\mathbf{r},t)=\psi_m(\mathbf{r})e^{-i\mu t/\hbar}$. We study the ground-state phase diagram of a uniform system (*i.e.*, $U=0$). Then, Eqs. (6.127)–(6.129) reduce to

$$(4q+2\tilde{\gamma}_0-\tilde{\mu})\,\psi_2+a\psi_{-2}^*=-\gamma_-\psi_1, \tag{6.132}$$

$$a^*\psi_2+(4q-2\tilde{\gamma}_0-\tilde{\mu})\,\psi_{-2}^*=-\gamma_-\psi_{-1}^*, \tag{6.133}$$

$$\gamma_+\psi_2+(q+\tilde{\gamma}_0-\tilde{\mu})\,\psi_1-a\psi_{-1}^*=-\frac{\sqrt{6}}{2}\,\gamma_-\psi_0, \tag{6.134}$$

$$-a^*\psi_1+(q-\tilde{\gamma}_0-\tilde{\mu})\,\psi_{-1}^*+\gamma_+\psi_{-2}^*=-\frac{\sqrt{6}}{2}\,\gamma_-\psi_0^*, \tag{6.135}$$

$$\tilde{\mu}\psi_0-a\psi_0^*=\frac{\sqrt{6}}{2}\,(r_+\psi_1+\gamma_-\psi_{-1}), \tag{6.136}$$

where $\tilde{\mu}\equiv\mu-c_0n$, $\tilde{\gamma}_0\equiv\gamma_0-p$, $\gamma_\pm\equiv c_1\langle f_\pm\rangle$, $\gamma_0\equiv c_1\langle f_z\rangle$, and $a\equiv c_2A/\sqrt{5}$. We may use the degree of gauge transformation (*i.e.*, the global phase) to make ψ_0 real. Furthermore, since the physical properties of the system are invariant under rotation about the quantization axis (*i.e.*, the z-axis), we may choose the coordinate system to make $\langle f_y \rangle = 0$, so that $\gamma_+ = \gamma_- \equiv \gamma$. A mean-field ground state can be obtained as a solution of Eqs. (6.132)–(6.136) under such simplifications.

Here, we consider the case of $\gamma = 0$, *i.e.*, the case of no transverse magnetization. Then, Eqs. (6.132)–(6.136) reduce to

$$(4q + 2\tilde{\gamma}_0 - \tilde{\mu})\,\psi_2 + a\psi_{-2}^* = 0, \tag{6.137}$$

$$a^*\psi_2 + (4q - 2\tilde{\gamma}_0 - \tilde{\mu})\,\psi_{-2}^* = 0, \tag{6.138}$$

$$(q + \tilde{\gamma}_0 - \tilde{\mu})\,\psi_1 - a\psi_{-1}^* = 0, \tag{6.139}$$

$$-a^*\psi_1 + (q - \tilde{\gamma}_0 - \tilde{\mu})\,\psi_{-1}^* = 0, \tag{6.140}$$

$$(\tilde{\mu} - a)\,\psi_0 = 0, \tag{6.141}$$

and the energy per particle is given by

$$\epsilon = \frac{1}{n}\sum_{m=-2}^{2}(-pm + qm^2)\,|\psi_m|^2 + \frac{c_0 n}{2} + \frac{c_1}{2n}\langle f_x\rangle^2 + \frac{c_2}{10n}\,|2\psi_2\psi_{-2} - 2\psi_1\psi_{-1} + \psi_0|^2. \tag{6.142}$$

If the determinant of the coefficient matrix of Eqs. (6.137) and (6.138)

$$D_2 \equiv (4q - \tilde{\mu})^2 - 4\tilde{\gamma}_0^2 - |a|^2 \tag{6.143}$$

is nonzero, we must have $\psi_2 = \psi_{-2} = 0$. In addition, if the determinant of the coefficient matrix of Eqs. (6.139) and (6.140)

$$D_1 \equiv (q - \tilde{\mu})^2 - \tilde{\gamma}_0^2 - |a|^2 \tag{6.144}$$

is nonzero, we obtain $\psi_1 = \psi_{-1} = 0$, and therefore, the order parameter of the ground state is given by $(0, 0, e^{i\chi_0}\sqrt{n}, 0, 0)$. The chemical potential is determined from (6.136) to be $\mu = c_0 n + c_2 n/5$ and $\langle f_z \rangle = 0$. The energy per particle is found from Eq. (6.124) to be $\epsilon_N = (c_0 + c_2/5)n/2$. This phase is referred to as the uniaxial nematic phase[1].

If $D_1 = 0$ and $D_2 \neq 0$, $\psi_2 = \psi_{-2} = 0$ and we may use Eqs. (6.134) and (6.135) to find $\langle f_z \rangle = p/(c_1 - c_2/5)$ and $\tilde{\mu} = q + c_2 n/5$. The solution of $D_2 = 0$ that is consistent with this $\tilde{\mu}$ is $\tilde{\mu} = q \pm \sqrt{\tilde{\gamma}_0^2 + |a|^2}$, where the plus (minus) sign corresponds to $c_2 > 0$ ($c_2 < 0$). It follows from Eq. (6.136) that $\psi_0 = 0$. The remaining components $\psi_{\pm 1}$ are determined so as to satisfy

[1] Note that for the spin-1 case, the corresponding state $(0, \sqrt{n}, 0)$ is also called polar because the order parameter $Y_1^0(\theta, \phi) \propto \cos\theta$ changes the sign under $\theta \to \pi - \theta$, that is, it has polarity. This is not the case for the spin-2 case because $Y_2^0(\theta, \phi) \propto 3\cos^2\theta - 1$.

$|\psi_1|^2 + |\psi_{-1}|^2 = n$ and $|\psi_1|^2 - |\psi_{-1}|^2 = \langle f_z \rangle$. The order parameter is thus given by $(0, e^{i\chi_1}\sqrt{(n+\langle f_z\rangle)/2}, 0, e^{i\chi_{-1}}\sqrt{(n-\langle f_z\rangle)/2}, 0)$.

If $D_1 \neq 0$ and $D_2 = 0$, $\psi_1 = \psi_{-1} = 0$. In addition, if $\tilde{\mu} \neq a$, we have $\psi_0 = 0$. The remaing components $\psi_{\pm 2}$ are determined to be $(e^{i\chi_2}\sqrt{(n+\langle f_z\rangle/2)/2}, 0, 0, 0, e^{i\chi_{-2}}\sqrt{(n+\langle f_z\rangle/2)/2})$, where $\tilde{\mu}$ and $\langle f_z \rangle$ are determined from Eqs. (6.132) and (6.133) to be $\tilde{\mu} = 4q + c_2 n/5$ and $\langle f_z \rangle = p/(c_1 - \frac{c_2}{20})$, respectively. This phase is referred to as the biaxial nematic phase or antiferromagnetic phase.

If $D_1 \neq 0$ and $D_2 = 0$ and $\tilde{\mu} = a$, we have $\psi_1 = \psi_{-1} = 0$, but $\psi_{\pm 2}$ and ψ_0 can be nonzero. They obey $|\psi_2|^2 + |\psi_{-2}|^2 + |\psi_0|^2 = n$ and $|\psi_2|^2 - |\psi_{-2}|^2 = \langle f_z \rangle/2$, and therefore, $|\psi_2|^2 = (n + \langle f_z\rangle/2 - \psi_0^2)/2$ and $|\psi_{-2}|^2 = (n - \langle f_z\rangle/2 - \psi_0^2)/2$, where μ and $\langle f_z \rangle$ are determined from Eqs. (6.132) and (6.133) by $\mu = 2q - \alpha^2/2q$ and $\langle f_z \rangle = (p+\alpha)/c_1$, where α is the real solution of the following equation:

$$\alpha^3 + p\alpha^2 + 4q\left[q + 2c_1(n - \psi_0^2)\right]\alpha + 4pq^2 = 0. \tag{6.145}$$

This phase is referred to as the cyclic phase.

Finally, let us consider the case of $D_1 = D_2 = 0$. By solving the simultaneous equations $D_1 = 0$ and $D_2 = 0$, we obtain

$$\tilde{\mu} = \frac{5q^2 - \tilde{\gamma}_0^2}{2q} = \pm\sqrt{4q^2 + |a|^2}. \tag{6.146}$$

When $q \neq 0$, $\tilde{\mu} \neq a$, we find from Eq. (6.141) that $\psi_0 = 0$. Then, $\gamma = 2\left(\psi_2^*\psi_1 + \psi_{-1}^*\psi_{-2}\right)$. We may use Eqs. (6.137)–(6.140) and (6.146) to show that

$$\gamma = 2c_1\psi_2^*\psi_1\left(1 - \frac{\tilde{\gamma}_0 + 3q}{\tilde{\gamma}_0 - 3q}\right) = 2c_1\psi_{-1}^*\psi_{-2}\left(1 - \frac{\tilde{\gamma}_0 - 3q}{\tilde{\gamma}_0 + 3q}\right) = 0. \tag{6.147}$$

Therefore, we must generally have $\psi_2\psi_1 = \psi_{-1}\psi_{-2} = 0$. To be consistent with Eqs. (6.137)–(6.140), we find that either $\psi_1 = \psi_{-2} = 0$ or $\psi_2 = \psi_{-1} = 0$ should hold. In the former case, the order parameter is given by $\left(e^{i\chi_2}\sqrt{(n+\langle f_z\rangle)/3}, 0, 0, e^{i\chi_{-1}}\sqrt{(2n-\langle f_z\rangle)/3}, 0\right)$ with $\langle f_z \rangle = (p-q)/c_1$. In the latter case, the solution is given by $\left(0, e^{i\chi_1}\sqrt{(2n+\langle f_z\rangle)/3}, 0, 0, e^{i\chi_{-2}}\sqrt{(n-\langle f_z\rangle)/3}\right)$ with $\langle f_z \rangle = (p+q)/c_1$. We refer to these two phases as the mixed phase.

The above results are summarized in Table 6.2. By comparing the energy of the ground-state, we can find the ground state phase diagram of a spin-2 BEC that is shown in Fig. 6.3 for a special case of $p = 0$ and $q < 0$ [Saito and Ueda (2005)].

Table 6.2 Possible phases for the ground state of a spin-2 BEC. F, BN, UN, C, and M indicate the ferromagnetic, biaxial nematic, uniaxial nematic, cyclic, and mixed phases, respectively. The biaxial nematic phase is also referred to as the antiferromagnetic phase. $\langle f_z\rangle/n, \mu$, and ε denote the polarization, chemical potential, and energy per particle. In the cyclic phase, α is a real solution of Eq. (6.145).

phase	order parameter	$\langle f_z\rangle$	μ	ε
F	$(e^{i\chi_2}\sqrt{n},0,0,0,0)$	$2n$	$4q+(c_0+\frac{c_2}{5})n$	$\frac{1}{2}(c_0+4c_1)n-2p+4q$
BN	$\left(e^{i\chi_2}\sqrt{\frac{1}{2}(n+\frac{\langle f_z\rangle}{2})},0,0,0,e^{i\chi_{-2}}\sqrt{\frac{1}{2}(n-\frac{\langle f_z\rangle}{2})}\right)$	$\frac{p}{c_1-\frac{c_2}{20}}$	$4q+(c_0+\frac{c_2}{5})n$	$4q+\frac{1}{2}(c_0+\frac{c_2}{5})n-\frac{3p^2}{2(c_1-\frac{c_2}{20})n}$
	$\left(0,e^{i\chi_1}\sqrt{\frac{1}{2}(n+\langle f_z\rangle)},0,e^{i\chi_{-1}}\sqrt{\frac{1}{2}(n-\langle f_z\rangle)},0\right)$	$\frac{p}{c_1-\frac{c_2}{5}}$	$q+(c_0+\frac{c_2}{5})n$	$q+\frac{1}{2}(c_0+\frac{c_2}{5})n-\frac{p^2}{2(c_1-\frac{c_2}{5})n}$
UN	$(0,0,e^{i\chi_0}\sqrt{n},0,0)$	0	$(c_0+\frac{c_2}{5})n$	$\frac{1}{2}(c_0+\frac{c_2}{5})n$
M	$\left(e^{i\chi_2}\sqrt{\frac{1}{3}(n+\langle f_z\rangle)},0,0,e^{i\chi_{-1}}\sqrt{\frac{1}{3}(2n-\langle f_z\rangle)},0\right)$	$\frac{p-q}{c_1}$	$2q$	$2q+\frac{c_0 n}{2}-\frac{(p-q)^2}{2c_1 n}$
	$\left(0,e^{i\chi_1}\sqrt{\frac{1}{3}(2n+\langle f_z\rangle)},0,0,e^{i\chi_{-2}}\sqrt{\frac{1}{3}(n-\langle f_z\rangle)}\right)$	$\frac{p+q}{c_1}$	$2q$	$2q+\frac{c_0 n}{2}-\frac{(p+q)^2}{2c_1 n}$
C	$\left(e^{i\chi_2}\sqrt{\frac{1}{2}(n+\frac{\langle f_z\rangle}{2}-\lvert\psi_0\rvert^2)},0,\psi_0,0,e^{i\chi_{-2}}\sqrt{\frac{1}{2}(n-\frac{\langle f_z\rangle}{2}-\lvert\psi_0\rvert^2)}\right)$	$\frac{p+\alpha}{c_1}$	$2q-\frac{\alpha^2}{2q}$	$4q+\frac{c_0 n}{2}-\frac{4q}{n}\lvert\psi_0\rvert^2+\frac{c_2}{10n}\lvert n$ $+\lvert\psi_0\rvert^2(e^{i(2\chi_0-\chi_2-\chi_{-2})}-1)\rvert^2$

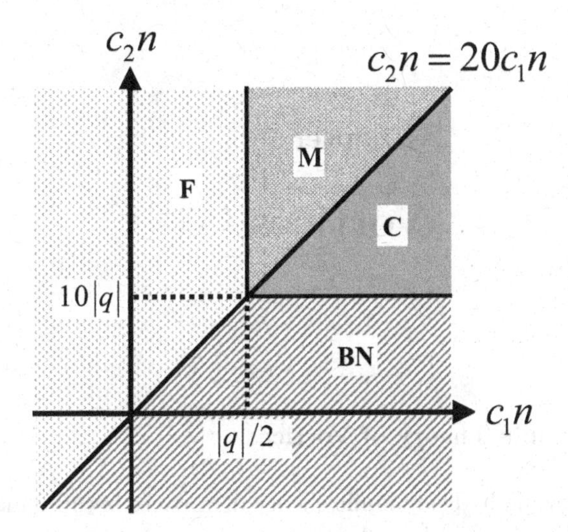

Fig. 6.3 Ground-state phase diagram of a spin-2 Bose–Einstein condensate for $p = 0$ and $q < 0$, where F, BN, C, and M indicate the ferromagnetic, biaxial nematic, cyclic, and mixed phases, respectively.

Chapter 7

Vortices

7.1 Hydrodynamic Theory of Vortices

We begin by reviewing hydrodynamic theory of vortices because many important features of vortices have hydrodynamic origins. Let $\rho(\mathbf{r},t)$ and $\mathbf{v}(\mathbf{r},t)$ be the mass density and the velocity field of a fluid at position $\mathbf{r}$ and time t. They obey the equation of continuity resulting from the conservation of mass:

$$\frac{\partial \rho}{\partial t} + \nabla(\rho \mathbf{v}) = 0. \tag{7.1}$$

An ideal fluid is the one with no viscosity. Furthermore, if there is no thermal conduction, the fluid is adiabatic and the velocity field changes according to the Euler equation

$$\frac{d\mathbf{v}}{dt} = \frac{\partial \mathbf{v}}{\partial t} + (\mathbf{v}\cdot\nabla)\mathbf{v} = -\frac{1}{\rho}\nabla p - \frac{1}{M}\nabla U, \tag{7.2}$$

where p, U, and M are the pressure, an external potential, and the mass of the constituent particle of the fluid. Since the entropy is conserved, $(1/\rho)\nabla p = dw$, where w is the enthalpy per unit mass, and the Euler equation can be written as

$$\frac{d\mathbf{v}}{dt} = \frac{\partial \mathbf{v}}{\partial t} + (\mathbf{v}\cdot\nabla)\mathbf{v} = -\nabla\left(w + \frac{U}{M}\right). \tag{7.3}$$

Using the identity

$$(\mathbf{v}\cdot\nabla)\mathbf{v} = \nabla\left(\frac{1}{2}\,\mathbf{v}^2\right) - \mathbf{v}\times \mathrm{rot}\mathbf{v},$$

we obtain the fundamental equation for the ideal isentropic fluid:

$$\frac{\partial \mathbf{v}}{\partial t} - \mathbf{v}\times\mathrm{rot}\mathbf{v} = -\nabla\left(\frac{1}{2}\,\mathbf{v}^2 + w + \frac{U}{M}\right). \tag{7.4}$$

When the flow is stationary (*i.e.*, $\partial \mathbf{v}/\partial t = 0$), Eq. (7.4) gives

$$\mathbf{v} \cdot \nabla \left(\frac{1}{2}\, \mathbf{v}^2 + w + \frac{U}{M} \right) = 0. \tag{7.5}$$

This result implies that the quantity $\mathbf{v}^2/2 + w + U/M$ does not change if we move along a streamline — a curve whose tangential line parallels the direction of $\mathbf{v}$ at the same point. We thus obtain Bernoulli's equation:

$$\frac{1}{2}\, \mathbf{v}^2 + w + \frac{U}{M} = \text{ constant along a streamline.} \tag{7.6}$$

When the mass density is constant, we may put $\omega = p/\rho$ in Eq. (7.6), obtaining

$$\frac{1}{2}\mathbf{v}^2 + \frac{p}{\rho} + \frac{U}{M} = \text{ constant along a streamline.} \tag{7.7}$$

The circulation of a fluid along a closed contour C is defined as the line integral of the velocity along C:

$$\Gamma \equiv \oint_C \mathbf{v} d\mathbf{r}. \tag{7.8}$$

By Stoke's theorem, the integral is converted into the surface integral over the area enclosed by C

$$\Gamma = \int_C \boldsymbol{\omega} d\mathbf{S}, \tag{7.9}$$

where

$$\boldsymbol{\omega} = \text{rot}\mathbf{v} \tag{7.10}$$

is called vorticity. Equation (7.9) shows that the circulation gives the total vorticity of the enclosed area. Because

$$\text{div}\boldsymbol{\omega} = 0, \tag{7.11}$$

the vortex cannot appear or disappear in the system; it can only close upon itself or terminate at the boundary. By the same token, a vortex ring can be created or annihilated in the system, but a line vortex cannot; it must enter or leave the system through its periphery. It follows from Eqs. (7.3) and (7.8) that

$$\begin{aligned} \frac{d\Gamma}{dt} &= \oint \frac{d\mathbf{v}}{dt}\, d\mathbf{r} + \oint \mathbf{v} d\mathbf{v} \\ &= -\oint \nabla \left(w + \frac{U}{M} \right) d\mathbf{r} + \frac{1}{2} \oint d(\mathbf{v}^2) = 0. \end{aligned} \tag{7.12}$$

This result implies that as a contour moves with the liquid, the circulation along the contour is conserved for isentropic flow (Kelvin's theorem). If the velocity field can be expressed as the gradient of a scalar potential φ

$$\mathbf{v} = \nabla\varphi, \tag{7.13}$$

the fluid is said to be irrotational because the vorticity defined in Eq. (7.10) vanishes throughout the system.

The conservation of circulation, together with the fact that the vortex cannot appear or disappear in the ideal fluids, lends a particular importance on the notion of vortices. Suppose that the system is contained in a cylinder of radius R and length ℓ and that the cylinder rotates at angular frequency Ω about the symmetry axis. When a line vortex is situated on the symmetry axis, the velocity of the fluid at a distance r from the center is given from Eq. (7.8) as

$$\mathbf{v} = \frac{\Gamma}{2\pi r}\mathbf{e}_\phi, \tag{7.14}$$

where $\mathbf{e}_\phi$ is the unit vector in the azimuthal direction. The vorticity of the line vortex is obtained from Eq. (7.9) as

$$\boldsymbol{\omega} = \Gamma\delta(x)\delta(y)\mathbf{e}_z, \tag{7.15}$$

where $\mathbf{e}_z$ is the unit vector in the z-direction. The kinetic energy of the fluid is thus given by

$$E = \int \frac{\rho}{2}v^2 d\mathbf{r} \simeq \frac{\rho\Gamma^2\ell}{4\pi}\ln\frac{R}{b}, \tag{7.16}$$

where b is the size of the vortex core. The creation of a vortex thus costs an energy. However, there is no minimum energy required to create a classical vortex because Γ can take on any small value. In contrast, there is a minimum energy cost to create a quantum vortex because the circulation is quantized as discussed in Sec. 7.2.

Some comments on the special role of a container is in order here. The surface of a solid-state container has atomic-scale roughness which allows the system to exchange energy and momentum with the container wall so that the system can reach thermodynamic equilibrium. Such an equilibration process occurs in a thin layer called boundary layer [Landau and Lifshitz (1987)]. The dynamics of the boundary layer is governed by the Navier–Stokes equation which is the Euler equation accompanied by the viscous term $(\eta/\rho)\nabla^2\mathbf{v}$, where η is the viscosity of the fluid. This boundary layer is responsible for thermal relaxation of the system, and therefore the thermal equilibrium of the system should be defined in the frame of reference of the container. In the case of gaseous BECs, however, the container is made up of an electromagnetic potential with no atomic-scale roughness, and does not act as a heat bath. To drive the system into rotation, we therefore need to break the axisymmetry of the potential by means of *e.g.*, dynamical phase imprinting using a binary BEC system [Matthews, *et al.* (1999)], laser-beam stirring [Madison, *et al.* (2000); Raman, *et al.* (2001)], or magnetic deformation of the potential [Hodby, *et al.* (2002)]. One can also create a vortex without breaking the axisymmetry of the potential by topological phase imprinting [Nakahara, *et al.* (2000); Leanhardt, *et al.* (2002)]. The observed vortex patterns [see also Fig. 7.3] are similar to those observed in rotating-bucket experiments of superfluid ^{4}He [Yarmchuk, *et al.* (1979)].

7.2 Quantized Vortices

As described in Sec. 1.3, a BEC is described by an order parameter or a macroscopic wave function ψ that emerges as a new thermodynamic variable when the system undergoes BEC. Let us decompose ψ into the amplitude A and the phase ϕ:

$$\psi(\mathbf{r}, t) = A(\mathbf{r}, t)\, e^{i\phi(\mathbf{r},t)}. \tag{7.17}$$

The mass density ρ and the current density $\mathbf{j}$ are given by

$$\rho(\mathbf{r}, t) = M\, |\psi(\mathbf{r}, t)|^2 = mA^2(\mathbf{r}, t), \tag{7.18}$$

$$\mathbf{j}(\mathbf{r}, t) = \frac{\hbar}{2i}\,(\psi^*\nabla\psi - \psi\nabla\psi^*) = \rho(\mathbf{r}, t)\,\frac{\hbar}{M}\nabla\phi(\mathbf{r}, t). \tag{7.19}$$

The superfluid velocity is defined as the ratio of the current density to the mass density:

$$\mathbf{v}_{\rm s}(\mathbf{r}, t) = \frac{\hbar}{M}\nabla\phi(\mathbf{r}, t). \tag{7.20}$$

Substituting this into Eq. (7.8) and assuming the single-valuedness of the condensate wave function, we obtain the quantization of circulation

$$\Gamma = \frac{\hbar}{M}\oint \nabla\phi d\mathbf{r} = \kappa n \quad (n = 0, \pm 1, \pm 2, \cdots), \tag{7.21}$$

where

$$\kappa \equiv \frac{h}{M} \tag{7.22}$$

is the quantum of circulation.

The vortex of a scalar BEC is characterized by the quantum number n in Eq. (7.21) which is referred to as the winding number or change of the vortex. A vortex with $|n| = 1$ (or 2) is called a singly quantized (or doubly quantized) vortex. Because the minimum value of $|n|$ for the vortex is one, the velocity field in Eq. (7.14) diverges at the origin. To prevent the kinetic energy from diverging, the amplitude of the wave function must vanish at the center of the vortex core where the phase of the wave function is arbitrary. In other words, the quantized vortex features the phase singularity that is characterized by the topological quantum number n. In contrast, classical vortices have no singularity because Γ can change continuously.

It is worthwhile to note that if $n \neq 0$, the superfluid velocity (7.14) exceeds the sound velocity (2.52) for $r < \xi$, where

$$\xi = \frac{1}{\sqrt{8\pi na}} \tag{7.23}$$

is the healing length. This affords an illustrative example of the fact that the superfluid is not destroyed if the velocity of the fluid exceeds the Landau critical velocity which in the present case is the sound velocity. Instead, the condensate wave function is suppressed near the vortex core according to [Ginzburg and Pitaevskii (1958)]

$$\psi \propto e^{in\phi} e^{ik_z z} \left(\frac{\rho}{\xi}\right)^n, \tag{7.24}$$

where $\rho \equiv \sqrt{x^2 + y^2}$. To show this, let us consider the GP equation (3.63) with $V(\mathbf{r}) = 0$. Expressing the Laplacian in cylindrical coordinates, we obtain

$$-\frac{\hbar^2}{2M}\left(\frac{\partial^2}{\partial\rho^2} + \frac{1}{\rho}\frac{\partial}{\partial\rho} + \frac{1}{\rho^2}\frac{\partial^2}{\partial\phi^2} + \frac{\partial^2}{\partial z^2}\right)\psi(\mathbf{r}) + U_0|\psi(\mathbf{r})|^2\psi(\mathbf{r}) = \mu\psi(\mathbf{r}). \tag{7.25}$$

A substitution of

$$\psi(\mathbf{r}) = e^{in\phi} e^{ik_z z} f(\rho) \tag{7.26}$$

in Eq. (7.25) gives

$$-\frac{\hbar^2}{2M}\left(\frac{\partial^2}{\partial\rho^2} + \frac{1}{\rho}\frac{\partial}{\partial\rho}\right) f(\rho) + \frac{\hbar^2 n^2}{2M\rho^2} f(\rho) + U_0 f^3(\rho) = (\mu - \varepsilon_{k_z}) f(\rho), \tag{7.27}$$

where $\varepsilon_{k_z} \equiv \hbar^2 k_z^2/(2M)$. The third term on the left-hand side arises from the centrifugal potential due to the vortex. We substitute $\rho = \xi\rho'$ in (7.27), obtaining

$$\left(-\frac{\partial^2}{\partial\rho'^2} - \frac{1}{\rho'}\frac{\partial}{\partial\rho'} + \frac{n^2}{\rho'^2}\right) f + f^3 = k_\rho^2 f, \tag{7.28}$$

where

$$k_\rho^2 \equiv \frac{2M\xi^2}{\hbar^2}(\mu - \varepsilon_{k_z}). \tag{7.29}$$

In a region far from the center, the centrifugal potential can be ignored and we have a stationary solution $f = k_\rho$ or

$$\psi(\mathbf{r}) = k_\rho e^{in\phi} e^{ik_z z}. \tag{7.30}$$

Near the center of the vortex, the terms f^3 and $k_\rho^2 f$ are negligible compared with the kinetic and centrifugal terms, so that the solution is given by $f \propto \rho^n$. The condensate wave function near the vortex core is therefore suppressed according to Eq. (7.24). This phenomenon is reminiscent of cavitation in a supersonic flow.

The energy of a singly quantized vortex per unit length is obtained by substitution of $\ell = 1$, $\Gamma = \kappa$ and $\rho = \rho_s$ in Eq. (7.16):

$$\epsilon = \frac{\rho_s \kappa^2}{4\pi} \ln \frac{R}{b}, \tag{7.31}$$

where ρ_s is the superfluid density and a numerical analysis gives $b \simeq 0.68\xi$ [Ginzburg and Pitaevskii (1958)]. The angular momentum per unit length is given by

$$L \simeq 2\pi \int_0^R \rho_s v_s r^2 dr = \frac{1}{2}\rho_s \kappa R^2 \tag{7.32}$$

The energy of a vortex with circulation $\Gamma = n\kappa$ is given by Eq. (7.31) with κ replaced by $n\kappa$. Because of the n^2 dependence, the quantized vortex with winding number n has n times as much energy as n singly quantized vortices. The vortex with $n > 1$ is therefore unstable against disintegration into singly quantized vortices.

When a singly quantized vortex resides at the center of the container, the free energy of the system at zero temperature is given by $F = \epsilon - L\Omega$. The critical frequency Ω_c for the thermodynamic vortex nucleation is obtained from the condition $F = 0$ as

$$\Omega_c = \frac{\kappa}{2\pi R^2} \ln \frac{R}{b}. \tag{7.33}$$

We thus find that the superfluid system does not rotate if the rotation frequency of the container is below Ω_c. This phenomenon is known as the Hess–Fairbank effect [Hess and Fairbank (1967)], and analogous to the Meissner effect in superconductors, where a magnetic field cannot penetrate into a superconductor if it is below a critical value. In the case of a gaseous Bose–Einstein condensate, the process of vortex nucleation is not explained by the thermodynamic argument alone because, as mentioned above, the trapping potential does not play a role of the heat bath. In this case, dynamical instabilities as well as the Landau (thermodynamic) instability trigger the nucleation process of vortices, and determine the critical frequency for vortex nucleation [Recati, *et al.* (2001); Sinha and Castin (2001); Kasamatsu and Tsubota (2003)].

A single vortex filament is known to exhibits an oscillatory helix excitation known as the Kelvin mode [Donnelly (1991)]. As shown in Eq. (7.31), the vortex filament has an energy per unit length which suggests the presence of tension in the filament and hence that of a restoring force against deviations from a straight line. Since a vortex has a circulation around the axis, the oscillatory mode of a vortex forms a helix with a quantized angular momentum opposite to the circulation around the vortex. Due to the translation symmetry in the z-direction, the Kelvin mode comprises a plane wave $\exp[i(kz-\omega t-\phi)]$ along the z-direction and rotates on the $x-y$ plane, where $-\phi$ describes the Kelvin mode with $m=-1$. The dispersion relation of the Kelvin mode is given by [Lifshitz and Pitaevskii (1991)]

$$\omega = \frac{\hbar k^2}{2M} \ln\left(\frac{1}{k\xi}\right). \tag{7.34}$$

The Kelvin mode in a harmonic potential is discussed by Svidzinsky and Fetter (2000).

When the rotation frequency exceeds the critical one, vortices enter the system. For uniform rotation $\boldsymbol{\Omega}=(0,0,\Omega)$, the velocity field is given by $\mathbf{v}_s = \boldsymbol{\Omega}\times\mathbf{r}$, so that the vorticity (7.10) is given by

$$\boldsymbol{\omega} = \mathrm{rot}\mathbf{v}_s = (0,0,2\Omega). \tag{7.35}$$

Since the vorticity gives the circulation per unit area and each singly quantized vortex carries a unit quantum of circulation κ, the vortex density is given by

$$n_v = \frac{2\Omega}{\kappa}. \tag{7.36}$$

This relation is referred to as Feynman's rule. The inverse of n_v gives the area per vortex. Equating n_v^{-1} to $\pi\ell_v^2$, we obtain a characteristic intervortex distance:

$$\ell_v = \sqrt{\frac{\kappa}{2\pi\Omega}} = \sqrt{\frac{\hbar}{M\Omega}}. \tag{7.37}$$

This length scale corresponds to the magnetic length

$$\ell_m = \sqrt{\frac{\hbar}{eB}} \tag{7.38}$$

which is the fundamental length scale of the quantizeed cyclotron motion of an electron in a magnetic field B.

The equation of motion for the superfluid velocity can be derived as follows. Using $\hat{\phi} = i\partial/\partial N$, the Heisenberg equation of motion for $\hat{\phi}$ reads

$$\hbar\,\frac{\partial\hat{\phi}}{\partial t} = \frac{\partial\hat{H}}{\partial N}. \tag{7.39}$$

Taking the expectation value of Eq. (7.39), we obtain

$$\hbar\,\frac{d\phi}{dt} = \frac{\partial E}{\partial N} = \mu. \tag{7.40}$$

This relation, combined with Eq. (7.20), leads to

$$\frac{d\mathbf{v}_\mathrm{s}}{dt} = \frac{1}{M}\nabla\mu. \tag{7.41}$$

The superfluid can thus be accelerated by the gradient of the chemical potential. At $T = 0$, we have $\nabla\mu = (1/n)\nabla p$, so that the superfluid is accelerated by a pressure gradient.

Consider now a closed contour C in the superfluid and take two arbitrary points 1 and 2 on C. Equation (7.21) implies that wherever a vortex with winding number n enters or leaves C, the line integral of $\nabla\phi$ along one segment of C between 1 and 2 differs from that along the other by $2\pi n$. On the other hand, the chemical potential difference between 1 and 2 is given from Eq. (7.40) by

$$\mu_1 - \mu_2 = \hbar\,\frac{d(\phi_1 - \phi_2)}{dt}. \tag{7.42}$$

Thus, a rapid change in phase difference, which is called phase slippage, caused by the drift of a vortex generates a chemical potential difference in superfluid. This makes a sharp contrast with an ideal classical fluid in which vortices cannot move across a streamline because the chemical potential is constant along the streamline [see Eq. (7.6)]. The distinction originates in the discrete nature of quantum vortices.

Let dn/dt be the net number of vortices per unit time that cross one segment of C between 1 and 2. It follows from Eq. (7.42) and $d(\phi_1 - \phi_2) = 2\pi dn$ that

$$\mu_1 - \mu_2 = h\,\frac{dn}{dt}. \tag{7.43}$$

Let the distance between 1 and 2 be infinitesimally small, say dx, and let dy be an infinitesimal segment perpendicular to dx. Dividing both sides of Eq. (7.43) by dx, we obtain

$$\frac{d\mu}{dx} = h\,\frac{dn}{dxdy}\,\frac{dy}{dt}. \tag{7.44}$$

The left-hand side of Eq. (7.44) is equal to $Md(\mathbf{v}_s)_x/dt$ [see Eq. (7.41)], while the right-hand side is equal to $h(M\omega/h)v_{\text{drift}}$, where $dn/(dxdy) = n_v = \omega/\kappa$ is the vortex density and $dy/dt = v_{\text{drift}}$ is the drift velocity of vortices. We thus obtain

$$\frac{d\mathbf{v}_s}{dt} = \mathbf{v}_{\text{drift}} \times \boldsymbol{\omega} + \frac{1}{M}\nabla\mu, \tag{7.45}$$

Where the last term is added to account for a pressure gradient [see Eq. (7.41)]. Equation (7.45) shows that when vortices drift, the superfluid is accelerated in the direction perpendicular to both $\boldsymbol{\omega}$ and the drift velocity. The corresponding force is called Magnus force whose physical origin can be explained as follows. Suppose that a vortex with vorticity $\boldsymbol{\omega}$ drifts to the left as shown in Fig. 7.1. Then the fluid flows faster at point A in Fig. 7.1 than at point B. According to Bernoulli's equation (7.7), the pressure is lower at A than at B, so that the vortex undergoes a force from the center toward A:

$$\mathbf{f}_{\text{Magnus}} = M\mathbf{v}_{\text{drift}} \times \boldsymbol{\omega}. \tag{7.46}$$

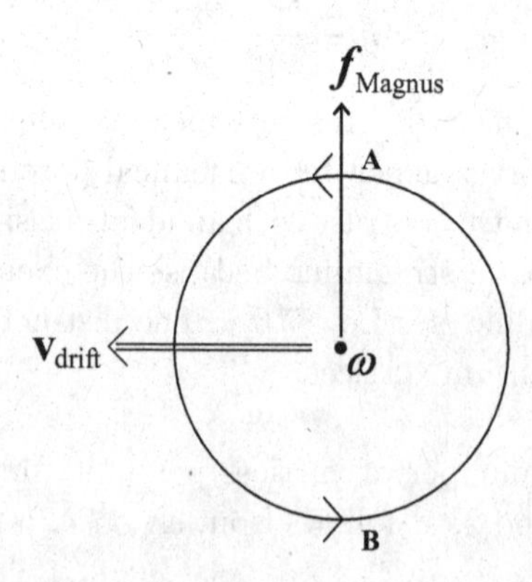

Fig. 7.1 Magnus force. When a vortex with vorticity $\boldsymbol{\omega}$ moves at velocity $\mathbf{v}_{\text{drift}}$, the Magnus force (7.46) exerts on the vortex.

7.3 Interaction Between Vortices

Let us consider two vortices with vorticities ω_1 and ω_2. When ω_1 and ω_2 are parallel as shown in Fig. 7.2 (a), the fluid flows faster at point A than at point B due to the flow caused by ω_2. It follows from Bernoulli's equation (7.7) that the two vorticies repel each other. In a similar manner, when ω_1 and ω_2 are antiparallel as in Fig. 7.2 (b), the two vortices attract each other.

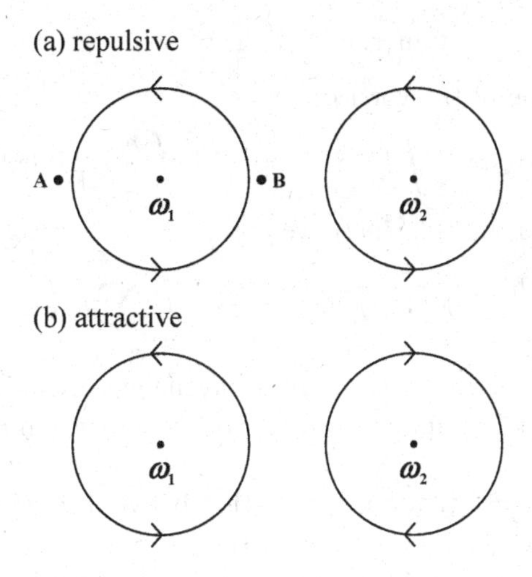

Fig. 7.2 Interaction between two vortices. It is repulsive or attractive, depending on whether their vorticities ω_1 and ω_2 are parallel (a) or antiparallel (b).

The interaction between vortices is, in general, more complex since they can bend flexibly. In two dimension, however, vortices become point objects and the interaction takes a simple form. Applying the Stokes theorem to Eq. (7.21), we find that

$$\mathrm{rot}\mathbf{v}_\mathrm{s}(\mathbf{r}) = \kappa \sum_i n_i \delta^{(2)}(\mathbf{r} - \mathbf{r}_i)\hat{\mathbf{e}}_z, \tag{7.47}$$

where $\delta^{(2)}(\mathbf{r})$ denotes the two-dimensional delta function, the z axis is taken to be perpendicular to the plane on which vortices reside, and n_i is the charge of a vortex at location $\mathbf{r}_i$. Let us introduce a stream function ψ such that

$$v_{\mathrm{s}x} = \frac{\partial \psi}{\partial y}, \quad v_{\mathrm{s}y} = -\frac{\partial \psi}{\partial x}. \tag{7.48}$$

Then Eq. (7.47) is reproduced if we choose

$$\psi(\mathbf{v}) = -\frac{\kappa}{2\pi}\sum_i n_i \ln\frac{|\mathbf{r}-\mathbf{r}_i|}{\xi}. \tag{7.49}$$

In fact,

$$(\mathrm{rot}\mathbf{v}_\mathrm{s})_z = \frac{\partial v_{\mathrm{s}y}}{\partial x} - \frac{\partial v_{\mathrm{s}x}}{\partial y} = -\Delta\psi = \kappa\sum_i n_i\delta^{(2)}(\mathbf{r}-\mathbf{r}_i), \tag{7.50}$$

where we used the formula

$$\Delta \ln|\mathbf{r}| = 2\pi\delta^{(2)}(\mathbf{r}). \tag{7.51}$$

The kinetic energy due to the vortices is

$$E = \int \frac{\rho_\mathrm{s}}{2}\mathbf{v}_\mathrm{s}^2 d\mathbf{r} = \int \frac{\rho_\mathrm{s}}{2}(\boldsymbol{\nabla}\psi)^2 d\mathbf{r} = -\int \frac{\rho_\mathrm{s}}{2}\psi\Delta\psi d\mathbf{r}. \tag{7.52}$$

Substituting Eq. (7.50) in Eq. (7.52), we obtain

$$E = -\frac{\kappa^2\rho_\mathrm{s}}{4\pi}\sum_{i\neq j} n_i n_j \ln\frac{|\mathbf{r}_i-\mathbf{r}_j|}{\xi} + E_\mathrm{c}\sum_i n_i^2, \tag{7.53}$$

where E_c is a core correction which arises from the fact that the amplitude of the condensate changes rapidly near the vortex core which makes (7.52) less precise.

Equation (7.53) shows that the interaction between vortices takes the same form as the Coulomb interaction in two dimensions. Comparing Eq. (7.53) with the Coulomb interaction in two dimensions, we find that $(\kappa^2\rho_\mathrm{s})^{-1}$ plays the role of a dielectric constant ε_0:

$$\varepsilon_0 = \frac{1}{\kappa^2\rho_\mathrm{s}}. \tag{7.54}$$

Consider, for example, two singly quantized vortices of the same sign ($n_1 = n_2 = 1$). According to what is stated at the beginning of this section, the interaction is repulsive. In fact, Eq. (7.53) shows that the energy of the system decreases logarithmically as the distance between the two vortices increases.

7.4 Vortex Lattice

7.4.1 *Dynamics of vortex nucleation*

When the rotation frequency of the container is above the critical frequency, vortices enter the system. All the vortices have the same sign of vorticity and therefore repel each other. In a stationary state, the vortices form a regular triangular lattice known as Abrikosov lattice [Abrikosov (1957)] in which vortices are placed equidistantly so as to minimize the repulsive interaction.

The dynamics of vortex lattice formation can be simulated with a GP equation that includes a phenomenological damping constant γ [Choi, *et al.* (1998); Tsubota, *et al.* (2002)]:

$$(i-\gamma)\hbar\frac{\partial\psi}{\partial t} = \left(-\frac{\hbar^2}{2M}\nabla^2 + V - \mathbf{\Omega}\cdot\mathbf{L} - \tilde{\mu}\right)\psi + \frac{4\pi\hbar^2 a}{M}|\psi|^2\psi, \quad (7.55)$$

where V is a trapping potential, $\mathbf{\Omega}$ is the angular frequency vector of the potential, $\mathbf{L} = -i\hbar\mathbf{r}\times\nabla$ is the angular momentum operator, and a is the s-wave scattering length. A time-dependent chemical potential $\tilde{\mu}$ is introduced to ensure the conservation of the number of particles. The damping term γ not only decreases energy but also attenuates the condensate wave function. The chemical potential $\tilde{\mu}$ is tuned at every instant of time so as to keep the number of particles constant.

In most numerical studies of vortex nucleation, a quadrupolar deformation is added to the trapping potential to drive the system into rotation:

$$V = \frac{m\omega_\perp^2}{2}[(1+\epsilon)x^2 + (1-\epsilon)y^2] + \frac{m\omega_z^2}{2}z^2, \quad (7.56)$$

where $|\epsilon| \ll 1$. In this case, the number of nucleated vortices must be even because the Hamiltonian is two-fold symmetric. To study nucleation dynamics of a single vortex, one must break the two-fold symmetry. An example of such a potential is given by

$$V = \frac{m\omega_\perp^2}{2}[(x-x_0)^2 + y^2] + \frac{m\omega_z^2}{2}z^2. \quad (7.57)$$

A stirring potential with a single off-center laser beam simulates the potential (7.57), while the potentail with a pair of symmetric laser beams simulates the potential (7.56). In either case, the system undergoes elliptic deformation with irrotational flow, followed by excitations of surface modes with high angular momenta, until at last vortices enter the system as illustrated in Fig. 7.3, where for each Ω, the left column shows the density profile and the right column shows the phase profile. The phase is displayed in a gray scale in which black and white shows 0 and 2π. Figure 7.3 (a) shows nucleation of a single vortex, while Figs. 7.3 (b) and (c) show formation of triangular lattices.

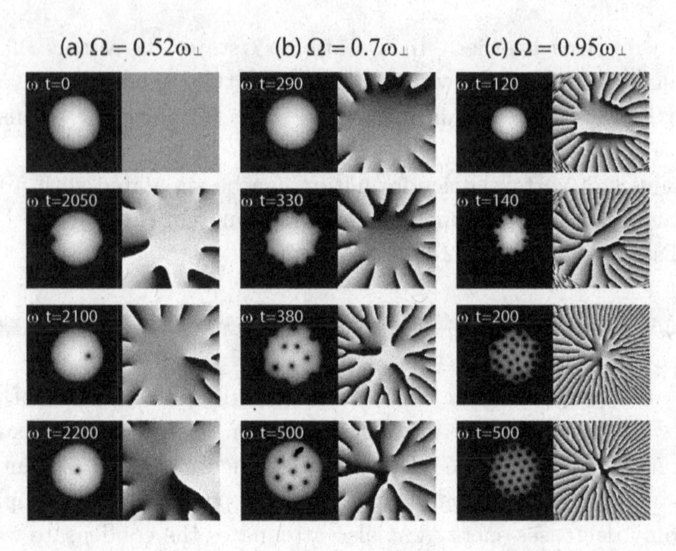

Fig. 7.3 Dynamics of vortex nucleation using potential (7.57) with $x_0 = 0.3a_{\text{ho}}$, where $a_{\text{ho}} = \sqrt{\hbar/(2m\omega_\perp)}$. For each Ω, the left and right columns show the density and phase profiles of the condensate. The strength of interaction is characterized with a dimensionless parameter $C = 4\sqrt{\pi\lambda}Na/a_{\text{ho}}$, where $\lambda \equiv \omega_z/\omega_\perp$. In the simulation, $C = 300$ is used. The field of view is 20×20 in (a) and (b), and 30×30 in (c) in units of a_{ho}. [Reproduced from Ueda and Saito (2007)]

An efficient vortex nucleation occurs when the rotation frequency hits the resonance frequency of the quadrupolar excitation which is $\omega_\perp/\sqrt{2}$ in the Thomas–Fermi limit. The surface modes with high angular momenta are exited because they cost little energy per angular momentum. In fact, the excitation energy of a surface mode with angular momentum $\hbar\ell$ is of the order of $\hbar\omega_\perp\sqrt{\ell}$ and the energy cost per angular momentum is $\hbar\omega_\perp/\sqrt{\ell}$. Such high-$\ell$ surface modes trigger vortices to enter the condensate as illustrate in Fig. 7.3

7.4.2 *Collective modes of a vortex lattice*

In the presence of vortices, the flow pattern becomes nonuniform. Reflecting the triangular configuration of the Abrikosov lattice, the restoring force of the system against deformation of the lattice becomes anisotropic which, in turn, affects collective modes of a vortex lattice.

Suppose that the system rotates at angular frequency vector $\mathbf{\Omega}$. Then, the average velocity of the fluid at position $\mathbf{r}$ is given by $\bar{\mathbf{v}} = \mathbf{\Omega} \times \mathbf{r}$. Decomposing the fluid velocity $\mathbf{v}$ at $\mathbf{r}$ into $\bar{\mathbf{v}}$ and $\mathbf{v}_{\text{local}} \equiv \mathbf{v} - \bar{\mathbf{v}}$, the kinetic energy of the fluid is written as

$$E = \int \frac{1}{2}\rho \mathbf{v}^2 d\mathbf{r} = \frac{I}{2}\mathbf{\Omega}^2 + \mathbf{\Omega} \cdot \int \rho \mathbf{r} \times \mathbf{v}_{\mathrm{local}} d\mathbf{r} + \frac{1}{2}\int \rho \mathbf{v}_{\mathrm{local}}^2 d\mathbf{r}, \quad (7.58)$$

where $I \equiv \int \rho r^2 dr$ is the classical moment of inertia. The angular momentum $\mathbf{L}$ of the system is given by

$$\mathbf{L} = \frac{\partial E}{\partial \mathbf{\Omega}} = E\mathbf{\Omega} + \int \rho \mathbf{r} \times \mathbf{v}_{\mathrm{local}} d\mathbf{r}. \quad (7.59)$$

When the container of the system rotates at frequency $\mathbf{\Omega}_0$, the equilibrium state of the system is determined by minimizing

$$E' = E - \mathbf{\Omega}_0 \cdot \mathbf{L}. \quad (7.60)$$

Substituting Eqs. (7.58) and (7.59) into Eq. (7.60), we obtain

$$E' = \frac{I}{2}(\mathbf{\Omega}^2 - 2\mathbf{\Omega}\mathbf{\Omega}_0) + (\mathbf{\Omega} - \mathbf{\Omega}_0) \cdot \int \rho \mathbf{r} \times \mathbf{v}_{\mathrm{local}} d\mathbf{r} + \frac{1}{2}\int \rho \mathbf{v}_{\mathrm{local}}^2 d\mathbf{r}. \quad (7.61)$$

We assume that $\mathbf{\Omega}$ and $\mathbf{\Omega}_0$ are parallel to each other and that vortices form a triangular lattice. We may then rewrite Eq. (7.61) as

$$E' = \frac{I}{2}(\Omega^2 - 2\Omega\Omega_0) + \int d^2 r n_{\mathrm{v}} [L_{\mathrm{v}}(\Omega - \Omega_0) + E_{\mathrm{v}}], \quad (7.62)$$

where n_{v} is the local density of vortices,

$$L_{\mathrm{v}} = \frac{\pi^2 \hbar^2 \rho}{4\sqrt{3}M^2 \Omega} \quad (7.63)$$

is the angular momentum per vortex [Tkachenko (1966, 1969); Baym and Chandler (1983)], and

$$E_{\mathrm{v}} = -\frac{\pi \hbar^2 \rho}{2M^2} \ln \frac{\Omega}{\Omega_{\mathrm{T}}} \quad (7.64)$$

is the energy per vortex, where $\Omega_{\mathrm{T}} \equiv \alpha \xi^{-2}$ with α being a constant depending on the geometry of the lattice. Minimizing (7.62) with respect to Ω gives

$$\Omega = \Omega_0 - \frac{1}{I}\int d^2 r n_{\mathrm{v}} \left(L_{\mathrm{v}} + \frac{E_{\mathrm{v}} - \frac{\pi \hbar^2 \rho}{2M^2}}{\Omega} \right). \quad (7.65)$$

When the size of the system is much larger than ξ, the second term can be ignored because it is of the order of $\hbar/(MR^2)$. We then find that $\Omega \simeq \Omega_0$ which implies that the vortex lattice rotates with the container.

Expanding $E - \Omega_0 L$ around $\Omega = \Omega_0$ up to second order, we obtain

$$E - \Omega_0 L \simeq (E - \Omega_0 L)|_{\Omega=\Omega_0} + \frac{I}{2}(\Omega - \Omega_0)^2 - \int d^2r \frac{\rho\hbar\Omega_0}{4M}\left(1 - \frac{\Omega}{\Omega_0}\right)^2. \tag{7.66}$$

The second term on the right-hand side is the kinetic energy for a uniform rotation of the system in the rotating frame of reference, while the last term can be interpreted as elastic energy $E_{\rm el}$ associated with a homogeneous displacement of the lattice by the amount

$$\boldsymbol{\epsilon} = \frac{1}{2}\left(1 - \frac{\Omega}{\Omega_0}\right)\mathbf{r}. \tag{7.67}$$

For a triangular lattice, there is also the contribution from shear deformation, so the elastic energy can be written as

$$E_{\rm el} = \int d^2r \left\{ 2C_1(\nabla\cdot\boldsymbol{\epsilon})^2 + C_2\left[\left(\frac{\partial \epsilon_x}{\partial x} - \frac{\partial \epsilon_y}{\partial y}\right)^2 + \left(\frac{\partial \epsilon_x}{\partial y} + \frac{\partial \epsilon_y}{\partial x}\right)^2\right]\right\}, \tag{7.68}$$

where C_1 is the compressibility modulus and C_2 is the shear modulus with $C_2 = -C_1 = \hbar\rho\Omega_0/(8M)$ [Tkachenko (1966, 1969); Baym and Chandler (1983)]. Tkachenko [Tkachenko (1966, 1969)] showed that the normal modes of an infinite triangular lattice involve an elliptical motion of the vortices around their equilibrium positions in a sense opposite to that of the uniform rotation with the dispersion relation given by

$$\omega_{\rm T}^2 = \frac{\hbar\Omega_0}{4M}\frac{c^2k^4}{4\Omega_0^2 + c^2k^2}. \tag{7.69}$$

For $\Omega \ll ck$, the system is incompressible and Eq. (7.69) gives

$$\omega_{\rm T} = \left(\frac{\hbar\Omega_0}{4M}\right)^{\frac{1}{2}} k. \tag{7.70}$$

On the other hand, for $\Omega \gg ck$, the system becomes compressible and the dispersion relation (7.69) becomes quadratic at the long-wavelength limit [Baym (2003); Cozzini, *et al.* (2004)] as

$$\omega_{\rm T} = \sqrt{\frac{\hbar}{16M\Omega_0}}ck^2, \tag{7.71}$$

where $c = \sqrt{gn/M}$ is the sound velocity. Furthermore, in the quantum-Hall regime, the dispersion relation is given by

$$\omega_{\rm T} = \frac{9}{4\pi^2\sqrt{10}}\frac{c^2k^2}{\Omega_0}. \tag{7.72}$$

The softening ($\omega_{\rm T} \propto k^2$) of the collective mode suggests the instability of the vortex lattice against a vortex liquid state. Apart from the Tkachenko mode, there is also the usual inertial mode whose dispersion relation is given by [Baym (2003); Cozzini, *et al.* (2004)]

$$\omega_{\rm I} = \sqrt{4\Omega_0^2 + c^2k^2}. \tag{7.73}$$

The Tkachenko mode was observed in liquid helium by Andereck and Glaberson [Andereck and Glaberson (1982)] and in gaseous BEC by Coddington *et al.* [Coddington, *et al.* (2003)]. The latter experiment was analyzed numerically by Mizushima *et al.* [Mizushima, *et al.* (2004)].

7.5 Fractional Vortices

A spinor condensate can exhibit both mass current and spin current, and consequently two kinds of circulation exist. Vortices in a spinor condensate are characterized with two winding numbers corresponding to mass circulation and spin circulation. Furthermore, the spinor condensate has a multicomponent order parameter that allows non trivial discrete symmetries and concomitant fractional vorteces.

By way of illustration, let us consider the polar phase of a spin-1 condensate. A general order parameter of this phase is given by Eq. (6.107). In the polar coordinates, the order parameter is proportional to

$$e^{i\chi}Y_1^0(\theta,\phi) = \sqrt{\frac{3}{4\pi}}e^{i\chi}\cos\theta \equiv \psi(\theta,\chi), \tag{7.74}$$

where θ and ϕ are the polar and azimuthal angles, and χ is the global (gauge) phase. This order parameter has a two-fold symmetry [see Fig. 7.4], and is invariant under spatial inversion $\theta \to \pi - \theta$ followed by a gauge transformation by π (*i.e.*, $\chi \to \chi + \pi$).

$$\psi(\theta,\chi) = \psi(\pi - \theta, \chi + \pi). \tag{7.75}$$

Since the single-valuedness of the wave function is met by the phase change of χ by π which is one half of the usual value of 2π, the system can possess a one-half vortex [Zhou (2001)].

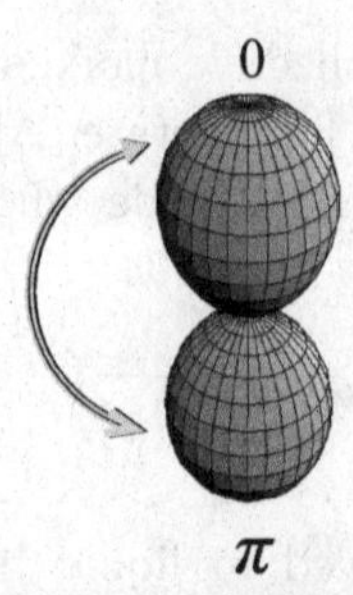

Fig. 7.4 Order parameter of the polar phase of a spin-1 BEC. It is two-fold symmetric and the upper and lower lobes have a phase difference by π due to a cosine factor in Eq. (7.74).

Consider next the cyclic phase of a spin-2 BEC. A general order parameter of this phase is given by

$$\begin{pmatrix} \psi_2 \\ \psi_1 \\ \psi_0 \\ \psi_{-1} \\ \psi_{-2} \end{pmatrix} = \frac{\sqrt{n}}{2} e^{i\chi} U(\alpha,\beta,\gamma) \begin{pmatrix} i \\ 0 \\ \sqrt{2} \\ 0 \\ i \end{pmatrix} \tag{7.76}$$

In the polar coordinates, the order parameter is proportional to

$$\begin{aligned} & e^{i\chi} \left\{ i \left[Y_2^2(\theta,\phi) + Y_2^{-2}(\theta,\phi) \right] + \sqrt{2} Y_2^0(\theta,\phi) \right\} \\ &= \sqrt{\frac{15}{8\pi}}\, e^{i\chi} \left[i \sin^2\theta \cos 2\phi + \frac{1}{\sqrt{3}} (3\cos^2\theta - 1) \right] \\ &\equiv \psi(\theta,\phi,\chi). \end{aligned} \tag{7.77}$$

Figure 7.5 shows the profile of the order parameter of the cyclic phase. It has a three-fold symmetry about the $(1,1,1)$ axis. This implies that the order parameter is invariant under a rotation through $2\pi/3$ (or $4\pi/3$) about the $(1,1,1)$ axis followed by a gauge transformation by $2\pi/3$ (or $4\pi/3$) which is one third (or two thirds) of 2π. The system can therefore hold a one-third (or two-thirds) vortex [Zhang, *et al.* (2006); Mäkelä (2006); Semenoff and Zhou (2007)].

7.6 Spin Current

The k-th component of the spin density is given by

$$s_k = \sum_{\alpha\beta} \psi_\alpha^* (\hat{s}_k)_{\alpha\beta} \psi_\beta \quad (k = x, y, z), \tag{7.78}$$

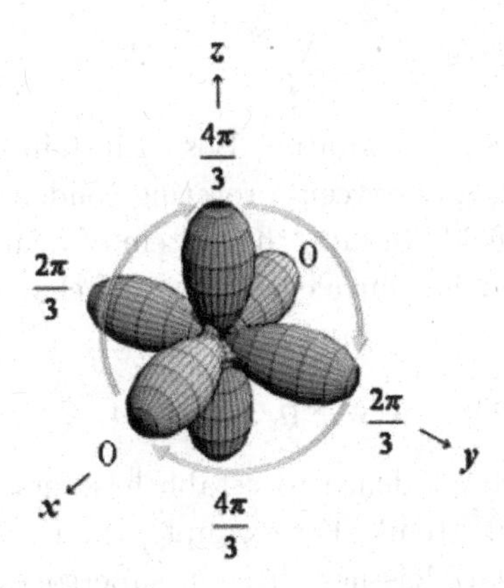

Fig. 7.5 Order parameter of the cyclic phase of a spin-2 BEC. The phase of each lobe is indicated.

where $\hat{s}_k$ is the k-component of spin matrices. The corresponding spin current density $\mathbf{j}^k_{\rm spin}$ is determined so as to meet the equation of continuity

$$\frac{\partial s_k}{\partial t} + \nabla \mathbf{j}^k_{\rm spin} = 0. \tag{7.79}$$

Substituting Eq. (7.78) into (7.79) and using (6.55), we obtain

$$\mathbf{j}^k_{\rm spin} = \frac{\hbar}{2Mi} \sum_{\alpha,\beta} \left[\psi^*_\alpha (\hat{s}_k)_{\alpha\beta} \nabla \psi_\beta - \psi_\beta (\hat{s}_k)_{\alpha\beta} \nabla \psi^*_\alpha\right]. \tag{7.80}$$

Consider first the polar-core vortex whose order parameter is given by

$$\psi_1 = e^{\pm i\theta + i\chi} f_1(r_\perp),\ \psi_0 = f_0(r_\perp),\ \psi_{-1} = e^{\mp i\theta - i\chi} f_1(r_\perp), \tag{7.81}$$

where $r_\perp \equiv \sqrt{x^2+y^2}$. For this vortex, the mass current (7.19) vanishes, while the spin currents are calculated to give

$$\mathbf{j}^x_{\rm spin} = \mathbf{j}^y_{\rm spin} = 0,\ \mathbf{j}^z_{\rm spin} = \frac{\hbar}{Mr_\perp} 2f_1^2(r_\perp)\hat{z}, \tag{7.82}$$

where $\hat{z} \equiv (0,0,1)$, and the factor $2f_1^2(r_\perp)$ shows the density of particles that contributes to $\mathbf{j}^z_{\rm spin}$. The z component of the spin velocity is therefore given by

$$\mathbf{v}^z_{\rm spin} = \frac{\hbar}{Mr_\perp}. \tag{7.83}$$

The corresponding spin circulation is obtained by integrating (7.83) around the z axis:

$$\oint \hat{z} \cdot \mathbf{v}^z_{\rm spin} d\ell = \frac{h}{M}. \tag{7.84}$$

7.7 Fast Rotating BECs

In considering the physics of a rotating Bose–Einstein condensate, it is useful to note a close analogy between a rotating condensate and a superconductor in a magnetic field. In fact, the system of neutral particles in a rotating frame of reference is equivalent to that of charged particles in a magnetic field through the correspondence:

$$\mathbf{p} - m\Omega \times \mathbf{r} \Longleftrightarrow \mathbf{p} - \frac{e}{2c}\mathbf{B} \times \mathbf{r}. \tag{7.85}$$

We can exploit this correspondence to establish several equivalent phenomena between the two systems. For example, the Hess–Fairbank effect in BEC corresponds to the Meissner effect in superconductivity, and the critical frequency Ω_c of vortex nucleation corresonds to the lower critical frequency of a type-II superconductor. What happens to the BEC when Ω gets larger? In type-II superconductors, the density of vortices increases with external magnetic field and the system becomes normal when the vortex cores touch each other. The critical magnetic field at which the superconducting state becomes normal defines the upper critical magnetic field. In contrast, in atomic-gas BEC, the vortex core size and the vortex lattice spacing adjust to each other so that no overlap of vortices occurs at high rotation frequency [Fischer and Baym (2003); Baym and Pethick (2004)]. Thus, there is no analogue in BEC for the upper critical frequency beyond which the system becomes normal. Instead, in the limit of fast rotation, the BEC is expected to become a strongly correlated system.

7.7.1 *Lowest Landau level approximation*

Consider a particle with mass M in a two-dimensional harmonic oscillator. The Hamiltonian $\hat{h}$ and the angular-momentum operator $\hat{\ell}$ of the particle are given by

$$\hat{h} = \frac{1}{2M}(\hat{p}_x^2 + \hat{p}_y^2) + \frac{M\omega^2}{2}(\hat{x}^2 + \hat{y}^2), \tag{7.86}$$

$$\hat{\ell} = \hat{x}\hat{p}_y - \hat{y}\hat{p}_x, \tag{7.87}$$

where $[\hat{x}, \hat{p}_x] = [\hat{y}, \hat{p}_y] = i\hbar$ and $[\hat{x}, \hat{p}_y] = [\hat{y}, \hat{p}_x] = 0$. To diagonalize $\hat{h}$ and $\hat{\ell}$, we first introduce

$$\hat{a}_x \equiv \sqrt{\frac{M\omega}{2\hbar}}\left(\hat{x} + i\frac{\hat{p}_x}{M\omega}\right), \quad \hat{a}_y \equiv \sqrt{\frac{M\omega}{2\hbar}}\left(\hat{y} + i\frac{\hat{p}_y}{M\omega}\right), \tag{7.88}$$

where $[\hat{a}_x, \hat{a}_x^\dagger] = [\hat{a}_y, \hat{a}_y^\dagger] = 1$ and $[\hat{a}_x, \hat{a}_y] = [\hat{a}_x, \hat{a}_y^\dagger] = 0$. In terms of $\hat{a}_x$ and $\hat{a}_y$, $\hat{h}$ and $\hat{\ell}$ are expressed as

$$\hat{h} = \hbar\omega(\hat{a}_x^\dagger \hat{a}_x + \hat{a}_y^\dagger \hat{a}_y + 1), \tag{7.89}$$

$$\hat{\ell} = i\hbar(\hat{a}_x \hat{a}_y^\dagger + \hat{a}_x^\dagger \hat{a}_y). \tag{7.90}$$

Since $\hat{h}$ and $\hat{\ell}$ commute, they can be diagonalized simultaneously:

$$\hat{h} = \hbar\omega(\hat{a}_+^\dagger \hat{a}_+ + \hat{a}_-^\dagger \hat{a}_- + 1), \tag{7.91}$$

$$\hat{\ell} = \hbar(\hat{a}_+^\dagger \hat{a}_+ - \hat{a}_-^\dagger \hat{a}_-), \tag{7.92}$$

where

$$\hat{a}_+ \equiv \frac{1}{\sqrt{2}}(\hat{a}_x - i\hat{a}_y), \quad \hat{a}_- \equiv \frac{1}{\sqrt{2}}(\hat{a}_x + i\hat{a}_y). \tag{7.93}$$

As can be seen from Eq. (7.92), $\hat{a}_+^\dagger$ and $\hat{a}_-^\dagger$ respectively increases and decreases the angular momentum of the system by $\hbar$. Denoting the eigenvalues of $\hat{a}_+^\dagger \hat{a}_+$ and $\hat{a}_-^\dagger \hat{a}_-$ as n_+ and n_-, respectively, and defining $n \equiv \min(n_+, n_-)$ and $m \equiv n_+ - n_-$, the energy eigenvalue ε is given by

$$\epsilon = \hbar\omega(2n + |m| + 1). \tag{7.94}$$

For a given angular momentum $\hbar m$, the minimum energy is attained with $n = 0$, that is, $n_- = 0$ if $m \geq 0$ and $n_+ = 0$ otherwise.

When the interparticle interaction per particle is much smaller than $\hbar\omega$, the total energy is given approximately by the sum of single-particle energies (7.94) of individual particles. Under the condition that the total angular momentum of the system is given, the total energy of the system is expected to be minimized if no two particles rotate in opposite directions. This assumption is referred to as the lowest Landau level (LLL) approximation.

When the system is rotated at angular frequency Ω, the energy of the system in the rotating frame of reference is given by

$$\epsilon' \equiv \hbar\big[(\omega - \Omega)n_+ + (\omega + \Omega)n_- + 1\big]. \tag{7.95}$$

When Ω approaches ω, the energy levels that have the same n_- are almost degenerate, while those that have different values of n_- are separated by an energy gap $\sim 2\hbar\omega$. Thus, the single-particle energy levels are grouped into nearly degenerate energy bands characterized by the Landau-level index n_-, as shown in Fig. 7.6.

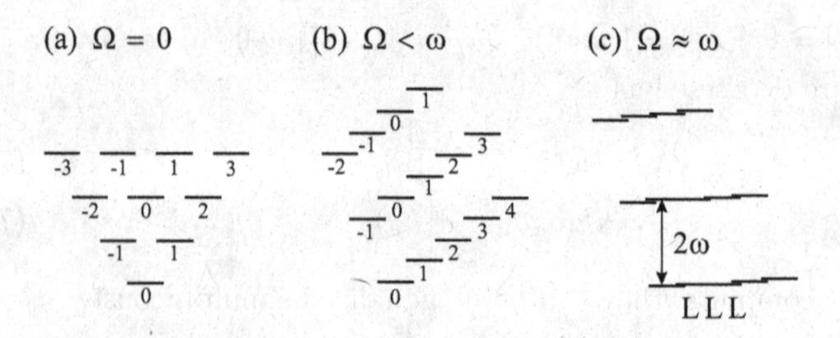

Fig. 7.6 Energy spectrum (7.95) for (a) $\Omega = 0$, (b) $\Omega < \omega$, and (c) $\Omega \to \omega$. The number below each level shows the magnetic quantum number.

To make the coordinates dimensionless, we measure x and y in units of $a_{\rm ho} = \sqrt{\hbar/(M\omega)}$, and introduce the complex variable $z \equiv x + iy$. In terms of z, a_+ and a_- are expressed as

$$\hat{a}_+ = \frac{z^*}{2} + \frac{\partial}{\partial z} = e^{-\frac{|z|^2}{2}} \frac{\partial}{\partial z} e^{\frac{|z|^2}{2}}, \tag{7.96}$$

$$\hat{a}_+^\dagger = \frac{z}{2} - \frac{\partial}{\partial z^*} = -e^{\frac{|z|^2}{2}} \frac{\partial}{\partial z^*} e^{-\frac{|z|^2}{2}}, \tag{7.97}$$

$$\hat{a}_- = \frac{z}{2} + \frac{\partial}{\partial z^*} = e^{-\frac{|z|^2}{2}} \frac{\partial}{\partial z^*} e^{\frac{|z|^2}{2}}, \tag{7.98}$$

$$\hat{a}_-^\dagger = \frac{z^*}{2} - \frac{\partial}{\partial z} = -e^{\frac{|z|^2}{2}} \frac{\partial}{\partial z} e^{-\frac{|z|^2}{2}}. \tag{7.99}$$

The ground-state wave function $\phi_0(z)$ is determined so as to satisfy $\hat{a}_+\phi_0(z) = \hat{a}_-\phi_0(z) = 0$. It follows from Eqs. (7.96) and (7.98) that

$$\phi_0(z) = \frac{1}{\sqrt{\pi}} e^{-\frac{|z|^2}{2}}. \tag{7.100}$$

The basis functions in the LLL approximation can be constructed as

$$\phi_m(z) = \frac{(\hat{a}_+^\dagger)^m}{\sqrt{m!}} \phi_0(z) = \frac{z^m}{\sqrt{\pi m!}} e^{-\frac{|z|^2}{2}} \quad (m \geq 0), \tag{7.101}$$

where ϕ_m's satisfy the orthonormal condition

$$\int d^2 z \phi_m^*(z) \phi_n(z) = \delta_{mn}, \tag{7.102}$$

where $d^2 z \equiv dxdy$. We thus find that in the LLL approximation, the wave function is a polynominal function of z and not of z^* apart from the factor $e^{-\frac{|z|^2}{2}}$.

7.7.2 *Mean field quantum Hall regime*

Consider what happens if we increase the rotation frequency Ω. Various phenomena can be expected, depending on the filling factor ν which is the ratio of the number of atoms N against the number of vortices N_v:

$$\nu = \frac{N}{N_v}. \tag{7.103}$$

When $\nu > 1000$, the behavior of the system can be well described by the GP mean-field theory.

Suppose that the system is confined in a harmonic potential with radial trapping frequency $\omega_\perp$. In a frame corotating with the container of the system, the state of the system at zero temperature is determined by minimizing

$$E' = E - \Omega L = \int d\mathbf{r} \left[\frac{1}{2M} |(\mathbf{p} - M\Omega \times \mathbf{r})\psi|^2 + \frac{M}{2}(\omega_\perp^2 - \Omega^2) r^2 |\psi|^2 + \frac{g}{2}|\psi|^4 \right]. \tag{7.104}$$

It is clear from this expression that as Ω approaches $\omega_\perp$, the condensate flattens out and becomes an effective two-dimensional system. When $1000 > \nu > 10$, the system enters a mean-field quantum Hall regime [Ho (2001)], in which the mean-field theory is still applicable but the state of the system can be well described in the LLL approximation [Butts and Rokhsar (1999)].

In the LLL approximation, the kinetic energy of the system becomes constant and equal to $\hbar\omega_\perp N$, where $N \equiv \int |\psi|^2 d\mathbf{r}$ is the number of particles of the system. Equation (7.104) therefore reduces to

$$E' = \hbar\omega_\perp N + \frac{M}{2}(\omega_\perp^2 - \Omega^2) \int d\mathbf{r}\, r^2 |\psi|^2 + \frac{g}{2} \int d\mathbf{r} |\psi|^4. \tag{7.105}$$

Because the wave function changes rapidly near the vortices, we introduce here an average density $\bar{n}$ over a distance larger than the inter-vortex spacing:

$$\overline{E'} = \hbar\omega_\perp N + \frac{M}{2}(\omega_\perp^2 - \Omega^2) \int d\mathbf{r} r^2 \bar{n} + \frac{g\beta}{2} \int d\mathbf{r} \bar{n}^2, \tag{7.106}$$

where $\beta \equiv \overline{n^2}/\bar{n}^2$ is the Abrikosov parameter which is estimated to be 1.16 for a triangular lattice [Abrikosov (1957); Kleiner, *et al.* (1964)]. Noting that $N = \int d\mathbf{r}\bar{n}$ and requiring that $\delta\overline{E'}/\delta\bar{n}$ be equal to the chemical potential μ, we obtain

$$\bar{n}(\mathbf{r}) = \bar{n}(0)\left(1 - \frac{r^2}{R^2}\right), \tag{7.107}$$

where

$$\bar{n}(0) \equiv \frac{\mu - \hbar\Omega}{g\beta}, \quad R \equiv \sqrt{\frac{2(\mu - \hbar\Omega)}{M(\omega_\perp^2 - \Omega^2)}}\,. \tag{7.108}$$

Thus, the coarse-grained density profile of a fast rotating BEC looks like a Thomas–Fermi distribution [Watanabe, *et al.* (2004)].

In the LLL, a useful relation holds between the vortex density $n_{\rm v}(\mathbf{r})$ and the particle density $n_{\rm LLL}(\mathbf{r})$ [Ho (2001); Watanabe, *et al.* (2004); Aftalion, *et al.* (2005)]. As discussed in Sec. 7.7.1, the single-particle wave function in the LLL approximation is a polynomial of $z = x + iy$ multiplied by a Gaussian function:

$$\psi_{\rm LLL}(z) = c\prod_i (z - \xi_i)e^{-\frac{|z|^2}{2d^2}}, \tag{7.109}$$

where c is the normalization constant, ξ_i denotes the location of a vortex and d is the width of the condensate. Since $n_{\rm LLL}(\mathbf{r}) = |\psi_{\rm LLL}(2)|^2$, where $\mathbf{r} = (x, y)$, we obtain

$$\nabla^2 \ln n_{\rm LLL}(\mathbf{r}) = -\frac{4}{d^2} + 4\pi \sum_i \delta^{(2)}(z - \xi_i), \tag{7.110}$$

where we use Eq. (7.51). Since the last term $\sum_i \delta^{(2)}(z - \xi_i)$ may be interpreted as the vortex density, we obtain

$$n_{\rm v}(\mathbf{r}) = \frac{1}{\pi d^2} + \frac{1}{4\pi}\nabla^2 \ln n_{\rm LLL}(\mathbf{r}). \tag{7.111}$$

When the particle density is constant, Eq. (7.111) reduces to the Feynman rule (7.36) with the identification $d = \sqrt{\hbar/(M\Omega)}$. The last term in Eq. (7.111) gives a correction to this rule. When the vortex density is constant, Eq. (7.111) predicts the Gaussian disribution of the particle density:

$$n_{\rm LLL}(\mathbf{r}) = n_{\rm LLL}(0)e^{-\frac{r^2}{\sigma^2}}, \tag{7.112}$$

where

$$\sigma^{-2} \equiv \pi\left(\frac{1}{\pi d^2} - n_{\rm v}\right). \tag{7.113}$$

In the trap system, however, the triangular lattice is slightly distorted at the periphery of the condensate, which results in a major change of the particle density distribution toward the TF distribution as shown in Eq. (7.107).

7.7.3 Many-body wave functions of a fast rotating BEC

When the interparticle interaction is much smaller than the single-particle energy-level spacing, the LLL approximation is applicable, and the many-body wave function of the system can be expanded in terms of states of the particles that rotate in the same sense as the rotation of the container. Let it be counterclockwise. Then, the many-body wave function can, in general, be constructed in the subspace spanned by $\phi_m(z)$ in Eq. (7.101):

$$\psi(z_1, z_2, \cdots, z_N) = f(z_1, z_2, \cdots, z_N)\exp\left(-\frac{1}{2}\sum_{k=1}^{N}|z_k|^2\right), \quad (7.114)$$

where $z_k = x_k + iy_k$ is the complex coordinate of the k-th particle and f is a symmetric polynomial of $z_1, z_2, \cdots, z_N$ because of the Bose symmetry. When the interparticle interaction is attaractive, the energy of the system can get lower if all particles bunch up and move together. The ground-state wave function can therefore be given by [Wilkin, *et al.* (1998)]

$$f(z_1, z_2, \cdots, z_N) \propto \left(\sum_{k=1}^{N} z_k\right)^L, \quad (7.115)$$

where $\hbar L$ is the total angular momentum of the system. The wave function (7.115) describes the situation in which all the angular momentum is carried by the center of mass of the particles.

On the other hand, when the interaction is repulsive and $0 \leq L \leq N$, the system gains energy if each individual particle carries a unit angular momentum. The ground-state wave function is therefore given by [Bertsch and Papenbrock (1999); Smith and Wilkin (2000); Ueda and Nakajima (2006)]

$$f(z_1, z_2, \cdots, z_N) \propto \sum_{1\leq i_1<i_2<\cdots<i_L\leq N} \tilde{z}_{i_1}\tilde{z}_{i_2}\cdots\tilde{z}_{i_L}, \quad (7.116)$$

where $\tilde{z}_i \equiv z_i - (1/N)\sum_{k=1}^{N} z_k$ and the sum runs over all possible choice of L atoms out of the N atoms.

When $\nu < 10$, the vortex lattice melts and a strongly correlated system is expected to emerge. In particular, when $L = N(N-1)$, we have [Cooper, *et al.* (2001); Paredes, *et al.* (2001); Sinova, *et al.* (2002); Regnault and Jolicoeur (2003); Nakajima and Ueda (2003)]

$$f(z_1, z_2, \cdots, z_N) \propto \prod_{i>j}(z_i - z_j)^2. \quad (7.117)$$

This state naturally circumvents the energy cost of repulsive interaction at $z_i = z_j$ while maintaining the Bose statistics. The interaction energy of this state is therefore zero. It is nothing but the Laughlin state with filling factor $\nu = 1/2$. The Laughlin state is incompressible in the sense that the (bulk) excitation spectrum is gapped at all momentum; however, the Laughlin state has gapless edge modes with the praticle-like excitations featuring fractional statistics with Abelian anyons [Prange and Girvin (1990)].

In the Laughlin state, the number of vortices per each particle is $N_{\rm v}^{(1)} = N - 1$. Similarly, we can envisage a state in which the number of vortices per particle is $N_{\rm v}^{(k)} = N/k - 1$, where k is a divisor of N. In a manner analogous to the Laughlin state (7.117), we construct the corresponding many-body wave function as follows:

$$f(z_1, z_2, \cdots, z_N) \propto \mathcal{S}\left[\prod_{i_1<j_1}^{N/k}(z_{i_1} - z_{j_1})^2 \cdots \prod_{i_k<j_k}^{N/k}(z_{i_k} - z_{j_k})^2\right], \quad (7.118)$$

where $\mathcal{S}$ is a symmetrization operator over k groups, each of which involves N/k particles. In Eq. (7.118), the $k = 2$ and $k \geq 3$ states are referred to as the Moore–Reed [Moore and Reed (1991)] and Reed–Rezayi [Reed and Rezayi (1999)] states, respectively. These strongly correlated states are predicted to appear in fast rotating Bose–Einstein condensates [Cooper, *et al.* (2001); Cooper (2008)].

Let us evaluate the contact interaction energy

$$V = 2\pi g \sum_{i\neq j} \delta^{(2)}(z_i - z_j) \quad (7.119)$$

for many-body wave functions (7.115), (7.116), and (7.117), where $\delta^{(2)}(z_i - z_j)$ is defined as

$$\delta^{(2)}(z_i - z_j) \equiv \int \frac{d^2k}{\pi^2} e^{i[k^*(z_i - z_j) + k(z_i^* - z_j^*)]} \quad (7.120)$$

with $d^2k \equiv d(\mathrm{Re}k)d(\mathrm{Im}k)$. Substituting $z_i = \hat{a}_{i+}^\dagger + \hat{a}_{i-}$ and $z_j = \hat{a}_{j+}^\dagger + \hat{a}_{j-}$ in the right-hand of Eq. (7.120), and using the Baker–Hausdorff formula

$$e^{\hat{A}+\hat{B}} = e^{-\frac{1}{2}[\hat{A},\hat{B}]} e^{\hat{A}} e^{\hat{B}}, \quad (7.121)$$

we can carry out the integration over k, obtaining

$$\delta^{(2)}(z_i - z_j) = \frac{1}{2\pi} \sum_{p,q,r,s=0}^{\infty} \delta_{p+s,q+r} \frac{(-1)^{(p+s)}(p+s)!}{2^{p+s}p!q!r!s!} \times (\hat{a}_{i+}^\dagger - \hat{a}_{j+}^\dagger)^p (\hat{a}_{i-}^\dagger - \hat{a}_{j-}^\dagger)^q (\hat{a}_{i+} - \hat{a}_{j+})^r (\hat{a}_{i-} - \hat{a}_{j-})^s. \quad (7.122)$$

In the LLL, $q = s = 0$ and therefore Eq. (7.122) is reduced to

$$\delta^{(2)}(z_i - z_j) = \frac{1}{2\pi} \sum_{m=0}^{\infty} \frac{(-1)^m}{2^m m!} (\hat{a}_{i+}^\dagger - \hat{a}_{j+}^\dagger)^m (\hat{a}_{i+} - \hat{a}_{j+})^m. \tag{7.123}$$

Expressing the right-hand side in terms of z_i and z_j using Eqs. (7.96) and (7.97), we obtain

$$\delta^{(2)}(z_i - z_j) = \frac{1}{2\pi} e^{\frac{|z_i|^2+|z_j|^2}{2}} \sum_{m=0}^{\infty} \frac{1}{2^m m!} \left(\frac{\partial}{\partial z_i^*} - \frac{\partial}{\partial z_j^*} \right)^m \times e^{-|z_i|^2-|z_j|^2} \left(\frac{\partial}{\partial z_i} - \frac{\partial}{\partial z_j} \right)^m e^{\frac{|z_i|^2+|z_j|^2}{2}}. \tag{7.124}$$

In the LLL, the polynominal part f of the many-body wave function (7.114) does not involve z_i^* and z_j^*. Equation (7.124) can then be used to show that

$$2\pi\delta^{(2)}(z_i - z_j) f(\cdots, z_i, \cdots, z_j, \cdots) \exp\left(-\frac{1}{2} \sum_{k=1}^{N} |z_k|^2 \right) = f\left(\cdots, \frac{z_i + z_j}{2}, \cdots, \frac{z_i + z_j}{2}, \cdots \right) \exp\left(-\frac{1}{2} \sum_{k=1}^{N} |z_k|^2 \right). \tag{7.125}$$

We use this result to calculate the eigenvalue of the many-body wave function (7.115), (7.116) and (7.117). First,

$$V \left(\sum_{k=1}^{N} z_k \right)^L \exp\left(-\frac{1}{2} \sum_{k=1}^{N} |z_k|^2 \right) = gN(N-1) \left(\sum_{k=1}^{N} z_k \right)^L \exp\left(-\frac{1}{2} \sum_{k=1}^{N} |z_k|^2 \right), \tag{7.126}$$

and therefore the eigenvalue is

$$E^{\text{attractive}} = \hbar\omega_\perp N + gN(N-1). \tag{7.127}$$

The wave function (7.116) for the repulsive interaction can also be shown to be invariant under the action of V with the eigenvalue:

$$E^{\text{repulsive}} = \hbar\omega_\perp N + gN\left(N - 1 - \frac{1}{2}\right). \tag{7.128}$$

For the $\nu = 1/2$ Laughlin state (7.117), the delta function obviously does not contribute at all. Therefore,

$$E^{\text{Laughlin}} = \hbar\omega_\perp N. \tag{7.129}$$

Chapter 8

Fermionic Superfluidity

8.1 Ideal Fermi Gas

We consider a system of ideal fermions with one-half spin and mass M. Unless otherwise specified, we assume that both spin-up and spin-down components share the same chemical potential μ and that they obey the Fermi distribution

$$f(\varepsilon) = \frac{1}{e^{\beta(\varepsilon-\mu)}+1}. \tag{8.1}$$

The chemical potential is determined such that the particle-number density is equal to a prescribed value n:

$$n = 2\int \frac{d^3k}{(2\pi)^3} f(\epsilon_k), \tag{8.2}$$

where the factor 2 before the integral accounts for the spin degrees of freedom and $\epsilon_k = \hbar^2k^2/2M$. Introducing the density of states per spin

$$D(\epsilon_k) \equiv \frac{4\pi k^2}{(2\pi)^3}\frac{dk}{d\epsilon_k} = \frac{Mk}{2\pi^2\hbar^2} = \frac{\sqrt{2M^3\epsilon_k}}{2\pi^2\hbar^3}, \tag{8.3}$$

Eq. (8.2) is rewritten as

$$n = 2\int_0^\infty d\varepsilon D(\varepsilon) f(\varepsilon). \tag{8.4}$$

By definition, at absolute zero, the chemical potential is equal to the Fermi energy E_{F}, and the Fermi distribution becomes a unit step function $f(\varepsilon) = \theta(E_{\mathrm{F}} - \varepsilon)$. It follows from Eq. (8.4) that

$$n = \frac{(2ME_{\mathrm{F}})^{\frac{3}{2}}}{3\pi^2\hbar^3} = \frac{p_{\mathrm{F}}^3}{3\pi^2\hbar^3}, \tag{8.5}$$

where $p_{\mathrm{F}} \equiv \sqrt{2ME_{\mathrm{F}}}$ is the Fermi momentum. The density of states per spin at the Fermi energy is given by

$$D(E_{\mathrm{F}}) = \frac{Mp_{\mathrm{F}}}{2\pi^2\hbar^3} = \frac{3n}{4E_{\mathrm{F}}}. \tag{8.6}$$

At finite temperature, various physical quantities can be calculated using the Sommerfeld expansion

$$\int_0^\infty g(\varepsilon)f(\varepsilon)d\varepsilon = \int_0^\mu g(\varepsilon)d\varepsilon + \frac{\pi^2}{6}(k_{\mathrm{B}}T)^2 g'(\mu) + O\left(\frac{k_{\mathrm{B}}T}{\mu}\right)^4. \tag{8.7}$$

To show this, we define $G(\varepsilon) \equiv \int_0^\varepsilon g(\varepsilon')d\varepsilon'$ and integrate the left-hand side of Eq. (8.7) by parts:

$$\int_0^\infty g(\varepsilon)f(\varepsilon)d\varepsilon = \int_0^\infty G(\varepsilon)\left(-\frac{\partial f(\varepsilon)}{\partial\varepsilon}\right)d\varepsilon.$$

Because $\partial f/\partial\varepsilon = -\beta f(1-f)$ deviates significantly from zero only in a small region ($\sim k_{\mathrm{B}}T$) around μ, we expand $G(\varepsilon)$ around $\varepsilon = \mu$:

$$\begin{aligned}\int_0^\infty g(\varepsilon)f(\varepsilon)d\varepsilon = \int_0^\infty &\left[G(\mu) + G'(\mu)(\varepsilon-\mu) + \frac{1}{2}G''(\mu)(\varepsilon-\mu)^2 + \cdots\right] \\ &\times\left(-\frac{\partial f(\varepsilon)}{\partial\varepsilon}\right)d\varepsilon.\end{aligned}$$

Here, the first term on the right-hand side is $G(\mu)$ because $f(0) \simeq 1$, and the second term vanishes because the integrand is an odd function in $\varepsilon-\mu$. (Because by assumption $\mu \gg k_{\mathrm{B}}T$, we may safely substitute the lower limit of integration with $-\infty$.) The third term can be calculated using

$$\int_{-\infty}^\infty (\varepsilon-\mu)^2\left(-\frac{\partial f(\varepsilon)}{\partial\varepsilon}\right)d\varepsilon = \frac{\pi^2}{3}(k_{\mathrm{B}}T)^2.$$

By combining these results, we obtain Eq. (8.7).

Applying Eq. (8.7) to Eq. (8.4), we find that

$$\begin{aligned}n &\simeq 2\int_0^\mu D(\varepsilon)d\varepsilon + \frac{\pi^2}{3}(k_{\mathrm{B}}T)^2 D'(\mu) \\ &\simeq 2\int_0^{E_{\mathrm{F}}} D(\varepsilon)d\varepsilon + 2(\mu - E_{\mathrm{F}})D(E_{\mathrm{F}}) + \frac{\pi^2}{3}(k_{\mathrm{B}}T)^2 D'(E_{\mathrm{F}}).\end{aligned}$$

The first term on the right-hand side is n. Thereore,

$$\mu = E_{\mathrm{F}} - \frac{\pi^2}{6}(k_{\mathrm{B}}T)^2\frac{D'(E_{\mathrm{F}})}{D(E_{\mathrm{F}})} = E_{\mathrm{F}}\left[1 - \frac{\pi^2}{12}\left(\frac{k_{\mathrm{B}}T}{E_{\mathrm{F}}}\right)^2\right]. \tag{8.8}$$

In a similar manner, the heat capacity at constant volume is calculated to be

$$c_v = \frac{\pi^2 k_{\rm B}^2 n}{2E_{\rm F}} T. \tag{8.9}$$

Because $E_{\rm F} \propto M$, the heat capacity is proportional to the mass of the atom.

When a uniform weak magnetic field is applied to the system, single-particle energy levels are shifted by μH ($-\mu H$) for up- (down-) spin atoms, where μ is the magnetic moment of the atom. Consequently, the system becomes magnetized with magnetization given by

$$M = \mu \sum_{\mathbf{k}} [f(\epsilon_k - \mu H) - f(\epsilon_k + \mu H)] \simeq -2\mu^2 H \sum_k \frac{\partial f(\epsilon_k)}{\partial \epsilon_k}. \tag{8.10}$$

When $k_{\rm B}T \ll E_{\rm F}$, we have $\partial f(\varepsilon)/\partial\varepsilon \simeq -\delta(\varepsilon - E_{\rm F})$ and

$$M = \frac{M p_{\rm F}}{\pi^2 \hbar^3} \mu^2 H = \chi_{\rm F} H, \tag{8.11}$$

where

$$\chi_{\rm F} = \frac{\mu^2 M p_{\rm F}}{\pi^2 \hbar^3} = 2\mu^2 D(E_{\rm F}) \tag{8.12}$$

is Pauli's paramagnetic susceptibility.

The Fermi distribution deviates significantly from the Maxwell–Boltzmann distribution when the temperature of the system is lower than the Fermi temperature $T_{\rm F}$ defined by

$$T_{\rm F} \equiv \frac{1}{k_{\rm B}} E_{\rm F} = \frac{p_{\rm F}^2}{2Mk_{\rm B}} = \frac{(3\pi^2)^{\frac{2}{3}} \hbar^2}{2Mk_{\rm B}} n^{\frac{2}{3}} \simeq 4.79 \frac{\hbar^2}{Mk_{\rm B}} n^{\frac{2}{3}}. \tag{8.13}$$

Below this temperature, the system undergoes Fermi degeneracy. Apart from the numerical coefficient, this temperature is the same as the transition temperature of Bose–Einstein condensation given in Eq. (1.19). This is because both Fermi degeneracy and Bose–Einstein condensation occur due to the indistinguishability of identical particles, the effect of which becomes prominent when the thermal de Broglie length (1.24) becomes comparable to the interparticle spacing $n^{-\frac{1}{3}}$. The larger numerical coefficient in Eq. (8.13) is due to the Pauli exclusion principle vis-a-vis bosonic enhancement.

When the system is confined in a harmonic potential, the density of states is given by Eq. (3.9). Keeping only the leading order term in $D(\varepsilon)$, the total number of particles and the energy of the system are given by

$$N = \frac{1}{(\hbar\bar{\omega})^3} \int_0^\infty d\varepsilon \frac{\varepsilon^2}{e^{\beta(\varepsilon-\mu)+1}}, \tag{8.14}$$

$$E = \frac{1}{(\hbar\bar{\omega})^3} \int_0^\infty d\varepsilon \frac{\varepsilon^3}{e^{\beta(\varepsilon-\mu)+1}}, \tag{8.15}$$

where $\bar{\omega} \equiv (\omega_x\omega_y\omega_z)^{1/3}$. At absolute zero, $\mu = E_{\rm F}$ and Eq. (8.14) gives

$$E_{\rm F} = (3N)^{1/3}\hbar\bar{\omega} \simeq 1.4N^{1/3}\hbar\bar{\omega}, \tag{8.16}$$

and Eq. (8.15) gives

$$E = \frac{E_{\rm F}^4}{4(\hbar\bar{\omega})^3} = \frac{3}{4}NE_{\rm F}. \quad \text{(harmonically trapped system)} \tag{8.17}$$

This result is to be compared with the uniform result:

$$E = \frac{3}{5}NE_{\rm F}. \quad \text{(uniform system)} \tag{8.18}$$

As in the uniform case, the Fermi temperature $T_{\rm F} \equiv E_{\rm F}/k_{\rm B}$ coincides with the BEC transition temperature (3.18), except for the numerical coefficient which is again larger for fermions than bosons.

In terms of the Fermi energy, the Thomas–Fermi radius R_i $(i = x, y, z)$ can be expressed as

$$R_i = \sqrt{\frac{2E_{\rm F}}{M\omega_i^2}} = (24N)^{\frac{1}{6}}\frac{\bar{\omega}}{\omega_i}d_0, \tag{8.19}$$

where $d_0 = \sqrt{\hbar/(M\bar{\omega})}$, and the Thomas–Fermi density distribution is given by

$$n(\mathbf{r}) = \frac{8N}{\pi^2 R_x R_y R_z}\left(1 - \frac{x^2}{R_x^2} - \frac{y^2}{R_y^2} - \frac{z^2}{R_z^2}\right)^{\frac{3}{2}}. \tag{8.20}$$

Using Eq. (8.20), the Fermi momentum is given by

$$p_{\rm F} = \hbar[3\pi^2 n(\mathbf{0})]^{\frac{1}{3}}, \tag{8.21}$$

which takes the same form as the uniform case (8.5). Note that the physical origin of the Thomas–Fermi density profile (8.20) of the ideal Fermi gas is the quantum kinetic pressure due to the Pauli exclusion principle, whereas the Thomas–Fermi distribution (3.66) for the Bose gas originates from the strong repulsive interaction that overwhelms the quantum pressure term.

8.2 Fermi Liquid Theory

The behavior of a Fermi system is fundamentally different depending on whether the interparticle interaction is attractive or repulsive. Here, we consider the case of repulsive interaction. The case of attractive interaction is discussed in the next section. Landau argued that unless phase transitions such as solidification and magnetic ordering occur, the system of interacting fermions, often referred to as the Fermi liquid, maintains the basic character of an ideal Fermi gas [Landau (1957)].

The basic concept of Landau's Fermi liquid theory is that if the interparticle interaction is switched on adiabatically, the momentum distribution of ideal fermions evolves into that of quasiparticles and that there is a one-to-one correspondence between the two spectra, as schematically illustrated in Fig. 8.1.

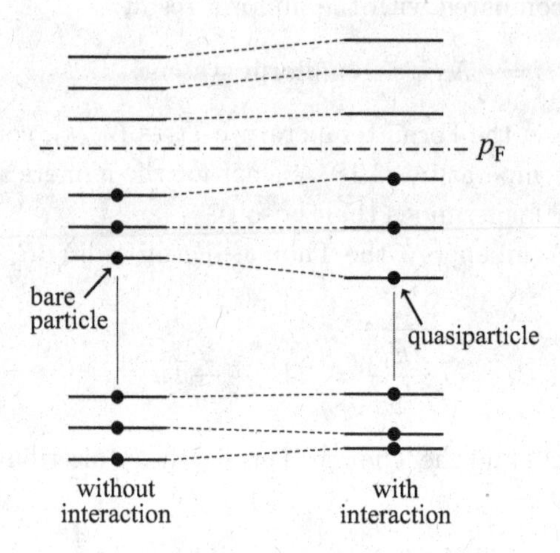

Fig. 8.1 Schematic of how the momentum distribution of an ideal Fermi gas (left) evolves into that of a Fermi liquid (right) as the interaction is adiabatically switched on.

The assumption of one-to-one correspondence, which is referred to as adiabatic continuity, enables us to label single-particle states of quasiparticles in terms of those of free fermions. In particular, quasiparticles obey the Fermi–Dirac statistics, whose momentum distribution function $n_{\mathbf{p}}$ satisfies the same relation as that in Eq. (8.2):

$$n = 2\int \frac{d^3p}{(2\pi\hbar)^3} n_{\mathbf{p}}. \tag{8.22}$$

The ground state of a Fermi liquid is one in which the quasiparticle levels are fully occupied below the Fermi surface with Fermi momentum $p_{\rm F}$ and empty above it. The relation between $p_{\rm F}$ and n is therefore the same as that of an ideal Fermi gas:

$$p_{\rm F} = \hbar(3\pi^2 n)^{\frac{1}{3}}. \tag{8.23}$$

The fact that the Fermi momentum is determined by the particle-number density alone allows us to introduce a formal parallelism between a Fermi liquid and an ideal Fermi gas. For example, when the system is isotropic, the dispersion relation $\varepsilon(p)$ of quasiparticles depends only on $p = |p|$. We can therefore expand $\varepsilon(p)$ around $p = p_{\rm F}$ to obtain

$$\varepsilon(p) = \mu + v_{\rm F}(p - p_{\rm F}), \tag{8.24}$$

where μ is the chemical potential that is equal to the Fermi energy at absolute zero, and

$$v_{\rm F} \equiv \left.\frac{\partial\varepsilon(p)}{\partial p}\right|_{p=p_{\rm F}} \equiv \frac{p_{\rm F}}{M^*} \tag{8.25}$$

is the Fermi velocity whose ratio against $p_{\rm F}$ defines the inverse effective mass $1/M^*$.

Many physical quantities of a Fermi liquid take the same form as those of an ideal Fermi gas, provided that M is replaced by M^*. For example, the density of states at the Fermi momentum is given by

$$D_{\rm F} = 2\sum_{\mathbf{p}} \delta(\varepsilon(\mathbf{p}) - \mu_{\rm F}) = 2\int \frac{d^3p}{(2\pi\hbar)^3} \frac{1}{\left|\frac{\partial\varepsilon(\mathbf{p})}{\partial p}\right|} \delta(p - p_{\rm F}) = \frac{M^* p_{\rm F}}{\pi^2\hbar^3}. \tag{8.26}$$

The distribution of quasiparticles at thermal equilibrium is determined by requiring that the entropy of the system

$$S = -2k_{\rm B}\sum_{\mathbf{p}} [n_{\mathbf{p}} \ln n_{\mathbf{p}} + (1 - n_{\mathbf{p}})\ln(1 - n_{\mathbf{p}})] \tag{8.27}$$

be maximal for the given total number of quasiparticles and total energy:

$$\delta N = 2\sum_{\mathbf{p}} \delta n_{\mathbf{p}} = 0,$$

$$\delta E = 2\sum_{\mathbf{p}} \varepsilon(\mathbf{p})\delta n_{\mathbf{p}} = 0,$$

where the factor 2 accounts for the spin degrees of freedom. Introducing Lagrange multipliers α and γ for N and E, respectively, the equilibrium distribution of quasiparticles should satisfy

$$k_{\rm B}^{-1}\delta S + \alpha\delta N + \gamma\delta E = -2\sum_{\mathbf{p}} \left(\ln\frac{n_{\mathbf{p}}}{1 - n_{\mathbf{p}}} - \alpha - \gamma\varepsilon(\mathbf{p})\right)\delta n_{\mathbf{p}} = 0.$$

Hence, we obtain $n_{\mathbf{p}} = (e^{-\beta\varepsilon(\mathbf{p})-\alpha} + 1)^{-1}$. By the standard argument of thermodynamics, α and γ are determined to be $\alpha = \mu/(k_{\rm B}T)$ and $\gamma = -1/(k_{\rm B}T)$. We therefore find that

$$n_{\mathbf{p}} = \frac{1}{e^{\beta(\varepsilon(\mathbf{p})-\mu)} + 1}. \tag{8.28}$$

The distribution of quasiparticles thus takes the same form as that of ideal fermions. However, it should be noted that $\varepsilon(\mathbf{p})$ depends on the distribution of all quasiparticles and it is therefore a functional of $n_{\mathbf{p}}$; Eq. (8.28) is actually a complicated implicit function for $n_{\mathbf{p}}$.

In the following discussions, we take into account the spin dependence of physical quantities and write the energy and distribution function of quasiparticles as $\varepsilon_\sigma(\mathbf{p})$ and $n_{\mathbf{p}\sigma}$, respectively. When quasiparticles are excited, $\varepsilon_\sigma(\mathbf{p})$ deviates from its value in the ground state $\varepsilon_\sigma^{(0)}(\mathbf{p})$ due to the influence of other excited quasiparticles $\delta n_{\mathbf{p}'\sigma'}$. When the number of excited quasiparticles is small, we may expect that $\varepsilon_\sigma(\mathbf{p}) - \varepsilon_\sigma^{(0)}(\mathbf{p})$ is given by the sum of individual contributions from excited quasiparticles:

$$\varepsilon_\sigma(\mathbf{p}) = \varepsilon_\sigma^{(0)}(\mathbf{p}) + \frac{1}{\Omega}\sum_{\mathbf{p}'\sigma'} f_{\sigma\sigma'}(\mathbf{p},\mathbf{p}')\delta n_{\mathbf{p}'\sigma'}, \tag{8.29}$$

where Ω is the volume of the system and $f_{\sigma\sigma'}(\mathbf{p},\mathbf{p}')$ is Landau's interaction function that characterizes the properties of the Fermi liquid.

As an illustrative simple example, we consider the Hartree–Fock approximation. In this approximation, the total energy is given by

$$E = E_0 + \frac{1}{2\Omega}\sum_{\mathbf{p}\mathbf{p}'\sigma\sigma'} \left[V(0) - V(\mathbf{p}-\mathbf{p}')\delta_{\sigma\sigma'}\right] n_{\mathbf{p}\sigma} n_{\mathbf{p}'\sigma'}, \tag{8.30}$$

where $V(\mathbf{p})$ is the Fourier transform of the interaction. In the sum of Eq. (8.30), the first term describes the Hartree (direct) interaction, whereas the second term describes the Fock (exchange) interaction, where the minus sign arises from the exchange of two fermions. The quasiparticle energy is calculated to be

$$\varepsilon_\sigma(\mathbf{p}) = \frac{\delta E}{\delta n_{\mathbf{p}\sigma}} = \varepsilon_\sigma^{(0)}(\mathbf{p}) + \frac{1}{\Omega}\sum_{\mathbf{p}'\sigma'} \left[V(0) - V(\mathbf{p}-\mathbf{p}')\delta_{\sigma\sigma'}\right] n_{\mathbf{p}'\sigma'}, \tag{8.31}$$

where

$$\varepsilon_\sigma^{(0)}(\mathbf{p}) = \frac{\delta E_0}{\delta n_{\mathbf{p}\sigma}}. \tag{8.32}$$

Comparing this with Eq. (8.29), we find that

$$f_{\uparrow\uparrow}(\mathbf{p},\mathbf{p}') = f_{\downarrow\downarrow}(\mathbf{p},\mathbf{p}') = V(0) - V(\mathbf{p}-\mathbf{p}'), \tag{8.33}$$

$$f_{\uparrow\downarrow}(\mathbf{p},\mathbf{p}') = f_{\downarrow\uparrow}(\mathbf{p},\mathbf{p}') = V(0). \tag{8.34}$$

We introduce the symmetric and antisymmetric parts of the interaction function as

$$f^{\mathrm{s}}(\mathbf{p},\mathbf{p}') \equiv \frac{1}{2}\left[f_{\uparrow\uparrow}(\mathbf{p},\mathbf{p}') + f_{\uparrow\downarrow}(\mathbf{p},\mathbf{p}')\right], \tag{8.35}$$

$$f^{\mathrm{a}}(\mathbf{p},\mathbf{p}') \equiv \frac{1}{2}\left[f_{\uparrow\uparrow}(\mathbf{p},\mathbf{p}') - f_{\uparrow\downarrow}(\mathbf{p},\mathbf{p}')\right]. \tag{8.36}$$

Substituting Eqs. (8.33) and (8.34) into Eqs. (8.35) and (8.36), we obtain

$$f^{\mathrm{s}}(\mathbf{p},\mathbf{p}') = V(0) - \frac{1}{2}V(\mathbf{p}-\mathbf{p}'), \tag{8.37}$$

$$f^{\mathrm{a}}(\mathbf{p},\mathbf{p}') = -\frac{1}{2}V(\mathbf{p}-\mathbf{p}'). \tag{8.38}$$

In many cases, only quasiparticles near the Fermi surface are important, where $|\mathbf{p}| = |\mathbf{p}'| = p_{\mathrm{F}}$ and $f^{\mathrm{s}}(\mathbf{p},\mathbf{p}')$ and $f^{\mathrm{a}}(\mathbf{p},\mathbf{p}')$ depend on $\mathbf{p}$ and $\mathbf{p}'$ only through their relative angle θ. We may therefore write them as $f^{\mathrm{s}}(\theta)$ and $f^{\mathrm{a}}(\theta)$. Furthermore, it is convenient to make them dimensionless:

$$F^{\mathrm{s}}(\theta) \equiv D_{\mathrm{F}} f^{\mathrm{s}}(\theta), \quad F^{\mathrm{a}}(\theta) \equiv D_{\mathrm{F}} f^{\mathrm{a}}(\theta), \tag{8.39}$$

where D_{F} is the density of state at the Fermi momentum defined in Eq. (8.26). Expanding these in terms of Legendre polynomials, we have

$$F^{\mathrm{s}}(\theta) = \sum_{\ell=0}^{\infty} P_\ell(\cos\theta) F_\ell^{\mathrm{s}}, \quad F^{\mathrm{a}}(\theta) = \sum_{\ell=0}^{\infty} P_\ell(\cos\theta) F_\ell^{\mathrm{a}}. \tag{8.40}$$

In general, the properties of the Fermi liquid are described by an infinite number of Landau parameters $F_\ell^{\mathrm{s,a}}$. In practice, only the first few of them are sufficient—a fact that makes Fermi liquid theory a versatile tool to relate a variety of experimental results to one another.

Here, we only cite a few important results, leaving their derivations to Appendix C. The ratio of the effective mass M^* to the bare mass M is given by

$$\frac{M^*}{M} = 1 + \frac{1}{3}F_1^{\mathrm{s}}. \tag{8.41}$$

The ratio of the ordinary (first) sound velocity c to the Fermi velocity v_{F} is given by

$$\frac{c}{v_{\mathrm{F}}} = \sqrt{\frac{1+F_0^{\mathrm{s}}}{3+F_1^{\mathrm{s}}}}. \tag{8.42}$$

In the limit of weak interaction ($F_0^{\mathrm{s}} \to 0,\ F_1^{\mathrm{s}} \to 0$), Eq. (8.42) gives $c = v_{\mathrm{F}}/\sqrt{3}$. The paramagnetic (or Pauli) spin susceptibility χ is independent of temperature and given by

$$\frac{\chi}{\chi_{\mathrm{F}}} = \frac{1}{1+F_0^{\mathrm{a}}}\frac{M^*}{M}, \tag{8.43}$$

where χ_{F} is the value for the ideal Fermi gas given in Eq. (8.12). The specific heat c_v and entropy density s can be calculated in a manner similar to the ideal Fermi gas. They are equal to each other, proportional to temperature, and given by Eq. (8.9) with M replaced by M^*:

$$c_v = s = \frac{M^* p_{\rm F}}{3\hbar^2} k_{\rm B}^2 T. \tag{8.44}$$

The positivity of the effective mass requires

$$1 + \frac{1}{3} F_1^{\rm s} > 0. \tag{8.45}$$

The positivity of the sound velocity requires (8.45) and

$$1 + F_0^{\rm s} > 0. \tag{8.46}$$

The positivity of the Pauli paramagnetic susceptibility requires

$$1 + F_0^{\rm a} > 0. \tag{8.47}$$

If this condition is not satisfied, the system may undergo a phase transition into an itinerant ferromagnetic state.

The instability criterion ($F^a < -1$) of the Fermi liquid can be restated for the case of a contact interaction as follows. In this case, $V(\mathbf{p})$ is independent of $\mathbf{p}$ and equal to $4\pi\hbar^2 a_{\rm s}/M^*$, and $F^a = -2p_{\rm F}a_{\rm s}/\pi$. Therefore, the condition of $F^a < -1$ gives

$$p_{\rm F} a_{\rm s} > \frac{\pi}{2}. \tag{8.48}$$

This is the Stoner criterion for itinerant ferromagnetism.

8.3 Cooper Problem

Cooper examined a situation in which two fermions are placed on a Fermi surface and found that if they undergo attractive interaction, they form a bound state irrespective of how weak the interaction is [Cooper (1956)]. This result is very different from the case of two fermions in free space; in the latter case, to form a bound state, the attractive interaction must be sufficiently strong to overcome the zero-point kinetic pressure.

8.3.1 *Two-body problem*

The Schrödinger equation for the wave function Ψ of two fermions with mass M interacting via a potential V is given by

$$-\frac{\hbar^2}{2M}\left(\frac{\partial^2}{\partial \mathbf{r}_1^2} + \frac{\partial^2}{\partial \mathbf{r}_2^2}\right)\Psi + V(\mathbf{r}_1 - \mathbf{r}_2)\Psi = E\Psi. \tag{8.49}$$

Introducing the center-of-mass coordinate $\mathbf{R} = (\mathbf{r}_1 + \mathbf{r}_2)/2$ and the relative coordinate $\mathbf{r} = \mathbf{r}_1 - \mathbf{r}_2$, Eq. (8.49) is rewritten as

$$-\frac{\hbar^2}{2M}\left(2\frac{\partial^2}{\partial \mathbf{r}^2} + \frac{1}{2}\frac{\partial^2}{\partial \mathbf{R}^2}\right)\Psi + V(\mathbf{r})\Psi = E\Psi. \tag{8.50}$$

Let us assume that Ψ is the product of the spin part χ, relative-coordinate part $\varphi(\mathbf{r})$, and center-of-mass part that we consider to be a plane wave $e^{i\mathbf{KR}}$:

$$\Psi = \chi\varphi e^{i\mathbf{KR}}. \tag{8.51}$$

Substituting this into Eq. (8.50), we obtain the Schrödinger equation for φ:

$$\left(-\frac{\hbar^2}{M}\frac{\partial^2}{\partial \mathbf{r}^2} + V(\mathbf{r})\right)\varphi(\mathbf{r}) = \tilde{E}\varphi(\mathbf{r}), \tag{8.52}$$

where $\tilde{E} \equiv E - \hbar^2K^2/4M$. Because the total wave function must be antisymmetric under the exchange of two fermions, φ must be symmetric under $\mathbf{r} \to -\mathbf{r}$ for a spin-singlet pair and antisymmetric for a spin-triplet one.

We solve Eq. (8.52) in Fourier space. Substituting Fourier expansions

$$\varphi(\mathbf{r}) = \frac{1}{\Omega}\sum_{\mathbf{k}} \varphi_{\mathbf{k}} e^{i\mathbf{kr}}, \quad V(\mathbf{r}) = \frac{1}{\Omega}\sum_{\mathbf{k}} V_{\mathbf{k}} e^{i\mathbf{kr}} \tag{8.53}$$

into Eq. (8.52), we obtain

$$2\epsilon_k\varphi_{\mathbf{k}} + \frac{1}{\Omega}\sum_{\mathbf{k}'} V_{\mathbf{k}-\mathbf{k}'}\varphi_{\mathbf{k}'} = \tilde{E}\varphi_{\mathbf{k}}, \tag{8.54}$$

where $\epsilon_k \equiv \hbar^2k^2/2M$. Hence, we obtain

$$\varphi_{\mathbf{k}} = \frac{1}{\tilde{E} - 2\epsilon_k}\frac{1}{\Omega}\sum_{\mathbf{k}'} V_{\mathbf{k}-\mathbf{k}'}\varphi_{\mathbf{k}'}, \tag{8.55}$$

where $\varphi_{-\mathbf{k}} = \varphi_{\mathbf{k}}$ for the spin-singlet pair and $\varphi_{-\mathbf{k}} = -\varphi_{\mathbf{k}}$ for the spin-triplet one. It follows from Eq. (8.55) that $V_{-\mathbf{k}} = V_{\mathbf{k}}$ ($V_{-\mathbf{k}} = -V_{\mathbf{k}}$) for the spin-singlet (spin-triplet) case.

We substitute

$$\varphi_{\mathbf{k}} = \sum_{\ell=0}^{\infty}\sum_{m=-\ell}^{\ell} \varphi_\ell(k) Y_\ell^m(\hat{\mathbf{k}}), \tag{8.56}$$

$$V_{\mathbf{k}-\mathbf{k}'} = 4\pi\sum_{\ell=0}^{\infty} V_\ell(k, k') \sum_{m=-\ell}^{\ell} Y_\ell^m(\hat{\mathbf{k}}) Y_\ell^{-m}(\hat{\mathbf{k}}') \tag{8.57}$$

in Eq. (8.55), where Y_ℓ^m are spherical harmonics with $\hat{\mathbf{k}} \equiv \mathbf{k}/k$ and $\hat{\mathbf{k}}' \equiv \mathbf{k}'/k'$. Then, we obtain

$$\sum_{\ell=0}^{\infty} \varphi_\ell(k) Y_\ell^m(\hat{\mathbf{k}}) = \frac{4\pi}{\tilde{E} - 2\epsilon_k} \frac{1}{\Omega} \sum_{\mathbf{k}'} \sum_{\ell m} \varphi_\ell(\hat{\mathbf{k}}') Y_\ell^m(\hat{\mathbf{k}}') \times \sum_{\ell' m'} Y_{\ell'}^{m'}(\hat{\mathbf{k}}) Y_{\ell'}^{-m'}(\hat{\mathbf{k}}'). \tag{8.58}$$

Rewriting the sum over $\mathbf{k}'$ as

$$\frac{1}{\Omega} \sum_{\mathbf{k}'} \to \int_0^\infty k'^2 dk' \int d\Omega_{\mathbf{k}'}, \tag{8.59}$$

where $\int d\Omega_{\mathbf{k}'}$ denotes the integral over the solid angle, and using the orthonormality condition

$$\int d\Omega_{\mathbf{k}'} Y_\ell^m(\hat{\mathbf{k}}') Y_{\ell'}^{-m'}(\hat{\mathbf{k}}') = \delta_{\ell\ell'} \delta_{mm'}, \tag{8.60}$$

Eq. (8.58) leads to

$$\varphi_\ell(k) = \frac{4\pi}{\tilde{E} - 2\epsilon_k} \int_0^\infty k'^2 dk' \varphi_\ell(k') V_\ell(k, k'). \tag{8.61}$$

If V_ℓ is nonzero only near the Fermi surface, we obtain

$$\varphi_\ell(k) = \frac{C_\ell}{\tilde{E} - 2\epsilon_k}, \tag{8.62}$$

where C_ℓ is a constant that is proportional to the coupling constant for the channel with orbital angular momentum ℓ and the density of states at the Fermi surface.

For the spin-singlet case with a delta-function interaction $V(\mathbf{r}) = -V\delta(\mathbf{r})$, we have $V_{\mathbf{k}} = -V$, and Eq. (8.55) leads to

$$1 = \frac{V}{\Omega} \sum_{\mathbf{k}} \frac{1}{2\epsilon_k - \tilde{E}}. \tag{8.63}$$

Converting the sum into an integral, we find that the right-hand side diverges for large k. This divergence is an artifact of using the delta function in three dimensions. Physically, there must be a cutoff k_c for k, and therefore, the sum over $k > k_c$ must be subtracted from the sum. We subtract $(V/\Omega) \sum_{\mathbf{k}} 1/(2\epsilon_k)$ from both sides of Eq. (8.63) and divide the resulting equation by $1 - (V/\Omega) \sum_{\mathbf{k}} 1/(2\epsilon_k)$:

$$1 = \frac{-V}{1 - \frac{V}{\Omega} \sum_{\mathbf{k}} \frac{1}{2\epsilon_k}} \frac{1}{\Omega} \sum_{\mathbf{k}} \left(\frac{1}{2\epsilon_k} - \frac{1}{2\epsilon_k - \tilde{E}} \right). \tag{8.64}$$

When a bound state is formed ($\tilde{E} < 0$), the summation on the right-hand side converges, giving

$$\frac{1}{\Omega}\sum_{\mathbf{k}}\left(\frac{1}{2\epsilon_k} - \frac{1}{2\epsilon_k - \tilde{E}}\right) = \frac{\sqrt{M|\tilde{E}|}}{4\pi\hbar^3}. \tag{8.65}$$

The coefficient in Eq. (8.64) can be interpreted as the zero-energy limit of the t-matrix. To understand this, we expand it as follows:

$$\frac{-V}{1 - \frac{V}{\Omega}\sum_{\mathbf{k}}\frac{1}{2\epsilon_k}} = -V\sum_{n=0}^{\infty}\left[\frac{1}{\Omega}\sum_{\mathbf{k}}\frac{1}{-2\epsilon_k}(-V)\right]^n. \tag{8.66}$$

Recalling $V_k = -V$, this is equivalent to the zero-energy limit of the t-matrix in free space defined by

$$T(E) = \sum_{n=0}^{\infty} V\left[G_0(E)V\right]^n, \tag{8.67}$$

where $G_0(E) = (E - H_0 + i0^+)^{-1}$ is the retarded Green's function with $\hat{H}_0$ being the free Hamiltonian. As schematically illustrated in Fig. 8.2, the t-matrix gives the contribution arising from all orders of binary collisions. If only the s-wave interaction is involved, as in the present case, $T(E = 0)$

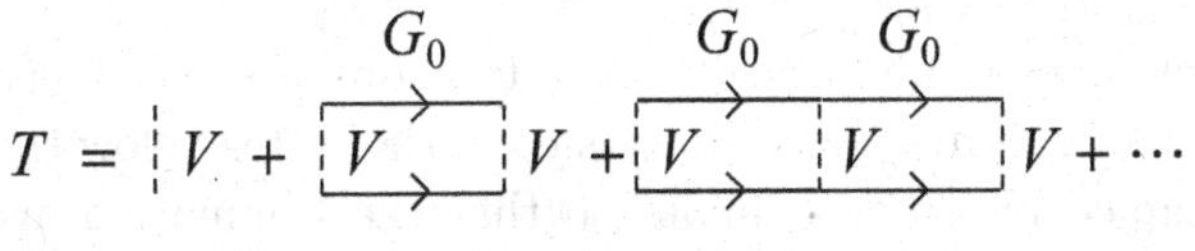

G_0 G_0 G_0

$T = V + V \; V + V \; V \; V + \cdots$

Fig. 8.2 Schematic illustration of the t-matrix (8.67), where G_0 describes the free propagation of two fermions (⇉) and V describes the interaction between them (⁞V).

is expressed in terms of the s-wave scattering length $a_{\rm s}$ as

$$T(E = 0) = \frac{-V}{1 - \frac{V}{\Omega}\sum_{\mathbf{k}}\frac{1}{2\epsilon_k}} = \frac{4\pi\hbar^2}{M}a_{\rm s}. \tag{8.68}$$

Substituting Eqs. (8.65) and (8.68) into Eq. (8.64), we obtain

$$1 = \frac{\sqrt{M|\tilde{E}|}}{\hbar M}a_{\rm s}. \tag{8.69}$$

Hence, the bound-state energy is given by

$$\tilde{E} = -\frac{\hbar^2}{Ma_{\rm s}^2}. \tag{8.70}$$

It is clear from Eq. (8.69) that the bound state can be formed in free space only if the sign of the s-wave scattering length is positive. The wave function for the bound state is calculated to give

$$\varphi(\mathbf{r}) \propto \frac{1}{\Omega}\sum_{\mathbf{k}} \frac{1}{2\epsilon_k + |\tilde{E}|} e^{i\mathbf{k}\mathbf{r}} = \frac{M}{2\pi\hbar^2}\frac{e^{-r/a_s}}{r}. \tag{8.71}$$

Therefore, the size of the dimer is of the same order as the s-wave scattering length.

It should be noted that $a_s > 0$ does not imply that the interaction potential is repulsive; it must be attractive and should, in fact, be sufficiently strong for the two atoms to bind together against the zero-point kinetic pressure. To see this, we explicitly calculate the sum in Eq. (8.68) by introducing the cutoff k_c for the wave number, transforming it into an integral. We then obtain

$$V = \frac{1}{D(\varepsilon_{k_c})\left(1 - \frac{\pi}{2k_c a_s}\right)} > \frac{1}{D(\varepsilon_{k_c})}, \tag{8.72}$$

where $D(\varepsilon_{k_c})$ is the density of states defined in Eq. (8.3). Recalling that $V(\mathbf{r}) = -V\delta(\mathbf{r})$, Eq. (8.72) implies that the attractive interaction must be stronger than $D^{-1}(\varepsilon_{k_c})$ for the two particles to form a bound state.

8.3.2 *Many-body problem*

The above result changes dramatically if two fermions are placed on a Fermi surface rather than in a free space. In this case, the states below the Fermi energy are occupied by other fermions so that the summation over $\mathbf{k}$ in Eq. (8.64) should be restricted to $k > k_F$. When the numbers of spin-up and spin-down atoms are equal, it appears reasonable to assume that pairing occurs most effectively when the center-of-mass momentum of the pair is zero ($\mathbf{K} = \mathbf{0}$). The sum in Eq. (8.64) is then calculated to give

$$\begin{aligned}\frac{1}{\Omega}\sum_{\mathbf{k}}\left(\frac{1}{2\varepsilon_{\mathbf{k}}} - \frac{1}{2\varepsilon_{\mathbf{k}} - E}\right) &= \int_{E_F}^{\infty} d\varepsilon D(\varepsilon)\left(\frac{1}{2\varepsilon} - \frac{1}{2\varepsilon - E}\right)\\ &= -\frac{1}{2}D(E_F)\sqrt{\frac{E}{2E_F}}\ln\frac{1+\sqrt{\frac{E}{2E_F}}}{1-\sqrt{\frac{E}{2E_F}}}.\end{aligned} \tag{8.73}$$

The coefficient in Eq. (8.64) is now expressed in terms of the s-wave scattering length at the Fermi energy:

$$\frac{-V}{1 - \frac{V}{\Omega}\sum_{k_F<k<k_c}\frac{1}{2\epsilon_k}} = \frac{4\pi\hbar^2}{M}a_s(E_F). \tag{8.74}$$

Substituting Eqs. (8.73) and (8.74) into Eq. (8.64), we obtain

$$1 = -\frac{k_F a_s(E_F)}{\pi}\sqrt{\frac{E}{2E_F}}\ln\frac{1+\sqrt{\frac{E}{2E_F}}}{1-\sqrt{\frac{E}{2E_F}}}. \tag{8.75}$$

This equation has a bound state solution ($E < 2E_F$) if and only if $a_s(E_F) < 0$ with the result

$$E = 2E_F - 8E_F \exp\left(-\frac{\pi}{k_F|a_s(E_F)|}\right). \tag{8.76}$$

We thus find that two fermions near the Fermi surface form a bound state irrespective of how weak the attractive interaction is. This is a nonperturbative effect since the scattering length appears in the exponent of Eq. (8.76) in the form of $1/|a_s|$. We also note that the exponent in Eq. (8.76) is inversely proportional to the density of states at the Fermi energy [see Eq. (8.109)]. Should the center-of-mass momentum of the pair be nonzero ($\mathbf{K} \neq 0$), the density of states available for the pair would reduce, and hence, the binding energy would decrease exponentially. This is the main reason why the pair state with $\mathbf{K} = 0$ is usually preferred.

The size d of the bound state can be estimated by equating the last term in Eq. (8.76) to $\hbar^2/Md^2$ with the result

$$d = \frac{2}{k_F}\exp\left(\frac{\pi}{k_F|a_s(E_F)|}\right) \gg k_F^{-1}. \tag{8.77}$$

Therefore, in the presence of the Fermi sea and in the weak-coupling limit, d is much larger than the average interparticle spacing $\sim k_F^{-1}$. This is in a sharp contrast with the case of a dimer in free space, where the size of the dimer is of the order of the scattering length [see Eq. (8.71)].

The relation between $a_s(E_F)$ and V can be found by performing the sum over $\mathbf{k}$ in Eq. (8.74):

$$V = \frac{1}{D(\varepsilon_{k_c})\left(1 - \frac{k_F}{k_c} + \frac{\pi}{2k_c|a_s(E_F)|}\right)}. \tag{8.78}$$

Because of the positive sign of the last term in the denominator, $V \to 0$ as $a_s(E_F) \to 0$. Therefore, a small negative a_s corresponds to a weak attractive interaction $V(\mathbf{r}) = -V\delta(\mathbf{r})$. The relation between $a_s(E_F)$ and its free-space value a_s can be obtained by rewriting the left-hand side of Eq. (8.74) as follows:

$$\frac{-V}{1-\frac{V}{\Omega}\sum\limits_{k_F<k<k_c}\frac{1}{2\epsilon_k}} = \frac{\dfrac{-V}{1-\frac{V}{\Omega}\sum\limits_{0<k<k_c}\frac{1}{2\epsilon_k}}}{1-\dfrac{-V}{1-\frac{V}{\Omega}\sum\limits_{0<k<k_c}\frac{1}{2\epsilon_k}}\frac{1}{\Omega}\sum\limits_{0<k<k_F}\frac{1}{2\epsilon_k}} = \frac{\frac{4\pi\hbar^2}{M}a_s}{1-\frac{2}{\pi}k_F a_s}.$$

Hence, we find that

$$a_{\rm s}(E_{\rm F}) = \frac{a_{\rm s}}{1 - \frac{2}{\pi} k_{\rm F} a_{\rm s}}. \tag{8.79}$$

Therefore, $a_{\rm s}(E_{\rm F}) \simeq a_{\rm s}$ if $k_{\rm F}|a_{\rm s}| \ll 1$, *i.e.*, if the interaction is weak.

8.4 Bardeen–Cooper–Schrieffer (BCS) Theory

The discussion in the preceding subsection suggests that if two fermions are placed near the Fermi surface, they form a bound state below the Fermi surface [see Eq. (8.76)]. However, at absolute zero, all states below the Fermi energy are occupied. A bound pair can only be formed if another pair is pushed outside the Fermi sea. Moreover, because there exist many fermions near the Fermi surface, such pairing of fermions should occur in a collective and self-consistent manner.

In fact, Bardeen, Cooper, and Schrieffer [Bardeen, *et al.* (1957)] showed that a degenerate Fermi gas with weak attractive interaction undergoes a second-order phase transition at a critical temperature below which ODLRO of Cooper pairs emerge.

It seems natural to assume that the zero-temperature ground state of the Fermi system takes the following form:

$$|\Psi_N\rangle = Z \left(\sum_{\mathbf{k}} \alpha_{\mathbf{k}} \hat{c}^\dagger_{\mathbf{k}\uparrow} \hat{c}^\dagger_{-\mathbf{k}\downarrow} \right)^{\frac{N}{2}} |\text{vac}\rangle, \tag{8.80}$$

where $|\text{vac}\rangle$ is the vacuum state, $\hat{c}^\dagger_{\mathbf{k}\sigma}$ is the creation operator of a fermion with wave vector $\mathbf{k}$ and spin $\sigma = \uparrow, \downarrow$, Z is a normalization constant, and $\alpha_{\mathbf{k}}$'s are the variational parameters that are determined so as to minimize the Hamiltonian of the system. We note that $|\Psi_N\rangle$ includes the normal state. In fact, if we choose $\alpha_{\mathbf{k}} = \theta(k_{\rm F} - |\mathbf{k}|)$, we obtain

$$\left(\sum_{\mathbf{k}} \theta(k_{\rm F} - |\mathbf{k}|) \hat{c}^\dagger_{\mathbf{k}\uparrow} \hat{c}^\dagger_{-\mathbf{k}\downarrow} \right)^{\frac{N}{2}} |\text{vac}\rangle \propto \prod_{k<k_{\rm F}} \hat{c}^\dagger_{\mathbf{k}\uparrow} \hat{c}^\dagger_{-\mathbf{k}\downarrow} |\text{vac}\rangle. \tag{8.81}$$

To proceed further with the calculation, it is convenient to switch from ψ_N to the coherent state representation

$$|\psi_{\rm BCS}\rangle = Z \exp\left(\sum_{\mathbf{k}} \alpha_{\mathbf{k}} \hat{c}^\dagger_{\mathbf{k}\uparrow} \hat{c}^\dagger_{-\mathbf{k}\downarrow} \right) |\text{vac}\rangle. \tag{8.82}$$

Expanding the exponential and using the Fermi statistics (*i.e.*, $\hat{c}^{\dagger\,2}_{\mathbf{k}\sigma} = 0$, etc.), we find that

$$|\psi_{\rm BCS}\rangle = \prod_{\mathbf{k}} \left(u_{\mathbf{k}} + v_{\mathbf{k}} \hat{c}^\dagger_{\mathbf{k}\uparrow} \hat{c}^\dagger_{-\mathbf{k}\downarrow}\right) |{\rm vac}\rangle, \tag{8.83}$$

where $u_{\mathbf{k}} = 1/\sqrt{1+|\alpha_k|^2}$ and $v_{\mathbf{k}} = \alpha_{\mathbf{k}}/\sqrt{1+|\alpha_k|^2}$, so that

$$|u_{\mathbf{k}}|^2 + |v_{\mathbf{k}}|^2 = 1. \tag{8.84}$$

The wave function (8.83) is referred to as the BCS wave function [Bardeen, *et al.* (1957)]. Once the BCS wave function is obtained, its number-conserving counterpart $|\psi_N\rangle$ can be obtained as follows:

$$|\psi_N\rangle \propto \int_0^{2\pi} d\varphi e^{-iN\varphi/2} \prod_{\mathbf{k}} \left(u_{\mathbf{k}} + e^{i\varphi} v_{\mathbf{k}} \hat{c}^\dagger_{\mathbf{k}\uparrow} \hat{c}^\dagger_{-\mathbf{k}\downarrow}\right) |{\rm vac}\rangle. \tag{8.85}$$

We shall determine the variational parameters $u_{\mathbf{k}}$ and $v_{\mathbf{k}}$ so as to minimize the expectation value of the Hamiltonian

$$\hat{H} = \sum_{\mathbf{k}\sigma} (\epsilon_k - \mu) \hat{c}^\dagger_{\mathbf{k}\sigma} \hat{c}_{\mathbf{k}\sigma} - \frac{V}{\Omega} \sum_{\mathbf{k}\mathbf{k}'\mathbf{k}''} \hat{c}^\dagger_{\mathbf{k}+\mathbf{k}''\uparrow} \hat{c}^\dagger_{\mathbf{k}'-\mathbf{k}''\downarrow} \hat{c}_{\mathbf{k}'\downarrow} \hat{c}_{\mathbf{k}\uparrow}, \tag{8.86}$$

where we introduce the chemical potential μ to set the average number of fermions to a prescribed value, and the interaction is chosen to be $V(\mathbf{r}) = -V\delta(\mathbf{r})$ $(V > 0)$. The expectation value of $\hat{H}$ over $|\psi_{\rm BCS}\rangle$ is calculated to give

$$\langle \hat{H} \rangle = 2\sum_{\mathbf{k}} \xi_k v_{\mathbf{k}}^2 - \frac{V}{\Omega} \sum_{\mathbf{k},\mathbf{k}'} u_{\mathbf{k}} v_{\mathbf{k}} u_{\mathbf{k}'} v_{\mathbf{k}'} - \frac{V}{4\Omega} \langle \hat{N} \rangle^2, \tag{8.87}$$

where $\xi_k \equiv \epsilon_k - \mu$ and

$$\langle \hat{N} \rangle = \sum_{\mathbf{k}\sigma} \langle \hat{c}^\dagger_{\mathbf{k}\sigma} \hat{c}_{\mathbf{k}\sigma} \rangle = 2\sum_{\mathbf{k}} v_{\mathbf{k}}^2 \tag{8.88}$$

is the average number of particles that we set to a constant value by appropriately choosing μ. We also note that $u_{\mathbf{k}}$ and $v_{\mathbf{k}}$ can be considered to be real as far as the ground-state properties are concerned.

The variational parameters $u_{\mathbf{k}}$ and $v_{\mathbf{k}}$ can be determined by requiring that

$$\frac{\partial}{\partial v_{\mathbf{k}}} \langle H \rangle = 0. \tag{8.89}$$

It follows from Eq. (8.89) and $u_{\mathbf{k}}^2 + v_{\mathbf{k}}^2 = 1$ that

$$2\xi_{\mathbf{k}} u_{\mathbf{k}} v_{\mathbf{k}} = \Delta(u_{\mathbf{k}}^2 - v_{\mathbf{k}}^2), \tag{8.90}$$

where

$$\Delta \equiv \frac{V}{\Omega}\sum_{\mathbf{k}} u_{\mathbf{k}} v_{\mathbf{k}} = \frac{V}{\Omega}\sum_{\mathbf{k}} \langle \psi_{\mathrm{BCS}} | \hat{c}^{\dagger}_{\mathbf{k}\uparrow} \hat{c}^{\dagger}_{-\mathbf{k}\downarrow} | \psi_{\mathrm{BCS}} \rangle \tag{8.91}$$

is called the order parameter of the BCS state and $\Delta/V = \sum_{\mathbf{k}} u_{\mathbf{k}} v_{\mathbf{k}}$ is called the pairing amplitude. We can solve Eq. (8.90), together with $u_{\mathbf{k}}^2 + v_{\mathbf{k}}^2 = 1$, to obtain

$$u_{\mathbf{k}} = \sqrt{\frac{E_k + \xi_k}{2E_k}}, \quad v_{\mathbf{k}} = \sqrt{\frac{E_k - \xi_k}{2E_k}}, \tag{8.92}$$

where

$$E_k = \sqrt{\xi_k^2 + \Delta^2} = \sqrt{(\epsilon_k - \mu)^2 + \Delta^2} \tag{8.93}$$

is the excitation energy of the system. Dividing both sides of Eq. (8.90) by ξ_k and summing over $\mathbf{k}$, we obtain the gap equation

$$1 = \frac{V}{2\Omega}\sum_{\mathbf{k}} \frac{1}{E_k}. \tag{8.94}$$

The right-hand side of Eq. (8.94) involves ultraviolet divergence that can be eliminated in a manner similar to what is done in deriving Eq. (8.64). That is, we first rewrite Eq. (8.94) as

$$1 = \frac{-V}{1 - \frac{V}{\Omega}\sum_{\mathbf{k}} \frac{1}{2\epsilon_k}} \frac{1}{\Omega}\sum_{\mathbf{k}} \left(\frac{1}{2\varepsilon_K} - \frac{1}{2E_k} \right) \tag{8.95}$$

and use Eq. (8.68) to eliminate the bare interaction V in favor of the experimentally measured s-wave scattering length a_{s}:

$$1 = \frac{4\pi\hbar^2 a_{\mathrm{s}}}{M} \frac{1}{\Omega}\sum_{\mathbf{k}} \left(\frac{1}{2\epsilon_k} - \frac{1}{2E_k} \right). \tag{8.96}$$

The gap equation (8.78), together with the number equation

$$N = 2\sum_{\mathbf{k}} v_{\mathbf{k}}^2 = \sum_{\mathbf{k}} \left(1 - \frac{\xi_k}{E_k} \right), \tag{8.97}$$

determines the energy gap Δ and the chemical potential μ. Converting the sums in Eqs. (8.96) and (8.97) into integrals, we obtain

$$\frac{\pi}{k_{\mathrm{F}} a_{\mathrm{s}}} = \frac{1}{\sqrt{E_{\mathrm{F}}}} \int_0^{\infty} d\varepsilon \sqrt{\varepsilon} \left(\frac{1}{\varepsilon} - \frac{1}{E} \right), \tag{8.98}$$

$$1 = \frac{3}{4\sqrt{E_{\mathrm{F}}^3}} \int_0^{\infty} d\varepsilon \sqrt{\varepsilon} \left(1 - \frac{\varepsilon - \mu}{E} \right), \tag{8.99}$$

where E is given in Eq. (8.93). The ground-state properties of the system can be investigated by solving Eqs. (8.98) and (8.99) for a given dimensionless parameter $k_F a_s$.

Let us consider the momentum distribution $n_{\mathbf{k}} = \langle \hat{c}^\dagger_{\mathbf{k}\sigma} \hat{c}^\dagger_{\mathbf{k}\sigma} \rangle = v^2_{\mathbf{k}}$. From Eqs. (8.92) and (8.93), we find that

$$n_{\mathbf{k}} = v^2_{\mathbf{k}} = \frac{1}{2}\left(1 - \frac{\epsilon_k - \mu}{\sqrt{(\epsilon_k - \mu)^2 + \Delta^2}}\right). \tag{8.100}$$

As shown below, in the BCS limit, where $a_s < 0$ and $k_F|a_s| \ll 1$, we have $\mu \approx \varepsilon_F \gg \Delta$, so that $n_{\mathbf{k}}$ is almost the same as the zero-temperature Fermi distribution $n_{\mathbf{k}} = \theta(\varepsilon_F - \epsilon_k)$ except for a narrow region of width $\sim\Delta$ around $\varepsilon_{\mathbf{k}} = \mu$ [see Fig. 8.3 (a)]. On the other hand, in the BEC limit where $a_s > 0$ and $k_F a_s \ll 1$, we have $\mu < 0$ and $|\mu| \gg \Delta$. Therefore, $n_{\mathbf{k}}$ is always much smaller than one and it is a monotonically decreasing function of ϵ_k [see Fig. 8.3 (b)]; in addition, the fermionic character disappears. The wavenum-

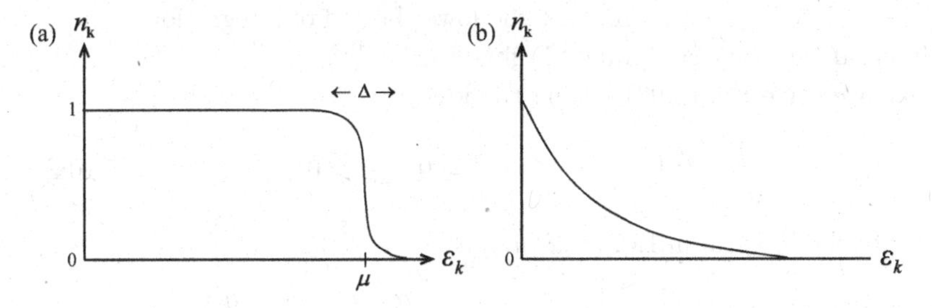

Fig. 8.3 Zero-temperature momentum distribution function $n_{\mathbf{k}}$ in the BCS limit (a) and BEC limit (b). In the BCS limit, $n_{\mathbf{k}} \ll 1$.

ber that minimizes the energy gap (8.93) decreases from $k = k_F$ in the BCS limit to $k = 0$ when the chemical potential changes sign and remains $k = 0$ toward the BEC limit. This suggests that the critical temperature in the BEC regime is governed not by gapped pair-breaking excitations but by gapless center-of-mass collective-mode excitations of Cooper pairs, as discussed in Sec. 8.7.

The BEC theory assumes that the system condenses in a single state characterized by the operator

$$\hat{b}^\dagger = \sum_{\mathbf{k}} \alpha_{\mathbf{k}} \hat{c}^\dagger_{\mathbf{k}\uparrow} \hat{c}^\dagger_{-\mathbf{k}\downarrow}, \tag{8.101}$$

where $\sum_{\mathbf{k}} |\alpha_{\mathbf{k}}|^2 = 1$. The commutation relation of this operator is calculated to be

$$[\hat{b}, \hat{b}^\dagger] = 1 - \sum_{\mathbf{k}} (\hat{c}^\dagger_{\mathbf{k}\uparrow} \hat{c}_{\mathbf{k}\uparrow} - \hat{c}^\dagger_{-\mathbf{k}\downarrow} \hat{c}_{-\mathbf{k}\downarrow}). \tag{8.102}$$

In the BEC limit, where $n_{\mathbf{k}\uparrow} \ll 1$ and $n_{\mathbf{k}\downarrow} \ll 1$, the right-hand side of Eq. (8.102) can be approximated by 1, and the system may be regarded as a collection of bosons.

8.5 BCS–BEC Crossover at $T = 0$

We solve Eqs. (8.98) and (8.99) in the BCS limit, where $(k_{\mathrm{F}}a_{\mathrm{s}})^{-1} \to -\infty$. Integrating the right-hand side of Eq. (8.99) by parts, we have

$$1 = \frac{\Delta^2}{2E_{\mathrm{F}}^{\frac{3}{2}}} \int_0^\infty d\varepsilon \frac{\varepsilon^{\frac{3}{2}}}{\left[(\varepsilon-\mu)^2+\Delta^2\right]^{\frac{3}{2}}}. \tag{8.103}$$

In the BCS limit, the attractive interaction is very weak, and we expect that μ is very close to E_{F} and $\Delta \ll \mu$. Therefore, the integrand exhibits a sharp peak at $\varepsilon \approx \mu$. We may therefore expand the numerator in Eq. (8.103) as $\varepsilon^{\frac{3}{2}} \simeq \mu^{\frac{3}{2}}\left[1+3(\varepsilon-\mu)/(2\mu)\right]$ and set the lower bound of integration as $-\infty$. The integral can then be evaluated, giving $\mu = E_{\mathrm{F}}$.

Next, we solve Eq. (8.98). Using the identity

$$\int_0^\infty d\varepsilon \left(\frac{1}{\sqrt{\varepsilon}} - \frac{\sqrt{\varepsilon}}{\varepsilon-\mu}\right) = 0 \quad (\mu > 0), \tag{8.104}$$

we rewrite the integral in Eq. (8.98) as

$$\begin{aligned}\int_0^\infty d\varepsilon\sqrt{\varepsilon}\left(\frac{1}{\varepsilon}-\frac{1}{E}\right) &= \int_0^\infty d\varepsilon\frac{\sqrt{\varepsilon}}{\varepsilon-\mu}\left(1-\frac{\varepsilon-\mu}{E}\right)\\ &= \int_0^\infty d\varepsilon\left(2\sqrt{\varepsilon}-\sqrt{\mu}\ln\left|\frac{(\sqrt{\varepsilon}+\sqrt{\mu})^2}{\varepsilon-\mu}\right|\right)\frac{\Delta^2}{E^3},\end{aligned} \tag{8.105}$$

where we perform integration by parts to derive the last equality. Because the attractive interaction is assumed to be very weak in the BCS limit, the energy gap Δ is much smaller than $\mu = E_{\mathrm{F}}$. Therefore, the integrand in Eq. (8.105) exhibits a sharp peak at $\varepsilon = \mu$ and each term can be evaluated as follows:

$$\int_0^\infty d\varepsilon\, 2\sqrt{\varepsilon}\frac{\Delta^2}{E^3} \simeq 2\Delta^2\sqrt{\mu}\int_{-\infty}^\infty dx\frac{1}{(x^2+1)^{\frac{3}{2}}} = 4\sqrt{\mu}, \tag{8.106}$$

$$\begin{aligned}\int_0^\infty d\varepsilon\,\sqrt{\mu}\ln\left|\frac{(\sqrt{\varepsilon}+\sqrt{\mu})^2}{\varepsilon-\mu}\right|\frac{\Delta^2}{E^3} &\simeq \int_0^\infty dx\ln\left|\frac{4\mu}{\Delta x}\right|\frac{2\sqrt{\mu}}{(x^2+1)^{\frac{3}{2}}}\\ &= 2\sqrt{\mu}\left(\frac{8\mu}{\Delta}\right).\end{aligned} \tag{8.107}$$

Substituting these results in Eq. (8.105) and substituting $\mu = E_F$, we obtain the expression for the energy gap in the BCS limit:

$$\Delta = \frac{8E_F}{e^2} e^{-\frac{\pi}{2k_F|a_s|}}. \tag{8.108}$$

The exponent in Eq. (8.108) can be written as

$$\frac{\pi}{2k_F a_s} = \frac{1}{D(E_F)U_0}, \tag{8.109}$$

where $D(E_F)$ is the density of states at the Fermi energy and $U_0 \equiv 4\pi\hbar^2 a_s/M$, the renormalized strength of interaction [see Eq. (8.74)]. The fact that the coupling constant U_0 appears in the denominator indicates that superfluidity in the BCS limit is a nonperturbative effect as a consequence of the Cooper instability discussed in Sec. 8.3.

Next, we consider the BEC limit ($k_F a_s \to 0^+$). In this case, μ is negative and $|\mu| \gg \Delta$, as shown below [see Eq. (8.117)]. We can therefore expand E as

$$E = (\varepsilon + |\mu|)\left[1 + \frac{\Delta^2}{(\varepsilon+|\mu|)^2}\right]^{\frac{1}{2}} \simeq (\varepsilon + |\mu|)\left[1 + \frac{\Delta^2}{2(\varepsilon+|\mu|)^2}\right]. \tag{8.110}$$

The integral in Eq. (8.103) can then be evaluated as

$$\begin{aligned}\int_0^\infty d\varepsilon \frac{\varepsilon^{\frac{3}{2}}}{[(\varepsilon+|\mu|)^2+\Delta^2]^{\frac{3}{2}}} &\simeq \int_0^\infty d\varepsilon \frac{\varepsilon^{\frac{3}{2}}}{(\varepsilon+|\mu|)^3} - \frac{3}{2}\Delta^2 \int_0^\infty d\varepsilon \frac{\varepsilon^{\frac{3}{2}}}{(\varepsilon+|\mu|)^3} \\ &= \frac{3\pi}{8|\mu|^{\frac{3}{2}}}\left(1 - \frac{3\Delta^2}{32\mu^2}\right).\end{aligned} \tag{8.111}$$

Substituting this in Eq. (8.103) and solving for Δ^2, we obtain

$$\Delta^2 = \frac{16}{3\pi} E_F^{\frac{3}{2}} |\mu|^{\frac{1}{2}} \left[1 + \frac{1}{2\pi}\left(\frac{E_F}{|\mu|}\right)^{\frac{3}{2}}\right]. \tag{8.112}$$

On the other hand, integrating the right-hand side of Eq. (8.98) by parts, we have

$$\frac{\pi}{k_F a_s} E_F^{\frac{1}{2}} = \int_0^\infty d\varepsilon 2\sqrt{\varepsilon} \frac{d}{d\varepsilon}\frac{\varepsilon}{E} = \int_0^\infty d\varepsilon \frac{2\sqrt{\varepsilon}(\varepsilon|\mu| + \mu^2 + \Delta^2)}{E^3}. \tag{8.113}$$

Substituting Eq. (8.110) in Eq. (8.113) and carrying out integration, we obtain

$$\mu = -\frac{E_F}{(k_F a_s)^2}\left[1 - \frac{1}{8}\left(\frac{\Delta}{|\mu|}\right)^2\right]. \tag{8.114}$$

Solving Eqs. (8.112) and (8.114), we find that

$$\mu = -\frac{\hbar^2}{2Ma_s^2} + \frac{\pi\hbar^2 a_s}{M} n, \tag{8.115}$$

$$\Delta = \frac{4}{\sqrt{3\pi}} \frac{E_F}{(3\pi^2 n a_s^3)^{\frac{1}{3}}} \left(1 + \frac{3\pi}{2} n a_s^3\right). \tag{8.116}$$

Taking the ratio of $|\mu|$ to Δ, we have

$$\frac{|\mu|}{\Delta} \simeq \frac{\sqrt{3\pi}}{4} \frac{1}{k_F a_s} = \frac{\sqrt{3\pi}}{4} \frac{1}{(3\pi^2 n a_s^3)^{1/3}} \gg 1; \tag{8.117}$$

our initial assumption ($|\mu| \gg \Delta$) is justified.

The chemical potential of a dimer is given by $\mu^{\text{dimer}} = 2\mu$ and it is given from Eq. (8.115) by

$$\mu^{\text{dimer}} = -\frac{\hbar^2}{Ma_s^2} + \frac{4\pi\hbar^2(2a_s)}{2M} n^{\text{dimer}}, \tag{8.118}$$

where $n^{\text{dimer}} = n/2$. The first term on the right-hand side is the binding energy of a dimer in free space, as given in Eq. (8.70), and the second term describes the Hartree energy due to the dimer-dimer interaction. The last term in Eq. (8.118) shows that the s-wave scattering length between dimers is found to be twice as large as the s-wave scattering length between fermions:

$$a^{\text{dimer}} = 2a_s. \tag{8.119}$$

A more precise four-body calculation gives $a^{\text{dimer}} = 0.6a_s$ [Petrov, *et al.* (2004)].

As can be seen from Eq. (8.114), $|\mu^{\text{dimer}}| \gg E_F$ implies $k_F a_s \ll 1$ or, equivalently, $n a_s^3 \ll 1$. Because the wave function of each dimer extends over a_s [see Eq. (8.71)], which is much shorter than the interparticle spacing $n^{-\frac{1}{3}}$, there is little overlap between dimers. Therefore, the Pauli exclusion principle plays a negligible role in the BEC limit and the system behaves like a collection of bosons.

In general, by solving Eqs. (8.98) and (8.99), we can obtain μ and Δ as functions of $(k_F a_s)^{-1}$, as schematically plotted in Fig. 8.4, where both μ and Δ change smoothly between the BCS limit ($(k_F a_s)^{-1} \to -\infty$) and the BEC limit ($(k_F a_s)^{-1} \to +\infty$) [Eagles (1969); Leggett (1980)]. We note that μ changes sign between the two limits, at which point the nature of the excitation gap also changes. In fact, from Eq. (8.93), we find that

$$E_{\text{gap}} = \begin{cases} \Delta & (\mu > 0), \\ \sqrt{\mu^2 + \Delta^2} & (\mu < 0). \end{cases} \tag{8.120}$$

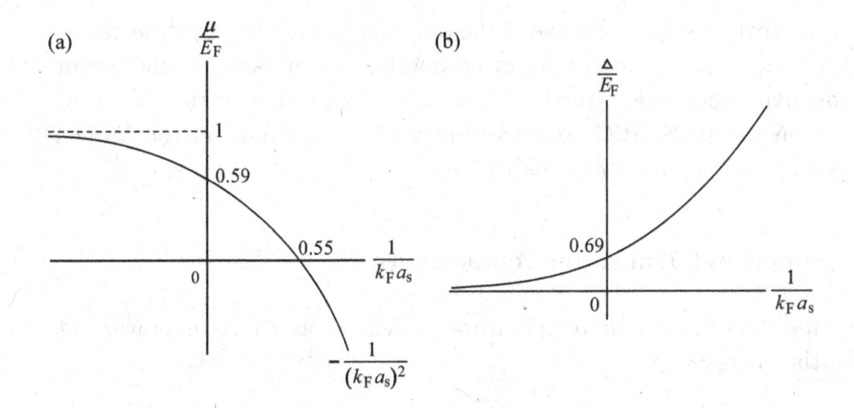

Fig. 8.4 Schematic plots of (a) $\mu/E_{\rm F}$ and (b) $\Delta/E_{\rm F}$ as functions of $(k_{\rm F}a_{\rm s})^{-1}$ obtained from Eqs. (8.98) and (8.99).

When $\mu = 0$, both Eqs. (8.98) and (8.99) can be solved analytically with the results

$$(k_{\rm F}a_{\rm s})^{-1} = \frac{4\Gamma\left(\frac{3}{4}\right)^2}{\pi^{\frac{4}{3}}\Gamma\left(\frac{1}{4}\right)^{\frac{2}{3}}} \simeq 0.553, \tag{8.121}$$

$$\frac{\Delta}{E_{\rm F}} = \frac{4\pi^{\frac{1}{3}}}{\Gamma\left(\frac{1}{4}\right)^{\frac{4}{3}}} \simeq 1.05. \tag{8.122}$$

On the other hand, at the unitary limit ($(k_{\rm F}a_{\rm s})^{-1} = 0$), Eqs. (8.98) and (8.99) can be solved numerically, giving

$$\frac{\Delta}{E_{\rm F}} \simeq 0.686, \tag{8.123}$$

$$\frac{\mu}{E_{\rm F}} \simeq 0.591. \tag{8.124}$$

In the region $|k_{\rm F}a_{\rm s}| \gtrsim 1$ or, equivalently, $n|a_{\rm s}|^3 \gtrsim 1$, the overlap between dimers is significant and we can no longer regard individual dimers as well-defined entities. Because no exchange effect is taken into account, the crossover theory described here cannot describe such a strongly interacting regime, in particular, the unitary limit at which $k_{\rm F}|a_{\rm s}|$ diverges. In fact, quantum Monte Carlo methods give $\Delta/E_{\rm F} \simeq 0.45 \pm 0.05$ [Carlson and Reddy (2008)] and $\mu/E_{\rm F} \simeq 0.43$ [Bulgac, *et al.* (2008)], which differ greatly from (8.123) and (8.124), respectively.

The crossover theory pressupposes that pairs of fermions constitute the fundamental building block that is assumed to be maintained throughout the entire regime, where the size of the pair increases as $(k_F a_s)^{-1}$ decreases from $+\infty$ to $-\infty$. It is an interesting question to investigate whether or not this hypothesis holds true in the unitary regime. A collection of review articles on the BCS–BEC crossover physics can be found in the Varena proceedings [Inguscio, *et al.* (eds.) (2008)].

8.6 Superfluid Transition Temperature

The superfluid transition temperature is defined as the temperature at which the energy gap

$$\Delta = \frac{V}{\Omega}\sum_{\mathbf{k}}\langle \hat{c}^\dagger_{\mathbf{k}\uparrow}\hat{c}^\dagger_{-\mathbf{k}\downarrow}\rangle \tag{8.125}$$

vanishes. To evaluate the pair amplitude $\langle \hat{c}^\dagger_{\mathbf{k}\uparrow}\hat{c}^\dagger_{-\mathbf{k}\downarrow}\rangle$ at finite temperature, we introduce the Bogoliubov quasiparticles whose creation and annihilation operators $\hat{\alpha}^\dagger_{\mathbf{k}\sigma}$ and $\hat{\alpha}_{\mathbf{k}\sigma}$ ($\sigma = \uparrow, \downarrow$) are related to the corresponding fermion operators by [Bogoliubov (1958); Valatin (1958)]

$$\hat{c}_{\mathbf{k}\uparrow} = u_{\mathbf{k}}\hat{\alpha}_{\mathbf{k}\uparrow} + v_{\mathbf{k}}\hat{\alpha}^\dagger_{-\mathbf{k}\downarrow}, \quad \hat{c}_{\mathbf{k}\downarrow} = u_{\mathbf{k}}\hat{\alpha}_{-\mathbf{k}\downarrow} - v_{\mathbf{k}}\hat{\alpha}^\dagger_{-\mathbf{k}\uparrow}, \tag{8.126}$$

where $u_{\mathbf{k}}$ and $v_{\mathbf{k}}$ are given in Eq. (8.92), and Bogoliubov quasiparticle operators are assumed to satisfy the following anticommutation relations:

$$\left\{\hat{\alpha}_{\mathbf{k}\sigma}, \hat{\alpha}^\dagger_{\mathbf{k}'\sigma'}\right\} = \delta_{\mathbf{k}\mathbf{k}'}\delta_{\sigma\sigma'}, \quad \left\{\hat{\alpha}_{\mathbf{k}\sigma}, \hat{\alpha}_{\mathbf{k}'\sigma'}\right\} = 0, \quad \left\{\hat{\alpha}^\dagger_{\mathbf{k}\sigma}, \hat{\alpha}^\dagger_{\mathbf{k}'\sigma'}\right\} = 0. \tag{8.127}$$

At thermal equilibrium, the expectation values of Bogoliubov quasiparticle operators are given by

$$\begin{aligned}&\langle \hat{\alpha}^\dagger_{\mathbf{k}\sigma}\hat{\alpha}_{\mathbf{k}'\sigma'}\rangle = \delta_{\mathbf{k}\mathbf{k}'}\delta_{\sigma\sigma'}f(E_k), \quad \langle \hat{\alpha}_{\mathbf{k}\sigma}\hat{\alpha}^\dagger_{\mathbf{k}'\sigma'}\rangle = \delta_{\mathbf{k}\mathbf{k}'}\delta_{\sigma\sigma'}(1 - f(E_k)),\\ &\langle \hat{\alpha}_{\mathbf{k}\sigma}\hat{\alpha}_{\mathbf{k}'\sigma'}\rangle = \langle \hat{\alpha}^\dagger_{\mathbf{k}\sigma}\hat{\alpha}^\dagger_{\mathbf{k}'\sigma'}\rangle = 0,\end{aligned} \tag{8.128}$$

where E_k is defined in Eq. (8.93) and

$$f(E) = \frac{1}{e^{\beta E} + 1}. \tag{8.129}$$

Substituting Eq. (8.126) into Eq. (8.125), we obtain

$$\Delta = \frac{V}{\Omega}\sum_{\mathbf{k}}\frac{\Delta}{2E_k}(1 - 2f(E)) = \frac{V}{\Omega}\sum_{\mathbf{k}}\frac{\Delta}{2E_k}\tanh\frac{\beta E_k}{2}. \tag{8.130}$$

For the system to have a non-vanishing energy gap Δ, the following gap equation must be satisfied:

$$1 = \frac{V}{\Omega}\sum_{\mathbf{k}}\frac{\tanh\frac{\beta E_k}{2}}{2E_k}. \tag{8.131}$$

To eliminate the bare coupling constant V in favor of the experimentally observable scattering length a_s, we subtract $(V/\Omega)\sum_{\mathbf{k}} 1/2\epsilon_k$ from both sides of Eq. (8.131) and divide the resultant equation by $1-(V/\Omega)\sum_{\mathbf{k}} 1/2\epsilon_k$. Finally, we use Eq. (8.68) to obtain

$$-1 = \frac{2\pi\hbar^2 a_s}{M\Omega}\sum_{\mathbf{k}}\left(\frac{\tanh\frac{\beta E_k}{2}}{E_k} - \frac{1}{\epsilon_k}\right). \tag{8.132}$$

Converting the sum into an integral, we obtain

$$-\frac{\pi}{k_F a_s} = \frac{1}{\sqrt{E_F}}\int_0^\infty d\varepsilon\sqrt{\varepsilon}\left(\frac{\tanh\frac{\beta E}{2}}{E} - \frac{1}{\varepsilon}\right). \tag{8.133}$$

This is the generalization of Eq. (8.98) to finite temperature.

The right-hand side of Eq. (8.133) involves the chemical potential μ and temperature β^{-1}. To determine them, we require another equation, namely, the number equation:

$$N = \sum_{\mathbf{k}\sigma}\langle \hat{c}^\dagger_{\mathbf{k}\sigma}\hat{c}_{\mathbf{k}\sigma}\rangle. \tag{8.134}$$

Substituting Eq. (8.126) into the right-hand side and using Eqs. (8.128) and (8.129), we obtain

$$N = \sum_{\mathbf{k}}\left(1 - \frac{\epsilon_k-\mu}{E_k}\tanh\frac{\beta E_k}{2}\right). \tag{8.135}$$

Converting the sum into an integral, we obtain

$$1 = \frac{3}{4E_F^{\frac{3}{2}}}\int_0^\infty d\varepsilon\sqrt{\varepsilon}\left(1 - \frac{\varepsilon-\mu}{E}\tanh\frac{\beta E}{2}\right). \tag{8.136}$$

This is the finite-temperature extension of Eq. (8.99). At the critical temperature $T_c \equiv (k_B\beta_c)^{-1}$, we have $\Delta = 0$ and $E = |\varepsilon - \mu|$, and Eqs. (8.133) and (8.136) reduce to

$$-\frac{\pi}{k_F a_s} = \frac{1}{\sqrt{E_F}}\int_0^\infty d\varepsilon\sqrt{\varepsilon}\left(\frac{\tanh\frac{\beta_c(\varepsilon-\mu)}{2}}{\varepsilon-\mu} - \frac{1}{\varepsilon}\right), \tag{8.137}$$

$$\begin{aligned}1 &= \frac{3}{4E_F^{\frac{3}{2}}}\int_0^\infty d\varepsilon\sqrt{\varepsilon}\left(1 - \tanh\frac{\beta_c(\varepsilon-\mu)}{2}\right)\\ &= \frac{\beta_c}{4E_F^{\frac{3}{2}}}\int_0^\infty d\varepsilon\frac{\varepsilon^{\frac{3}{2}}}{\cosh^2\frac{\beta_c(\varepsilon-\mu)}{2}}.\end{aligned} \tag{8.138}$$

Let us first consider the BCS limit ($(k_F a_s)^{-1} \to -\infty$). In this case, the integrand in Eq. (8.138) deviates significantly from zero only around

$\varepsilon = \mu > 0$. We thus replace $\varepsilon^{\frac{3}{2}}$ and the lower bound of integration with $\mu^{\frac{3}{2}}$ and $-\infty$, respectively. On evaluating the integral, we obtain $\mu = E_{\rm F}$. Substituting $\mu = E_{\rm F}$ in Eq. (8.137), using Eq. (8.104), and performing integration by parts, we obtain

$$-\frac{\pi}{k_{\rm F}a_{\rm s}} = \frac{\beta_{\rm c}E_{\rm F}}{2}\int_0^\infty dx \frac{2\sqrt{x} + \ln\left|\frac{\sqrt{x}-1}{\sqrt{x}+1}\right|}{\cosh^2\frac{\beta_{\rm c}E_{\rm F}}{2}(x-1)}. \tag{8.139}$$

Because the attractive interaction is very weak in the BCS limit, $\beta_{\rm c}E_{\rm F} \gg 1$, and the integrand deviates significantly from zero only within a narrow region around $x = 1$. We therefore expand $\sqrt{x}$ as $\sqrt{x} \simeq 1 + (x-1)/2$ and replace the lower bound of integration with $-\infty$. Using the formula

$$\int_0^\infty \frac{\ln x}{\cosh^2 x} = \ln\frac{\pi}{4} - \gamma, \tag{8.140}$$

where $\gamma = 0.577216\cdots$ is Euler's constant, we obtain the superfluid transition temperature in the BCS limit:

$$k_{\rm B}T_{\rm c} = \frac{8e^{\gamma-2}}{\pi}E_{\rm F}e^{-\frac{\pi}{2k_{\rm F}|a_{\rm s}|}} \simeq 0.613E_{\rm F}e^{-\frac{\pi}{2k_{\rm F}|a_{\rm s}|}}. \tag{8.141}$$

The ratio of the pair-breaking energy gap 2Δ at absolute zero (Eq. (8.108)) to $k_{\rm B}T_{\rm c}$ gives a universal constant independent of material parameters:

$$\frac{2\Delta(T=0)}{k_{\rm B}T_{\rm c}} = \frac{2\pi}{e^\gamma} \simeq 3.53. \tag{8.142}$$

In fact, there is a substantial many-body correction to the critical temperature in Eq. (8.141) as discussed in Sec. 8.8.

8.7 BCS–BEC Crossover at $T \neq 0$

Next, we consider the BEC limit ($(k_{\rm F}a_{\rm s})^{-1} \to +\infty$). In this limit, $\mu < 0$, and integrating the right-hand side of Eq. (8.137) by parts, we obtain

$$-\frac{\pi}{k_{\rm F}a_{\rm s}} = -\pi\sqrt{\frac{|\mu|}{E_{\rm F}}} - \frac{\beta_{\rm c}|\mu|^{\frac{3}{2}}}{\sqrt{E_{\rm F}}}\int_0^\infty dx \frac{\sqrt{x} - {\rm Arctan}\sqrt{x}}{\cosh^2\frac{\beta_{\rm c}|\mu|}{2}(x+1)}. \tag{8.143}$$

Because $\beta_{\rm c}|\mu| \to \infty$ in the BEC limit, the integral is negligible, giving

$$\mu = -\frac{\hbar^2}{2Ma_{\rm s}^2}. \tag{8.144}$$

Substituting this into the last term in Eq. (8.138) and noting that $\cosh\beta_{\rm c}/(\varepsilon + |\mu|)/2 \simeq e^{\beta_{\rm c}(\varepsilon+|\mu|)/2}/2$, we obtain

$$k_{\rm B}T_{\rm c} \sim \frac{2|\mu|}{3\ln\frac{|\mu|}{E_{\rm F}}}. \tag{8.145}$$

On physical grounds, this result does not appear to be correct because as we discussed in the preceding section, the system should behave like a collection of tightly bound dimers, so that T_c is expected to be close to that of an ideal Bose gas with mass $2M$.

The origin of the discrepancy lies in the fact that T_c is determined by different physical mechanisms in the BCS and BEC limits. In the BCS limit where the strength of fermion–fermion interaction is weak, T_c is determined by pair breaking, *i.e.*, the disappearance of the energy gap, and therefore, the gap equation (8.132) should provide the correct result. However, in the BEC limit, no pair breaking occurs because each pair is tightly bound. Therefore, T_c should be determined by the onset of Bose–Einstein condensation of the center of mass of the pairs. Once such pair fluctuations [see Fig. 8.5] are taken into account, the correct T_c is obtained in both the BEC and the BCS limits and smoothly interpolated between them [Nozières and Schmitt-Rink (1985)].

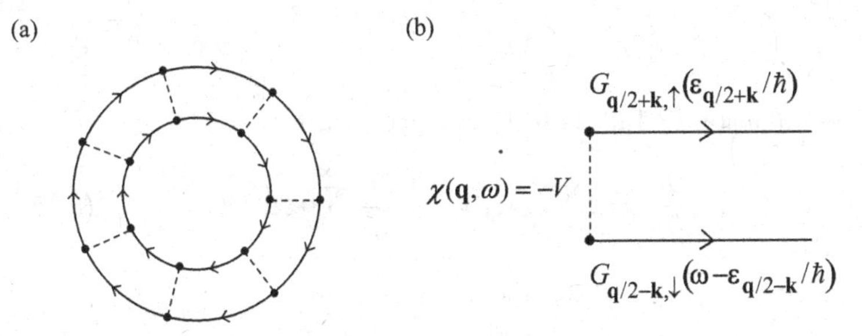

Fig. 8.5 (a) Bubble diagram describing pair fluctuations. (b) A building block of the bubble diagram.

To incorporate the effect of pair fluctuations into theory, it is convenient to calculate the thermodynamic potential for Hamiltonian $\hat{H} = \hat{H}_0 + \hat{H}_1$, where

$$\hat{H}_0 = \sum_{\mathbf{k},\sigma} \xi_k \hat{c}^\dagger_{\mathbf{k}\sigma} \hat{c}_{\mathbf{k}\sigma}, \tag{8.146}$$

$$\hat{H}_1 = -\frac{V}{\Omega} \sum_{\mathbf{k},\mathbf{k}',\mathbf{k}''} \hat{c}^\dagger_{\mathbf{k}+\mathbf{k}''\uparrow} \hat{c}^\dagger_{\mathbf{k}'-\mathbf{k}''\downarrow} \hat{c}_{\mathbf{k}'\downarrow} \hat{c}_{\mathbf{k}\uparrow}. \tag{8.147}$$

We define the expectation value of an arbitrary observable $\hat{O}$ as

$$\langle \hat{O} \rangle_0 = \frac{\mathrm{Tr}(e^{-\beta \hat{H}_0} \hat{O})}{\mathrm{Tr}(e^{-\beta \hat{H}_0})}. \tag{8.148}$$

The thermodynamic potential can be calculated as [Kubo, *et al.* (1992)]

$$\Omega = -\frac{1}{\beta} \ln \mathrm{Tr} e^{-\beta \hat{H}} = \Omega_0 + \Omega_1, \tag{8.149}$$

where

$$\Omega_0 = -\frac{1}{\beta} \ln \mathrm{Tr} e^{-\beta \hat{H}_0} = -\frac{2}{\beta} \sum_{\mathbf{k}} \ln(1 + e^{-\beta \xi_k}) \tag{8.150}$$

and

$$\Omega_1 = -\frac{1}{\beta} \left[\left\langle T_\tau \exp\left(-\frac{1}{\hbar} \int_0^{\beta\hbar} d\tau \hat{H}_1(\tau) \right) \right\rangle_0^{\mathrm{con}} - 1 \right]. \tag{8.151}$$

In Eq. (8.151), T_τ is the time-ordering operator that places operators from left to right in descending order in imaginary time τ, $\langle \cdots \rangle_0^{\mathrm{con}}$ indicates that only connected diagrams are to be considered and that the average is to be taken over $\hat{\rho}_0 = e^{-\beta \hat{H}_0} / \mathrm{Tr} e^{-\beta \hat{H}_0}$, and

$$\hat{H}_1(\tau) \equiv e^{\frac{1}{\beta} \hat{H}_0 \tau} H_1 e^{-\frac{1}{\beta} \hat{H}_0 \tau}. \tag{8.152}$$

The number equation (8.138) is to be replaced with

$$N = -\frac{\partial \Omega}{\partial \mu} = 2 \sum_{\mathbf{k}} f(\xi_{\mathbf{k}}) + \frac{1}{\beta} \frac{\partial}{\partial \mu} \sum_{M=1}^{\infty} \Omega_1^{(M)}, \tag{8.153}$$

where f is defined in Eq. (8.129) and

$$\Omega_1^{(M)} = \left\langle T_\tau \left(\frac{1}{\beta\hbar} \int_0^{\beta\hbar} d\tau \hat{H}_1(\tau) \right)^m \right\rangle_0^{\mathrm{con}}. \tag{8.154}$$

This term is calculated in Appendix D with the result

$$\Omega_1^{(M)} = \sum_{\mathbf{k},n} \frac{1}{M} \left[V \Pi(\mathbf{k}, i\omega_n) \right]^m e^{i\omega_n 0^+}, \tag{8.155}$$

where $\omega_m \equiv 2\pi M/(\beta\hbar)$, M runs over all integers, and

$$\Pi(\mathbf{k}, i\omega_n) = -\frac{1}{\Omega} \sum_{\mathbf{q}} \frac{1 - f(\xi_{\mathbf{q}+\frac{\mathbf{k}}{2}}) - f(-\xi_{\mathbf{q}+\frac{\mathbf{k}}{2}})}{i\omega_n - \xi_{\mathbf{q}+\frac{\mathbf{k}}{2}} - \xi_{-\mathbf{q}+\frac{\mathbf{k}}{2}}}. \tag{8.156}$$

Substituting Eq. (8.155) into Eq. (8.153), we obtain

$$N = 2 \sum_{\mathbf{k}} f(\xi_k) + \frac{1}{\beta} \frac{\partial}{\partial \mu} \sum_{\mathbf{k},n} e^{i\omega_n 0^+} \ln \left[1 - V \Pi(\mathbf{k}, i\omega_n) \right]. \tag{8.157}$$

To eliminate the bare coupling constant V in favor of the s-wave scattering length a_s through Eq. (8.68), we rewrite

$$1 - V\Pi = \left(1 - \frac{V}{\Omega}\sum_{\mathbf{k}}\frac{1}{2\epsilon_k}\right)\left[1 + \frac{-V}{1 - \frac{V}{\Omega}\sum_{\mathbf{k}}\frac{1}{2\epsilon_k}}\left(\Pi - \frac{1}{\Omega}\sum_{\mathbf{k}}\frac{1}{2\epsilon_k}\right)\right]$$

$$= \left(1 - \frac{V}{\Omega}\sum_{\mathbf{k}}\frac{1}{2\epsilon_k}\right)\left[1 + \frac{4\pi\hbar^2 a_s}{M}\left(\Pi - \frac{1}{\Omega}\sum_{\mathbf{k}}\frac{1}{2\epsilon_k}\right)\right]. \tag{8.158}$$

Substituting Eq. (8.158) into Eq. (8.157) and noting that $\ln\left(1 - \frac{V}{\Omega}\sum_{\mathbf{k}}\frac{1}{2\epsilon_k}\right)$ vanishes upon differentiation with respect to μ, we obtain

$$N = 2\sum_{\mathbf{k}} f(\xi_k) - \frac{1}{\beta}\frac{\partial}{\partial\mu}\sum_{\mathbf{k},n} e^{i\omega_n 0^+} \times \ln\left[1 + \frac{4\pi\hbar^2 a_s}{M}\left(\Pi(\mathbf{k}, i\omega_n) - \frac{1}{\Omega}\sum_{\mathbf{k}}\frac{1}{2\epsilon_k}\right)\right]. \tag{8.159}$$

In the BCS limit, the chemical potential μ is positive and the first term on the right-hand side is dominant. In the BEC limit, μ is negative and very large, and therefore, the first term is negligible. To evaluate the second term, we substitute $f = 0$ in Π, obtaining

$$\Pi(\mathbf{k}, i\omega_n) - \frac{1}{\Omega}\sum_{\mathbf{k}}\frac{1}{2\epsilon_k} \cong -\frac{M^{\frac{3}{2}}}{4\pi\hbar^3}\sqrt{\frac{\hbar^2 k^2}{4M} - 2\mu - i\hbar\omega_n}. \tag{8.160}$$

Substituting Eq. (8.160) into Eq. (8.159) and carrying out differentiation with respect to μ, we obtain

$$N = 2\sum_{\mathbf{k}} f(\xi_k) - \frac{1}{\beta}\sum_{\mathbf{k},n} e^{i\omega_n 0^+} \times \frac{1}{\frac{\hbar}{\sqrt{M}a_s}\sqrt{\frac{\hbar^2k^2}{4M} - 2\mu - i\hbar\omega_n} - \left(\frac{\hbar^2k^2}{4M} - 2\mu - i\hbar\omega_n\right)}. \tag{8.161}$$

Because $\mu \simeq -\hbar^2/(2Ma_s^2) \to -\infty$ in the BEC limit, we expand the square root in the denominator as

$$\frac{\hbar}{\sqrt{M}a_s}\sqrt{\frac{\hbar^2k^2}{4M} - 2\mu - i\hbar\omega_n} \simeq \sqrt{\frac{2\hbar^2|\mu|}{Ma_s^2}} + \frac{1}{2}\left(\frac{\hbar^2k^2}{4M} - i\hbar\omega_n\right). \tag{8.162}$$

We thus obtain

$$N = 2\sum_{\mathbf{k}} f(\xi_k) - \frac{2}{\beta}\sum_{\mathbf{k},n} e^{i\omega_n 0^+}\frac{1}{i\hbar\omega_n - \frac{\hbar^2k^2}{4M} + \mu^{\mathrm{dimer}}}, \tag{8.163}$$

where

$$\mu^{\text{dimer}} = -2|\mu|\left(1 - \sqrt{\frac{\hbar^2}{2|\mu|ma_s^2}}\right)$$

is the chemical potential of dimers that vanishes in the limit of $\mu \to -\hbar^2/(2Ma_s^2)$. The sum over n in Eq. (8.163) can be carried out using the formula

$$\sum_{n=-\infty}^{\infty} e^{i\omega_n 0^+} \frac{1}{i\hbar\omega_n - \xi_p} = -\frac{\beta}{e^{\beta_i \xi_p} - 1} \equiv -\beta n_{\text{B}}(\xi_p).$$

Substituting $\epsilon_k^{\text{dimer}} = \hbar^2 k^2/(4M)$, we find that

$$N = 2\sum_{\mathbf{k}} f(\xi_k) + 2\sum_{\mathbf{k}} n_{\text{B}}(\epsilon_k^{\text{dimer}} - \mu^{\text{dimer}}). \tag{8.164}$$

This result shows that the total number of particles is equal to the number of fermions and twice the number of dimers. The superfluid transition temperature T_c and the chemical potential μ can be determined by solving the gap equation (8.132) and the number equation (8.135) self-consistently. At unitarity, we find that $T_c/T_F = 0.224$; this is to be compared with $T_c/T_F = 0.157 \pm 0.007$ obtained by a quantum Monte Carlo method [Burovski, *et al.* (2006)].

8.8 Gor'kov–Melik–Barkhudarov Correction

Gor'kov and Melik-Barkhudarov [Gor'kov and Melik-Barkhudarov (1961); Pethick and Smith (2002)] pointed out that the strength of interaction is renormalized by the presence of other particles and that in the BCS limit, T_c is lowered by a factor of approximately two.

Figure 8.6 (a) illustrates the direct (first-order) scattering from a pair state $(\mathbf{p}, -\mathbf{p})$ to another $(\mathbf{p}', -\mathbf{p}')$, and Figs. 8.6 (b) and (c) show the second-order processes. In Fig. 8.6 (b), the momentum of a particle first changes from $\mathbf{p}$ to $\mathbf{p}+\mathbf{p}'+\mathbf{p}'$ by exciting a particle with momentum $-\mathbf{p}'$ and a hole with momentum $\mathbf{p}''$. Then, the particle with momentum $\mathbf{p}+\mathbf{p}'+\mathbf{p}''$ causes the particle with momentum $-\mathbf{p}$ to recombine with the hole with momentum $\mathbf{p}''$ by the second scattering. Consequently, the pair $(\mathbf{p}, -\mathbf{p})$ is scattered into $(\mathbf{p}', -\mathbf{p}')$. Figure 8.6 (c) shows a similar second-order process. According to the second-order perturbation theory, the correction of these two processes to U_0 is given by

$$\begin{aligned}\delta U &= \frac{U_0^2}{\Omega}\sum_{\mathbf{p}''}\left[\frac{(1 - f(\xi_{\mathbf{p}+\mathbf{p}'+\mathbf{p}''}))f(\xi_{\mathbf{p}''})}{\epsilon_{\mathbf{p}+\mathbf{p}'+\mathbf{p}''} - \epsilon_{\mathbf{p}''}} + \frac{f(\xi_{\mathbf{p}+\mathbf{p}'+\mathbf{p}''})(1 - f(\xi_{\mathbf{p}''}))}{\epsilon_{\mathbf{p}''} - \epsilon_{\mathbf{p}+\mathbf{p}'+\mathbf{p}''}}\right] \\ &= U_0^2 L(|\mathbf{p}+\mathbf{p}'|),\end{aligned} \tag{8.165}$$

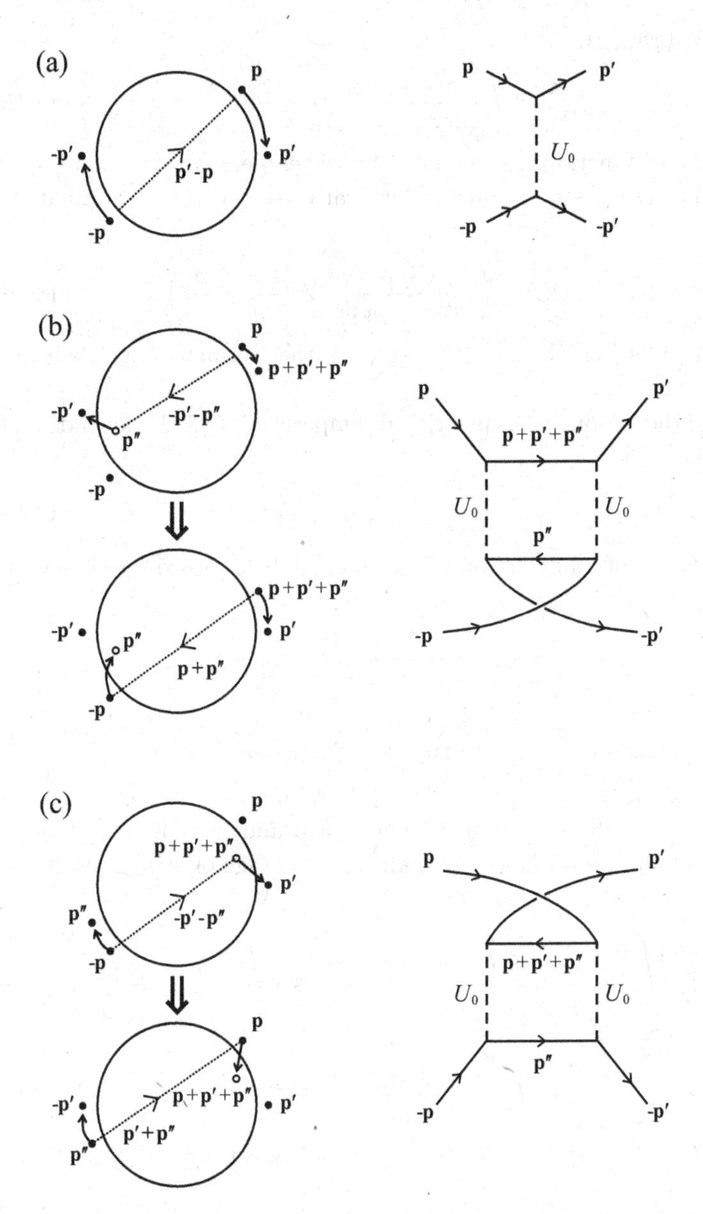

Fig. 8.6 (a) First-order and (b)(c) second-order pair scattering from $(\mathbf{p}, -\mathbf{p})$ to $(\mathbf{p}', -\mathbf{p}')$. In the left column, the filled and open circles indicate particles and holes, respectively. The dotted lines indicate interactions whose strength is given by $U_0 = 4\pi\hbar^2 a_s/M$. The right column shows the corresponding Feynman diagrams, where the arrows pointing toward the light (left) indicate propagation of particles (holes).

where $U_0 = 4\pi\hbar^2 a/M$ and

$$L(p) = \int \frac{d\mathbf{p}'}{(2\pi\hbar)^3} \frac{f(\xi_{\mathbf{p}'}) - f(\xi_{\mathbf{p}+\mathbf{p}'})}{\epsilon_{\mathbf{p}+\mathbf{p}'} - \epsilon_{\mathbf{p}'}} \tag{8.166}$$

is the Lindhard function. At sufficiently low temperatures, we may substitute $f(\xi_{\mathbf{p}'}) = \theta(E_{\mathrm{F}} - \epsilon_{\mathbf{p}'})$ and the integral in Eq. (8.166) is evaluated to give

$$L(p) = D(E_{\mathrm{F}}) \left(\frac{1}{2} + \frac{4p_{\mathrm{F}}^2 - p^2}{8p_{\mathrm{F}}p} \ln \left| \frac{2p_{\mathrm{F}} + p}{2p_{\mathrm{F}} - p} \right| \right), \tag{8.167}$$

where $D(E_{\mathrm{F}})$ is the density of states at the Fermi energy defined in Eq. (8.6).

Because the exponent of the critical temperature (8.141) depends on U_0 as

$$\frac{\pi}{2k_{\mathrm{F}}a_{\mathrm{s}}} = \frac{1}{D_{\mathrm{F}}U_0}, \tag{8.168}$$

the modification of U_0 by δU due to the second-order processes changes the exponent by

$$\begin{aligned} A(p) &= \frac{1}{D_{\mathrm{F}}(U_0 + \delta U)} - \frac{1}{D_{\mathrm{F}}U_0} \simeq -\frac{\delta U}{D_{\mathrm{F}}U_0^2} \\ &= -\frac{1}{2} - \frac{4p_{\mathrm{F}}^2 - p^2}{8p_{\mathrm{F}}p} \ln \left| \frac{2p_{\mathrm{F}} + p}{2p_{\mathrm{F}} - p} \right| . \end{aligned} \tag{8.169}$$

Now, p in Eq. (8.169) is equal to $|\mathbf{p} + \mathbf{p}'|$. Assuming that $|\mathbf{p}| = |\mathbf{p}'| = p_{\mathrm{F}}$, $p = 2p_{\mathrm{F}} \cos\frac{\theta}{2}$, where θ is the angle between $\mathbf{p}$ and $\mathbf{p}'$. The net change in the exponent can be obtained by taking the average of $A(p)$ over all solid angles:

$$\frac{1}{4\pi} \int d\Omega \cdots = \frac{1}{2} \int_0^{\pi} \sin\theta d\theta \cdots = \frac{1}{2p_{\mathrm{F}}^2} \int_0^{2p_{\mathrm{F}}} p dp \cdots . \tag{8.170}$$

Therefore,

$$\frac{1}{2p_{\mathrm{F}}^2} \int_0^{2p_{\mathrm{F}}} A(p) p dp = -\ln(4e)^{\frac{1}{3}}. \tag{8.171}$$

Combined with Eq. (8.141), T_{c} is suppressed by a factor of $(4e)^{\frac{1}{3}} \simeq 2.22$, giving

$$k_{\mathrm{B}}T_{\mathrm{c}} = \left(\frac{2}{e}\right)^{\frac{7}{3}} \frac{e^{\gamma}}{\pi} E_{\mathrm{F}} e^{-\frac{\pi}{2k_{\mathrm{F}}|a_{\mathrm{s}}|}} \simeq 0.277 E_{\mathrm{F}} e^{-\frac{\pi}{2k_{\mathrm{F}}|a_{\mathrm{s}}|}}. \tag{8.172}$$

The physics behind the suppression of the critical temperature is the medium effect that screens the bare interaction.

8.9 Unitary Gas

A dilute gas is a system of particles in which the average interparticle distance $n^{-1/3}$ is much greater than the range of interaction r_0:

$$n^{-1/3} \gg r_0. \quad \text{(diluteness condition)} \tag{8.173}$$

Moreover, if the scattering length is much larger than $n^{-1/3}$, the system is said to be a unitary gas:

$$a_s \gg n^{-1/3} \gg r_0. \quad \text{(condition for unitary gas)} \tag{8.174}$$

In this case, the properties of the system are expected to be universal in that they are described in terms of some universal functions of the density n and temperature T alone.

The partial-wave analysis of scattering theory gives the s-wave scattering amplitude as

$$f_0(k) = \frac{e^{2i\delta_0(k)} - 1}{2ik} = \frac{1}{k \cot \delta_0(k) - ik}, \tag{8.175}$$

where $\delta_0(k)$ is the s-wave phase shift. The corresponding scattering cross section σ_0 is given by

$$\sigma_0 = 4\pi |f_0|^2 = \frac{4\pi}{k^2} \sin^2 \delta_0(k). \tag{8.176}$$

Therefore, the scattering cross section becomes maximal when

$$\delta_0(k) = \left(n + \frac{1}{2}\right)\pi, \quad n = 0, \pm 1, \pm 2, \cdots. \tag{8.177}$$

The limit is referred to as the unitary limit.

In the zero-range limit (*i.e.*, $kr_0 \ll 1$), the first term in the denominator of Eq. (8.175) can be expanded as follows:

$$k \cot \delta_0(k) = -a_s^{-1} + \frac{1}{2}k^2 r_{\text{eff}} + O\big(k(kr_{\text{eff}})^2\big), \tag{8.178}$$

where the s-wave scattering length a_s is defined by

$$a_s \equiv -\lim_{k\to 0} \frac{\tan \delta_0(k)}{k} \tag{8.179}$$

and r_{eff} is the effective range which is equal to $2a/3$ for the pseudo-potential (2.18). Substituting Eq. (8.178) into Eq. (8.175), we obtain

$$f_0(k) = \frac{-1}{a_s^{-1} - \frac{1}{2}k^2 r_{\text{eff}} + ik}. \tag{8.180}$$

When $k|a_s| \ll 1$ and $kr_{\rm eff} \lesssim 1$, Eq. (8.180) gives $f_0(k) \simeq -a_s$ and the cross section is given by $\sigma_0 = 4\pi a_s^2$. In contrast, in the unitary limit (8.177), Eqs. (8.179) and (8.175) show that the scattering length diverges and that the scattering amplitude is given by

$$f_0(k) = \frac{i}{k} \quad \text{(unitary limit)}. \tag{8.181}$$

A unitary gas is realized if both the zero-range limit and the infinite scattering-length limit are satisfied. Because in this case the scattering amplitude does not depend on a_s, the only relevant length scales are the average interparticle distance $n^{-1/3}$ and the thermal de Broglie length $\lambda_{\rm th} = h/\sqrt{2\pi M k_B T}$. Therefore, all thermodynamic quantities become universal functions of these quantities, or equivalently, the Fermi energy E_F and the temperature [Ho (2004)]. Here, we show some examples.

At absolute zero, the ground-state energy per particle E/N is, by assumption, proportional to E_F and it is expressed as

$$\frac{E}{N} = \xi\frac{3}{5}E_F = (1+\beta)\frac{3}{5}E_F = (1+\beta)\frac{3(3\pi^2)^{\frac{2}{3}}\hbar^2}{10M}n^{\frac{2}{3}}, \tag{8.182}$$

where $3E_F/5$ is the ground-state energy per particle for the ideal Fermi gas [see Eq. (8.18)] and the proportionality constant $\xi = 1+\beta$ is a universal number called Bertsch's parameter. The same parameter describes the equation of state as

$$P = -\frac{\partial E}{\partial V} = (1+\beta)\frac{2}{5}E_F n = (1+\beta)\frac{(3\pi^2)^{\frac{2}{3}}\hbar^2}{5M}n^{\frac{5}{3}} \tag{8.183}$$

and the chemical potential μ as

$$\mu = \frac{\partial E}{\partial N} = \sqrt{1+\beta}E_F. \tag{8.184}$$

Consequently, the sound velocity is given by

$$c = \sqrt{\frac{1}{M}\frac{\partial P}{\partial n}} = \sqrt{1+\beta}\frac{v_F}{\sqrt{3}}. \tag{8.185}$$

It follows from Eqs. (8.182) and (8.183) that the energy density $e \equiv E/V = En/N$ is related to pressure P as

$$e = \frac{3}{2}P. \tag{8.186}$$

This relation is the same as that of an ideal gas. The quantum Monte Carlo calculations give the value of β as [Carlson, *et al.* (2003); Astrakharchik, *et al.* (2004); Carlson and Reddy (2005)]

$$\beta = -0.58 \pm 0.01. \tag{8.187}$$

When the system is confined in a potential $U(\mathbf{r})$, the mechanical balance gives

$$\nabla P(\mathbf{r}) + n(\mathbf{r})\nabla U(\mathbf{r}) = 0. \tag{8.188}$$

When $U(\mathbf{r}) \propto r^k$, we have $\mathbf{r} \cdot \nabla U(\mathbf{r}) = kU(\mathbf{r})$, and therefore,

$$\int d\mathbf{r}\ \mathbf{r} \cdot \nabla P = -k \int d\mathbf{r}\ nU. \tag{8.189}$$

The right-hand side gives $-kN\langle U\rangle$, whereas the left-hand side gives

$$\int d\mathbf{r}\ \mathbf{r} \cdot \nabla P = -3 \int d\mathbf{r}\ P = -2 \int d\mathbf{r}\ e, \tag{8.190}$$

where Eq. (8.186) is used. Because $\int d\mathbf{r}\ e = E^{\mathrm{total}} - N\langle U\rangle$, where E^{total} is the total energy of the system, we obtain the following virial theorem at unitarity [Thomas, *et al.* (2005)]:

$$\frac{E^{\mathrm{total}}}{N} = \frac{k+2}{2}\langle U\rangle. \tag{8.191}$$

At the unitary limit, the thermodynamic quantities are expected to be universal functions of the density n and the reduced temperature $\theta \equiv T/T_{\mathrm{F}}$ [Ho (2004); Ho and Mueller (2004)]. For example, the internal energy E, Helmholtz free energy F, chemical potential μ, and entropy S are expressed as

$$E = NE_{\mathrm{F}} f_E(\theta), \tag{8.192}$$

$$F = NE_{\mathrm{F}} f_F(\theta), \tag{8.193}$$

$$\mu = E_{\mathrm{F}} f_\mu(\theta), \tag{8.194}$$

$$S = Nk_{\mathrm{B}} f_S(\theta), \tag{8.195}$$

respectively. The universal functions are related to each other by the following thermodynamic relations:

$$f_E(\theta) = f_F(\theta) - \theta f_{\mathrm{F}}'(\theta), \tag{8.196}$$

$$f_\mu(\theta) = \frac{5}{3} f_F(\theta) - \frac{2}{3}\theta f_{\mathrm{F}}'(\theta), \tag{8.197}$$

$$f_S(\theta) = -f_{\mathrm{F}}'(\theta). \tag{8.198}$$

Therefore, it is sufficient to know only one among them. These universal functions have been measured based on the local density approximation [Horikoshi, *et al.* (2010); Nascimbène, *et al.* (2010)].

The properties concerning superfluidity are also expected to be universal. According to quantum Monte Carlo methods, the superfluid transition temperature T_c is given by [Bulgac, *et al.* (2006); Burovski, *et al.* (2006)]

$$\frac{T_c}{T_F} = 0.157 \pm 0.007, \tag{8.199}$$

and the pairing gap at $T = 0$ is given by [Carlson and Reddy (2008)]

$$\frac{\Delta}{E_F} = 0.45 \pm 0.05. \tag{8.200}$$

The critical region at unitarity is expected to be large because the system is strongly interacting. The issue of the pseudogap above T_c has not yet been fully understood. These subjects raise interesting problems that have profound implications for other strongly correlated systems such as high-T_c superconductors.

8.10 Imbalanced Fermi Systems

In the preceding sections, we assumed that the numbers of spin-up and spin-down atoms are equal. In atomic gases, the spin relaxation is often slow and there are many situations in which the number of spin-up atoms $N_\uparrow$ and that of spin-down atoms $N_\downarrow$ are conserved separately. Then, we can define chemical potentials $\mu_\uparrow$ and $\mu_\downarrow$ for spin-up and spin-down atoms, respectively.

Let us define the spin polarization as

$$P \equiv \frac{N_\uparrow - N_\downarrow}{N_\uparrow + N_\downarrow}, \tag{8.201}$$

where $P = 0$ and $P = \pm 1$ correspond to the balanced and spin-polarized systems, respectively. Suppose that the spin-up and spin-down atoms undergo attractive interaction at absolute zero and that there are no interactions between the same-spin atoms. The ground state of the system is superfluid at $P = 0$ and normal at $P = 1$. Therefore, there must exist a limit P_c above which the system undergoes a quantum phase transition to the normal state. This limit is referred to as the Chandrasekhar–Clogston (CC) limit [Chandrasekhar (1962); Clogston (1962)].

On physical grounds, the CC limit is achieved when the chemical potential difference between spin-up and spin-down atoms becomes comparable to the pairing gap:

$$h \equiv \frac{\mu_\uparrow - \mu_\downarrow}{2} = \alpha\Delta, \tag{8.202}$$

where Δ is the energy, α is a constant of the order of unity, and we assume $h \geq 0$ without loss of generality. The free-energy difference between the superfluid and the normal states is given by

$$F_{\rm n} - F_{\rm s} = \frac{1}{2} D(E_{\rm F}) \Delta^2. \tag{8.203}$$

On the other hand, the kinetic-energy cost of the imbalance is given by

$$2D(E_{\rm F}) \int_0^h \mu d\mu = D(E_{\rm F}) h^2. \tag{8.204}$$

Equating (8.203) with (8.204), we obtain $\alpha = 1/\sqrt{2}$.

At unitarity, quantum Monte Carlo calculations give $P_{\rm c} = 0.39$ for a uniform system [Bedaque, *et al.* (2003)] and $P_{\rm c} = 0.77$ for a harmonically trapped system [Lobo, *et al.* (2006)]. The larger value of $P_{\rm c}$ for the latter system is attributed to the higher peak density at the trap center. The measurement of $P_{\rm c}$ for a trapped system [Shin, *et al.* (2006)] exhibits excellent agreement with theory.

The new degrees of freedom of the polarization raise many new questions and motivate further investigation of old issues such as the Fulde–Ferrell–Larkin–Ovchinnikov (FFLO) state [Fulde and Ferrell (1964); Larkin and Ovchinnikov (1964)] and the Sarma or breached-pair state [Sarma (1963); Liu and Wilczek (2003)]. When the chemical-potential difference exceeds the CC limit, the gap equation admits no spatially uniform solution; however, the system may maintain superfluidity by making the energy gap vary in space, where the order parameter varies in space as $\Delta \propto e^{i\mathbf{g}\mathbf{r}}$ (Fulde–Ferrell state) or $\Delta \propto \cos \mathbf{g} \cdot \mathbf{r}$ (Larkin–Ovchinnikov state). This implies that Cooper pairs have nonzero center-of-mass momentum. The time-reversal and translational symmetries are broken in the FF and LO states, respectively. In the Sarma state, the chemical-potential difference exceeds the energy gap; nonetheless, Cooper pairs with zero center-of-mass momentum are formed by occupying a region of the momentum space that is separated by unpaired fermions. The Sarma state is therefore gapless.

To investigate the properties of an imbalanced Fermi system, we consider the Hamiltonian

$$H = \sum_{\mathbf{k}\sigma} (\xi_{\mathbf{k}} - h\sigma_z) \hat{c}^\dagger_{\mathbf{k}\sigma} \hat{c}_{\mathbf{k}\sigma} - \frac{V}{\Omega} \sum_{\mathbf{k}\mathbf{k}'\mathbf{k}''} \hat{c}^\dagger_{\mathbf{k}+\mathbf{k}''\uparrow} \hat{c}^\dagger_{\mathbf{k}'-\mathbf{k}''\downarrow} \hat{c}_{\mathbf{k}'\downarrow} \hat{c}_{\mathbf{k}\uparrow}, \tag{8.205}$$

where h is defined in Eq. (8.202), σ_z is the Pauli matrix, and

$$\xi_k \equiv \epsilon_k - \frac{\mu_\uparrow + \mu_\downarrow}{2} \tag{8.206}$$

with $\epsilon_k = \hbar^2 k^2/(2M)$. Because the only new term in the Hamiltonian (8.205) as compared to the balanced case (8.86) is the term $h\sigma_z$ that shifts the chemical potential, the gap equation for the imbalanced system is modified from Eq. (8.130) as

$$\Delta = \frac{V}{\Omega}\sum_{\mathbf{k}} \frac{\Delta}{2E_{\mathbf{k}}}\left[1 - f(E_{\mathbf{k}} - h) - f(E_{\mathbf{k}} + h)\right], \tag{8.207}$$

where

$$E_{\mathbf{k}} = \sqrt{\xi_k^2 + \Delta^2}. \tag{8.208}$$

At absolute zero, we have $f(E_{\mathbf{k}} + h) = 0$. Furthermore, if $h \leq \Delta$, the gap equation (8.207) reduces to the usual one

$$1 = \frac{V}{2}\int_0^{E_c} \frac{d\xi}{\sqrt{\xi^2 + \Delta_0^2}} = \frac{V}{2}\ln\frac{2E_c}{\Delta_0}, \tag{8.209}$$

where $\Delta_0 = 1.76 k_B T_c$ is the energy gap for the balanced system [see Eq. (8.142)] and E_c, the cut-off energy. On the other hand, if $h > \Delta$, the lower bound of the integral in Eq. (8.209) is replaced with $\sqrt{h^2 - \Delta^2}$ and we obtain

$$1 = \frac{V}{2}\ln\frac{2E_c}{h + \sqrt{h^2 - \Delta^2}}. \tag{8.210}$$

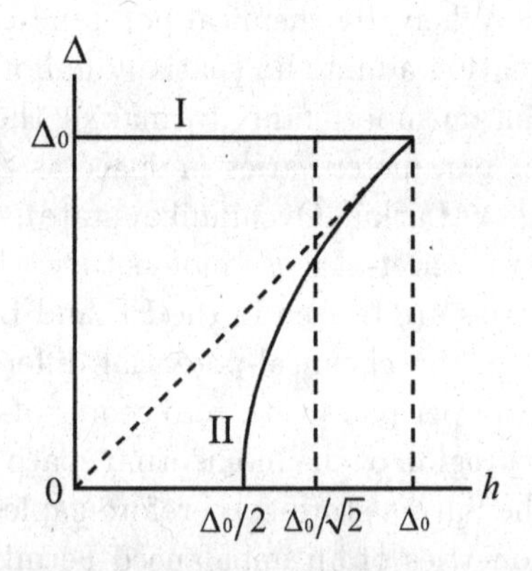

Fig. 8.7 Two branches of the solution of Eq. (8.207) at $T = 0$, where $\Delta_0 = 1.76 k_B T_c$ is the energy gap for the balanced case ($h = 0$). For $\Delta > h$, the solution is $\Delta = \Delta_0$, whereas for $\Delta < h$, it is $\Delta = \sqrt{2\Delta_0(h - \Delta_0/2)}$.

Comparing Eqs. (8.209) and (8.210), we find that the energy gap for the imbalanced system with $h > \Delta$ is given by

$$\Delta = \sqrt{2\Delta_0(h - \Delta_0/2)}. \tag{8.211}$$

Note that this solution is obtained for $h > \Delta$; however, this is a regime beyond the CC limit. The free-energy argument made above shows that in this regime, the normal-state energy is lower than the superfulid-state energy, and thus, the superfluid state in this regime, if it exists, can only be metastable. The h-dependence of the energy gap is shown in Fig. 8.7.

8.11 *P*-Wave Superfluid

8.11.1 *Generalized pairing theory*

Let $\hat{c}^{\dagger}_{\mathbf{k}\alpha}$ be the creation operator of a spin-1/2 fermion with wave vector $\mathbf{k}$ and spin projection $\alpha = \uparrow, \downarrow$. The pairing amplitudes of a general pairing state with zero center-of-mass momentum are defined as

$$\psi_{\alpha\beta}(\mathbf{k}) = \langle \hat{c}_{\mathbf{k}\alpha}\hat{c}_{-\mathbf{k}\beta}\rangle. \tag{8.212}$$

First, we study some general properties of the pairing amplitudes under a rotation in spin space. The rotation matrix that describes a rotation through an infinitesimal angle $\delta\theta$ about a unit vector $\mathbf{n}$ is given by

$$R = 1 + \frac{i}{2}\mathbf{n}\cdot\boldsymbol{\sigma}\delta\theta, \tag{8.213}$$

where $\boldsymbol{\sigma} = (\sigma_x, \sigma_y, \sigma_z)$ is the vector of Pauli matrices

$$\sigma_x = \begin{pmatrix} 0 & 1 \\ 1 & 0 \end{pmatrix}, \quad \sigma_y = \begin{pmatrix} 0 & -i \\ i & 0 \end{pmatrix}, \quad \sigma_z = \begin{pmatrix} 1 & 0 \\ 0 & -1 \end{pmatrix}. \tag{8.214}$$

Under the rotation (8.213), $\psi_{\alpha\beta}$ transforms as

$$\psi'_{\alpha\beta} = R_{\alpha\gamma}R_{\beta\delta}\psi_{\gamma\delta} = R_{\alpha\gamma}\psi_{\gamma\delta}{}^{t}R_{\delta\beta}, \tag{8.215}$$

where t denotes the matrix transpose. Because ${}^{t}\boldsymbol{\sigma} = -\sigma_y\boldsymbol{\sigma}\sigma_y$, we obtain

$$\begin{aligned} \psi'_{\alpha\beta} &= \left(1 + \frac{i}{2}\mathbf{n}\cdot\boldsymbol{\sigma}\delta\theta\right)_{\alpha\gamma}\psi_{\gamma\delta}\left(1 - \frac{i}{2}\mathbf{n}\cdot\sigma_y\boldsymbol{\sigma}\sigma_y\delta\theta\right)_{\delta\beta} \\ &= \left[\Psi + \frac{i}{2}\mathbf{n}\cdot(\boldsymbol{\sigma}\Psi\sigma_y - \Psi\sigma_y\boldsymbol{\sigma})\sigma_y\delta\theta\right]_{\alpha\beta}, \end{aligned} \tag{8.216}$$

where we ignore the term of the order of $(\delta\theta)^2$ and define the 2×2 matrix $\Psi \equiv (\psi_{\alpha\beta})$. Thus,

$$\Psi' = \Psi + \frac{i}{2}\mathbf{n}\cdot(\boldsymbol{\sigma}\Psi\sigma_y - \Psi\sigma_y\boldsymbol{\sigma})\sigma_y\delta\theta. \tag{8.217}$$

Under a rotation in spin space, the spin-singlet term Ψ^{singlet} and spin-triplet term Ψ^{triplet} transform as a scalar and a vector, respectively. By inspection, we obtain

$$\Psi^{\text{singlet}} = A_0(\mathbf{k}) i\sigma_y, \tag{8.218}$$

$$\Psi^{\text{triplet}} = \mathbf{A}(\mathbf{k}) \cdot i\boldsymbol{\sigma}\sigma_y, \tag{8.219}$$

where $\mathbf{A} = (A_x, A_y, A_z)$. In fact, substituting Eq. (8.218) in Eq. (8.217), we find that $\Psi' = \Psi$. In addition, substituting Eq. (8.219) in Eq. (8.217), we have

$$\begin{aligned}
\mathbf{n} \cdot (\boldsymbol{\sigma}\Psi\sigma_y - \Psi\sigma_y\boldsymbol{\sigma}) &= i(\mathbf{n}\cdot\boldsymbol{\sigma}) \cdot (\mathbf{A}\cdot\boldsymbol{\sigma}) - i(\mathbf{A}\cdot\boldsymbol{\sigma}) \cdot (\mathbf{n}\cdot\boldsymbol{\sigma}) \\
&= i\sum_{jk} n_j A_k [\sigma_j, \sigma_k] = i\sum_{jk} n_j A_k 2i\epsilon_{jk\ell}\sigma_\ell \\
&= -2(\mathbf{n}\times\mathbf{A})\cdot\boldsymbol{\sigma}.
\end{aligned}$$

Thus, we find that Ψ^{triplet} indeed transforms under rotation like a vector:

$$\Psi' = \Psi - i(\mathbf{n}\times\mathbf{A})\cdot\boldsymbol{\sigma}\sigma_y\delta\theta. \tag{8.220}$$

Due to the anticommutation relations of $\hat{c}_{\mathbf{k}\alpha}$, $\psi_{\alpha\beta}(\mathbf{k})$ must change its sign under the exchange of two particles:

$$\psi_{\beta\alpha}(-\mathbf{k}) = -\psi_{\alpha\beta}(\mathbf{k}). \tag{8.221}$$

It follows that the orbital part must satisfy

$$A_0(-\mathbf{k}) = A_0(\mathbf{k}), \quad \text{(spin singlet)} \tag{8.222}$$

$$\mathbf{A}(-\mathbf{k}) = -\mathbf{A}(\mathbf{k}). \quad \text{(spin triplet)} \tag{8.223}$$

Noting that $i\sigma_x\sigma_y = -\sigma_z$, $i\sigma_y^2 = i$, and $i\sigma_z\sigma_y = \sigma_x$, we find that the most general pairing amplitude can be written as

$$\begin{aligned}
\Psi &= A_0\begin{pmatrix} 0 & 1 \\ -1 & 0 \end{pmatrix} + A_x\begin{pmatrix} -1 & 0 \\ 0 & 1 \end{pmatrix} + A_y\begin{pmatrix} i & 0 \\ 0 & i \end{pmatrix} + A_z\begin{pmatrix} 0 & 1 \\ 1 & 0 \end{pmatrix} \\
&= \begin{pmatrix} -A_x + iA_y & A_z + A_0 \\ A_z - A_0 & A_x + iA_y \end{pmatrix}.
\end{aligned} \tag{8.224}$$

Next, we introduce the superfluid gap function

$$\Delta_{\alpha\beta}(\mathbf{k}) \equiv \sum_{\mathbf{k}'\alpha'\beta'} V_{\alpha\beta,\alpha'\beta'}(\mathbf{k},\mathbf{k}')\psi_{\alpha'\beta'}(\mathbf{k}'), \tag{8.225}$$

where $V_{\alpha\beta,\alpha'\beta'}(\mathbf{k},\mathbf{k}')$ is the matrix element of the interaction Hamiltonian between fermions:

$$\hat{V} = \frac{1}{2\Omega}\sum_{\mathbf{k}\mathbf{k}'\mathbf{q}}\sum_{\alpha\beta\alpha'\beta'} V_{\alpha\beta,\alpha'\beta'}(\mathbf{k},\mathbf{k}')\hat{c}^\dagger_{-\mathbf{k}+\frac{\mathbf{q}}{2},\beta}\hat{c}^\dagger_{\mathbf{k}+\frac{\mathbf{q}}{2},\alpha}\hat{c}_{\mathbf{k}'+\frac{\mathbf{q}}{2},\alpha'}\hat{c}_{-\mathbf{k}'+\frac{\mathbf{q}}{2},\beta'}. \tag{8.226}$$

The matrix element satisfies

$$V_{\beta\alpha,\alpha'\beta'}(-\mathbf{k},\mathbf{k}') = V_{\alpha\beta,\beta'\alpha'}(\mathbf{k},-\mathbf{k}') = -V_{\alpha\beta,\alpha'\beta'}(\mathbf{k},\mathbf{k}') \tag{8.227}$$

due to the anticommutation relations of fermion operators and

$$V_{\alpha'\beta',\alpha\beta}(\mathbf{k}',\mathbf{k}) = V_{\alpha\beta,\alpha'\beta'}(\mathbf{k},\mathbf{k}')^* \tag{8.228}$$

due to the Hermiticity of $\hat{V}$.

When the interaction conserves spin, its matrix element takes the form

$$\begin{aligned} V_{\alpha\beta,\alpha'\beta'}(\mathbf{k},\mathbf{k}') &= \frac{1}{2}(\delta_{\alpha\alpha'}\delta_{\beta\beta'} - \delta_{\alpha\beta'}\delta_{\alpha'\beta})V^{(\mathrm{e})}(\mathbf{k},\mathbf{k}') \\ &\quad + \frac{1}{2}(\delta_{\alpha\alpha'}\delta_{\beta\beta'} + \delta_{\alpha\beta'}\delta_{\alpha'\beta})V^{(\mathrm{o})}(\mathbf{k},\mathbf{k}'), \end{aligned} \tag{8.229}$$

where $V^{(\mathrm{e})}$ and $V^{(\mathrm{o})}$ respectively describe the even and odd parts:

$$V^{(\mathrm{e})}(-\mathbf{k},\mathbf{k}') = V^{(\mathrm{e})}(\mathbf{k},-\mathbf{k}') = V^{(\mathrm{e})}(\mathbf{k},\mathbf{k}'), \tag{8.230}$$

$$V^{(\mathrm{o})}(-\mathbf{k},\mathbf{k}') = V^{(\mathrm{o})}(\mathbf{k},-\mathbf{k}') = -V^{(\mathrm{o})}(\mathbf{k},\mathbf{k}'). \tag{8.231}$$

Substituting Eq. (8.229) in Eq. (8.225), we obtain

$$\begin{aligned} \Delta_{\alpha\beta}(\mathbf{k}) &= \frac{1}{2}\sum_{\mathbf{k}'}\Big\{[\psi_{\alpha\beta}(\mathbf{k}') - \psi_{\beta\alpha}(\mathbf{k}')]V^{(\mathrm{e})}(\mathbf{k},\mathbf{k}') \\ &\quad + [\psi_{\alpha\beta}(\mathbf{k}') + \psi_{\beta\alpha}(\mathbf{k}')]V^{(\mathrm{o})}(\mathbf{k},\mathbf{k}')\Big\}. \end{aligned} \tag{8.232}$$

The first term on the right-hand side describes the spin-singlet pairing

$$\hat{\Delta}^{\mathrm{singlet}}(\mathbf{k}) = \Delta(\mathbf{k}) i\sigma_y, \tag{8.233}$$

where

$$\Delta(\mathbf{k}) = \sum_{\mathbf{k}'} V^{(\mathrm{e})}(\mathbf{k},\mathbf{k}')A_0(\mathbf{k}'), \tag{8.234}$$

$$A_0(\mathbf{k}') = \frac{1}{2}(\psi_{\uparrow\downarrow}(\mathbf{k}') - \psi_{\downarrow\uparrow}(\mathbf{k}')) = A_0(-\mathbf{k}), \tag{8.235}$$

whereas the second term describes the spin-triplet pairing

$$\hat{\Delta}^{\mathrm{triplet}}(\mathbf{k}) = \mathbf{\Delta}(\mathbf{k}) \cdot i\boldsymbol{\sigma}\sigma_y, \tag{8.236}$$

where

$$\mathbf{\Delta}(\mathbf{k}) = \sum_{\mathbf{k}'} V^{(\mathrm{o})}(\mathbf{k},\mathbf{k}')\mathbf{A}(\mathbf{k}') \equiv (\Delta_x, \Delta_y, \Delta_z), \tag{8.237}$$

$$A_x = -\frac{\psi_{\uparrow\uparrow} - \psi_{\downarrow\downarrow}}{2}, \quad A_y = \frac{\psi_{\uparrow\uparrow} + \psi_{\downarrow\downarrow}}{2i}, \quad A_z = \frac{\psi_{\uparrow\downarrow} + \psi_{\downarrow\uparrow}}{2}. \tag{8.238}$$

Because $\mathbf{A}(\mathbf{k})$ transforms like a vector under rotation in spin space, so does $\mathbf{\Delta}(\mathbf{k})$. From Eq. (8.236),

$$\begin{aligned}\hat{\Delta}^{\text{triplet}} \cdot (\hat{\Delta}^{\text{triplet}})^{\dagger} &= \sum_{j,k} \Delta_j \sigma_j \sigma_y^2 \sigma_k \Delta_k^* \\ &= \sum_j \Delta_j \Delta_j^* + \sum_{j,k,\ell} i\epsilon_{jkl} \Delta_j \Delta_k^* \sigma_\ell \\ &= \mathbf{\Delta} \cdot \mathbf{\Delta}^* + i(\mathbf{\Delta} \times \mathbf{\Delta}^*) \cdot \boldsymbol{\sigma}. \end{aligned} \tag{8.239}$$

By taking the quantization axis along the direction of $\mathbf{\Delta} \times \mathbf{\Delta}^*$, we find that the eigenvalues of the energy gap for the spin-triplet state are given by

$$\sqrt{|\mathbf{\Delta}|^2 \pm |\mathbf{\Delta} \times \mathbf{\Delta}^*|^2}. \tag{8.240}$$

The pairing state with $\mathbf{\Delta} \times \mathbf{\Delta}^* = 0$ $(\neq 0)$ is referred to as unitary (nonunitary). In other words, if the vector $\mathbf{\Delta}$ is real up to an overall phase, the state is referred to as unitary. For unitary states, the energy gap is given by $|\Delta(\mathbf{k})|$ and the excitation spectrum is

$$E_{\mathbf{k}} = \sqrt{\xi_k^2 + |\Delta(\mathbf{k})|^2}. \tag{8.241}$$

The properties of the nonunitary state can be illustrated by considering the case of $\Delta_z = 0$. In this case,

$$\mathbf{\Delta} \times \mathbf{\Delta}^* = \left(0, 0, \frac{|\Delta_{\uparrow\uparrow}|^2 - |\Delta_{\downarrow\downarrow}|^2}{2i}\right); \tag{8.242}$$

and therefore, the energy gap depends on the direction of the spin.

When the interaction is invariant under simultaneous rotation of $\mathbf{k}$ and $\mathbf{k}'$, $V(k,k')$ can be expanded as

$$V(k,k') = \sum_{\ell=0}^{\infty} (2\ell+1) V_\ell(k,k') P_\ell(\hat{\mathbf{k}} \cdot \hat{\mathbf{k}}'), \tag{8.243}$$

where P_ℓ is the Legendre polynomial and $\hat{\mathbf{k}} = \mathbf{k}/k$. The gap equation (8.130) for the s-wave ($\ell = 0$) interaction is now replaced by

$$\Delta_{\alpha\beta}(\mathbf{k}) = -\frac{1}{\Omega} \sum_{\mathbf{k}'} V_\ell(k,k') \frac{\Delta_{\alpha\beta}(\mathbf{k}')}{2E_{\mathbf{k}',\alpha}} \left(1 - 2f(E_{\mathbf{k}',\alpha})\right). \tag{8.244}$$

Assuming that V_ℓ is nonzero only near the Fermi surface and constant there, we may exclude it from the sum, and we obtain

$$1 = -D(E_{\mathrm{F}}) V_\ell \int_0^{\omega_c} d\xi \frac{\tanh \frac{\beta_c \xi}{2}}{\xi}. \tag{8.245}$$

Thus, the critical temperature $T_c = (k_B \beta_c)^{-1}$ is determined for ℓ for which V_ℓ is most attractive.

8.11.2 *Spin-triplet p-wave states*

As shown in Eq. (8.236), the order parameter of a spin-triplet p-wave state is described by the order parameter $\mathbf{\Delta}(\mathbf{k})$ that behaves like a vector under rotation in spin space. When the orbital part is p-wave, $\mathbf{\Delta}(\mathbf{k})$ should also behave like a vector under rotation in orbital space. Thus, we obtain

$$\Delta_\mu(\mathbf{k}) = \sum_j d_{\mu j}\hat{k}_j, \tag{8.246}$$

where $\mu = x, y, z$ refer to the spin space and $j = x, y, z$, to the orbital space. Because $d_{\mu j}$ is, in general, complex, the order parameter of the spin-triplet p-wave superfluid involves $2 \times 3 \times 3 = 18$ degrees of freedom, and it exhibits a very rich variety of phases [Vollhardt and Wölfle (1990)]. Here, we show some typical examples.

BW state

The most symmetric state is the BW state [Balian and Werthamer (1963)]. A representation of the order parameter is given by

$$d_{\mu j} = e^{i\phi}\Delta_{\rm B}\delta_{\mu j}, \tag{8.247}$$

where ϕ is the global phase. In the absense of the dipole–dipole interaction, the free energy is invariant under rotation of the spin relative to the orbital angular momentum. Thus, the general order parameter of the BW state is given by

$$d_{\mu j} = e^{i\phi}\Delta_{\rm B}R_{\mu j}, \tag{8.248}$$

where $R_{\mu j}$ is a 3×3 rotation matrix. Because

$$\sum_\mu |\Delta_\mu(\mathbf{k})|^2 = \Delta_{\rm B}^2,$$

we find that the energy gap of the BW state is isotropic, as shown in Fig. 8.8 (a). The BW state describes the B phase of a superfluid ^{3}He.

ABM state

The order parameter of the ABM state [Anderson and Morel (1961); Anderson and Brinkman (1973)] is given by

$$d_{\mu j} = \Delta_{\rm A}\hat{d}_\mu(\hat{m}_j + i\hat{n}_j), \tag{8.249}$$

where $\hat{\mathbf{d}}$, $\hat{\mathbf{m}}$, and $\hat{\mathbf{n}}$ are unit vectors with $\hat{\mathbf{m}} \perp \hat{\mathbf{n}}$. Substituting Eq. (8.249) in Eq. (8.246), we obtain

$$\Delta_\mu(\mathbf{k}) = \Delta_{\rm A}\hat{d}_\mu(\hat{\mathbf{k}}\cdot\hat{\mathbf{m}} + i\hat{\mathbf{k}}\cdot\hat{\mathbf{n}}). \tag{8.250}$$

To calculate the energy gap, we introduce $\hat{\boldsymbol{\ell}} = \hat{\mathbf{m}} \times \hat{\mathbf{n}}$ and define $\hat{\mathbf{k}} = (\sin\theta\cos\phi,\ \sin\theta\sin\phi,\ \cos\theta)$ in the frame of reference $(\hat{\mathbf{m}},\ \hat{\mathbf{n}},\ \hat{\boldsymbol{\ell}})$, as shown in Fig. 8.9. Then,

$$\sum_{\mu} |\Delta_\mu(\mathbf{k})|^2 = \Delta_{\mathrm{A}}^2 \sin^2\theta. \tag{8.251}$$

Thus, the energy gap is anisotropic and has point nodes at $\theta = 0$ and π, as shown in Fig. 8.8 (b). If we choose $\hat{\mathbf{m}}$ and $\hat{\mathbf{n}}$ along the x- and y-directions, respectively, from Eq (8.250), we find that $\Delta_\mu(\mathbf{k}) \propto \hat{k}_x + i\hat{k}_y \propto Y_1^1(\hat{\mathbf{k}})$, implying that the orbital angular momentum of every Cooper pair points in the z-direction. The ABM state therefore exhibits orbital ferromagnetism and breaks time-reversal symmetry. The ABM phase describes the A phase of a superfluid ^{3}He.

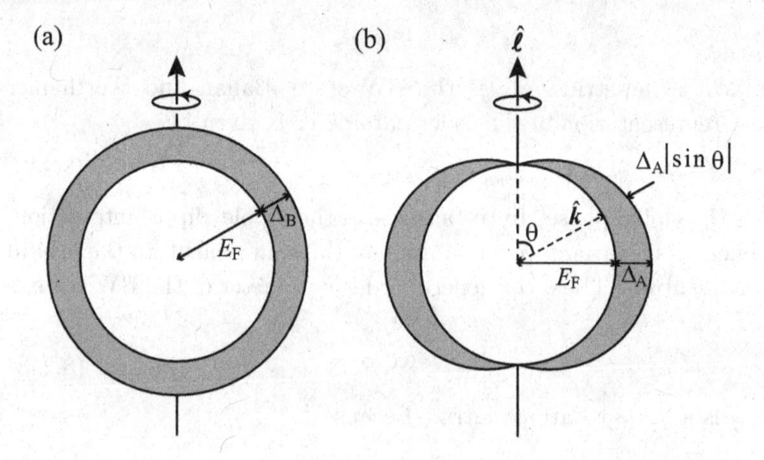

Fig. 8.8 The energy gap of the BW state (a) and the ABM state (b).

Polar state

The order parameter of a polar state is given by

$$d_{\mu j} = \Delta_{\mathrm{P}} \hat{d}_\mu \hat{\ell}_j. \tag{8.252}$$

Substituting this in Eq. (8.246), we obtain

$$\Delta_\mu(\mathbf{k}) = \Delta_{\mathrm{P}} \hat{d}_\mu \hat{\mathbf{k}} \cdot \hat{\boldsymbol{\ell}} \tag{8.253}$$

and

$$\sum_{\mu} |\Delta_\mu(\mathbf{k})|^2 = \Delta_{\mathrm{P}}^2 \cos^2\theta. \tag{8.254}$$

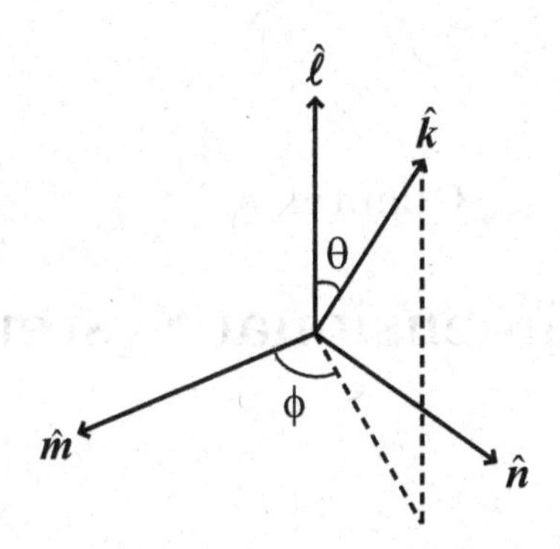

Fig. 8.9 A triad of vectors $\hat{\boldsymbol{\ell}}$, $\hat{\mathbf{m}}$, $\hat{\mathbf{n}}$ which are orthogonal to each other and provide a frame of reference in orbital space.

Therefore, the polar state has a nodal line on the equator ($\theta = \pi/2$). Because $\Delta_\mu(\mathbf{k}) \propto \hat{k}_z \propto Y_1^0(\hat{\mathbf{k}})$, the polar state is nonmagnetic with respect to the orbit.

If a magnetic Feshbach resonance is utilized to form spin-triplet Cooper pairs, the spin of every fermion is polarized along the direction of the magnetic field. In the case of ^{40}K [Ticknor, *et al.* (2004)], the magnetic dipole–dipole interaction (DDI) is significant and thus the polar state is favored, whereas in the case of ^{6}Li [Zhang, *et al.* (2004); Schunck, *et al.* (2005)], the DDI is not strong enough to stabilize the polar state.

Chapter 9

Low-Dimensional Systems

9.1 Non-interacting Systems

The Bose–Einstein condensation of particles into the zero-momentum state is disturbed by the presence of low-energy excitations. In particular, in a uniform infinite system, BEC does not exist in one and two dimensions at nonzero temperatures [Hohenberg (1967)] and in one dimension even at zero temperature [Pitaevskii and Stringari (1991)]. On the other hand, BEC can exist in one and two dimensions for a finite system [Widom (1968); Bagnato and Kleppner (1991)] and this fact has important implications in atomic-gas BECs. We begin by proving the Hohenberg theorem:

Theorem. *A Bose gas in a uniform infinite system with dimension* $d \leq 2$ *does not exhibit BEC at finite temperature.*

We first prove this theorem for the case of an ideal Bose gas because although the argument is simple, it captures the essential feature of the underlying physics. The proof for the case of an interacting Bose gas is given in the next section. The chemical potential μ in a d-dimensional ideal gas of N bosons is determined so as to satisfy the number equation:

$$\sum_{\mathbf{k}} \frac{1}{e^{\beta(\epsilon_k - \mu)} - 1} = N, \tag{9.1}$$

where $\beta \equiv 1/(k_B T)$ and $\epsilon_k \equiv \hbar^2 k^2/2m$. Suppose that μ vanishes at a nonzero temperature $T_c \equiv 1/(k_B \beta_c) > 0$. Equation (9.1) then leads to

$$\int_0^\infty d\epsilon \frac{\rho(\epsilon)}{e^{\beta_c \epsilon} - 1} = N, \tag{9.2}$$

where $\rho(\epsilon)$ is the density of states (DOS), which is given by $\rho(\epsilon) \propto \epsilon^{d/2-1}$, for

$$\sum_{\mathbf{k}} \cdots \Rightarrow \frac{V}{(2\pi)^d} \int d^d k \cdots \propto \int k^{d-1} dk \propto \int \epsilon^{\frac{d}{2}-1} d\epsilon.$$

Then, for $T < T_c$, a macroscopic number of particles

$$N_0 = N - \int_0^\infty d\epsilon \frac{\rho(\epsilon)}{e^{\beta\epsilon} - 1}$$

condense into the state with $\epsilon = 0$. However, if $d \leq 2$, μ does not vanish because the integral in Eq. (9.2) diverges for small ϵ. This implies that μ does not vanish for $d \leq 2$. Therefore, BEC does not occur for $d \leq 2$ at finite temperature for a uniform infinite system. The physics behind this absence of BEC is the proliferation of low-lying thermal excitations (infrared divergence) that destroy ODLRO.

In three dimensions, the DOS is proportional to $\sqrt{\epsilon}$ and rapidly decreases as $\epsilon \to 0$; therefore, a macroscopic occupation of the lowest-energy state is energetically favorable. In contrast, in one dimension, the DOS is proportional to $1/\sqrt{\epsilon}$ and diverges as $\epsilon \to 0$; therefore, it is difficult to occupy the zero-energy state by a macroscopic number of particles. The situation is marginal in two dimensions, where the DOS is independent of ϵ. In this case, there is no ODLRO; however, a quasi long-range order associated with a topological quantum phase transition can emerge, as discussed in Sec. 9.4.

These results change dramatically for a spatially confined system. In fact, under appropriate conditions, such a system can exhibit BEC even for $d \leq 2$ [Bagnato and Kleppner (1991)]. This is because the volume that is occupied by a nonuniform system, in general, depends on its energy ϵ, and this dependence alters the low-energy behavior of the DOS in favor of BEC. For example, when the system is confined in a power-law potential

$$V(r) \propto r^\eta, \tag{9.3}$$

there is a contribution to the DOS from the volume L^d such that

$$\rho(\epsilon) \propto L^d \epsilon^{\frac{d}{2}-1} \propto \epsilon^{\frac{d}{\eta}+\frac{d}{2}-1} \tag{9.4}$$

because particles with energy ϵ can spatially extend only over an extent of $L \propto \epsilon^{1/\eta}$. Then, a condition under which the integral in Eq. (9.2) converges, and hence, BEC can exist is given by $d/\eta + d/2 - 1 > 0$, that is,

$$d > \frac{2\eta}{\eta + 2} = \begin{cases} 1 \text{ if } \eta = 2 \text{ (parabolic potential)}, \\ \frac{2}{3} \text{ if } \eta = 1 \text{ (linear potential)}. \end{cases} \tag{9.5}$$

The transition temperature of BEC monotonically decreases with an increase in η ($> 2/3$) in 1D; however, it has a broad maximum around $\eta = 2$ in 2D [Bagnato and Kleppner (1991)]. According to Eq. (9.5), a noninteracting system confined in a harmonic potential undergoes BEC in 2D; however, the situation appears marginal in 1D. In fact, the 1D system in a harmonic trap undergoes BEC due to the discrete nature of energy levels, as shown in Sec. 3.2.

9.2 Hohenberg–Mermin–Wagner Theorem

Let us now consider the case of interacting bosons. The absence of the $U(1)$ gauge symmetry breaking in one and two dimensions at nonzero temperature can be shown based on Bogoliubov's inequality [Bogoliubov (1962); Wagner (1966); Mermin and Wagner (1966)]:

$$\langle\{\hat{A},\hat{A}^\dagger\}\rangle\langle[\hat{B}^\dagger,[\hat{H},\hat{B}]]\rangle \geq 2k_{\rm B}T|\langle[\hat{A},\hat{B}]\rangle|^2, \tag{9.6}$$

where $\hat{A}$ and $\hat{B}$ are arbitrary operators, $\hat{H}$ is the Hamiltonian of the system, $\{\hat{A},\hat{A}^\dagger\} \equiv \hat{A}\hat{A}^\dagger + \hat{A}^\dagger\hat{A}$, and $\langle\cdots\rangle$ denotes the thermal average

$$\langle\hat{O}\rangle \equiv \sum_m p_m\langle m|\hat{O}|m\rangle,\ p_m \equiv \frac{e^{-\beta E_m}}{{\rm Tr}e^{-\beta\hat{H}}}, \tag{9.7}$$

where $|m\rangle$ and E_m are the eigenstate and eigenvalue, respectively of $\hat{H}$, *i.e.*, $(\hat{H}|m\rangle = E_m|m\rangle)$. To prove (9.6), we define

$$(\hat{A},\hat{B}) \equiv {\sum_{m,n}}'\langle m|\hat{A}|n\rangle^*\langle m|\hat{B}|n\rangle\frac{p_m - p_n}{E_n - E_m}, \tag{9.8}$$

where $\sum'_{m,n}$ denotes the summation over m and n with $m \neq n$. Substituting $x = \beta(E_m - E_n)/2$ in $\tanh x/x \leq 1$, we obtain

$$\frac{\tanh\frac{\beta}{2}(E_m - E_n)}{\frac{\beta}{2}(E_m - E_n)} = \frac{2}{\beta(p_m + p_n)}\frac{p_m - p_n}{E_n - E_m} \leq 1.$$

Hence,

$$0 < \frac{p_m - p_n}{E_n - E_m} \leq \frac{\beta}{2}(p_m + p_n). \tag{9.9}$$

It follows from this and definition (9.8) that

$$\begin{aligned}(\hat{A},\hat{A}) &= {\sum_{m,n}}'|\langle m|\hat{A}|n\rangle|^2\frac{p_m - p_n}{E_n - E_m} \\ &\leq \frac{\beta}{2}\sum_{m,n}|\langle m|\hat{A}|n\rangle|^2(p_m + p_n) = \frac{\beta}{2}\langle\{\hat{A},\hat{A}^\dagger\}\rangle.\end{aligned} \tag{9.10}$$

We substitute $\hat{C} \equiv [\hat{B}^\dagger,\hat{H}]$ into the Schwartz' inequality [1]

[1] For an arbitrary complex number α, we have

$$\begin{aligned}(\alpha\hat{A}+\hat{C},\alpha\hat{A}+\hat{C}) &= |\alpha|^2(\hat{A},\hat{A}) + \alpha^*(\hat{A},\hat{C}) + \alpha(\hat{A},\hat{C})^* + (\hat{C},\hat{C}) \\ &= (\hat{A},\hat{A})\left|\alpha + \frac{(\hat{A},\hat{C})}{(\hat{A},\hat{A})}\right|^2 + \frac{(\hat{A},\hat{A})(\hat{C},\hat{C}) - |(\hat{A},\hat{C})|^2}{(\hat{A},\hat{A})} \geq 0.\end{aligned}$$

Because this inequality holds for arbitrary α, we obtain (9.11).

$$(\hat{A},\hat{A})(\hat{C},\hat{C}) \geq |(\hat{A},\hat{C})|^2. \tag{9.11}$$

Then, $(\hat{C},\hat{C})$ and $(\hat{A},\hat{C})$ are evaluated as

$$\begin{aligned}(\hat{C},\hat{C}) &= \sum_{m,n}{}' \langle m|\hat{C}|n\rangle^* \langle m|[\hat{B}^\dagger,\hat{H}]|n\rangle \frac{p_m - p_n}{E_n - E_m} \\ &= \sum_{m,n} \langle n|\hat{C}^\dagger|m\rangle\langle m|\hat{B}^\dagger|n\rangle(p_m - p_n) \\ &= \langle[\hat{B}^\dagger,\hat{C}^\dagger]\rangle = \langle[\hat{B}^\dagger,[\hat{H},\hat{B}]]\rangle,\end{aligned} \tag{9.12}$$

$$\begin{aligned}(\hat{A},\hat{C}) &= \sum_{m,n}{}' \langle m|\hat{A}|n\rangle^* \langle m|[\hat{B}^\dagger,\hat{H}]|n\rangle \frac{p_m - p_n}{E_n - E_m} \\ &= \sum_{m,n} \langle n|\hat{A}^\dagger|m\rangle\langle m|\hat{B}^\dagger|n\rangle(p_m - p_n) \\ &= \langle[\hat{B}^\dagger,\hat{A}^\dagger]\rangle = \langle[\hat{A},\hat{B}]\rangle^*.\end{aligned} \tag{9.13}$$

Substituting (9.10), (9.12), and (9.13) into (9.11) proves (9.6).

Now, we substitute $\hat{A} = \hat{a}^\dagger_{\mathbf{p}}$ and $\hat{B} = \hat{\rho}_{\mathbf{p}} \equiv \sum_{\mathbf{k}} a^\dagger_{\mathbf{k}} a_{\mathbf{k}+\mathbf{p}}$ in (9.6). Then,

$$\begin{aligned}\langle\{\hat{A},\hat{A}^\dagger\}\rangle &= 2n_{\mathbf{p}} + 1, \ n_{\mathbf{p}} \equiv \langle\hat{a}^\dagger_{\mathbf{p}}\hat{a}_{\mathbf{p}}\rangle, \\ [\hat{B}^\dagger,[\hat{H},\hat{B}]] &= [\hat{\rho}^\dagger_{\mathbf{p}},[\hat{H},\hat{\rho}_{\mathbf{p}}]] = \frac{\mathbf{p}^2}{M}N, \quad (f\text{-sum rule})^2 \\ [\hat{A},\hat{B}] &= -\hat{a}^\dagger_0.\end{aligned} \tag{9.14}$$

Substituting these results into Bogoliubov's inequality (9.6), we obtain

$$n_{\mathbf{p}} \geq \frac{Mk_{\mathrm{B}}T}{p^2}\frac{|\langle\hat{a}_0\rangle|^2}{N} - \frac{1}{2}. \tag{9.15}$$

It follows from Eq. (9.15) that in one and two dimensions, the summation $\sum_{\mathbf{p}} n_{\mathbf{p}}$ diverges unless $\langle\hat{a}_0\rangle = 0$. Thus, there is no breaking of the $U(1)$ gauge symmetry in one and two dimensions at finite temperatures.

At absolute zero, $\langle\hat{a}_0\rangle$ can take a nonzero value in two dimensions, whereas it cannot in one dimension. To show this, we use a different in-

[2]See Appendix E for proof.

equality [3]

$$\langle\{\hat{A}^\dagger, \hat{A}\}\rangle\langle\{\hat{B}^\dagger, \hat{B}\}\rangle \geq |\langle[\hat{A}^\dagger, \hat{B}]\rangle|^2. \tag{9.16}$$

Substituting $\hat{A} = \hat{a}_{\mathbf{p}}$ and $\hat{B} = \hat{\rho}_{\mathbf{p}}$ into (9.16), we have

$$\begin{aligned}\langle\{\hat{A}^\dagger, \hat{A}\}\rangle &= 2n_{\mathbf{p}} + 1,\\ \langle\{\hat{B}^\dagger, \hat{B}\}\rangle &= 2\langle\hat{\rho}_{\mathbf{p}}\hat{\rho}_{\mathbf{p}}^\dagger\rangle \equiv S(\mathbf{p}), \text{ (static structure factor)}\\ [\hat{A}^\dagger, \hat{B}] &= -\hat{a}_0^\dagger.\end{aligned} \tag{9.17}$$

Hence [Pitaevskii and Stringari (1991); Stringari (1995)],

$$n_{\mathbf{p}} \geq \frac{|\langle\hat{a}_0\rangle|^2}{2S(\mathbf{p})} - \frac{1}{2}. \tag{9.18}$$

In the long-wavelength limit ($p \to 0$), we can show that [see Appendix E]

$$S(\mathbf{p}) \leq \frac{p}{Mc}N,$$

where c is the velocity of sound. Thus,

$$n_{\mathbf{p}} \geq \frac{Mc}{2}\frac{|\langle\hat{a}_0\rangle|^2}{pN} - \frac{1}{2}. \tag{9.19}$$

It follows from Eq. (9.19) that the sum $\sum_{\mathbf{p}} n_{\mathbf{p}}$ converges in two dimensions but diverges logarithmically in one dimension unless $\langle\hat{a}_0\rangle$ vanishes. We therefore conclude that even at absolute zero, the $U(1)$ gauge symmetry does not break down in one dimension but it does so in two dimensions. Because the DOS diverges for $\epsilon \to 0$ in one dimension, quantum fluctuations are enhanced, thus destroying ODLRO.

The Hohenberg–Mermin–Wagner theorem shows the absence of the $U(1)$ gauge symmetry breaking in low dimensions. However, as emphasized in Sec. 1.5, BEC can exist in the absence of the $U(1)$ symmetry breaking. Nevertheless, it is generally considered based on the Hohenberg–Mermin–Wagner theorem and by replacing $|\langle\hat{a}_0\rangle|^2$ with n_0—the number of

[3] The proof of (9.16) is as follows.

$$\langle(\alpha\hat{A} + i\hat{B})^\dagger(\alpha\hat{A} + i\hat{B})\rangle = |\alpha|^2\langle\hat{A}^\dagger\hat{A}\rangle + i\alpha^*\langle\hat{A}^\dagger\hat{B}\rangle - i\alpha\langle\hat{B}^\dagger\hat{A}\rangle + \langle\hat{B}^\dagger\hat{B}\rangle \geq 0$$
$$\langle(\alpha\hat{A} - i\hat{B})(\alpha\hat{A} - i\hat{B})^\dagger\rangle = |\alpha|^2\langle\hat{A}\hat{A}^\dagger\rangle - i\alpha^*\langle\hat{B}\hat{A}^\dagger\rangle + i\alpha\langle\hat{A}\hat{B}^\dagger\rangle + \langle\hat{B}\hat{B}^\dagger\rangle \geq 0$$

Adding both sides, we have

$$\begin{aligned}&|\alpha|^2\langle\{\hat{A}^\dagger, \hat{A}\}\rangle - i\alpha^*\langle[\hat{A}, \hat{B}^\dagger]\rangle^* + i\alpha\langle[\hat{A}, \hat{B}^\dagger]\rangle + \langle\{\hat{B}, \hat{B}^\dagger\}\rangle\\ &= \langle\{\hat{A}^\dagger, \hat{A}\}\rangle\left|\alpha - i\frac{\langle[\hat{A}, \hat{B}^\dagger]\rangle^*}{\langle\{\hat{A}^\dagger, \hat{A}\}\rangle}\right|^2 + \frac{\langle\{\hat{A}^\dagger, \hat{A}\}\rangle\langle\{\hat{B}^\dagger, \hat{B}\}\rangle - |\langle[\hat{A}^\dagger, \hat{B}]\rangle|^2}{\langle\{\hat{A}^\dagger, \hat{A}\}\rangle} \geq 0\end{aligned}$$

Because this inequality holds for arbitrary α, we obtain (9.16).

Bose–condensed atoms—that BEC cannot exist in a uniform infinite system in two dimensions at nonzero temperatures and in one dimension down to zero temperature. While this replacement is usually assumed, it is of interest to clarify the logical gap and validate the theorem without invoking the notion of the $U(1)$ symmetry breaking.

9.3 Two-Dimensional BEC at Absolute Zero

The two-dimensional system is special at least in two respects. First, BEC exists only at absolute zero. Second, the DOS is constant and independent of energy, implying that the critical region extends over all temperature below the Berezinskii–Kosterlitz–Thouless transition temperature, as discussed in the next section. Here, a true condensate with ODLRO exists at absolute zero and with increasing temperature up to the transition temperature, a quasicondensate with quasi long-range order persists.

We examine the properties of a two-dimensional BEC in a uniform system. The average number of excited particles is given by

$$N'_T = \sum_{\mathbf{k}\neq 0} \langle \hat{a}^\dagger_{\mathbf{k}} \hat{a}_{\mathbf{k}} \rangle. \tag{9.20}$$

Substituting the Bogoliubov transformations (2.40) in Eq. (9.20), we obtain

$$N'_T = \sum_{k\neq 0} \left(\frac{\alpha_k^2}{1-\alpha_k^2} + \frac{1+\alpha_k^2}{1-\alpha_k^2} \frac{1}{e^{\beta E_k} - 1} \right), \tag{9.21}$$

where

$$E_k = \sqrt{\epsilon_k(\epsilon_k + 2U_0 n)} \tag{9.22}$$

is the energy of a Bogoliubov quasiparticle with $\epsilon_k = \hbar^2 k^2/(2M)$. At $T = 0$, the last term in Eq. (9.21) vanishes and only the term arising from quantum depletion remains nonvanishing:

$$N'_{T=0} = \sum_{k\neq 0} \frac{\alpha_k^2}{1-\alpha_k^2} = \frac{S}{(2\pi)^2} \int d^2k \frac{\alpha_k^2}{1-\alpha_k^2}, \tag{9.23}$$

where S is the area of the system. Changing the variable of integration to x defined in Eq. (2.42), we have

$$N'_{T=0} = \frac{MnU_0S}{\pi\hbar^2} \int_0^\infty dx x \frac{\alpha_k^2}{1-\alpha_k^2}. \tag{9.24}$$

Performing integration by parts and eliminating x in favor of α_k using Eq. (2.44), we obtain

$$\begin{aligned} N'_{T=0} &= \frac{MnU_0S}{\pi\hbar^2}\int_0^\infty dx \frac{x^2}{2}\frac{2\alpha_k}{(1-\alpha_k^2)^2}\frac{d\alpha_k}{dx} \\ &= \frac{MnU_0S}{2\pi\hbar^2}\int_0^\infty d\alpha_k \frac{1}{(1+\alpha_k)^2} \\ &= \frac{MnU_0S}{2\pi\hbar^2}. \end{aligned} \tag{9.25}$$

The ratio of quantum depletion to the total number of particles $N = nS$ is

$$\frac{N'_{T=0}}{N} = \frac{MU_0}{2\pi\hbar^2}. \tag{9.26}$$

The right-hand side is of the order of the interaction energy per particle $\epsilon_{\rm int} = nU_0$ to the kinetic energy $\epsilon_{\rm kin} = \hbar^2 n/2M$. Thus, when the interparticle interaction is weak ($\epsilon_{\rm int}/\epsilon_{\rm kin} \ll 1$), the fraction of quantum depletion is small, and thus, most particles are Bose–Einstein condensed at absolute zero, and the system possesses ODLRO.

At nonzero temperature, the second term in Eq. (9.21) is proportional to k^{-2} for small k and it logarithmically diverges in two dimensions. Thus, there is no genuine BEC, and hence, no ODLRO. However, the single-particle density matrix decays only algebraically and a quasi long-range order survives at low tempearture, as discussed in the next section.

9.4 Berezinskii–Kosterlitz–Thouless Transition

9.4.1 *Universal jump*

The Hohenberg–Mermin–Wagner theorem precludes a long-range order in one and two dimensions at finite temperature. This implies that low-dimensional systems do not exhibit conventional order–disorder phase transitions. On the other hand, it can be shown that the correlation function of the phase of the order parameter decays only algebraically at sufficiently low temperature which, in two dimensions, leads to a topological phase transition known as the Berezinskii–Kosterlitz–Thouless transition [Berezinskii (1970, 1971); Kosterlitz and Thouless (1973); Kosterlitz (1974)].

A two-dimensional system described by a scalar order parameter is equivalent to a planar spin model. The topological excitation of this system is a singly quantized vortex. The energy cost ϵ required to create a singly quantized vortex is given by Eq. (7.31). On the other hand, the entropy of the system S is given by

$$S = k_{\rm B} \ln \left(\frac{R}{\xi} \right)^2 = 2k_{\rm B} \ln \frac{R}{\xi}, \tag{9.27}$$

where ξ is the vortex core radius and the constant term is ignored. The change in free energy due to the creation of a vortex is therefore given by

$$\Delta F = \epsilon - TS = \left(\frac{\rho_{\rm s} \kappa^2}{4\pi} - 2k_{\rm B}T \right) \ln \frac{R}{\xi} + \text{constant}, \tag{9.28}$$

where $\rho_{\rm s}$ is the superfluid mass density and $\kappa \equiv h/M$, the quantum of circulation. If the temperature of the system T is larger than

$$T_{\rm BKT} = \frac{\rho_{\rm s} \kappa^2}{8\pi k_{\rm B}} = \frac{\pi \hbar^2}{2Mk_{\rm B}} n_{\rm s}, \tag{9.29}$$

where $n_{\rm s} \equiv \rho_{\rm s}/M$ is the superfluid density, there will exist a proliferation of vortices that destroy the superfluidity. If $T < T_{\rm BKT}$, on the other hand, the vortex creation is suppressed and superfluidity is maintained. In terms of the thermal de Broglie length $\lambda_{\rm th} = h/\sqrt{2\pi M k_{\rm B} T_{\rm BKT}}$, the condition (9.29) is expressed as

$$n_{\rm s} \lambda_{\rm th}^2 = 4. \tag{9.30}$$

We note that the BKT transition temperature $T_{\rm BKT}$ is substantially lower than the temperature of quantum degeneracy

$$T_d = \frac{2\pi \hbar^2}{M} n, \tag{9.31}$$

where n is the 2D particle density [Prokof'ev, *et al.* (2001); Prokof'ev and Svistunov (2002)]. We also note that at $T_{\rm BKT}$, the ratio of $\rho_{\rm s}$ to $k_{\rm B}T_{\rm BKT}$ jumps from 0 to

$$\frac{\rho_{\rm s}}{k_{\rm B} T_{\rm BKT}} = \frac{8\pi}{\kappa^2}. \tag{9.32}$$

Because the right-hand side depends only on the fundamental unit of circulation and does not depend on the material properties, this phenomenon is referred to as a universal jump [Nelson and Kosterlitz (1977)]. This implies that for the superflow to persist stably, $\rho_{\rm s}$ must satisfy

$$\rho_{\rm s} \geq \frac{8\pi}{\kappa^2} k_{\rm B} T_{\rm BKT}. \tag{9.33}$$

9.4.2 *Quasi long-range order*

Next, we calculate the correlation function of the phase of the order parameter:

$$\psi(\mathbf{r}) = \psi_0 e^{i\phi(\mathbf{r})}. \tag{9.34}$$

For the sake of simplicity, we assume the amplitude ψ_0 to be constant. The change in phase leads to a cost in the kinetic energy:

$$H_{\text{KE}} = \int d\mathbf{r} \frac{\hbar^2}{2M} |\nabla\psi|^2 = \frac{J}{2} \int d\mathbf{r} |\nabla\phi|^2, \tag{9.35}$$

where

$$J = \frac{\hbar^2}{M} |\psi_0|^2. \tag{9.36}$$

The correlation function is given by

$$\begin{aligned} \langle \psi^*(\mathbf{r})\psi(0)\rangle &= |\psi_0|^2 \left\langle e^{-i(\phi(\mathbf{r})-\phi(0))} \right\rangle \\ &= |\psi_0|^2 \exp\left[-\frac{1}{2} \langle (\phi(\mathbf{r}) - \phi(0))^2 \rangle \right], \end{aligned} \tag{9.37}$$

where we use the fact that the canonical distribution for H_{KE} is Gaussian and that only the quadratic term remains nonvanishing in the cumulant expansion for the Gaussian distribution. Expanding $\phi(\mathbf{r})$ in a Fourier series

$$\phi(\mathbf{r}) = \sum_{\mathbf{k}} \phi_{\mathbf{k}} e^{i\mathbf{k}\mathbf{r}}, \tag{9.38}$$

Eq. (9.35) becomes

$$H_{\text{KE}} = \frac{J}{2} \sum_{\mathbf{k}} \mathbf{k}^2 |\phi_{\mathbf{k}}|^2. \tag{9.39}$$

This result can be used to show that

$$\langle \phi_{\mathbf{k}} \phi_{\mathbf{k}'} \rangle = \frac{\text{Tr}\left(e^{-\beta H_{\text{KE}}} \phi_{\mathbf{k}} \phi_{\mathbf{k}'} \right)}{\text{Tr} e^{-\beta H_{\text{KE}}}} = \frac{1}{\beta J k^2} \delta_{\mathbf{k},-\mathbf{k}'}. \tag{9.40}$$

Thus,

$$\begin{aligned} \langle (\phi(\mathbf{r}) - \phi(0))^2 \rangle &= \sum_{\mathbf{k},\mathbf{k}'} (e^{i\mathbf{k}\mathbf{r}} - 1)(e^{i\mathbf{k}'\mathbf{r}} - 1) \langle \phi_{\mathbf{k}} \phi_{\mathbf{k}'} \rangle \\ &= \frac{2}{\beta J} \sum_{\mathbf{k}} \frac{1 - \cos(\mathbf{k} \cdot \mathbf{r})}{k^2}. \end{aligned} \tag{9.41}$$

In the limit of large r, the rapidly oscillating cosine term does not contribute to the sum and we obtain

$$\langle (\phi(\mathbf{r}) - \phi(0))^2 \rangle = \frac{2}{\beta J} \frac{2\pi}{(2\pi)^2} \int_{r^{-1}}^{\xi^{-1}} \frac{dk}{k} = \frac{1}{\pi \beta J} \ln \frac{r}{\xi}. \tag{9.42}$$

Substituting this into Eq. (9.37), we obtain

$$\langle\psi^*(\mathbf{r})\psi(0)\rangle = |\psi_0|^2 \left(\frac{r}{\xi}\right)^{-\eta}, \tag{9.43}$$

where

$$\eta = \frac{1}{2\pi\beta J} = \frac{k_B T}{2\pi J} = \frac{1}{|\psi_0|^2\lambda_{\rm th}^2}. \tag{9.44}$$

Thus, the order parameter decays algebraically and the system is said to possess a quasi long-range order. The algebraic decay of the correlation function at all temperatures below $T_{\rm BKT}$ implies that the entire region below the transition temperature constitutes a critical line.

At the BKT transition temperature, we substitute Eqs. (9.32) and (9.36) in Eq. (9.44) to obtain

$$\eta = \frac{\rho_s}{4M|\psi_0|^2}. \tag{9.45}$$

If we identify $\rho_s = M|\psi_0|^2$, we find that $\eta = 1/4$. In accordance with Eq. (9.33), for the superflow to be stable, η must satisfy

$$\eta \leq \frac{1}{4}. \tag{9.46}$$

However, the relationship between ρ_s and $|\psi_0|^2$ is not simple. We refer to [Prokof'ev, *et al.* (2001)] for a quantum Monte Carlo study.

9.4.3 *Renormalization-group analysis*

Let $p(r)$ be the probability density of a pair of vortices that are separated by distance r. Assuming that vortex pairs are excited thermally, $p(r+dr)$ and $p(r)$ are related to each other by

$$p(r+dr) = p(r)e^{-\beta\Delta E}, \tag{9.47}$$

where ΔE is the work required to increase the separation from r to $r+dr$. Because the energy of a vortex pair is given from (7.53) as

$$E(r) = 2\pi\kappa^2\rho_s \ln\frac{r}{\xi} + 2E_c, \tag{9.48}$$

we have

$$\Delta E = \frac{dE(r)}{dr}dr = \frac{2}{\varepsilon r}dr, \tag{9.49}$$

where $\varepsilon \equiv (\pi\kappa^2\rho_s)^{-1}$ is the "dielectric constant" of the medium. Bearing in mind the fact that thermally excited vortex pairs that are smaller than r

screen the interaction, we regard ε as a function of r. Substituting Eq. (9.49) into Eq. (9.47), we obtain

$$\frac{dp(r)}{dr} = -\frac{\beta}{\varepsilon(r)r}p(r). \tag{9.50}$$

On the other hand, the variation of $\varepsilon(r)$ is related to $\chi(r)$—the polarizability of a vortex pair of separation r—as

$$\varepsilon(r+dr) - \varepsilon(r) = 4\pi\chi(r) \cdot p(r)2\pi r dr. \tag{9.51}$$

Here, $\chi(r)$ is estimated to be

$$\chi(r) \propto \langle r^2 \rangle = \frac{\int_r^\infty r'^3 e^{-\beta E(r')} dr'}{\int_r^\infty r' e^{-\beta E(r')} dr'} = r^2 \frac{\pi K - 1}{\pi K - 2}, \tag{9.52}$$

where $K \equiv \beta\kappa^2\rho_s$. The polarizability diverges at $K = 2/\pi$, below which vortex pairs can dissociate freely. Here, we use a more precise value $\chi(r) = \beta r^2/2$ [Kosterlitz and Thouless (1973)] to obtain

$$\frac{d\varepsilon(r)}{dr} = 4\pi^2\beta r^3 p(r). \tag{9.53}$$

Introducing dimensionless variables

$$\ell \equiv \ln\frac{r}{\xi}, \tag{9.54}$$

$$K(\ell) \equiv \frac{\beta}{\pi\varepsilon(r)}, \tag{9.55}$$

$$y(\ell) \equiv r^2\sqrt{p(r)}\Big|_{r=\xi e^\ell}, \tag{9.56}$$

and keeping only the leading-order terms in small y, we obtain the Kosterlitz recursion relations [Kosterlitz (1974)]:

$$\frac{dx}{d\ell} = 4\pi^3 y^2 + O(y^4), \tag{9.57}$$

$$\frac{dy}{d\ell} = 2xy + O(y^3), \tag{9.58}$$

where

$$x \equiv \frac{2}{\pi K} - 1. \tag{9.59}$$

If follows from Eqs. (9.57) and (9.58) that

$$x^2 - 4\pi^2 y^2 = \text{constant}. \tag{9.60}$$

This equation shows how x and y change with an increase in ℓ, as shown in Fig. 9.1.

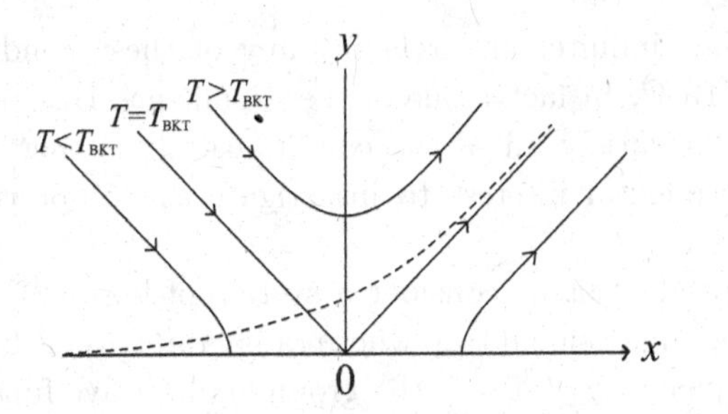

Fig. 9.1 Renormalization-group flow diagram, where x is a shifted dimensionless superfluid density and y is proportional to the fugacity of vortices. The arrow shows how x and y change with an increase in the length scale ℓ. The dashed curve shows the initial conditions as a function of temperature.

We begin with $x(\ell = \xi)$ and $y(\ell = \xi)$ that represent the unscreened properties of the system and draw the trajectory $(x(\xi), y(\xi))$ as a function of temperature (dashed curve in Fig. 9.1). The point at which this trajectory intersects the separatrix $(y = -x/2\pi)$ gives $T_{\rm BKT}$ and the superfluid density $K(\infty)$.

At $T = T_{\rm BKT}$, the constant in Eq. (9.60) is zero and the renormalization-group flow diagram follows the separatrix,

$$y = -\frac{x}{2\pi} \sim \frac{1}{4\pi \ln \frac{r}{\xi}}, \tag{9.61}$$

and we reproduce the universal jump:

$$K = \frac{2}{\pi}. \tag{9.62}$$

At $T = T_{\rm BKT}$, we obtain

$$p(r) \simeq \frac{y^2}{r^4} \simeq \frac{1}{r^4(\ln r)^2}. \tag{9.63}$$

The BKT transition was observed in helium-4 by Rudnick [Rudnick (1978)] and Bishop and Reppy [Bishop and Reppy (1978, 1980)], and in a ^{87}Rb BEC by Hadzibabic *et al.* [Hadzibabic, *et al.* (2006); Hadzibabic, *et al.* (2008)].

9.5 Quasi One-Dimensional BEC

The Hohenberg–Mermin–Wagner theorem precludes BEC in one-dimensional infinite systems. However, real systems are neither strictly one-dimensional nor infinite, and relaxing any of these conditions often works in favor of BEC. In fact, none of the atomic-gas BEC experiments carried out thus far satisfy all of the conditions. It is therefore of theoretical and experimental interest to investigate the properties of quasi one-dimensional BEC.

To be more specific, let us consider a system of bosons that are confined in a quasi one-dimensional trap whose radial direction is harmonically confined with frequency $\omega_\perp$. Then, the ground-state wave function of the noninteracting bosons has a radial width

$$d_\perp = \sqrt{\frac{\hbar}{M\omega_\perp}}. \tag{9.64}$$

The correlation length (or healing length) ξ_c is obtained by equating the chemical potential μ with the zero-point kinetic energy:

$$\xi_c = \frac{\hbar}{\sqrt{M\mu}} = \frac{\hbar}{\sqrt{Mn\tilde{g}}}, \tag{9.65}$$

where n is the particle density and $\tilde{g}$, the coupling constant which, in free space, would be related to the s-wave scattering length a via $\tilde{g} = 4\pi\hbar^2 a/M$. The condition of tight confinement in the radial direction is then given by

$$\xi_c \gg d_\perp \Leftrightarrow \hbar\omega_\perp \gg \mu. \tag{9.66}$$

We define the dimensionless strength of interaction, γ, as the ratio of the strength of interaction $n\tilde{g}$ to the kinetic energy $\sim \hbar^2 n^2/M$:

$$\gamma = \frac{M\tilde{g}}{\hbar^2 n}. \tag{9.67}$$

Thus, in one dimension, the interaction becomes stronger for lower density. To understand this phenomenon, let us consider a scattering problem of two particles with mass M interacting via a delta function $\tilde{g}\delta(z)$, where z is the relative coordinate between the two particles. The Schrödinger equation for this problem is

$$\left[-\frac{\hbar^2}{M}\frac{d^2}{dz^2} + \tilde{g}\delta(z)\right]\psi(z) = E\psi(z), \tag{9.68}$$

where $E = \hbar^2 k^2/M$ is the incident energy. In one dimension, there are only transmission and reflection, and the general solution is given by

$$\psi(z) = \begin{cases} e^{ikz} + re^{-ikz} & \text{for } z < 0; \\ te^{ikz} & \text{for } z > 0, \end{cases} \tag{9.69}$$

where r and t are the reflection and transmission coefficients which are calculated to be

$$r = -\left(1 - \frac{2i\hbar^2 k}{M\tilde{g}}\right)^{-1}, \quad t = \left(1 + \frac{iM\tilde{g}}{2\hbar^2 k}\right)^{-1}. \tag{9.70}$$

The lower density implies smaller k, which leads to smaller t. In particlu-lar, in the low-density limit with $k \to 0$, we have $t \to 0$, indicating that the system becomes an impenetrable Tonks-Girardeau gas as discussed in Sec. 9.6.

Let r_{eff} be an effective range of interaction. If the condition

$$d_\perp \gg r_{\text{eff}} \tag{9.71}$$

is satisfied, the scattering occurs in three dimensions and we may use the three-dimensional scattering length. In the following discussions, we assume that conditions (9.66) and (9.71) are both satisfied. Then, the dynamics of the system occurs predominantly in the z-direction, whereas the microscopic scattering occurs in three dimensions. Thus, the coupling constant $\tilde{g}$ can be expressed as a function of a and $d_\perp$. In the case of a harmonic confinement, the single-particle energy spectrum in the radial direction is quantized in units of $\hbar\omega_\perp$, and such a discreteness modifies the coupling constant $\tilde{g}$ as [Olshanii (1998)]

$$\tilde{g} = \frac{g}{2\pi d_\perp^2} \frac{1}{1 - C\frac{a}{d_\perp}}, \tag{9.72}$$

where $g = 4\pi\hbar^2 a/M$ and $C = -\zeta(1/2)/\sqrt{2} \simeq 1.0326$ with $\zeta(x)$ being the Riemann zeta function. Equation (9.72) implies that a 1D effective interaction

$$V_{\text{1D}}(z) = \tilde{g}\delta(z) \tag{9.73}$$

exhibits a resonant structure called the confinement-induced resonance (CIR) which may be interpreted as a Feshbach resonance of a bound state with the lowest transverse mode. As a consequence of the CIR, the effective interaction can be attractive ($\tilde{g} < 0$) for positive $a(> d_\perp/C)$. The CIR was observed by H. Moritz *et al.* [Moritz, *et al.* (2005)].

For the sake of simplicity, we first assume that the system is uniform in the z-direction and periodic with periodicity L. Such a situation can be realized using a quasi one-dimensional torus trap. Within the Bogoliubov approximation, we may use the results described in Sec. 2.4 to find the number of non-condensed atoms at $T = 0$:

$$N' = 2\sum_{m=1}^{\infty} \sinh^2\theta_m = 2\sum_{m=1}^{\infty} \frac{(\tilde{g}n)^2}{\epsilon_m(\epsilon_m + 2\tilde{g}n)}, \tag{9.74}$$

where $gN = \tilde{g}n$ and ϵ_m is defined in Eq. (2.104) with $L = 2\pi R$. Taking the sum in Eq. (9.74), we obtain

$$N' = \frac{\pi^2}{12}\tilde{\gamma} + \frac{\pi}{4}\sqrt{\tilde{\gamma}}\coth\left(\pi\sqrt{\tilde{\gamma}}\right) + \frac{1}{4}, \tag{9.75}$$

where

$$\tilde{\gamma} \equiv \frac{2\tilde{g}n}{\hbar\omega_c} = \frac{N^2}{\pi^2}\gamma. \tag{9.76}$$

For $\tilde{\gamma} \gg 1$, the first term on the right-hand side of Eq. (9.75) dominates and we have

$$\frac{N'}{N} \simeq \frac{\pi^2}{12}\frac{\tilde{\gamma}}{N} = \frac{\pi^2}{6}\frac{\tilde{g}n}{\hbar\omega_c N}. \tag{9.77}$$

The assumption of weakly interacting bosons is self-consistent if $N'/N \ll 1$, and then, we have a genuine BEC. When the circumference L of the torus increases, $\tilde{\gamma}$ increases and the assumption of weak interaction breaks down when the right-hand side of Eq. (9.77) becomes of the order of one.

A similar conclusion holds when the axial direction is confined by a harmonic potential as well. Let ω_z be the frequency of the axial trap. Then, the 1D condensate at $T = 0$ obeys

$$\left(-\frac{\hbar^2}{2M}\frac{d^2}{d_z^2} + \frac{M\omega_z^2}{2}z^2 + g|\psi_0|^2\right)\psi_0 = \mu\psi_0. \tag{9.78}$$

In the Thomas–Fermi regime (*i.e.*, $\mu \gg \hbar\omega_z$), ψ_0 is given by

$$\psi_0(z) = \sqrt{\frac{\mu}{g}\left(1 - \frac{z^2}{R_{\rm TF}^2}\right)}\,\theta(R_{\rm TF} - |z|), \tag{9.79}$$

where

$$R_{\rm TF} \equiv \sqrt{\frac{2\mu}{m\omega_z^2}}$$

is the Thomas–Fermi radius. The chemical potential μ is determined from

$$N = \int_{-R_{\rm TF}}^{R_{\rm TF}} |\psi_0(z)|^2 d_z, \tag{9.80}$$

giving

$$\mu = \frac{\hbar\omega_z}{2}\left(\frac{3N\alpha}{2}\right)^{\frac{2}{3}}, \tag{9.81}$$

where

$$\alpha \equiv \frac{M\tilde{g}d_z}{\hbar^2} = \gamma(d_z n)$$

with $d_z = \sqrt{\hbar/(M\omega_z)}$ distinguishes various regimes of the quasi 1D Bose gas [Petrov, *et al.* (2004)]. The system is in the Thomas–Fermi regime ($\mu \gg \hbar\omega_z$) for $N\alpha \gg 1$ and the Gaussian regime ($\mu \ll \hbar\omega_z$) for $N\alpha \ll 1$. When $\alpha \gg 1$ and $N \ll \alpha^2$, the system is in the Tonks–Girardeau regime, as discussed in the next section [Cladé, *et al.* (2009)].

9.6 Tonks–Girardeau Gas

When the repulsive interaction becomes very strong, the one-dimensional wave function begins to develop antibunching correlation reminescent of the Fermi statistics. Let ξ_a be the correlation length of the antibunching. Equating the energy cost of the interaction $\sim \tilde{g}\xi_a^{-1}$ with the zero-point kinetic energy $\hbar^2/(M\xi_a^2)$, we obtain

$$\xi_a = \frac{\hbar^2}{M\tilde{g}} = \frac{1}{\gamma n} = \frac{d_z}{\alpha}. \tag{9.82}$$

The Tonks–Girardeau regime is realized when ξ_a becomes much smaller than the average interparticle distance n^{-1}, *i.e.*, $\xi_a n = \gamma^{-1} \ll 1$. In this regime, the repulsive interaction is so strong that the probability of two bosons being found at the same place is negligible. Thus, the many-body wave function of the Bose system, $\psi_{\rm B}(x_1, x_2, \cdots, x_N; t)$, vanishes when any two coordinates coincide:

$$\psi_{\rm B}(x_1, x_2, \cdots, x_N; t) = 0 \ \text{ if } \ x_i = x_j \ \text{ for all } \ 1 \le i < j \le N. \tag{9.83}$$

This property is exactly shared by the many-body wave function of the Fermi system, $\psi_{\rm A}(x_1, x_2, \cdots, x_N)$, for the same Hamiltonian. On the other hand, for the exchange of any pair of coordinates, $\psi_{\rm A}$ changes its sign and $\psi_{\rm B}$ does not. Furthermore, if the interaction is of a contact type, the effect of interaction enters the problem only through the boundary condition (9.83). Hence, the energy of a system of impenetrable bosons becomes purely kinetic.

In this case, one can introduce the one-to-one mapping between $\psi_{\rm B}$ and $\psi_{\rm A}$, known as the Bose–Fermi mapping [Girardeau (1960); Girardeau (1965); Yukalov and Girardeau (2005)]:

$$\psi_{\rm B}(x_1, x_2, \cdots, x_N; t) = A(x_1, x_2, \cdots, x_N)\psi_{\rm F}(x_1, x_2, \cdots, x_N; t), \tag{9.84}$$

where

$$A(x_1, x_2, \cdots, x_N) = \prod_{i>j}^{N} \mathrm{sign}(x_1 - x_j)$$

is a unit antisymmetric function. Note that the Bose–Fermi mapping holds true not only for the ground state but also for time-dependent nonequilibrium situations. The Tonks–Girardeau model was originally envisaged by Tonks [Tonks (1936)] as a classical statistical-mechanical problem of hard spheres and later extended by Girardeau [Girardeau (1960)] as an effective 1D system of impenetrable hard-core bosons. The Tonks–Girardeau gas has been realized experimentally [Paredes, *et al.* (2004); Kinoshita, *et al.* (2004)].

When the system is in the ground state, the Bose–Fermi mapping theorem takes a particularly simple form. In the ground state, the many-body wave function of a Bose system is nodeless and can be considered to be real [see Sec. 1.6], and therefore, Eq. (9.84) reduces to

$$\psi_{\mathrm{B}}(x_1, x_2, \cdots, x_N) = |\psi_{\mathrm{F}}(x_1, x_2, \cdots, x_N)|. \tag{9.85}$$

This relationship is very useful for calculating $\psi_{\mathrm{B}}(x_1, x_2, \cdots, x_N)$ because ψ_{F} is given by a Slater determinant of the N lowest single-particle wave functions $\phi_n(x)$:

$$\psi_{\mathrm{F}}(x_1, x_2, \cdots, x_N) = \frac{1}{\sqrt{N!}} \det\phi_n(x_m), \tag{9.86}$$

where $\det\phi_n(x_m)$ is the determinant of the $N \times N$ matrix $\{\phi_n(x_m); n, m = 1, 2, \cdots, N\}$. It follows from Eqs.(9.85) and (9.86) that the single-particle density distribution is given by

$$\begin{aligned} \rho(x) &= N \int |\psi_{\mathrm{B}}(x_1, x_2, \cdots, x_N)|^2 dx_2 \cdots dx_N \\ &= \sum_{n=1}^{N} |\phi_n(x)|^2. \end{aligned} \tag{9.87}$$

A simplest example is a system of N free bosons subject to the periodic boundary condition with period L. Then, the single-particle wave functions are plane waves $\phi_n(x) = e^{2\pi i n x/L}$ $(n = 1, 2, \cdots, N)$ and the Slater determinant (9.86) becomes a Vandermonde matrix; thus,

$$\psi_{\mathrm{B}}(x_1, x_2, \cdots, x_N) \propto \prod_{i>j} |\sin[\pi(x_i - x_j)/L]|. \tag{9.88}$$

Let us next consider a 1D gas in a harmonic potential

$$V(x) = \frac{M\omega^2}{2} x^2. \tag{9.89}$$

Then, the single-particle wave function is given by

$$\phi_n(x) = \frac{1}{4\sqrt{\pi}\sqrt{2^n n!d}} H_n\left(\frac{x}{d}\right) e^{-\frac{x^2}{2d^2}}, \tag{9.90}$$

where $d \equiv \sqrt{\hbar/(M\omega)}$ and H_n is the nth Hermitian polynomial. Substituting Eq. (9.90) into Eq. (9.86) and using Eq. (9.85), we find that

$$\psi_{\mathrm{B}}(x_1, x_2, \cdots, x_N) \propto \left(\prod_{i>j}^{N} |x_i - x_j|\right) \exp\left(-\sum_{n=1}^{N} \frac{x_n^2}{2d^2}\right). \tag{9.91}$$

It can be shown that the largest eigenvalue of the single-particle density matrix corresponding to Eq. (9.91) is of the order of $\sqrt{N}$ [Forrester, *et al.* (2003)], indicating that there is no BEC in the Tonks–Girardeau regime. However, the system does exhibit a sharp peak near zero momentum [Girardeau and Wright (2001); Lapeyre, *et al.* (2002)] that is markedly different from the broad momentum distribution of the Fermi system.

9.7 Lieb–Liniger Model

When the radial confinement $d_\perp = \sqrt{\hbar/(M\omega_\perp)}$ is much tighter than the correlation length $\xi_c = \hbar/\sqrt{M\mu}$, *i.e.*, when $d_\perp \ll \xi_c$ or $\hbar\omega_\perp \gg \mu$, the system may be regarded as effectively one-dimensional with an effective coupling constant given by

$$g_{\rm 1D} = \frac{\hbar^2}{Md_\perp^2}a = \hbar\omega_\perp a, \tag{9.92}$$

where $a \ll d_\perp$ is assumed. Furthermore, if the condition $\xi_a \ll n^{-1}$ is satisfied, the Tonks–Girardeau model can be applied; otherwise, the system is strongly interacting but the overlap of the wave functions between bosons occurs. Such a strongly correlated system of bosons can be described by the Lieb–Liniger model:

$$H = -\sum_{i=1}^{N} \frac{\hbar^2}{2M}\frac{\partial^2}{\partial x_i^2} + 2g_{\rm 1D}\sum_{i>j}\delta(x_i - x_j), \tag{9.93}$$

where the Tonks–Girardeau model is obtained in the limit $g_{\rm 1D} \to \infty$. This model was introduced by Lieb and Liniger [Lieb and Liniger (1963)] and all the eigenvalues and eigenfunctions can, in principle, be solved exactly. This model predicts two types of solutions [Lieb (1963)], one of which gives the low-lying spectrum consistent with the Bogoliubov theory. The other solution corresponds to a soliton branch [Ishikawa and Takayama (1980)] or a yrast state—the lowest-energy solution for a given angular momentum [Kanamoto, *et al.* (2010)]. For a given one-dimensional density $n_{\rm 1D}$, the ratio of the mean-field interaction energy to the kinetic energy is given by

$$\gamma = \frac{2g_{\rm 1D}n_{\rm 1D}}{\hbar^2 n_{\rm 1D}^2/(2M)} = \frac{4a}{n_{\rm 1D}d_\perp^2}. \tag{9.94}$$

Thus, γ is inversely proportional to the particle density as opposed to the 3D case.

The delta-function interaction in Eq. (9.93) serves as the boundary conditions for the many-body wave function $\psi_B(x_1, x_2, \cdots, x_N)$ such that

$$\left(\frac{\partial}{\partial x_i} - \frac{\partial}{\partial x_j}\right) \psi_B \big|_{x_i = x_j + 0^+} = g_{1D} \psi_B \big|_{x_i = x_j} . \tag{9.95}$$

Because ψ_B is symmetric under the exchange of any two of its coordinates, it is sufficient to find the solution for $0 \leq x_1 \leq x_2 \leq \cdots \leq x_N \leq L$. It can be shown that all eigenfunctions of the Hamiltonian (9.93) can be written in the Bethe ansatz form [Dorlas (1993)]:

$$\psi_B(x_1, x_2, \cdots, x_N) = \sum_p a(p) \exp\left(i \sum_{j=1}^{N} k_{pj} x_j\right), \tag{9.96}$$

where $k_1, k_2, \cdots, k_N$ are real numbers such that $k_1 < k_2 < \cdots < k_N$ and the sum is taken over all $N!$ permutations of $1, 2, \cdots, N$. It follows from the Schrödinger equation $H\psi_B = E\psi_B$ that the eigenvalue E and the coefficients $a(p)$ are given by

$$E = \sum_{i=1}^{N} k_i^2, \tag{9.97}$$

$$a(p) = \prod_{1 \leq i < j \leq N} \left(1 + \frac{i g_{1D}}{k_{pi} - k_{pj}}\right). \tag{9.98}$$

Substituting Eq. (9.98) back into Eq. (9.96), the many-body wave function is obtained in terms of $k_1, k_2, \cdots, k_N$. By imposing the periodicc boundary conditions

$$\begin{aligned} &\psi_B(x_1, \cdots, x_i + L, \cdots, x_N) \\ &= \psi_B(x_1, \cdots, x_i, \cdots, x_N) (i = 1, 2, \cdots, N), \end{aligned} \tag{9.99}$$

we obtain a set of N equations that determine $k_1, k_2, \cdots, k_N$:

$$k_i L = 2\pi n_i - 2 \sum_{j=1}^{N} \arctan\left(\frac{k_i - k_j}{g_{1D}}\right) \quad (i = 1, 2, \cdots, N), \tag{9.100}$$

where $n_1 < n_2 < \cdots < n_N$ are integers (half-integers) when N is odd (even).

The set of equations (9.100) can be solved numerically for a small number of particles. In the thermodynamic limit ($L, N \to \infty$ with $\rho = N/L$ fixed), the dimensionless ground-state energy per particle

$$e(g_{1D}) \equiv \frac{2m}{\hbar^2 \rho^2} \frac{E}{N} \tag{9.101}$$

can be calculated numerically [Lieb and Liniger (1963)], and it is consistent with the Bogoliubov spectrum when g_{1D} is small. Conversely, when $g_{1D} = \infty$, we have

$$e(\infty) = \frac{\pi^2}{3}, \tag{9.102}$$

which corresponds to the Tonks–Girardeau model.

The exact solution of the Lieb–Liniger model at finite temperature was obtained by Yang and Yang [Yang and Yang (1969)], and the zero-temperature solution for attractive bosons was found by McGuire [McGuire (1964); Calogero and Degasperis (1975)]. For extensive reviews, we suggest [Bloch, *et al.* (2008); Giorgini, *et al.* (2008)].

Chapter 10

Dipolar Gases

10.1 Dipole–Dipole Interaction

10.1.1 *Basic properties*

The interaction between two dipoles $\mathbf{d}_1$ and $\mathbf{d}_2$ that are separated by $\mathbf{r}$ [see Fig. 10.1 (a)] is described by

$$V_{\rm dd}(\mathbf{r}) = c_{\rm dd}\frac{\hat{\mathbf{d}}_1 \cdot \hat{\mathbf{d}}_2 - 3(\hat{\mathbf{d}}_1 \cdot \hat{\mathbf{r}})(\hat{\mathbf{d}}_2 \cdot \hat{\mathbf{r}})}{r^3}, \tag{10.1}$$

where $\hat{\mathbf{d}}_i \equiv \mathbf{d}_i/|\mathbf{d}_i|$, $\hat{\mathbf{r}} \equiv \mathbf{r}/|\mathbf{r}|$, and

$$c_{\rm dd} = \begin{cases} \frac{d_1 d_2}{4\pi\epsilon_0}, & \text{(electric dipoles);} \\ \frac{\mu_0 d_1 d_2}{4\pi} & \text{(magnetic dipoles),} \end{cases} \tag{10.2}$$

with ϵ_0 and μ_0 being the dielectric constant and magnetic permeability, respectively. As can be seen from Eq. (10.1), the dipole–dipole interaction (DDI) is long-ranged and anisotropic, and it can be attractive or repulsive depending on the relative position and orientation of the dipoles. To understand this, let us consider the simplest case in which the dipoles are polarized as illustrated in Fig. 10.1 (b). Then, Eq. (10.1) reduces to

$$V_{\rm dd}(\mathbf{r}) = c_{\rm dd}\frac{1 - 3\cos^2\theta}{r^3}. \tag{10.3}$$

Hence, the DDI is most repulsive at $\theta = \pi/2$ [Fig. 10.1 (c)] and most attractive at $\theta = 0$ [Fig. 10.1 (d)], and the sign changes at the magic angle

$$\theta_{\rm m} = \cos^{-1}\frac{1}{\sqrt{3}} \simeq 54.7^\circ. \tag{10.4}$$

At low temperatures, the DDI exhibits some unique scattering properties. In general, for a potential that decays as r^{-n} at long distances, the

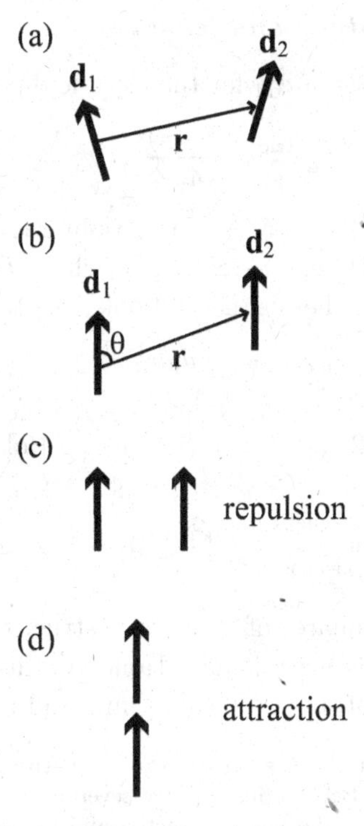

Fig. 10.1 Interaction between two dipoles $\mathbf{d}_1$ and $\mathbf{d}_2$ that are separated by $\mathbf{r}$. (a) Unpolarized case; (b)polarized case; (c) side-by-side alignment ($\theta = \pi/2$) for which the interaction is most repulsive; and (d) head-to-tail alignment ($\theta = 0$) for which the interaction is most attractive.

scattering phase shift $\delta_\ell(k)$ for the ℓth partial wave behaves in the low-energy limit $k \to 0$ as [Landau and Lifshitz (1999)]

$$\delta_\ell(k) \propto \begin{cases} k^{2\ell+1} \text{ for } \ell < \frac{n-3}{2}; \\ k^{n-2} \text{ for } \ell \geq \frac{n-3}{2}. \end{cases} \tag{10.5}$$

For the case of a van der Waals potential with $n = 6$, $\delta_0(k) \propto k$ (s-wave), $\delta_1(k) \propto k^3$ (p-wave), and $\delta_\ell(k) \propto k^4$ for $\ell \geq 2$. Therefore, the s-wave scattering dominates in the low-energy limit. However, for the DDI with $n = 3$, $\delta_\ell(k) \propto k$ for all ℓ, implying that all partial waves contribute to the low-energy scattering. Moreover, the anisotropy of the DDI induces coupling between different partial waves, as discussed in Sec. 10.1.3.

10.1.2 *Order of magnitude and length scale*

A typical order of magnitude of c_{dd} for the electric dipole is

$$c_{\mathrm{dd}}^{\mathrm{electric}} = \frac{(ea_{\mathrm{B}})^2}{4\pi\epsilon_0}, \tag{10.6}$$

where $a_{\mathrm{B}} \equiv 4\pi\epsilon_0\hbar^2/(m_{\mathrm{e}}e^2) \simeq 0.53\text{Å}$ is the Bohr radius, with m_{e} being the mass of the electron. Here, $ea_{\mathrm{B}} \simeq 2.54D$, where D is the debye with $D \simeq 3.34 \times 10^{-30}\mathrm{C}\cdot\mathrm{m}$. For the magnetic dipole, c_{dd} is given by

$$c_{\mathrm{dd}}^{\mathrm{magnetic}} = \frac{\mu_0\mu_{\mathrm{B}}^2}{4\pi}, \tag{10.7}$$

where $\mu_{\mathrm{B}} \equiv e\hbar/(2m_{\mathrm{e}}) \simeq 9.27 \times 10^{-24}\mathrm{J}\cdot\mathrm{T}^{-1}$ is the Bohr magneton. Their ratio

$$\frac{c_{\mathrm{dd}}^{\mathrm{magnetic}}}{c_{\mathrm{dd}}^{\mathrm{electric}}} = \frac{\alpha^2}{4} \sim 10^{-5} \tag{10.8}$$

is proportional to the square of the fine structure constant $\alpha \equiv e^2/(4\pi\epsilon_0\hbar c) \simeq 1/137$ and is very small. Table 10.1 lists the electric and magnetic dipole moments of some typical atomic and molecular species.

Table 10.1 Dipole moment and the corresponding length scale a_{dd} defined in Eq. (10.9) in units of the Bohr radius a_{B} for several atomic and molecular species, where μ_{B} and D are the Bohr magneton and the debye, respectively. $\varepsilon_{\mathrm{dd}}$ defined in Eq. (10.9) gives the ratio of the dipole–dipole interaction to the contact interaction. The electron spin and orbital angular momentum (AM) are measured in units of $\hbar$.

species	dipole moment	$a_{\mathrm{dd}}/a_{\mathrm{B}}$	$\varepsilon_{\mathrm{dd}}$	electron spin	orbital AM
^{87}Rb	$1\mu_{\mathrm{B}}$	0.7	0.007	1/2	0
^{52}Cr	$6\mu_{\mathrm{B}}$	16	0.16	3	0
^{166}Er	$7\mu_{\mathrm{B}}$	67		1	5
^{162}Dy	$10\mu_{\mathrm{B}}$	132		2	6
KRb	$0.6D$	2000			
ND_3	$1.5D$	3600			
HCN	$3.0D$	24000			

By equating the characteristic energy scale of the DDI $\sim c_{\mathrm{dd}}/r^3$ with the zero-point kinetic energy $\sim \hbar^2/(Mr^2)$, where M is the mass of the particle, we obtain a typical length scale a_{dd} for the DDI:

$$a_{\mathrm{dd}} \equiv \frac{c_{\mathrm{dd}}M}{3\hbar^2}, \tag{10.9}$$

where the numerical factor is introduced so that a uniform condensate becomes unstable against dipolar collapse for $a_{\rm dd} > a$. The ratio of $a_{\rm dd}$ to the s-wave scattering length a gives an estimate of the relative strength of the DDI against the contact interaction:

$$\varepsilon_{\rm dd} \equiv \frac{a_{\rm dd}}{a} = \frac{4\pi c_{\rm dd}}{3U_0}, \tag{10.10}$$

where $U_0 = 4\pi\hbar^2 a/M$ is the coupling constant for the s-wave contact interaction. This ratio is 0.007 for ^{87}Rb and 0.16 for ^{52}Cr [Griesmaier, *et al.* (2005)]; it is expected to be much greater for polar molecules because of the relationship (10.8) [see Table 10.1 for examples].

It is important to note that even when the DDI is weak, $\varepsilon_{\rm dd}$ can be made large by reducing a by means of a Feshbach resonance. Near the Feshbach resonance, the scattering length a is modified as

$$a = a_{\rm bg}\left(1 - \frac{\Delta}{B - B_0}\right), \tag{10.11}$$

where $a_{\rm bg}$ is the background scattering length; B_0, the location of the resonance; and Δ, its width. By setting $B = B_0 + \Delta$, we can quench a completely. Using this technique, magnetic DDI-dominated condensates have been created in ^{52}Cr [Lahaye, *et al.* (2007)] and ^{7}Li [Pollack, *et al.* (2009)].

10.1.3 *D-wave nature*

An important feature of the DDI is the d-wave nature of the relative orbital angular momentum between the interacting particles. This can be understood by rewriting Eq. (10.1) as follows:

$$V_{\rm dd}(\mathbf{r}) = c_{\rm dd} \sum_{\mu,\nu=x,y,z} (\hat{\mathbf{d}}_1)_\mu Q_{\mu\nu}(\mathbf{r})(\hat{\mathbf{d}}_2)_\nu, \tag{10.12}$$

where the kernel of the interaction

$$Q_{\mu\nu}(\mathbf{r}) \equiv \frac{\delta_{\mu\nu} - 3\hat{r}_\mu\hat{r}_\nu}{r^3} = \sqrt{\frac{6\pi}{5}}\frac{1}{r^3} \times \begin{bmatrix} \sqrt{\frac{2}{3}}Y_2^0 - Y_2^2 - Y_2^{-2} & i(Y_2^2 - Y_2^{-2}) & Y_2^1 - Y_2^{-1} \\ i(Y_2^2 - Y_2^{-2}) & \sqrt{\frac{2}{3}}Y_2^0 + Y_2^2 + Y_2^{-2} & -i(Y_2^1 + Y_2^{-1}) \\ Y_2^1 - Y_2^{-1} & -i(Y_2^1 + Y_2^{-1}) & -2\sqrt{\frac{2}{3}}Y_2^0 \end{bmatrix} \tag{10.13}$$

is a traceless symmetric tensor of rank 2 and the arguments of spherical harmonics are $Y_\ell^m(\theta,\phi)$ with $\hat{\mathbf{r}} \equiv \mathbf{r}/r = (\sin\theta\cos\phi,\ \sin\theta\sin\phi,\ \cos\theta)$. Because the integral

$$\int d\Omega (Y_L^M)^* Y_2^{m'} Y_\ell^m \tag{10.14}$$

is nonzero only if $M = m + m'$ and $L = \ell$ or $\ell \pm 2$, the DDI connects states with (ℓ, m) to those with $(\ell, m + m')$ or $(\ell \pm 2, m + m')$, where $Y_2^{m'}$ represents the spherical harmonics that appear in Eq. (10.13). In a special case in which all dipoles are polarized, the DDI involves only Y_2^0, and therefore,

$$V_{\rm dd}(\mathbf{r}) = c_{\rm dd} Q_{\rm zz}(\mathbf{r}) = c_{\rm dd} \frac{1 - 3\cos^2\theta}{r^3} = -\sqrt{\frac{16\pi}{5}} c_{\rm dd} \frac{Y_2^0(\hat{\mathbf{r}})}{r^3}. \tag{10.15}$$

In this case, the DDI conserves the projected orbital angular momentum along the quantization axis, connecting states of (ℓ, m) with those of (ℓ, m) or $(\ell \pm 2, m)$.

Due to the d-wave nature of the DDI, it is not invariant under individual rotations of the spin or the orbital angular momentum, and neither the total spin nor the total orbital angular momentum of the system is conserved. However, the DDI is invariant under a simultaneous rotation of the spin and orbital angular momentum, so that the total angular momentum is conserved.

10.1.4 *Tuning the dipole–dipole interaction*

The strength and sign of the DDI can be altered by the application of a time-dependent electric or magnetic field [Giovanazzi, *et al.* (2002)]. Suppose that the dipoles are made to oscillate in time as [see Fig 10.2]

$$\hat{\mathbf{d}}_1 = \hat{\mathbf{d}}_2 = (\sin\phi\cos\Omega t,\ \sin\phi\sin\Omega t,\ \cos\phi). \tag{10.16}$$

Substituting Eq. (10.16) and $\hat{\mathbf{r}} = (\sin\theta,\ 0,\ \cos\theta)$ in Eq. (10.1) and averaging the result over one period of oscillations, we obtain

$$\overline{V_{\rm dd}(\mathbf{r}, t)} = c_{\rm dd} \frac{1 - 3\cos^2\theta}{r^3} \frac{3\cos^2\phi - 1}{2}. \tag{10.17}$$

By adjusting ϕ, the DDI can be tuned as compared to Eq. (10.3) by a factor of $(3\cos^2\phi - 1)/2$ that varies between 1 for $\phi = 0$ and $-1/2$ for $\phi = \pi/2$. The DDI can also be quenched completely if ϕ is set to be the magic angle θ_m defined in Eq. (10.4).

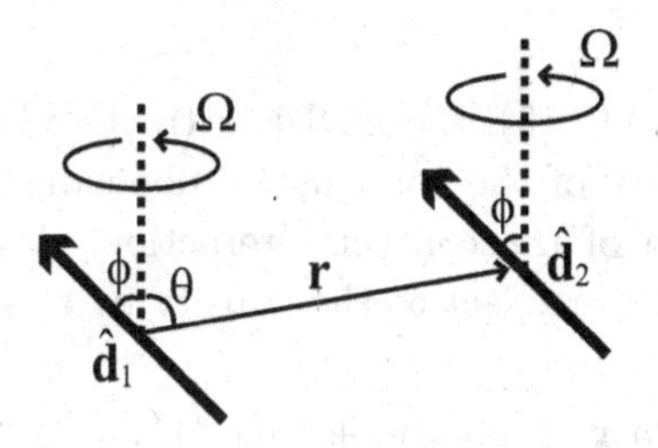

Fig. 10.2 Tuning of the dipole–dipole interaction by the application of a rotating external field with frequency Ω.

10.2 Polarized Dipolar BEC

10.2.1 *Nonlocal Gross–Pitaevskii equation*

When the dipoles are fully polarized, a Bose–Einstein condensate is described with a scalar order parameter $\psi(\mathbf{r}, t)$. However, as compared to the Gross–Pitaevskii equation (GPE) for a scalar BEC, the GPE now involves a nonlocal term that describes the DDI:

$$V_{\rm dd}(\mathbf{r}, t) = c_{\rm dd} \int d\mathbf{r}' \frac{1 - 3\cos^2\theta}{|\mathbf{r} - \mathbf{r}'|^3} |\psi(\mathbf{r}', t)|^2, \tag{10.18}$$

where θ is indicated in Fig. 10.1 (b) with $\mathbf{r}$ replaced by $\mathbf{r} - \mathbf{r}'$. The dynamics of the system is therefore described by the nonlinear GPE:

$$i\hbar \frac{\partial}{\partial t}\psi(\mathbf{r}, t) = \left[-\frac{\hbar^2}{2M}\nabla^2 + V(\mathbf{r}) + U_0|\psi(\mathbf{r}, t)|^2 + V_{\rm dd}(\mathbf{r}, t) \right] \psi(\mathbf{r}, t), \tag{10.19}$$

where $V(\mathbf{r})$ is an external potential.

We decompose $\psi(\mathbf{r}, t)$ into the amplitude and phase

$$\psi(\mathbf{r}, t) = \sqrt{n(\mathbf{r}, t)}e^{i\phi(\mathbf{r}, t)}, \tag{10.20}$$

where $n(\mathbf{r}, t)$ is the particle density. Substituting Eq. (10.20) into Eq. (10.19) and comparing the real and imaginary parts separately, we obtain

$$M\frac{\partial \mathbf{v}_{\rm s}}{\partial t} = -\boldsymbol{\nabla}\left[\frac{M}{2}\mathbf{v}_{\rm s}^2 + V(\mathbf{r}) + U_0 n + V_{\rm dd} - \frac{\hbar^2}{2M\sqrt{n}}\nabla^2\sqrt{n} \right], \tag{10.21}$$

$$\frac{\partial n}{\partial t} + \boldsymbol{\nabla}(n\mathbf{v}_{\rm s}) = 0, \tag{10.22}$$

where $\mathbf{v}_{\rm s}$ is the superfluid velocity defined in Eq. (1.84) and the last term in Eq. (10.21) is a quantum pressure term that originates from the kinetic energy.

10.2.2 *Stability*

As can be seen from Eq. (10.15), the DDI is attractive for $\theta < \theta_m$, which may lead to the instability of the condensate, depending on the trap geometry and the strength of the contact interaction. We investigate the stability of the system with respect to the density by expanding it around the equilibrium value $n_0(\mathbf{r})$:

$$n(\mathbf{r},t) = n_0(\mathbf{r}) + \delta n(\mathbf{r},t). \tag{10.23}$$

Substituting this into Eq. (10.22) and keeping only those terms that are linear in δn and $\mathbf{v}_\mathrm{s}$, we have

$$\frac{\partial \delta n}{\partial t} + n_0 \boldsymbol{\nabla} \mathbf{v}_\mathrm{s} = 0. \tag{10.24}$$

Differentiating both sides of this equation with respect to t and using Eq. (10.21), we obtain

$$\frac{\partial^2 \delta n}{\partial t^2} = \frac{n_0}{M}\boldsymbol{\nabla}^2\left[V(\mathbf{r}) + U_0 n + V_\mathrm{dd}\right] - \frac{\hbar^2 n_0}{2M^2}\nabla^2\left(\frac{1}{\sqrt{n}}\nabla^2\sqrt{n}\right), \tag{10.25}$$

where

$$V_\mathrm{dd}(\mathbf{r},t) = c_\mathrm{dd}\int d\mathbf{r}'\frac{1-3\cos^2\theta}{|\mathbf{r}'-\mathbf{r}|^3}n(\mathbf{r}',t). \tag{10.26}$$

When the system is uniform with $V=0$, n_0 is constant and Eqs. (10.25) and (10.26) become

$$\frac{\partial^2 \delta n}{\partial t^2} = \frac{n_0}{M}\nabla^2(U_0\delta n + V_\mathrm{dd}) - \frac{\hbar^2}{4M^2}\nabla^4\delta n, \tag{10.27}$$

$$V_\mathrm{dd}(\mathbf{r},t) = c_\mathrm{dd}\int d\mathbf{r}'\frac{1-3\cos^2\theta}{|\mathbf{r}-\mathbf{r}'|^3}\delta n(\mathbf{r}',t), \tag{10.28}$$

where Eq. (10.28) does not involve n_0 because it vanishes upon integration over θ. Substituting $\delta n = \mathrm{const}\cdot e^{i(\mathbf{kr}-\omega t)}$ in Eq. (10.27), we obtain

$$E_\mathbf{k}^2 \equiv (\hbar\omega)^2 = \epsilon_k^2 + 2n_0U_0\epsilon_k + 2n_0\epsilon_k\tilde{V}_\mathrm{dd}(\mathbf{k}), \tag{10.29}$$

where $\epsilon_k \equiv \hbar^2k^2/(2M)$ and

$$\tilde{V}_\mathrm{dd}(\mathbf{k}) = c_\mathrm{dd}\int d\mathbf{r}\frac{1-3\cos^2\theta}{r^3}e^{i\mathbf{kr}} = c_\mathrm{dd}\int_0^\infty\frac{dr}{r}\int d\Omega(1-3\cos^2\theta)e^{i\mathbf{kr}} \tag{10.30}$$

with $d\Omega = \sin\theta d\theta d\phi$. The integral over the solid angle can be carried out, giving [1]

$$\int d\Omega(1-3\cos^2\theta)e^{i\mathbf{kr}} = -4\pi(1-3\cos^2\theta_\mathbf{k})j_2(kr), \tag{10.31}$$

[1] This can be shown as follows. We expand $e^{i\mathbf{kr}}$ as

$$e^{i\mathbf{kr}} = 4\pi\sum_{l=0}^{\infty} i^l j_l(kr)\sum_{m=-l}^{l} Y_l^m(\hat{\mathbf{k}})^* Y_l^m(\hat{\mathbf{r}}).$$

where $\theta_{\mathbf{k}}$ is the angle between $\mathbf{k}$ and the direction of the dipole (*i.e.*, $\cos\theta_k = k_z/k$) and j_2 is the 2nd spherical Bessel function. The remaining integration over r can be carried out by using the formula

$$\int_0^\infty \frac{j_2(kr)}{r} dr = \frac{1}{3}. \tag{10.32}$$

Thus, we obtain

$$\tilde{V}_{\rm dd}(\mathbf{k}) = -\frac{4\pi}{3} c_{\rm dd}(1 - 3\cos^2\theta_{\mathbf{k}}). \tag{10.33}$$

Substituting Eq. (10.33) in Eq. (10.29), we obtain the excitation spectrum for a polarized uniform dipolar BEC:

$$E_{\rm k} = \sqrt{\epsilon_k\{\epsilon_k + 2n_0 U_0[1 + \varepsilon_{\rm dd}(3\cos^2\theta_{\mathbf{k}} - 1)]\}}, \tag{10.34}$$

where $\varepsilon_{\rm dd}$ is defined in Eq. (10.10). From this result, we find that the system is stable if $1 \geq \varepsilon_{\rm dd} \geq -1/2$; otherwise, the system would suffer dynamical instabilities because $E_{\rm k}$ becomes imaginary for some $\mathbf{k}$ and $\theta_{\mathbf{k}}$.

When the system is confined in a trapping potential, it becomes more stable because the spatial confinement gives rise to the zero-point kinetic pressure that helps counterbalance the attractive part of the DDI. Here, we consider the case of an axisymmetric harmonic potential:

$$V(\mathbf{r}) = \frac{M}{2}\left[\omega_\perp^2(x^2 + y^2) + \omega_z^2 z^2\right], \tag{10.35}$$

where the dipoles are assumed to be polarized along the z-direction. As can be seen from Figs. 10.1 (c) and (d), the BEC is more stable in an oblate potential ($\ell \equiv \sqrt{\omega_\perp/\omega_z} < 1$) than in a prolate one ($\ell > 1$) [Santos, *et al.* (2000)]. For a certain range of parameters of ℓ and a close to the unstable regime, the system is predicted to exhibit a biconcave shape in which the density distribution exhibits a local minimum at the trap center [Ronen, *et al.* (2007)]. The collapsing dynamics of a spin-polarized dipolar condensate has been observed to exhibit a cloverleaf pattern that reflects the d-wave nature of the DDI and is in excellent agreement with numerical simulations based on the nonlocal GPE (10.19) combined with three-body loss [Lahaye, *et al.* (2008)].

Substitute this and $1 - 3\cos^2\theta = -\sqrt{16\pi/5}Y_2^0(\hat{\mathbf{r}})$ into the left-hand side of Eq. (10.31), and using the orthnormality relation

$$\int d\Omega Y_l^m(\hat{\mathbf{r}})Y_2^0(\hat{\mathbf{r}}) = \delta_{l2}\delta_{m0},$$

we obtain the right-hand side of Eq. (10.31).

10.2.3 *Thomas–Fermi limit*

When the s-wave contact interaction is strongly repulsive so that the size of the condensate is much larger than the size of the ground-state wave function of the harmonic potential, the kinetic term can be ignored, and the GPE reduces to the Thomas–Fermi equation

$$V(\mathbf{r}) + U_0|\psi(\mathbf{r})|^2 + V_{\text{dd}}(\mathbf{r}) = \mu, \tag{10.36}$$

where μ is the chemical potential and $V_{\text{dd}}(\mathbf{r})$ is given in Eq. (10.18). Equation (10.36) is a nonlocal integral equation for the particle density $n(\mathbf{r}) = |\psi(\mathbf{r})|^2$. When $V(\mathbf{r})$ is parabolic, Eq. (10.36) admits an analytic solution [O'Dell, *et al.* (2004); Eberlein, *et al.* (2005)]. To understand this, we use the formula

$$\frac{\delta_{\mu\nu} - 3\hat{r}_\mu \hat{r}_\nu}{r^3} = -\nabla_\mu \nabla_\nu \frac{1}{r} - \frac{4\pi}{3}\delta_{\mu\nu}\delta(\mathbf{r}) \tag{10.37}$$

to rewrite Eq. (10.18) as

$$V_{\text{dd}}(\mathbf{r}) = -c_{\text{dd}}\left[\frac{4\pi}{3}n(\mathbf{r}) + \frac{\partial^2}{\partial z^2}\Phi(\mathbf{r})\right], \tag{10.38}$$

where $n(\mathbf{r}) = |\psi(\mathbf{r})|^2$ and

$$\Phi(\mathbf{r}) = \int d\mathbf{r}' \frac{n(\mathbf{r}')}{|\mathbf{r}' - \mathbf{r}|}. \tag{10.39}$$

Then, Eq. (10.36) reduces to

$$V(\mathbf{r}) + U_0(1 - \varepsilon_{\text{dd}})n(\mathbf{r}) - \frac{3U_0}{4\pi}\varepsilon_{\text{dd}}\frac{\partial^2}{\partial z^2}\Phi(\mathbf{r}) = \mu, \tag{10.40}$$

which is an integro-differential equation for $n(\mathbf{r})$. In the absence of the DDI ($\varepsilon_{\text{dd}} = 0$), the solution of Eq. (10.40) is an inverted parabola, as shown in Sec. 3.5. Remarkably, the density distribution remains to be an inverted parabola in the presence of the DDI with the aspect ratio altered by the DDI. In fact, operating ∇^2 on both sides of Eq. (10.40) and noting the fact that Φ satisfies the Poisson equation

$$\nabla^2\Phi(\mathbf{r}) = -4\pi n(\mathbf{r}), \tag{10.41}$$

we obtain

$$\nabla^2 V(\mathbf{r}) + U_0(1 - \varepsilon_{\text{dd}})\nabla^2 n(\mathbf{r}) + 3U_0\varepsilon_{\text{dd}}\frac{\partial^2}{\partial z^2}n(\mathbf{r}) = 0. \tag{10.42}$$

Because $\nabla^2 V(\mathbf{r})$ is constant, the sum of the remaining terms in Eq. (10.42) must also be constant. This is possible if $n(\mathbf{r})$ is quadratic. Thus, we obtain

$$n(\mathbf{r}) = n(0)\left(1 - \frac{x^2 + y^2}{R_\perp^2} - \frac{z^2}{R_z^2}\right)\theta\left(1 - \frac{x^2 + y^2}{R_\perp^2} - \frac{z^2}{R_z^2}\right), \tag{10.43}$$

where $\theta(x)$ is the unit-step function. Substituting Eqs. (10.35) and (10.43) in Eq. (10.42), we obtain

$$\frac{1}{R_z^2} = \frac{1}{1+2\varepsilon_{\rm dd}}\left[\frac{M(2\omega_\perp^2+\omega_z^2)}{2n(0)U_0} - \frac{2(1-\varepsilon_{\rm dd})}{R_\perp^2}\right]. \tag{10.44}$$

In general, the remaining parameter $R_\perp$ can be determined numerically by solving Eq. (10.40). On the other hand, when $\varepsilon_{\rm dd} \ll 1$, it is sufficient to find the solution to the first order in $\varepsilon_{\rm dd}$. In this case, we may substitute $n(\mathbf{r}) = [\mu - V(\mathbf{r})]/U_0$ in Eq. (10.39) because Φ is already multiplied by $\varepsilon_{\rm dd}$ in Eq. (10.42). For an isotropic trap $V = M\omega^2 r^2/2$, the integration can be carried out straightforwardly, giving

$$\Phi(\mathbf{r}) = \frac{\pi\mu}{U_0 R^2}\left(R^4 - \frac{2}{3}R^2 r^2 + \frac{r^4}{5}\right). \tag{10.45}$$

Substituting this in Eq. (10.40), we obtain

$$n(\mathbf{r}) = \frac{\mu}{U_0}\left(1 - \frac{x^2+y^2}{R_\perp^2} - \frac{z^2}{R_z^2}\right)\theta\left(1 - \frac{x^2+y^2}{R_\perp^2} - \frac{z^2}{R_z^2}\right), \tag{10.46}$$

where

$$R_\perp = R\left(1 - \frac{\varepsilon_{\rm dd}}{5}\right), \quad R_z = R\left(1 + \frac{2\varepsilon_{\rm dd}}{5}\right), \tag{10.47}$$

with $R = \sqrt{2\mu/(M\omega^2)}$. These results indicate that the condensate becomes increasingly prolate with an increase in the DDI. This is because with an increase in the DDI, the particles are more likely to align in a head-to-tail configuration to lower the energy of the system. For the case of a magnetic DDI, such an effect of magnetostriction has been observed in a chromium BEC [Stuhler, *et al.* (2005); Lahaye, *et al.* (2007)].

A more rigorous method is to carry out the integration in Eq. (10.39) with Eq. (10.43) directly [Eberlein, *et al.* (2005)], with the result

$$R_\perp = \kappa R_z = \left\{\frac{15U_0 N\kappa}{4\pi M\Omega_\perp^2}\left[1 + \varepsilon_{\rm dd}\left(\frac{3\kappa^2 f(\kappa)}{2(1-\kappa^2)} - 1\right)\right]\right\}^{\frac{1}{5}}, \tag{10.48}$$

where

$$f(\kappa) = \frac{1+2\kappa^2}{1-\kappa^2} - \frac{3\kappa^2}{(1-\kappa^2)^{3/2}}\mathrm{arctanh}\sqrt{1-\kappa^2} \tag{10.49}$$

and κ is a real positive root of the following equation:

$$3\kappa\varepsilon_{\rm dd}\left(\frac{(\omega_z^2+2\omega_\perp^2)f(\kappa)}{2\omega_\perp^2(1-\kappa^2)} - 1\right) + (\varepsilon_{\rm dd}-1)\left(\kappa^2 - \frac{\omega_z^2}{\omega_\perp^2}\right) = 0. \tag{10.50}$$

When the trapping potential is isotropic ($\omega_\perp = \omega_z$) and the DDI is weak ($\varepsilon_{\rm dd} \ll 1$), κ must be close to one. Substituting $\kappa = 1 - x$ in Eq. (10.49), we obtain $f(\kappa) = 4x/5 + O(x^2)$. Substituting this in Eq. (10.50), we find that $x \simeq 3\varepsilon_{\rm dd}/5$, and hence, $R_\perp/R_z \simeq 1 - 3\varepsilon_{\rm dd}/5$ is in agreement with Eq. (10.47).

10.2.4 *Quasi two-dimensional systems*

Quasi two-dimensional dipolar condensates exhibit numerous unique phenomena. Here, by quasi two-dimension, we imply that the system is tightly confined in one direction and free or only weakly confined in the other two directions. For the sake of clarity, we assume that the system is tightly confined in the z-direction with l_z being a characteristic width of the confinement, and that the dipoles are polarized in the z-direction, unless otherwise stated.

10.2.4.1 *Roton–maxon spectrum*

We first consider the excitation spectrum. When the wave number k of the excitation is much smaller than l_z^{-1}, the excitations are predominantly phonons that propagate on the x-y plane. In this regime, most dipoles align in the side-by-side configuration and the DDI is repulsive, leading to an enhancement in the compressibility, and hence, the velocity of sound. The dispersion relation of this phonon part is linear, as shown in Fig. 10.3.

When the wave number of the excitation increases, atoms are more likely to align in the head-to-tail configuration along the z-direction so as to lower the energy of the system. Consequently, the dispersion relation tends to bend downward and eventually reaches the maximum, before beginning to decrease. The excitation at the (local) maximum of the dispersion relation is referred to as the maxon.

With a further increase in k, the spectrum eventually approaches that of a free particle with a mean-field offset. En route to this free part, the dispersion relation reaches a local minimum. The excitation around the minimum is called the roton in analogy with the case of superfluid helium-4. Thus, the dispersion relation exhibits the roton–maxon spectrum, as illustrated in Fig. 10.3 [Santos, *et al.* (2003)]. In contrast to the case of the superfluid helium-4, where the interatomic interaction is isotropic, the spin-polarized dipolar condensate exhibits the roton–maxon spectrum as a consequence of the anisotropic nature of the DDI.

10.2.4.2 *Soliton*

A scalar Bose–Einstein condensate cannot support a stable soliton in two dimensions. This is because both the kinetic energy and the interaction energy scale are inversely proportional to the square of the length scale of the system. Thus, they cannot balance at a particular length scale and

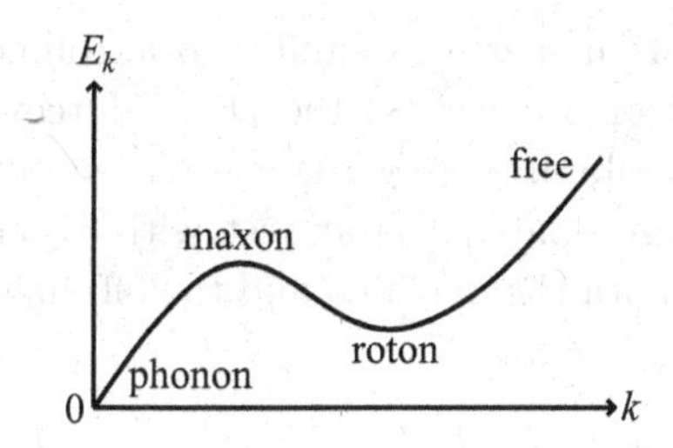

Fig. 10.3 Dispersion relation of a quasi two-dimensional condensate with the DDI.

the system collapses or diffuses away, depending on the strength of the contact interaction. With the DDI, the system can support a stable soliton if $\varepsilon_{\rm dd} < 0$ [Pedri and Santos (2005)]. The negative sign of $\varepsilon_{\rm dd}$ can be realized by the method described in Sec. 10.1.4. We assume that the system is trapped in an axisymmetric harmonic potential with the axial and radial confinement frequencies given by ω_z and $\omega_\perp$, respectively, and consider a Gaussian trial wave function for the soliton condensate:

$$\Psi(\mathbf{r}) = \frac{1}{\sqrt{\pi^{3/2}R_\perp^2 R_z}} \exp\left(-\frac{x^2+y^2}{2R_\perp^2} - \frac{z^2}{2R_z^2}\right). \tag{10.51}$$

Then, the energy E of the system, which is given by

$$E = \int d\mathbf{r}\Psi^*(\mathbf{r}) \left[-\frac{\hbar^2}{2M}\nabla^2 + V(\mathbf{r}) + V_{\rm dd}(\mathbf{r})\right] \Psi(\mathbf{r}), \tag{10.52}$$

is calculated to give

$$\frac{E}{\hbar\omega_z} = \frac{\pi}{\kappa^2}\left\{2\pi + \tilde{g}[1 - \varepsilon_{\rm dd} f(\kappa)]\right\}, \tag{10.53}$$

where $\tilde{g} \equiv U_0/(\sqrt{2\pi}\hbar\omega_z\ell_z^3)$ and $f(\kappa)$ is given by Eq. (10.49). Because $f(\kappa)$ ranges from $f(0) = -1$ to $f(\infty) = 2$, the energy can have a minimum if $\varepsilon_{\rm dd}\tilde{g} < 2\pi + \tilde{g} < -2\varepsilon_{\rm dd}\tilde{g}$. This is possible if $\varepsilon_{\rm dd} < 0$. The physics behind the stabilization of the soliton is the dependence of the DDI on the size of the system, which originates in the long-range nature of the DDI.

10.2.4.3 *Quantum ferrofluidity*

A ferrofluid is a colloidal mixture of fine ferromagnetic particles suspended in a carrier fluid. While the system is paramagnetic and does not exhibit ferromagnetism by itself, it has a high magnetic susceptibility and develops various instabilities such as the Rosensweig instability under an external magnetic field. A binary mixture of dipolar and nonmagnetic condensates is predicted to exhibit various patterns similar to ferrofluids [Saito, *et al.* (2009)] due to the anisotropic nature of the DDI. Moreover, because the system can support a superfluid, there exists a situation in which a density pattern such as a hexagonal pattern is at rest with a persistent current flowing around it. Thus, both the diagonal and the off-diagonal long-range order can coexist in such systems.

10.3 Spinor-Dipolar BEC

When the system is confined in an optical trap, atomic spins acquire dynamical degrees of freedom and they are coupled via the DDI to the orbital degrees of freedom. Suppose that the condensate consists of atoms with spin f. The second-quantized form of the DDI is described by

$$\hat{V}_{\rm dd} = \frac{c_{\rm dd}}{2} \iint d\mathbf{r} d\mathbf{r}' \sum_{\mu,\nu=x,y,z} : \hat{f}_\mu(\mathbf{r}) Q_{\mu\nu}(\mathbf{r}-\mathbf{r}') \hat{f}_\nu(\mathbf{r}') :, \tag{10.54}$$

where the integration kernel $Q_{\mu\nu}$ is given by (10.13), and $\hat{f}_\mu(\mathbf{r})$ is given in terms of the spin-f matrix f_μ as

$$\hat{f}_\mu(\mathbf{r}) = \sum_{m,n=-f}^{f} \hat{\psi}_m^\dagger(\mathbf{r}) (f_\mu)_{mn} \hat{\psi}_n(\mathbf{r}). \tag{10.55}$$

The coupling constant $c_{\rm dd}$ is given by $c_{\rm dd} = \mu_0 (g_F \mu_B)^2/(4\pi)$, where g_F is the Landé g-factor. The corresponding Gross–Pitaevskii equations involve nonlocal terms due to the DDI:

$$\begin{aligned} i\hbar \frac{\partial \psi_m}{\partial t} &= \left[-\frac{\hbar^2 \nabla^2}{2M} + V(\mathbf{r}) - pm + qm^2 \right] \psi_m + c_0 n \psi_m \\ &+ \sum_{n,m',n'=-f}^{f} \sum_{S=0,2,\cdots,2f} g_S \langle mn|P_S|m'n'\rangle \psi_n^*(\mathbf{r}) \psi_{m'}(\mathbf{r}) \psi_{n'}(\mathbf{r}) \\ &+ \sum_{\mu=x,y,z} \sum_{n=-f}^{f} B_{\rm eff}^\mu(\mathbf{r}) g\mu_B (f_\mu)_{mn} \psi_n(\mathbf{r}), \end{aligned} \tag{10.56}$$

where $g_S = 4\pi\hbar^2 a_s/M$ is the strength of the spin-exchange interaction for the total spin S channel and P_S, the projection operator onto the total spin S state. In terms of the Clebsch–Gordan coefficients, the matrix elements of P_S are expressed as $\langle mn|P_S|m'n'\rangle = \sum_{M=-S}^{S} \langle fmfn|SM\rangle\langle SM|fm'fn'\rangle$. In the last term of Eq. (10.56),

$$B_{\rm eff}^\mu(\mathbf{r}) = \frac{c_{\rm dd}}{g\mu_B} \sum_\nu \int d\mathbf{r}' Q_{\mu\nu}(\mathbf{r}-\mathbf{r}') f_\nu(\mathbf{r}) \tag{10.57}$$

plays the role of an effective magnetic field at $\mathbf{r}$ produced by the surrounding magnetic dipoles.

10.3.1 *Einstein–de Haas effect*

Because the DDI conserves the total (spin plus orbital) angular momentum and its projection J_z onto the quantization axis, the order parameter ψ_m can be written as

$$\psi_m(r, \varphi, z) = e^{i(J_z - m)\varphi}\eta_m(r, z), \tag{10.58}$$

where (r, φ, z) are the cylindrical coordinates and η_m is a function of r and z only. Now, suppose that spins are initially fully polarized by an external magnetic field, and the magnetic field is quenched. The spin angular momentum will then be transferred to the orbital angular momentum because the DDI produces an effective magnetic field (10.57) that causes the Larmor precession of atomic spins around the local dipole field. Consequently, the system begins to rotate spontaneously. This phenomenon is known as the Einstein–de Haas effect [Kawaguchi, *et al.* (2006); Santos and Pfau (2006)].

10.3.2 *Flux closure and ground-state circulation*

Due to the long-range nature of the DDI, the spin texture tends to arrange itself to minimize the energy of the DDI. To show this, we use Eq. (10.37) to rewrite Eq. (10.54) as follows:

$$\hat{V}_{\rm dd} = \frac{c_{\rm dd}}{2}\iint d\mathbf{r}d\mathbf{r}'\frac{:[\nabla\cdot\hat{\mathbf{f}}(\mathbf{r})][\nabla\cdot\hat{\mathbf{f}}(\mathbf{r}')]:}{|\mathbf{r}-\mathbf{r}'|} - \frac{2\pi c_{\rm dd}}{3}\int d\mathbf{r} :\hat{\mathbf{f}}(\mathbf{r})^2: . \tag{10.59}$$

The last term on the right-hand side implies that the energy of the DDI is lowered by maximizing the local magnetization. This is because the larger $\langle\hat{\mathbf{f}}\rangle$ implies that the dipoles align themselves in a head-to-tail configuration that lowers the energy of the DDI. On the other hand, the first term suggests that the energy of the DDI decreases if the local magnetization satisfies $\nabla\cdot\hat{\mathbf{f}} = 0$. This condition is referred to as the flux-closure relation because it requires the magnetic flux to close upon itself, which in turn reduces the energy of the magnetic field. Thus, we expect that in the presence of the DDI, the system tends to develop a magnetic domain structure whose characteristic size is given by the dipole healing length

$$\xi_{\rm dd} = \frac{\hbar}{\sqrt{2Mc_{\rm dd}n}}. \tag{10.60}$$

Now, due to the spin-gauge symmetry discussed in Sec. 6.3.3, a nonuniform spin texture caused by the domain structure produces a mass current. Therefore, a spinor-dipolar condensate with $c_1 < 0$ can have a mass current in the ground state [Kawaguchi, *et al.* (2006b)].

In the presence of an external magnetic field B, say, in the z-direction, the magnetic diples undergo Larmor precessions around the magnetic field and the DDI is partially cancelled. To analyze this problem, it is convenient to work in the frame of reference corotating with the Larmor precession, where

$$\hat{f}_{\pm}^{(\mathrm{rot})} \equiv \hat{f}_x^{(\mathrm{rot})} \pm i\hat{f}_y^{(\mathrm{rot})} = e^{\pm i\omega_{\mathrm{L}}t}\hat{f}_{\pm}, \quad \hat{f}_z^{(\mathrm{rot})} = \hat{f}_z. \tag{10.61}$$

Then, the DDI becomes

$$\begin{aligned}\hat{V}_{\mathrm{dd}} = -\sqrt{\frac{6\pi}{5}}\frac{c_{\mathrm{dd}}}{2}\int d\mathbf{r}\int d\mathbf{r}'\frac{1}{|\mathbf{r}-\mathbf{r}'|^3} : \Bigg\{ & \frac{Y_2^0(\hat{\mathbf{r}})}{\sqrt{6}} \\ & \times\left[4\hat{f}_z^{(\mathrm{rot})}(\mathbf{r})\hat{f}_z^{(\mathrm{rot})}(\mathbf{r}') - \hat{f}_+^{(\mathrm{rot})}(\mathbf{r})\hat{f}_-^{(\mathrm{rot})}(\mathbf{r}') - \hat{f}_-^{(\mathrm{rot})}(\mathbf{r})\hat{f}_+^{(\mathrm{rot})}(\mathbf{r}')\right] \\ & + e^{-i\omega_{\mathrm{L}}t}Y_2^{-1}(\hat{\mathbf{r}})\left[\hat{f}_+^{(\mathrm{rot})}(\mathbf{r})\hat{f}_z^{(\mathrm{rot})}(\mathbf{r}') + \hat{f}_z^{(\mathrm{rot})}(\mathbf{r})\hat{f}_+^{(\mathrm{rot})}(\mathbf{r}')\right] \\ & - e^{i\omega_{\mathrm{L}}t}Y_2^{1}(\hat{\mathbf{r}})\left[\hat{f}_-^{(\mathrm{rot})}(\mathbf{r})\hat{f}_z^{(\mathrm{rot})}(\mathbf{r}') + \hat{f}_z^{(\mathrm{rot})}(\mathbf{r})\hat{f}_-^{(\mathrm{rot})}(\mathbf{r}')\right] \\ & + e^{-2i\omega_{\mathrm{L}}t}Y_2^{-2}(\hat{\mathbf{r}})\hat{f}_+^{(\mathrm{rot})}(\mathbf{r})\hat{f}_+^{(\mathrm{rot})}(\mathbf{r}') \\ & + e^{2i\omega_{\mathrm{L}}t}Y_2^{2}(\hat{\mathbf{r}})\hat{f}_-^{(\mathrm{rot})}(\mathbf{r})\hat{f}_-^{(\mathrm{rot})}(\mathbf{r}')\Bigg\} :,\end{aligned} \tag{10.62}$$

where $\hat{\mathbf{r}} \equiv (\mathbf{r}-\mathbf{r}')/|\mathbf{r}-\mathbf{r}'|$. When the Zeeman energy dominates the DDI, *i.e.*, $\hbar\omega_{\mathrm{L}} \gg c_{\mathrm{dd}}n$, the spin dynamics due to the dipolar interaction proceeds much slower than the Larmor precession, and the time-dependent terms are cancelled upon time averaging. Thus,

$$\begin{aligned}\overline{\hat{V}}_{\mathrm{dd}} = -\frac{c_{\mathrm{dd}}}{4}\int d\mathbf{r}\int d\mathbf{r}'\frac{|\mathbf{r}-\mathbf{r}'|^2 - 3(z-z')^2}{|\mathbf{r}-\mathbf{r}'|^5} \\ \times : \left\{\hat{\mathbf{f}}^{(\mathrm{rot})}(\mathbf{r})\cdot\hat{\mathbf{f}}^{(\mathrm{rot})}(\mathbf{r}') - 3\hat{f}_z^{(\mathrm{rot})}(\mathbf{r})\hat{f}_z^{(\mathrm{rot})}(\mathbf{r}')\right\} : .\end{aligned} \tag{10.63}$$

Accordingly, the effective DDI is given by

$$\overline{\hat{V}}_{\mathrm{dd}} = -\frac{c_{\mathrm{dd}}}{4}\sum_{\mu\nu}\int d\mathbf{r}\int d\mathbf{r}' : \hat{\mathbf{f}}^{(\mathrm{rot})}(\mathbf{r})\bar{Q}_{\mu\nu}(\mathbf{r}-\mathbf{r}')\hat{\mathbf{f}}^{(\mathrm{rot})}(\mathbf{r}') :, \tag{10.64}$$

where the time-averaged integration kernel is given by

$$\bar{Q}_{\mu\nu}(\mathbf{r}) = \frac{1-3(\hat{\mathbf{r}}\cdot\hat{\mathbf{B}})^2}{r^3}(\delta_{\mu\nu} - 3\hat{B}_\mu\hat{B}_\nu). \tag{10.65}$$

We note that the time-averaged interaction separately conserves the projected total spin angular momentum and the projected relative orbital angular momentum on the symmetry axis. However, the long-range and anisotropic nature of the dipolar interaction are still maintained as can be seen from Eq. (10.63). In fact, the magnetic DDI plays a role under an external magnetic field as large as 100 mG [Kawaguchi, *et al.* (2007)].

The field of dipolar gases is still witnessing rapid developments. For further reading, we suggest two review articles [Baranov (2008); Lahaye, *et al.* (2009)].

Chapter 11

Optical Lattices

11.1 Optical Potential

11.1.1 *Optical trap*

An optical dipole trap is an indispensable tool for trapping atoms in cases in which a magnetic trap cannot be used. For example, when we utilize a magnetic Feshbach resonance, we must change the magnetic field to control the scattering length; thus, atoms must be trapped by other means. Another example is cases in which atoms are high-field seekers (*i.e.*, their energy levels decrease with an increase in the magnetic field), because Earnshaw's theorem forbits a static local maximum of the magnetic field in free space.

An optical trap is a potential created by the ac Stark effect; this is a second-order process of the electric dipole interaction between an atom and an electric field. The electric dipole interaction is described by

$$\hat{V} = -\hat{\mathbf{d}} \cdot \hat{\mathbf{E}}(\mathbf{r}), \tag{11.1}$$

where $\hat{\mathbf{d}}$ is the electric dipole moment of the atom and $\hat{\mathbf{E}}(\mathbf{r})$, the electric field at the location of the atom. Because the atomic energy levels are not equally spaced, we can tune the frequency of a laser such that a transition is induced between two selected levels that we call the ground state $|\mathrm{g}\rangle$ and the excited state $|\mathrm{e}\rangle$. By assumption, these two states form a complete set:

$$|\mathrm{g}\rangle\langle\mathrm{g}| + |\mathrm{e}\rangle\langle\mathrm{e}| = \hat{I}, \tag{11.2}$$

where $\hat{I}$ is the identity operator. This relationship can be used to rewrite $\hat{\mathbf{d}}$ as

$$\hat{\mathbf{d}} = \hat{I}\hat{\mathbf{d}}\hat{I} = \langle\mathrm{e}|\hat{\mathbf{d}}|\mathrm{g}\rangle\hat{\sigma}^\dagger + \langle\mathrm{g}|\hat{\mathbf{d}}|\mathrm{e}\rangle\hat{\sigma}, \tag{11.3}$$

where $\hat{\sigma}^\dagger \equiv |\mathrm{e}\rangle\langle \mathrm{g}|$ and $\hat{\sigma} \equiv |\mathrm{g}\rangle\langle \mathrm{e}|$ are the raising and lowering operators of the atom, respectively. We note that $\langle e|\hat{\mathbf{d}}|e\rangle = \langle g|\hat{\mathbf{d}}|g\rangle = 0$ because they change sign under spatial inversion. We assume that the laser field is described by a single mode with frequency ω_L:

$$\hat{\mathbf{E}}(\mathbf{r}) = i\sqrt{\frac{\hbar\omega_\mathrm{L}}{2\epsilon_0}}[\hat{a}\mathbf{u}(\mathbf{r}) - \hat{a}^\dagger\mathbf{u}^*(\mathbf{r})], \tag{11.4}$$

where $\mathbf{u}(\mathbf{r})$ is a spatial mode function that is determined by the boundary conditions for the laser field. Substituting Eqs. (11.3) and (11.4) in Eq. (11.1) and using the rotating wave approximation (*i.e.*, dropping the terms $\hat{a}\hat{\sigma}$ and $\hat{a}^\dagger\hat{\sigma}^\dagger$), we obtain

$$\hat{V} = g\hat{a}\hat{\sigma}^\dagger + g^*\hat{a}^\dagger\hat{\sigma}, \tag{11.5}$$

where

$$g = -i\sqrt{\frac{\hbar\omega_\mathrm{L}}{2\epsilon_0}}\langle \mathrm{e}|\hat{\mathbf{d}}|\mathrm{g}\rangle\mathbf{u}(\mathbf{r}) \tag{11.6}$$

is the coupling constant of the electric dipole interaction. The simplified interaction (11.5) between the two-level atom and the single-mode electric field is referred to as the Jaynes–Cummings interaction.

We consider a situation in which the laser field is detuned from the transition frequency ω_0 of the atom by $\Delta = \omega_\mathrm{L} - \omega_0$. If $|\Delta|$ is much larger than the linewidth of the absorption peak, the atom can absorb a photon only virtually during a time interval $\sim|\Delta|^{-1}$ before it returns to the ground state by emitting a photon. According to the second-order perturbation theory, the energy of the combined atom–photon system changes by

$$V_\mathrm{dipole}(\mathbf{r}) = \frac{|\langle \mathrm{e}|\langle\alpha|\hat{V}|\alpha\rangle|\mathrm{g}\rangle|^2}{\hbar\Delta}, \tag{11.7}$$

where the state of the laser field is assumed to be a coherent state $|\alpha\rangle = e^{-|\alpha|^2/2}e^{\alpha\hat{a}^\dagger}|0\rangle$. Substituting Eq. (11.5) in Eq. (11.7), we obtain

$$V_\mathrm{dipole}(\mathbf{r}) = \frac{\omega_\mathrm{L}}{2\epsilon_0}\frac{|\langle \mathrm{e}|\hat{\mathbf{d}}|\mathrm{g}\rangle|^2}{\Delta}I(\mathbf{r}), \tag{11.8}$$

where $I(\mathbf{r}) \equiv |\mathbf{u}(\mathbf{r})\alpha|^2$ is the intensity of the laser light. The decay rate of the excited state (Einstein's A coefficient) is given by

$$\Gamma = \frac{\omega_0^3}{3\pi\epsilon_0\hbar c^3}|\langle \mathrm{e}|\hat{\mathbf{d}}|\mathrm{g}\rangle|^2. \tag{11.9}$$

In terms of Γ, $V_{\text{dipole}}(\mathbf{r})$ is expressed as

$$V_{\text{dipole}}(\mathbf{r}) = \frac{3\pi c^3 \hbar \Gamma}{2\omega_0^2 \Delta} I(\mathbf{r}), \tag{11.10}$$

where we ignore the terms of the order of $|\Delta/\omega_0| \ll 1$.

This result implies that the optical dipole potential can be attractive or repulsive depending on whether the laser field is red-detuned ($\Delta < 0$) or blue-detuned ($\Delta > 0$). When the electric field changes spatially, so does the dipole potential, and therefore, the atom experiences a dipole force given by

$$\mathbf{F}_{\text{dipole}}(\mathbf{r}) = -\nabla V_{\text{dipole}}(\mathbf{r}). \tag{11.11}$$

As a consequence, atoms are attracted to the location with stronger (weaker) laser intensity when $\Delta < 0$ ($\Delta > 0$).

The optical dipole potential is not purely conservative but accompanied by a nonzero dissipation due to spontaneous emission. According to the Kramers–Kronig relationship, the shift in the energy level must be accompanied by energy dissipation. In the present example, the ac Stark shift must be accompanied by real absorption of photons and an ensuing spontaneous emission with decay rate Γ_{abs}. The ratio of the former against the latter, $V_{\text{dipole}}/(\hbar\Gamma_{\text{abs}})$, is given by the ratio of the real and imaginary parts of the scattering amplitude $\propto 1/(\Delta + i\Gamma/2)$ and is therefore given by $\Gamma/(\Delta)$. Thus, the spontaneous emission loss is suppressed by increasing the detuning and strengthening the laser intensity.

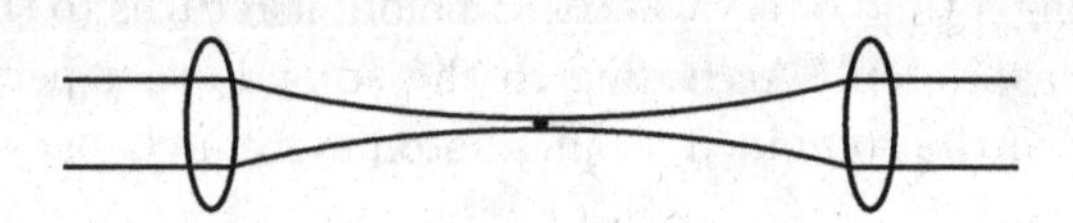

Fig. 11.1 Schematic illustration of a dipole potential. A laser beam incident from the left is focused by a pair of lenses. A typical intensity profile is given by Eq. (11.12) which produces a harmonic potential near the center (indicated by the dot), where atoms are trapped.

Figure 11.1 shows a schematic diagram of an optical potential. An incident laser is focused along the z-direction by a pair of lenses. If a laser intensity profile is Gaussian, it is described by

$$I(r, z) = \frac{2P}{\pi w^2(z)} e^{-\frac{2r^2}{w^2(z)}}, \tag{11.12}$$

where P is the total laser power; r, the radial distance from the z-axis; and $w(z) = w_0\sqrt{1 + (z/z_{\text{R}})^2}$, the $1/e^2$ radius with w_0 and $z_{\text{R}} = \pi w_0^2/\lambda$ being the beam waist and Rayleigh length, respectively [Bloch, *et al.* (2008)]. Here, the Rayleigh length is the distance along the z-direction from the beam center to the location at which the cross-sectional area is twice as large as that at the center. Expanding Eq. (11.12) around $r = z = 0$, we obtain

$$I(r,z) = \frac{2P}{\pi w_0^2}\left(1 - \frac{z^2}{z_{\rm R}^2} - \frac{2r^2}{w_0^2}\right). \tag{11.13}$$

Thus, the optical dipole potential is harmonic around the trap center.

11.1.2 *Optical lattice*

An optical lattice is a periodic potential that is created by a standing wave of laser beams. When two counterpropagating laser beams with wavelength λ interfere, they create a standing wave of period $\lambda/2$. If the two laser beams have a Gaussian profile as in Eq. (11.12), the interference leads to a periodic potential of the form

$$V(r,z) \simeq -V_0 e^{-\frac{2r^2}{w^2(z)}} \sin^2(kz), \tag{11.14}$$

where V_0 is the maximum depth of the lattice potential and $k = 2\pi/\lambda$. The period of the lattice can be increased by tilting the two beams relative to each other [Peil, *et al.* (2003); Hadzibabic, *et al.* (2004)]. A schematic illustration of a one-dimensional optical lattice is shown in Fig. 11.2(a), where each potential has a quasi two-dimensional disk.

A two-dimensional optical lattice can be created by using two pairs of counterpropagating laser beams with orthogonal polarizations, as shown in Fig. 11.2(b). When the confinement potential is tight, each lattice potential is a quasi one-dimensional tube, that is, a quantum wire. In this case, atoms can move in one direction; such quasi one-dimensional systems of neutral atoms have been realized [Greiner, *et al.* (2001); Moritz, *et al.* (2003); Kinoshita, *et al.* (2004); Paredes, *et al.* (2004); Tolra, *et al.* (2004)].

A three-dimensional optical lattice can be created by using three pairs of counterpropagating laser beams with orthogonal polarizations, as shown in Fig. 11.2(c), where each potential is a quasi zero-dimensional dot, that is, a quantum dot, and all the quantum dots are arrayed in a three-dimensional periodic lattice. In the case of a deep potential, the confinement potential of each dot can be approximated by a tighly confined harmonic potential with a trapping frequency $\omega_{\rm trap}$ that is typically a few tens of kilohertz. Thus, the energy level spacing of each dot, $\hbar\omega_{\rm trap} = 2E_{\rm r}\sqrt{V_0/E_{\rm r}}$, can be increased considerably relative to the recoil energy $E_{\rm r} = \hbar^2k^2/(2M)$, and

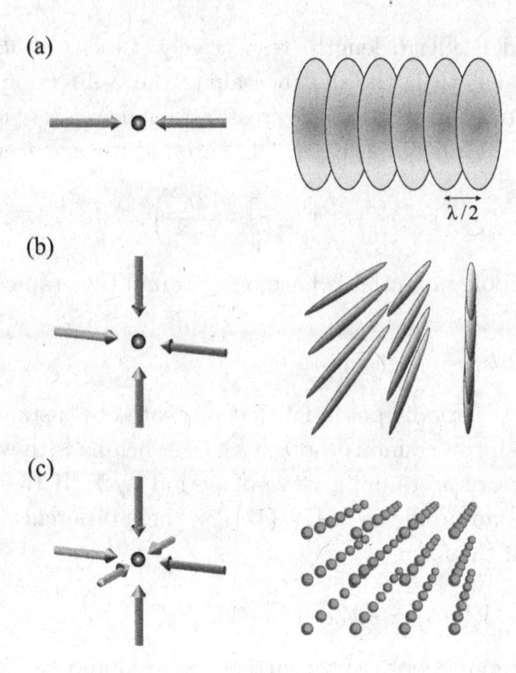

Fig. 11.2 One-dimensional (a), two-dimensional (b), and three-dimensional (c) optical lattices. The left column shows the configurations of laser beams and the right column shows the created optical potentials. The polarizations for orthogonal pairs of beams are usually taken to be perpendicular to each other to avoid interference between orthogonal beams that leads to heating.

an atom lies in the lowest-energy level of the dot. For a red-detuned optical lattice made from Gaussian beams of $1/e^2$ radii w_i $(i = x, y, z)$, the lattice potential is given by

$$V(\mathbf{r}) = -\sum_{i=x,y,z} V_i e^{-2\frac{r^2-i^2}{w_i^2}} \sin^2(ki). \tag{11.15}$$

Near the center of the trap, we may approximate this as a sum of the rapidly oscillating sinusoidal potentials and a slowly varying three-dimensional harmonic potential in which the $\sin^2(ki)$ terms are averaged:

$$V(\mathbf{r}) = \sum_{i=x,y,z} \left[-V_i \sin^2(ki) + \frac{M\omega_i^2}{2} i^2 \right], \tag{11.16}$$

where

$$\omega_i = \sqrt{\frac{4}{M}\left(\sum_{j=x,y,z}\frac{V_j}{w_j^2} - \frac{V_i}{w_i^2}\right)}. \tag{11.17}$$

In the above discussions, it is tacitly assumed that the detuning Δ is much greater than the fine-structure splittings of an atom. Then, the dipole force is the same for all magnetic sublevels. If Δ is decreased, it is possible to make the lattice potential dependent on the spin state of the atom because different fine-structure levels can make unequal contributions to the lattice potential. For example, consider two laser beams that counterpropagate in the z-direction and have linear polarizations on the $x-y$ plane that tilt from each other by an angle θ. Then the right-going and left-going electric fields are given by

$$\mathbf{E}_{\rm R} = E[\mathbf{e}_x \cos(\theta/2) + \mathbf{e}_y \sin(\theta/2)]e^{ikz}, \tag{11.18}$$

$$\mathbf{E}_{\rm L} = E[\mathbf{e}_x \cos(\theta/2) - \mathbf{e}_y \sin(\theta/2)]e^{-ikz}, \tag{11.19}$$

where $\mathbf{e}_x$ and $\mathbf{e}_y$ are the polarization vector in the x- and y-directions, respectively. Then the superposed state is

$$\mathbf{E} = E\left[\mathbf{e}_+ \cos(kz - \theta/2) + \mathbf{e}_- \cos(kz + \theta/2)\right], \tag{11.20}$$

where $\mathbf{e}_\pm \equiv \mathbf{e}_x \pm i\mathbf{e}_y$. Thus, the standing wave consists of two sublattices made from right (+) and left (-) circularly polarized light with optical potentials given by $V_\pm(z,\theta) = V_0\cos^2(kz \mp \theta/2)$. Thus, by changing θ, the relative separation of the two potentials $\Delta z = \theta\lambda/(2\pi)$ can be tuned. Such a state-selective optical lattice can be used as a versatile tool to manipulate spin-dependent transport and quantum gates.

A lattice geometry can be manipulated by tuning the angles of laser beams. For example, a triangular lattice can be created if three pairs of laser beams are tilted relative to each other by $2\pi/3$ [Becker, *et al.*]. By utilizing two optical lattices with different periods, an array of double-well potentials can also be created [Sebby-Strabley, *et al.* (2006); Anderlini, *et al.* (2007); Fölling, *et al.* (2007)]. By making frequencies of the two counterpropagating beams slightly different, the standing wave can be set in motion and plays a role of an atomic conveyor belt that transport the trapped atoms.

An optical lattice is placed in a vacuum chamber and very well isolated from the environment. Neutral atoms in an optical lattice are robust against decoherence. Moreover, a large number of atoms can be initialized and manipulated in a highly controlled manner. All these features make this system an ideal playground to test predictions and explore new phenomena in condensed matter physics, quantum information, and precision measurement [see [Bloch (2008)] for a review].

11.2 Band Structure

11.2.1 *Bloch theorem*

An optical lattice that is made from three orthogonal pairs of counter-propagating beams is a simple cubic lattice, where each lattice point $\mathbf{R}$ is expressed in terms of the unit vectors $\mathbf{a}_i$ as

$$\mathbf{R} = n_1\mathbf{a}_1 + n_2\mathbf{a}_2 + n_3\mathbf{a}_3, \tag{11.21}$$

where n_i's are integers, and 1, 2, and 3 refer to the x-, y-, and z-direction, respectively. For lattice vector $\mathbf{R}$, we may introduce the reciprocal lattice vector $\mathbf{K}$ such that it satisfies the following relation:

$$e^{i\mathbf{K}\cdot\mathbf{R}} = 1 \Longleftrightarrow \mathbf{K}\cdot\mathbf{R} = 2\pi \times \text{integer}. \tag{11.22}$$

The three basis vectors, $\mathbf{b}_j$ $(j = 1, 2, 3)$, for reciprocal lattice vectors satisfy

$$\mathbf{a}_i \cdot \mathbf{b}_j = 2\pi\delta_{ij} \quad (i, j = 1, 2, 3), \tag{11.23}$$

and they are expressed in terms of the fundamental lattice vectors $\mathbf{a}_i$ as

$$\mathbf{b}_1 = \frac{2\pi(\mathbf{a_2}\times\mathbf{a_3})}{\mathbf{a_1}\cdot(\mathbf{a_2}\times\mathbf{a_3})}, \ \mathbf{b}_2 = \frac{2\pi(\mathbf{a_3}\times\mathbf{a_1})}{\mathbf{a_1}\cdot(\mathbf{a_2}\times\mathbf{a_3})}, \ \mathbf{b}_3 = \frac{2\pi(\mathbf{a_1}\times\mathbf{a_2})}{\mathbf{a_1}\cdot(\mathbf{a_2}\times\mathbf{a_3})}. \tag{11.24}$$

The single-particle wave function Ψ of a particle in an optical lattice obeys the following Schrödinger equation:

$$\left[-\frac{\hbar^2}{2M}\nabla^2 + V(\mathbf{r})\right]\Psi(\mathbf{r}) = \varepsilon\Psi(\mathbf{r}), \tag{11.25}$$

where $V(\mathbf{r})$ is the potential of an optical lattice that is periodic with the period of the fundamental lattice vectors:

$$V(\mathbf{r} + \mathbf{a}_i) = V(\mathbf{r}) \quad (i = 1, 2, 3). \tag{11.26}$$

According to the Bloch theorem [Ashcroft and Mermin (1976)], the solution of Eq. (11.25) takes the following form:

$$\Psi_{n\mathbf{k}}(\mathbf{r}) = e^{i\mathbf{k}\mathbf{r}}u_{n\mathbf{k}}(\mathbf{r}), \tag{11.27}$$

where $\mathbf{k}$ is the wave vector, n, the band index, and $u_{n\mathbf{k}}(\mathbf{r})$, a periodic function with the same periodicity as $V(\mathbf{r})$:

$$u_{n\mathbf{k}}(\mathbf{r} + \mathbf{a}_i) = u_{n\mathbf{k}}(\mathbf{r}). \tag{11.28}$$

Thus, for any lattice vector $\mathbf{R}$,

$$\Psi_{n\mathbf{k}}(\mathbf{r}+\mathbf{R}) = e^{i\mathbf{kR}}\Psi_{n\mathbf{k}}(\mathbf{r}). \tag{11.29}$$

The Bloch theorem states that the wave function of a particle in a periodic potential is the product of a plane wave $e^{i\mathbf{kr}}$ and a periodic function with the period of the fundamental lattice vectors.

To prove the Bloch theorem, we introduce the translation operator $\hat{T}_{\mathbf{R}}$ that translates the coordinate by $\mathbf{R}$. Because by assumption, the Hamiltonian is periodic with the periodicity of $\mathbf{R}$, we obtain

$$\hat{T}_{\mathbf{R}}\hat{H}(\mathbf{r})\Psi(\mathbf{r}) = \hat{H}(\mathbf{r}+\mathbf{R})\Psi(\mathbf{r}+\mathbf{R}) = \hat{H}(\mathbf{r})\hat{T}_{\mathbf{R}}\Psi(\mathbf{r}). \tag{11.30}$$

Because this result holds true for arbitrary $\Psi(\mathbf{r})$, $\hat{T}_{\mathbf{R}}$ commutes with $\hat{H}$ and it is possible to find a simultaneous eigenstate of $\hat{H}$ and $\hat{T}_{\mathbf{R}}$:

$$\hat{H}\Psi(\mathbf{r}) = \varepsilon\Psi(\mathbf{r}), \quad \hat{T}_{\mathbf{R}}\Psi(\mathbf{r}) = c(\mathbf{R})\Psi(\mathbf{r}). \tag{11.31}$$

The last equation implies that the eigenfunction of the Schrödinger equation (11.25) satisfies

$$\Psi(\mathbf{r}+\mathbf{R}) = c(\mathbf{R})\Psi(\mathbf{r}). \tag{11.32}$$

On the other hand, from $\hat{T}_{\mathbf{R}}\hat{T}_{\mathbf{R}'} = \hat{T}_{\mathbf{R}+\mathbf{R}'}$, we obtain $c(\mathbf{R}+\mathbf{R}') = c(\mathbf{R})c(\mathbf{R}')$. It follows then that for $\mathbf{R}$ in Eq. (11.21) we obtain

$$c(\mathbf{R}) = c(\mathbf{a}_1)^{n_1}c(\mathbf{a}_2)^{n_2}c(\mathbf{a}_3)^{n_3}. \tag{11.33}$$

Because the wave function is normalized, Eq. (11.32) gives $|c(\mathbf{R})| = 1$. We may therefore substitute $c(\mathbf{a}_i) = e^{2\pi i k_i}$ in Eq. (11.33), where k_i is real, to obtain

$$c(\mathbf{R}) = e^{2\pi i(k_1n_1+k_2n_2+k_3n_3)} = e^{i\mathbf{k}\cdot\mathbf{R}}, \tag{11.34}$$

where $\mathbf{k} = k_1\mathbf{b}_1 + k_2\mathbf{b}_2 + k_3\mathbf{b}_3$. Thus, we prove Eq. (11.27).

We note that the wave function $\Psi_{n\mathbf{k}}$ is not an eigenstate of the momentum. In fact,

$$\hat{\mathbf{p}}\Psi_{n\mathbf{k}} = \frac{\hbar}{i}\vec{\nabla}(e^{i\mathbf{k}\cdot\mathbf{r}}u_{n\mathbf{k}}(\mathbf{r})) = \hbar\mathbf{k}\Psi_{n\mathbf{k}} + e^{i\mathbf{kr}}\frac{\hbar}{i}\vec{\nabla}u_{n\mathbf{k}}(\mathbf{r}) \neq \alpha\Psi_{n\mathbf{k}}(\mathbf{r}). \tag{11.35}$$

This is a conseqnence of the fact that the system is not invariant under an arbitrary space translation but invariant only under discrete translations of lattice vectors. Thus, the momentum of a particle in an optical lattice is conserved only up to an arbitrariness of the reciprocal lattice vectors. In this sense, $\hbar\mathbf{k}$ is referred to as a crystal momentum or quasi-momentum.

The Bloch function is periodic not only in real space but also in momentum space.

$$\Psi_{n,\mathbf{k}+\mathbf{K}}(\mathbf{r}) = \Psi_{n,\mathbf{k}}(\mathbf{r}), \quad \varepsilon_{n,\mathbf{k}+\mathbf{K}} = \varepsilon_{n,\mathbf{k}}. \tag{11.36}$$

In fact, $\Psi_{n,\mathbf{k}}(\mathbf{r})$ can be expanded as

$$\Psi_{n,\mathbf{k}}(\mathbf{r}) = \sum_{\mathbf{K}'} c_{n,\mathbf{k}-\mathbf{K}'} e^{i(\mathbf{k}-\mathbf{K}')\mathbf{r}}. \tag{11.37}$$

Hence,

$$\Psi_{n,\mathbf{k}+\mathbf{K}}(\mathbf{r}) = \sum_{\mathbf{K}'} c_{n,\mathbf{k}+\mathbf{K}-\mathbf{K}'} e^{i(\mathbf{k}+\mathbf{K}-\mathbf{K}')}. \tag{11.38}$$

By substituting $\mathbf{K}'$ by $\mathbf{K}' + \mathbf{K}$, the right-hand side becomes

$$\sum_{\mathbf{K}'} c_{n,\mathbf{k}-\mathbf{K}'} e^{i(\mathbf{k}-\mathbf{K}')} = \Psi_{n,\mathbf{k}}(\mathbf{r}).$$

Hence, the assertion is proved.

11.2.2 *Brillouin zone*

Let the number of lattice sites in the i-direction be N_i ($i = 1, 2, 3$). The first Brillouin zone is the discrete set of the wave vectors that are expressed in terms of reciprocal lattice vectors $\mathbf{b}$ as

$$\mathbf{k} = \sum_{j=1}^{3} \frac{m_j}{N_j} \mathbf{b}_j, \tag{11.39}$$

where $-N_j/2 < m_j \le N_j/2$ if N_j is even and $-(N_j - 1)/2 \le m_j \le (N_j - 1)/2$ if N_j is odd. The volume of the first Brillouin zone is given by

$$v_{\mathrm{B}} = \mathbf{b}_1 \cdot (\mathbf{b}_2 \times \mathbf{b}_3) = \frac{(2\pi)^3}{v}, \tag{11.40}$$

where $v = \mathbf{a}_1 \cdot (\mathbf{a}_2 \times \mathbf{a}_3)$ is the volume of the primitive cell in the direct lattice. The number of states in the first Brillouin zone is equal to the number of unit cells $N_{\mathrm{cell}} = \Omega/v$ in the system, where Ω is the volume of the system, because

$$\frac{\Omega}{(2\pi)^3} \sum_{\mathbf{k}} 1 = \frac{\Omega}{(2\pi)^3} v_{\mathrm{B}} = N_{\mathrm{cell}}. \tag{11.41}$$

The energy spectrum specified by the band index n, $\varepsilon_{n,\mathbf{k}}$, is called the energy band. Because it is periodic with respect to $\mathbf{k}$, it forms an energy band, as shown in Fig. 11.3. The Brillouin zone of a 2D simple cubic optical lattice was observed in a time-of-flight experiment [Greiner, *et al.* (2001)].

The properties of a Fermi system are classified according to how each energy band is occupied by atoms.

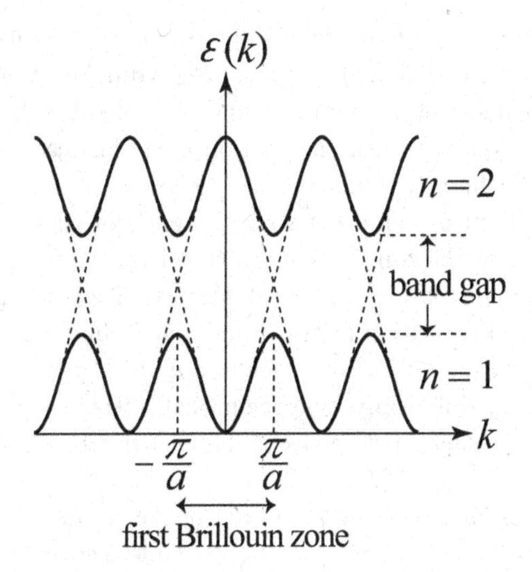

Fig. 11.3 Schematic band structure in a one-dimensional periodic potential, where a is the lattice spacing and n, the band index.

- If all states up to an energy band, say $n = 1$, are fully occupied and the states above it are empty, there exists an excitation energy gap called the band gap $E_{\rm gap}$ [see Fig. 11.3]. If $k_{\rm B}T \ll E_{\rm gap}$, the system is a band insulator. This is a single-particle effect that is independent of interparticle interactions.
- If a band is only partially occupied, there exists no energy gap for transfer of particles, and the system is metallic.
- If the hopping of a particle from one site to others is prohibited by interactions, the system is called a Mott insulator.

11.2.3 *Bloch oscillations*

Suppose that an external force $\mathbf{F}$ acts on atoms in an optical lattice. Then, the wave vector $\mathbf{k}$ changes in time according to the equation of motion:

$$\hbar \frac{d\mathbf{k}}{dt} = \mathbf{F}. \tag{11.42}$$

If $\mathbf{F}$ is constant, $\mathbf{k}$ increases linearly in time. On the other hand, the group velocity is given as

$$\mathbf{v}_{n,\mathbf{k}} = \frac{1}{\hbar}\frac{\partial \epsilon_{n,\mathbf{k}}}{\partial \mathbf{k}}. \tag{11.43}$$

Because $\epsilon_{n,\mathbf{k}}$ is periodic in $\mathbf{k}$ [see Fig. 11.3], the group velocity oscillates periodically as $\mathbf{k}$ increases linearly in time. Accordingly, atoms undergo periodic motion in real space. This phenomenon is called Bloch oscillations which make a sharp contrast with the motion of a classical particle that will be accelerated indefinitely under a constant force.

The physical mechanism behind the phenomenon of Bloch oscillations is the Bragg reflection at the zone boundary, that is, when atoms reach the zone boundary, say, at $k = \pi/a$ in Fig. 11.3, they receive recoil momentum $-2\pi/a$ from the crystal and jump to the other zone boundary $-\pi/a$ of the first Brillouin zone. (Note that quasimomenta π/a and $-\pi/a$ are physically equivalent because they differ only by a reciprocal lattice vector). It is the mixing of forward and backward waves that gives rise to energy gaps at zone boundaries.

In metals, Bloch oscillations usually do not occur because electrons will be scattered by impurities before they make a complete cycle of the oscillations. In a Bose–Einstein condensate of ^{87}Rb loaded in a one-dimensional optical lattice, Bloch oscillations were observed [Morsch, *et al.* (2001)]. In this case, there is no impurity scattering, but the Laudau–Zener tunneling to a higher band occurs for a strong drive, which leads to a damping of the Bloch oscillations.

11.2.4 *Wannier function*

When the lattice potential is deep, the tunneling probability to the neighboring sites is small and the atomic wave function is expected to localize at the lattice site. In such a situation, it is convenient to express the wave function in terms of a set of Wannier functions that are localized near potential valleys. Because the Bloch wave function $\Psi_{n,\mathbf{k}}(\mathbf{r})$ is periodic in $\mathbf{k}$ with periodicity of reciprocal lattice vectors, it can be expanded in a Fourier series as follows:

$$\Psi_{n,\mathbf{k}}(\mathbf{r}) = \sum_{\mathbf{R}} w_n(\mathbf{r} - \mathbf{R}) e^{i\mathbf{k}\cdot\mathbf{R}}, \tag{11.44}$$

where $\mathbf{R}$ runs over all lattice vectors. The inverse Fourier transform of this equation gives the Wannier function:

$$w_n(\mathbf{r} - \mathbf{R}) = \frac{1}{v_{\mathrm{B}}} \int d\mathbf{k} \Psi_{n,\mathbf{k}}(\mathbf{r}) e^{-i\mathbf{k}\cdot\mathbf{R}}, \tag{11.45}$$

where v_{B} is given in Eq. (11.40), and the integral is over the first Brillouin zone. This can be verified if we substitute Eq. (11.45) in the right-hand side of Eq. (11.44) and use the following relation:

$$\sum_{\mathbf{R}} e^{i(\mathbf{k}-\mathbf{k}')\cdot\mathbf{R}} = N_{\text{cell}}\delta_{\mathbf{k},\mathbf{k}'} = \frac{(2\pi)^3}{v}\delta(\mathbf{k}-\mathbf{k}'). \tag{11.46}$$

The Wannier functions satisfy the orthonormality conditions:

$$\int d\mathbf{r} w_n^*(\mathbf{r}-\mathbf{R})w_{n'}(\mathbf{r}-\mathbf{R}') = \frac{1}{v_{\text{B}}}\delta_{n,n'}\delta_{\mathbf{R},\mathbf{R}'}. \tag{11.47}$$

Thus, the Wannier funtions at different lattice sites or in different bands are orthogonal to one another, and they provide a convenient set of basis functions to describe a system of interacting particles.

11.3 Bose–Hubbard Model

11.3.1 *Bose–Hubbard Hamiltonian*

The Bose–Hubbard model was originally proposed to investigate a superfluid system of interacting bosons in the presence of disorder [Fisher, *et al.* (1989)]. It was theoretically proposed [Jaksch, *et al.* (1989)] and experimentally demonstrated [Greiner, *et al.* (2002)] that the model can be applied to ultracold atoms in an optical lattice.

We start with a second-quantized Hamiltonian with a contact s-wave interaction:

$$\hat{H} = \int d\mathbf{r}\hat{\Psi}^\dagger(\mathbf{r})\left[-\frac{\hbar^2}{2M}\nabla^2 + V(\mathbf{r})\right]\hat{\Psi}(\mathbf{r}) + \frac{U_0}{2}\int d\mathbf{r}\hat{\Psi}^{\dagger 2}(\mathbf{r})\hat{\Psi}^2(\mathbf{r}), \tag{11.48}$$

where $\hat{\Psi}(\mathbf{r})$ is the boson field operator that satisfies the canonical commutation relations (2.33) and $V(\mathbf{r})$, the lattice potential including a weak confinement potential if needed [see Eq. (11.16)]. We consider a situation in which the lattice potential is so deep that the energy gap between the first and second bands is much larger than the thermal and mean-field interaction energies per particle. Then, we can restrict ourselves to the lowest band $n = 0$; the field operator can be expanded in terms of the Wannier functions as

$$\hat{\Psi}(\mathbf{r}) = \sum_i w_0(\mathbf{r}-\mathbf{R}_i)\hat{b}_i, \tag{11.49}$$

where i denotes a lattice point and $\hat{b}_i$, the annihilation operator for bosons at lattice point i. Substituting Eq. (11.49) in Eq. (11.48), we obtain

$$\hat{H} = \sum_{ij} J_{ij}\hat{b}_i^\dagger\hat{b}_j + \sum_{ijkl} U_{ijkl}\hat{b}_i^\dagger\hat{b}_j^\dagger\hat{b}_k\hat{b}_l, \tag{11.50}$$

where

$$J_{ij} = \int d\mathbf{r}\, w_0^*(\mathbf{r}-\mathbf{R}_i)\left[-\frac{\hbar^2}{2M}\nabla^2 + V(\mathbf{r})\right] w_0(\mathbf{r}-\mathbf{R}_j), \tag{11.51}$$

$$U_{ijkl} = \frac{U_0}{2}\int d\mathbf{r}\, w_0^*(\mathbf{r}-\mathbf{R}_i)w_0^*(\mathbf{r}-\mathbf{R}_j)w_0(\mathbf{r}-\mathbf{R}_k)w_0(\mathbf{r}-\mathbf{R}_l). \tag{11.52}$$

When the potential is deep, the Wannier function $w_0(\mathbf{r}-\mathbf{R}_i)$ is sharply localized at lattice point $\mathbf{R}_i$ and the overlap of the Wannier functions for different lattice sites is expected to be small. We therefore keep up to the nearest-neighbor terms of J_{ij} and only the diagonal terms U_{iiii} in the Hamiltonian. We also decompose J_{ij} into its diagonal and non-diagonal parts: $J_{ij} \equiv \epsilon_i\delta_{ij} - J(1-\delta_{ij})$. Then, Eq. (11.50) reduces to

$$\hat{H} = -J\sum_{\langle ij\rangle}\hat{b}_i^\dagger\hat{b}_j + \frac{U}{2}\sum_i \hat{n}_i(\hat{n}_i - 1) + \sum_i \epsilon_i\hat{n}_i, \tag{11.53}$$

where $\hat{n}_i \equiv \hat{b}_i^\dagger\hat{b}_i$, $U \equiv U_{iiii}$, $\langle ij\rangle$ denotes the nearest-neighbor pair, and ϵ_i is an on-site one-body potential; it can also describe the effect of disorder. In the following, we consider the case of $\epsilon_i = 0$ unless otherwise stated.

The free part of the Hamiltonian (11.53) can be diagonalized as follows. We express $\hat{b}_j$ in a Fourier series:

$$\hat{b}_j = \frac{1}{\sqrt{N_{\rm cell}}}\sum_{\mathbf{k}}\hat{c}_{\mathbf{k}}e^{i\mathbf{k}\cdot\mathbf{R}_j}. \tag{11.54}$$

Then, the first term on the right-hand side of Eq. (11.53) is diagonalized as

$$\hat{H}^{\rm free} = \sum_{\mathbf{k}} E_{\mathbf{k}}\hat{c}_{\mathbf{k}}^\dagger\hat{c}_{\mathbf{k}}, \tag{11.55}$$

where

$$E_{\mathbf{k}} = -2J\sum_{i=1}^{3}\cos(\mathbf{k}\cdot\mathbf{a}_i). \tag{11.56}$$

11.3.2 *Superfluid–Mott-insulator transition*

The Bose–Hubbard model allows both superfluid and Mott insulator phases. The properties of the system can be studied by looking at the eigenvalues of the single-particle density matrix defined by

$$\rho_{ij} \equiv {\rm Tr}(\hat{\rho}\hat{b}_i^\dagger\hat{b}_j) = \langle\hat{b}_i^\dagger\hat{b}_j\rangle, \tag{11.57}$$

where $\hat{\rho}$ is the density operator of the system. For a given ρ_{ij}, we write down an eigenvalue equation:

$$\sum_{j=1}^{M} \rho_{ij}\psi_j = \sum_{j=1}^{M} \langle \hat{b}_i^\dagger \hat{b}_j \rangle \psi_j = \lambda \psi_i, \tag{11.58}$$

where λ is an eigenvalue, $(\psi_1, \psi_2, \cdots, \psi_M)$ is the corresponding eigenvector, and i runs over all lattice points $i = 1, 2, \cdots, M$.

If the potential is so deep that inter-site hopping is negligible (*i.e.*, $J = 0$), the system has only local correlations: $\langle \hat{b}_i^\dagger \hat{b}_j \rangle \propto \delta_{ij}$. In this case, Eq. (11.58) gives

$$\lambda = \langle \hat{b}_i^\dagger \hat{b}_i \rangle = \frac{N}{M}. \tag{11.59}$$

Thus, the eigenvalue is equal to the average number of particles per site and there is no BEC. An unnormalized state vector for such a state is given by

$$|\Psi\rangle = \sum_{1 \le i_1, < \cdots < i_N \le M} \hat{b}_{i_1}^\dagger \hat{b}_{i_2}^\dagger \cdots \hat{b}_{i_N}^\dagger |\mathrm{vac}\rangle, \tag{11.60}$$

where $i_1, \cdots, i_N$ are the lattice sites that are occupied by a single particle.

An apparently similar but physically very different state occurs at unit filling $N = M$ and in the limit of $U/J \gg 1$. In this case, no two particles can occupy the same site because of a large energy cost U. As a result, although the hopping matrix element J is nonzero, particle transport is prohibited by the interaction. This state is called the Mott insulator. The state of the Mott insulator is given by Eq. (11.60) with $N = M$; the eigenvalue of the single-particle density matrix is 1 and there is no BEC.

In the opposite limit of $U/J \ll 1$, we expect that the wave function of each particle spreads over the entire system. In this case, it is reasonable to assume that the unnormalized state vector is given by

$$|\Psi\rangle = \left(\sum_{i=1}^{M} \hat{b}_i^\dagger \right)^N |\mathrm{vac}\rangle. \tag{11.61}$$

In this case, the matrix element of the single-particle density matrix is calculated to give

$$\rho_{ij} = \frac{\langle \Psi | \hat{b}_i^\dagger \hat{b}_j | \Psi \rangle}{\langle \Psi | \Psi \rangle} = \frac{N}{M}. \tag{11.62}$$

Because ρ_{ij} does not vanish in the large limit of $|i-j|$, the system possesses ODLRO. The eigenvalue is found from Eq. (11.58) to be $\lambda = N$. Thus, the system undergoes BEC.

11.3.3 *Phase diagram*

We first consider the case with $J = 0$. In this case, the ground-state phase is determined by minimizing the on-site energy

$$\begin{aligned}\epsilon(n) &= -\mu n + \frac{U}{2}n(n-1) \\ &= \frac{U}{2}\left\{\left[n - \left(\frac{\mu}{U} + \frac{1}{2}\right)\right]^2 - \left(\frac{\mu}{U} + \frac{1}{2}\right)^2\right\}, \end{aligned} \tag{11.63}$$

where μ is the chemical potential. The minimum-energy configuration is $n = 0$ if $\mu < 0$ and $n = [\mu/U + 1]$ otherwise, where $[x]$ denotes the maximum integer that does not exceed x. Conversely, for a given n with $J = 0$, μ/U can take on the value $n - 1 \leq \mu/U < n$. For nonzero J, we expect that the allowed range of μ/U becomes narrower because the system can gain kinetic energy if particles are allowed to hop to the neighboring sites. With increasing J, the kinetic energy eventually dominates over the interaction energy and the system undergoes a phase transition to a superfluid state, as schematically illustrated in Fig. 11.4(a). Thus, for each $n > 0$, there exists a finite region in the $\mu - J$ plane in which the number of bosons per site is fixed. Within each lobe, the system is a Mott insulator; outside it, it is a superfluid.

The superfluid–Mott-insulator (SF–MI) transition is continuous in any dimension because the d-dimensional Bose–Hubbard model can be mapped into a (d+1)-dimensional XY model [Fisher, *et al.* (1989)]. This mapping suggests that in one dimension, the SF–MI transition is analogous to the Kosterlitz–Thouless transition with a jump in the superfluid density at the transition point [Kühner, *et al.* (2000); Kollath, *et al.* (2004)].

The characteristic feature of the Mott insulating phase is the energy gap for particle-transfer excitations, that is, the energy required to add or remove a particle to or from a lattice site. For a given point in a Mott lobe, the excitation energy is given by the minimum distance to the phase boundary in the μ-direction because the particle-transfer is allowed only outside the lobe. The Mott phase is incompressible because n is fixed, and therefore, the compressibility (4.59) vanishes.

We next start from a point in the superfluid phase in which J/U is sufficiently large and the average number of particles $\bar{n} = 1 + \epsilon$ is slightly greater than 1. If we fix $\bar{n}$ and decrease J/U, then the state cannot penetrate the Mott lobe with $n = 1$; instead, it traces the dashed curve shown in Fig 11.4(a) and maintains superfluidity. A similar situation applies to

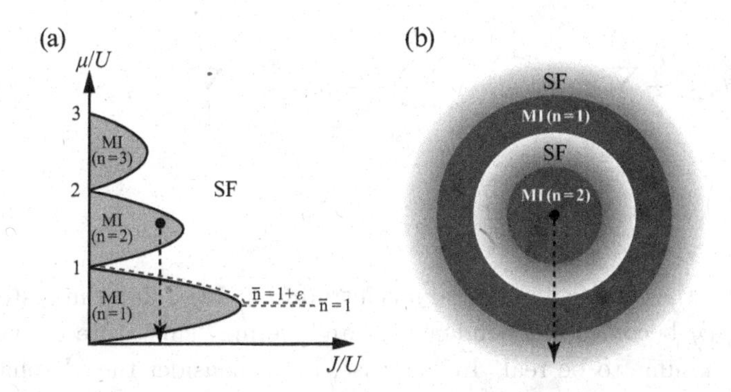

Fig. 11.4 Phase diagram of the Bose–Hubbard model for a uniform system (a) and in a harmonic trap (b), where SF and MI stand for superfluid and Mott insulator, respectively, and n is the number of particles per site. Here, μ, J, and U are the chemical potential, hopping amplitude, and strength of the on-site interaction, respectively.

the case of $\bar{n} = 1 - \epsilon$. Thus, an SF–MI transition occurs in the uniform system only at commensurate filling, that is, when $\bar{n}$ is an integer.

In a real system, an optical lattice potential is not flat but has a weak harmonic confinement, as indicated by Eq. (11.16). As a result, the density of particles is maximal at the center of the trap and decreases as one moves toward the edge of the trapping potential. Now, suppose that at the center of the trap, the system is a Mott insulator with $n = 2$. As one moves toward the edge of the potential, the phase of the system changes according to the dashed arrow shown in Fig. 11.4(a). Therefore, the system has concentric rings of alternating Mott insulator and superfluid phases, as shown in Fig. 11.4(b).

11.3.4 *Mean-field approximation*

In the Bose–Hubbard Hamiltonian, terms on different sites are coupled via the hopping term—the first term on the right-hand side of Eq. (11.53). If this term is decoupled as described below, the Hamiltonian can be decomposed into a sum of local Hamiltonians. Such an apprximation is called the decoupling approximation [van Oosten, *et al.* (2001)]. In this approximation, the hopping term is approximated as

$$\hat{b}_i^\dagger \hat{b}_j \simeq \phi_i^* \hat{b}_j + \hat{b}_i^\dagger \phi_j - \phi_i^* \phi_j, \tag{11.64}$$

where $\phi_i = \langle \hat{b}_i \rangle$ and the last term is subtracted to avoid double counting. Then, the Hamiltonian (11.53) is written as

$$\hat{H} = \hat{H}_0 + \hat{H}_1, \tag{11.65}$$

where

$$\hat{H}_0 = \frac{U}{2}\sum_i \hat{n}_i(\hat{n}_i - 1) + \sum_i (\epsilon_i - \mu)\hat{n}_i + J\sum_{\langle ij\rangle} \phi_i^*\phi_j \tag{11.66}$$

and

$$\hat{H}_1 = -J\sum_{\langle ij\rangle}(\phi_i^*\hat{b}_j + \hat{b}_i^\dagger\phi_j). \tag{11.67}$$

In Eq. (11.66), the chemical potential μ is introduced. For a uniform system with $\epsilon_i = 0$, ϕ_i becomes independent of i and we may substitute ϕ_i with ϕ, which we assume to be real. In this case, let us consider the MI phase and calculate the change ΔE in the ground-state energy by applying a second-order perturbation theory.

$$\Delta E = \sum_{\mathrm{e}} \frac{|\langle \mathrm{e}|\hat{H}_1|\mathrm{g}\rangle|^2}{E_{\mathrm{g}} - E_{\mathrm{e}}}, \tag{11.68}$$

where E_{g} is the ground-state energy of the Hamiltonian $\hat{H}_0$; it is given by $E_{\mathrm{g}} = zJN_{\mathrm{s}}\phi^2/2$ if $\mu < 0$ and $E_{\mathrm{g}} = N_{\mathrm{s}}[Un(n-1)/2 - \mu n + zJ\phi^2/2]$ if $n - 1 < \mu/U < n$, where N_{s} is the number of lattice sites and z is the coordination number (i.e., the number of nearest-neighbor lattice points). Because $\hat{H}_1$ creates or annihilates one particle from the ground state, $E_{\mathrm{e}} = Un(n+1)/2 - \mu(n+1) + (N_{\mathrm{s}} - 1)[Un(n-1)/2 - \mu n] + zJN_{\mathrm{s}}\phi^2/2$ in the former case and $E_{\mathrm{e}} = U(n-1)(n-2)/2 - \mu(n-1) + (N_{\mathrm{s}} - 1)[Un(n-1)/2 - \mu n] + zJN_{\mathrm{s}}\phi^2/2$ in the latter case. Thus,

$$\Delta E = \left(-\frac{n+1}{Un - \mu} + \frac{n}{U(n-1) - \mu}\right) z^2J^2\phi^2. \tag{11.69}$$

The ground-state energy of the interacting system is therefore given by

$$E_{\mathrm{g}}(\phi) = \text{constant} + A\phi^2 + O(\phi^4), \tag{11.70}$$

where $A = zJ + \Delta E/\phi^2$. According to the Ginzburg–Landau theory, the phase boundary is determined by the condition $A = 0$, from which we obtain

$$\frac{\mu_\pm}{U} = \frac{1}{2}\left[2n - 1 - \frac{zJ}{U} \pm \sqrt{\left(\frac{zJ}{U}\right)^2 - (2n+1)\frac{zJ}{U} + \frac{1}{4}}\right], \tag{11.71}$$

where $\mu_\pm$ corresponds to the upper and lower boundaries of the Mott lobe. The width of the lobe for a given J/U is given by

$$\Delta\left(\frac{\mu}{U}\right) = \sqrt{\left(\frac{zJ}{U}\right)^2 - (2n+1)\frac{zJ}{U} + \frac{1}{4}}, \tag{11.72}$$

which vanishes at $z(J/U)_c = 2n+1-2\sqrt{n(n+1)}$. The right-hand side of this critical value approaches $1/(4n)$ for large n. In particular, for $\bar{n}=1$, the superfluid–Mott insulator transition occurs at [Fisher, *et al.* (1989); Sheshadri, *et al.* (1993); Freericks and Monien (1996); van Oosten, *et al.* (2001)]

$$\left(\frac{U}{J}\right)_c = (3+2\sqrt{2})z \simeq 5.8z. \tag{11.73}$$

This result agrees reasonably with the result $(U/J)_c = 29.34$ of quantum Monte Carlo calculations for a simple cubic lattice ($z=6$) [Capogrosso-Sansone, *et al.* (2007)]. The decoupling approximation holds for $zJ/U \ll 1$; it also holds for $zJ/U \gg 1$ because the superfluid phase is well approximated by the mean field ϕ.

The decoupling approximation is closely related to the so-called Gutzwiller mean-field approach [Rokhsar and Kotliar (2001)] in which the following variational wave function is assumed:

$$|\Psi_{\rm GW}\rangle = \prod_i \sum_{n=0}^{n_{\max}} f_n^{(i)} |n\rangle_i, \tag{11.74}$$

where $|n\rangle_i$ denotes the Fock state at the i-th lattice site, $n_{\max}$ is the maximum number of bosons per site, and $f_n^{(i)}$'s play the role of variational parameters that are normalized to unity: $\sum_n |f_n^{(i)}|^2 = 1$. By requiring that

$$\langle \Psi_{\rm GW}| i(\partial/\partial t) - \hat{H} |\Psi_{\rm GW}\rangle \tag{11.75}$$

be stationary with respect to $f_n^{(i)}$, we obtain the equation of motion for $f_n^{(i)}$ [Jaksch, *et al.* (2002)]:

$$i\frac{df_n^{(i)}}{dt} = \frac{U}{2}n(n-1)f_n^{(i)} - J\sqrt{n+1}\Phi_i^* f_{n+1}^{(i)} - J\sqrt{n}\Phi_i f_{n-1}^{(i)}, \tag{11.76}$$

where

$$\Phi_i = \sum_j{}' \langle \hat{b}_j \rangle = \sum_j{}' \sum_n \sqrt{n} f_{n-1}^{(i)*} f_n^{(j)}, \tag{11.77}$$

where the sum over j is taken over the neighboring sites of i. Equation (11.76) can be used to investigate the mean-field dynamics of the Bose–Hubbard model.

11.3.5 *Supersolid*

Penrose and Onsager argued that the ground state of a many-body system exhibits ODLRO based on the assumption that there is no long-range configurational (*i.e.*, diagonal) order [Penrose and Onsager (1956)]. They also argued that a perfect crystal at absolute zero does not exhibit ODLRO. More than a decade later, Andreev and Lifshitz [Andreev and Lifshitz (1969)] suggested the possibility that zero-point fluctuations generate vacancies in a solid that undergo BEC so that the diagonal and off-diagonal orders can exist simultaneously. Today, supersolidity is defined as a state of matter in which the diagonal long-range order (DLRO) coexists with ODLRO. DLRO implies a long-range configurational order, whereas ODLRO implies a long-range phase coherence. Hence, if a system that exhibits a crystalline order also exhibits superfluidity, it is a supersolid.

Optical lattices simulate lattice-based supersolid models. The Hamiltonian of a hard-core model is given by

$$\hat{H} = -J \sum_{\langle ij \rangle} \hat{b}_i^\dagger \hat{b}_j + V \sum_{\langle ij \rangle} \hat{n}_i \hat{n}_j - \mu \sum_i \hat{n}_i, \tag{11.78}$$

where μ is the chemical potential. Because the hard-core bosons prohibit double occupancy, the model can be mapped to that of spin-1/2 magnets [Matsuda and Tsuneto (1970)]. In fact, by using the correspondence

$$\hat{b}_i \leftrightarrow \hat{S}_i^+, \quad \hat{b}_i^\dagger \leftrightarrow \hat{S}_i^-, \quad \hat{n}_i \leftrightarrow \hat{S}_i^z - \frac{1}{2}, \tag{11.79}$$

where $\hat{S}_i^\pm \equiv \hat{S}_x \pm i\hat{S}_y$, we find that the Hamiltonian (11.78) is equivalent to the spin-1/2 anisotropic Heisenberg model:

$$\hat{H} = \sum_{\langle ij \rangle} [J_\parallel \hat{S}_i^z \hat{S}_j^z + J_\perp (\hat{S}_i^x \hat{S}_j^x + \hat{S}_i^y \hat{S}_j^y)] - h \sum_i \hat{S}_i^z, \tag{11.80}$$

where $J_\parallel \equiv V$, $J_\perp \equiv -J$, and $h \equiv \mu - zV/2$, with z being the coordination number of the lattice. In this representation, the order parameter of the BEC is described by the transverse magnetization, *i.e.*, $\langle \hat{S}_i^+ \rangle = \langle \hat{b}_i \rangle$, and the compressibility corresponds to the magnetic susceptibility $\chi = \partial M_z / \partial h$. A supersolid phase is identified with the one in which a transverse magnetization coexists with a periodic modulation of the longitudinal magnetization [Liu and Fisher (1973)]. A mean-field analysis of the triangular and kagome lattices indicates the existence of supersolid phases [Murthy, *et al.* (1997)], whereas the evidence of supersolidity for a triangular lattice near commensurate filling due to frustration was presented using quantum Monte Carlo calculations [Wessel and Troyer (2005); Boninsegni and Prokof'ev (2005)].

Chapter 12

Topological Excitations

Topological excitations in a superfluid are singular defects of a superfluid order parameter. The unique feature of topological excitations lies in the fact that they can move freely in space without changing their characteristics and that they are robust against external perturbations. The properties of topological excitations are independent of material-specific parameters and charecterized by a set of integers called topological charges. Topological excitations are best classified in terms of homotopy theory. This theory gives what type of topological excitations can exist in which type of order parameter manifolds, and describes what happens if two topological defects coexist and how they coalesce or disintegrate. In this chapter, we present a brief introduction of homotopy theory and touch upon a rich variety of topological excitations in superfluid systems.

12.1 Homotopy Theory

12.1.1 *Homotopic relation*

To describe topological defects, we develop a mathematical tool with which to identify objects that can transform into each other in a continuous manner as belonging to the same equivalent class. Such a classification can be carried out using homotopy theory that classifies continuous maps. Let $f : X \to Y$ and $g : X \to Y$ be two continuous maps between topological spaces X and Y. These two maps are called homotopic and denoted as $f \sim g$ if there exists a continuous map

$$F : X \times [0,1] \to Y \tag{12.1}$$

such that $F(x,0) = f(x)$ and $F(x,1) = g(x)$ for $^\forall x \in X$. Viewed as a function of $t, F(x,t)$ gives a continuous family of maps from X to Y; F is called the homotopy for f and g.

The homotopic relation satisfies the following three conditions that are necessary and sufficient for this relation to be an equivalence one. Let $f, g, h : X \to Y$ be continuous maps. Then

$$\begin{array}{lll} \text{(i)} & f \sim f & \text{(symmetric);} \\ \text{(ii)} & \text{If } f \sim g, \text{ then } g \sim f & \text{(reflexive);} \\ \text{(iii)} & \text{If } f \sim g \text{ and } g \sim h, \text{ then } f \sim h & \text{(transitive).} \end{array} \tag{12.2}$$

Condition (i) can be shown by taking $F(x,t) = f(x)$. To show (ii), let $F(x,t)$ be the homotopy for $f \sim g$. Then, $F(x, 1-t)$ gives the homotopy for $g \sim f$. To show condition (iii), let $F : X \times I \to Y$ and $G : X \times I \to Y$ be the homotopies for $f \sim g$ and $g \sim h$, respectively, where $I \equiv [0,1]$. Then,

$$H(x,t) \equiv \begin{cases} F(x,2t) & 0 \le t \le \frac{1}{2} \\ G(x,2t-1) & \frac{1}{2} \le t \le 1 \end{cases} \tag{12.3}$$

gives the homotopy for $f \sim h$.

Let f and g be maps for $X \to Y$ and $Y \to Z$, respectively. Then, the composite map $g \cdot f : X \to Z$ is defined as $g(f(x))$. Let f, f' and g, g' be mappings for $X \to Y$ and $Y \to Z$, respectively. If $f \sim f'$ and $g \sim g'$, then $g \cdot f \sim g' \cdot f'$. To show this, let $F : X \times I \to Y$ and $G : Y \times I \to Z$ be homotopies for f, f' and g, g', respectively. Then, $H(x,t) \equiv G(F(x,t),t)$ defines a continuous map $H : X \times I \to Z$ such that

$$\begin{aligned} H(x,0) &= G(f(x),0) = g \cdot f(x), \\ H(x,1) &= G(f'(x),1) = g' \cdot f'(x). \end{aligned} \tag{12.4}$$

Thus, $g \cdot f$ and $g' \cdot f'$ are homotopic under H.

A constant map $c : X \to Y$ is defined as a map whose image is an arbitrary fixed element of Y. The identity map $i_d : X \to X$ is defined as a map such that $i_d(x) = x$ for $^{\forall}x$. A topological space X is said to be contractible if the identity map i_d is homotopic to a constant map c:

$$X \text{ is contractible} \iff i_d \sim c. \tag{12.5}$$

If a topological space X is contractible, any continuous map $f : X \to Y$ is homotopic to a constant map. In fact, if X is contractible, there exists some homotopy $F : X \times I \to X$ such that $F(x,0) = x$ and $F(x,1) = \text{constant} \equiv x_0 \in X$ for $^{\forall}x \in X$. Then, a homotopy for $f \sim c$ is given by $G(x,t) \equiv f(F(x,t))$, for $G(x,0) = f(F(x,0)) = f(x)$ and $G(x,1) = f(F(x,1)) = f(x_0) = \text{constant} \in Y$.

12.1.2 *Fundamental group*

Let a continuous map $\varphi : I \to X$ be called a path whose initial point and terminal point are given by $\varphi(0)$ and $\varphi(1)$, respectively. The inverse φ^{-1} of φ is defined as

$$\varphi^{-1}(s) \equiv \varphi(1-s). \tag{12.6}$$

If the terminal point of a path φ_1 coincides with the initial point of another path φ_2, *i.e.*, $\varphi_1(1) = \varphi_2(0)$, we can define the product of the two paths $\varphi_1 \cdot \varphi_2$ as

$$\varphi_1 \cdot \varphi_2(t) \equiv \begin{cases} \varphi_1(2s) & 0 \le s \le \frac{1}{2} \\ \varphi_2(2s-1) & \frac{1}{2} \le s \le 1. \end{cases} \tag{12.7}$$

A loop ℓ is a special path such that the initial and terminal points coincide:

$$\ell(0) = \ell(1) \equiv x_0, \tag{12.8}$$

where x_0 is called the base point. Let $\ell_1(s)$ and $\ell_2(s)$ be two loops sharing the same base point. If there exists a homotopy $F : I \times I \to X$ such that

$$\begin{aligned} F(s,0) &= \ell_1(s), \\ F(s,1) &= \ell_2(s), \\ F(0,t) &= F(1,t) = x_0, \end{aligned} \tag{12.9}$$

then ℓ_1 and ℓ_2 are said to be homotopic. The constant loop ℓ_c is the one in which $\ell_c(s) = x_0$ for $^\forall s \in I$. We can classify all the loops that share the same base point into a set of equivalent classes called homotopy classes:

$$[\ell_1], [\ell_2], \cdots\cdots, \tag{12.10}$$

where $[\ell_i]$ comprises all loops that are homotopic to ℓ_i. For each homotopy class $[\ell_i]$, we may choose an arbitrary loop in $[\ell_i]$ to represent the homotopy class because all elements belonging to the same class are homotopic to each other.

The product of two homotopy classes $[\ell_1]$ and $[\ell_2]$ is defined as

$$[\ell_1] \cdot [\ell_2] = [\ell_1 \cdot \ell_2], \tag{12.11}$$

where $\ell_1 \cdot \ell_2$ denotes the product of two loops in which ℓ_1 is first traversed and then ℓ_2 is traversed as shown in Fig. 12.1. Members of $[\ell_1 \cdot \ell_2]$, which are homotopic to $\ell_1 \cdot \ell_2$, need not return to the base point x_0 en route to the terminal point [see the dashed curve in Fig. 12.1]. With definition (12.11),

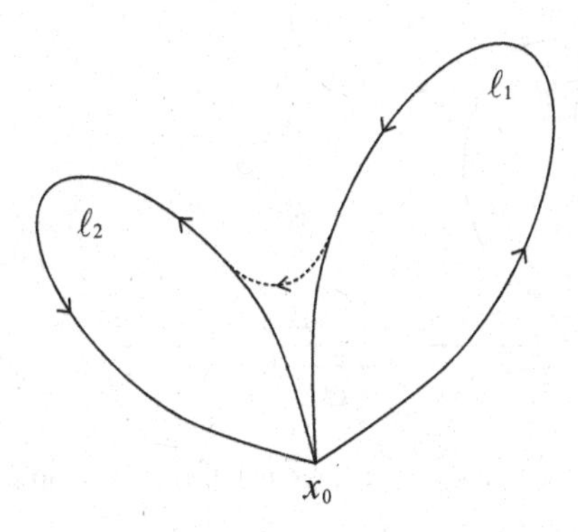

Fig. 12.1 A product of two loops $\ell_1 \cdot \ell_2$ in which ℓ_1 is first traversed and then ℓ_2 is traversed. Loops homotopic to $\ell_1 \cdot \ell_2$ need not return to the base point x_0 en route to the terminal point, as indicated by the dashed curve.

a set of homotopy classes form a group. In fact, they satisfy the associative law

$$([\ell_1] \cdot [\ell_2]) \cdot [\ell_3] = [\ell_1] \cdot ([\ell_2] \cdot [\ell_3]); \tag{12.12}$$

those loops that are homotopic to the constant loop c form the identity element $[c]$:

$$[c] \cdot [\ell] = [\ell] \cdot [c] = [\ell] \text{ for any } [\ell]; \tag{12.13}$$

the inverse $[\ell^{-1}]$ of $[\ell]$ is the homotopy class that consists of inverse loops [see Eq. (12.6)] of $[\ell]$:

$$[\ell^{-1}] \cdot [\ell] = [\ell] \cdot [\ell^{-1}] = [c]. \tag{12.14}$$

The group defined above is called the fundamental group or the first homotopy group and denoted as $\pi_1(X, x_0)$.

Two points x_0 and x_1 in X are said to be arcwise connected if there exists a continuous map $f : I \to X$ such that $f(0) = x_0$ and $f(1) = x_1$. A topological space X is said to be arcwise connected if any two pairs in X are arcwise connected. Then, $\pi_1(X, x_0)$ and $\pi_1(X, x_1)$ are isomorphic to each other:

$$\pi_1(X, x_0) \cong \pi_1(X, x_1). \tag{12.15}$$

In fact, by assumption, there exists a path α that connects x_1 to x_0 [see Fig. 12.2]. Then, for every loop ℓ that has the base point at x_0, loop $\alpha \cdot \ell \cdot \alpha^{-1}$ has the base point at x_1. This correspondence can be utilized to introduce a map from $\pi_1(X, x_0)$ to $\pi_1(X, x_1)$ via

$$[\ell] \to [\alpha \cdot \ell \cdot \alpha^{-1}]. \tag{12.16}$$

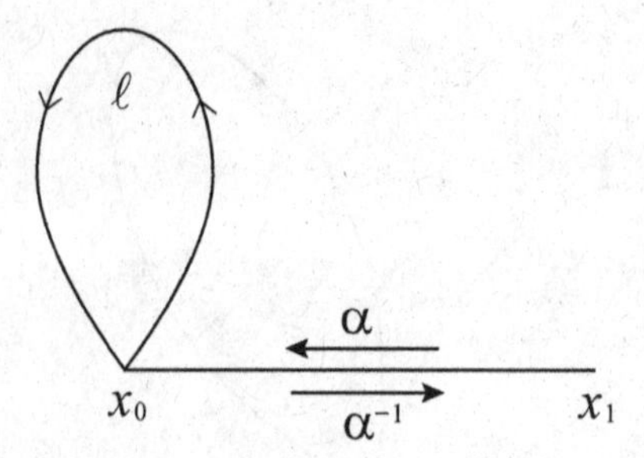

Fig. 12.2 For every loop ℓ that has the base point at x_0, the loop $\alpha \cdot \ell \cdot \alpha^{-1}$ has the base point at x_1.

This map is a homomorphism, because the product $[\ell_1] \cdot [\ell_2]$ is mapped into

$$\begin{aligned}[\ell_1] \cdot [\ell_2] = [\ell_1 \cdot \ell_2] &\to [\alpha \cdot \ell_1 \cdot \ell_2 \cdot \alpha^{-1}] \\ &= [(\alpha \cdot \ell_1 \cdot \alpha^{-1}) \cdot (\alpha \cdot \ell_2 \cdot \alpha^{-1})] \\ &= [\alpha \cdot \ell_1 \cdot \alpha^{-1}] \cdot [\alpha \cdot \ell_2 \cdot \alpha^{-1}]. \end{aligned} \tag{12.17}$$

To prove the isomorphism (12.15), we must show that the mapping in (12.16) is a bijection. Since the map has the inverse $[\ell'] \to [\alpha^{-1} \cdot \ell' \cdot \alpha] \in \pi_1(X, x_0)$ for any $[\ell'] \in \pi_1(X, x_1)$, the map (12.16) is surjective. If $[\ell'_1]$ and $[\ell'_2]$ are two different homotopy classes of $\pi_1(X, x_1)$, their inverses $[\alpha^{-1} \cdot \ell'_1 \cdot \alpha]$ and $[\alpha^{-1} \cdot \ell'_2 \cdot \alpha]$ are different classes of $\pi_1(X, x_0)$, because if they belong to the same class, there exists some homotopy $F : I \times I \to X$ such that $F(s, 0) = \alpha^{-1} \cdot \ell'_1 \cdot \alpha$ and $F(s, 1) = \alpha^{-1} \cdot \ell'_2 \cdot \alpha$. Then, $\alpha \cdot F(s, t) \cdot \alpha^{-1}$ is a homotopy for ℓ'_1 and ℓ'_2; this contradicts our initial assumption. By *reductio ad absurdum*, the mapping (12.16) is injective, and hence, it is bijective. Unless otherwise stated, we assume that X is arcwise connected and denote $\pi_1(X, x_0)$ simply as $\pi_1(X)$ because all fundamental groups that have different base points will be isomorphic to each other. If $\pi_1(X)$ consists only of the identity element, the topological space X is called simply connected.

Let the n-sphere S^n be defined as

$$S^n = \left\{(x_1, x_2, \cdots, x_{n+1}) \in \mathbb{R}^{n+1} | x_1^2 + x_2^2 + \cdots + x_{n+1}^2 = 1\right\}, \tag{12.18}$$

where S^1 is a unit circle and S^2, a unit sphere. Because $\pi_1(S^1)$ concerns a mapping from a loop to the unit circle, it is isomorphic to the additive group of integers:

$$\pi_1(S^1) \cong \mathbb{Z}, \tag{12.19}$$

where each element of integers corresponds to the number of times the unit circle is wound by the map. Because any loop on S^2 can be contracted to a point, we have

$$\pi_1(S^2) \cong \{[c]\}. \tag{12.20}$$

This is also denoted simply as $\pi_1(S^2) = 0$.

The fundamental group of the direct product of two arcwise connected topological spaces X and Y is isomorphic to the direct product of their individual fundamental groups:

$$\pi_1(X \times Y) \cong \pi_1(X) \times \pi_1(Y). \tag{12.21}$$

In fact, each element of $\pi_1(X \times Y)$ is expressed as $[(\ell_X, \ell_Y)]$, where ℓ_X and ℓ_Y are loops belonging to X and Y, respectively, and each pair of loops defines a map $(\ell_X, \ell_Y)\colon I \to X \times Y$. If $\ell_X \sim \ell'_X$ and $\ell_Y \sim \ell'_Y$, then $(\ell_X, \ell_Y) \sim (\ell'_X, \ell'_Y)$, and therefore, there exists a one-to-one correspondence between $[(\ell_X, \ell_Y)]$ and $([\ell_X], [\ell_Y])$, which proves (12.21).

12.1.3 *Higher homotopy groups*

We consider a map $\varphi_n : I^n \to X$, where I^n is the unit n-cube:

$$I^n = \{(s_1, s_2, \cdots, s_n) |\, 0 \le s_i \le 1 \text{ for } i = 1, 2, \cdots, n\}. \tag{12.22}$$

Let ∂I^n be the boundary of I^n defined by

$$\partial I^n = \left\{(s_1, s_2, \cdots, s_n) |\, s_i = 0 \text{ or } 1 \text{ for } {}^{\exists}i\right\}. \tag{12.23}$$

Similar to the case of π_1, we require that all points on the boundary be mapped to a single base point x_0; then, φ_n is called an n-loop at x_0.

Two maps α and β are said to be homotopic to each other and denoted as $\alpha \sim \beta$ if there exists a continuous map $H : I^n \times I \to X$ such that

$$\begin{aligned} H(s_1, s_2, \cdots, s_n, 0) &= \alpha(s_1, s_2, \cdots, s_n), \\ H(s_1, s_2, \cdots, s_n, 1) &= \beta(s_1, s_2, \cdots, s_n), \\ H(s_1, s_2, \cdots, s_n, t) &= x_0 \text{ for } {}^{\forall}t \in I \text{ if } (s_1, s_2, \cdots, s_n) \in \partial I^n. \end{aligned} \tag{12.24}$$

The homotopic relation allows us to classify all n-loops at x_0 into homotopy classes $[\alpha]$.

The inverse α^{-1} of α is defined by

$$\alpha^{-1}(s_1, s_2, \cdots, s_n) = \alpha(1 - s_1, s_2, \cdots, s_n). \tag{12.25}$$

The product of two n-loops α and β is defined by

$$\alpha \cdot \beta = \begin{cases} \alpha(2s_1, s_2, \cdots, s_n) & 0 \le s_1 \le \frac{1}{2} \\ \beta(2s_1 - 1, s_2, \cdots, s_n) & \frac{1}{2} \le s_1 \le 1. \end{cases} \tag{12.26}$$

It can be seen that the homotopy classes $[\alpha], [\beta], \cdots$ form a group called the n-th homotopy group $\pi_n(X, x_0)$, where the multiplication law is defined by

$$[\alpha] \cdot [\beta] \equiv [\alpha \cdot \beta]. \tag{12.27}$$

Similarly to Eq. (12.15), if X is arcwise connected, two homotopy groups $\pi_n(X, x_0)$ and $\pi_n(X, x_1)$ are isomorphic to each other, and therefore, we may omit the base point and write $\pi_n(X, x_0)$ simply as $\pi_n(X)$. A relationship similar to (12.21) holds for the n-th homotophy group:

$$\pi_n(X \times Y) \cong \pi_n(X) \times \pi_n(Y). \tag{12.28}$$

12.2 Order Parameter Manifold

12.2.1 *Isotropy group*

Let ψ be an order parameter of the system and let G be a Lie group whose arbitrary element g transforms ψ in such a manner to keep the free-energy functional invariant. We also introduce the isotropy group H whose arbitrary element h makes ψ invariant, that is, $h\psi = \psi$. The order parameter manifold M is defined as the coset of H in G:

$$M = G/H. \tag{12.29}$$

For example, in s-wave superconductors, superfluid helium-4, or spin-polarized gaseous BECs, a global phase change $\psi \to e^{i\phi}\psi$ does not change the free energy of the system. Thus, $G = U(1)$. In this case, only the identity element makes ψ invariant. Thus, $H = \{1\}$ and $M = U(1)$.

For the case of spinor BECs in the absence of an external magnetic field, the free energy is invariant under the global gauge transformation and an orbitrary rotation in spin space. Thus,

$$G = U(1)_\phi \times SO(3)_\mathbf{S}, \tag{12.30}$$

where the subscripts ϕ and $\mathbf{S}$ indicate that the group operation acts on the gauge (*i.e.*, the global phase) and spin, respectively. In the presence of a magnetic field, the isotropy of space is broken; however, the rotation about the direction, say the z-axis, of the magnetic field still keeps the free energy invariant. Thus,

$$G = U(1)_\phi \times U(1)_{S_z}, \tag{12.31}$$

where $U(1)_{S_z}$ describes the group of rotations of spin about the z-axis.

The isotropy group of a spinor BEC is a subgroup of G in Eq. (12.30); therefore, we can list possible candidates without investigating individual phases. The rotation group $SO(3)$ is the group of rotations in three dimensions, where each rotation is specified by the rotation axis (determined by the polar and azimuthal angles) and the angle of rotation about the axis. The group is therefore characterized by three continuous parameters. The $SO(3)$ group has a continuous subgroup that is the $U(1)$ group of rotations about a symmetry axis, and the following five discrete subgroups:

- $\mathbb{C}_n$: the cyclic group of rotations through angle $2\pi k/n$, where $k = 0, 1, 2, \cdots, n-1$.
- $\mathbb{D}_n$: the dihedral group consisting of $\mathbb{C}_n$ and n reflections about the axes symmetrically placed on a plane perpendicular to the symmetry axis of the $\mathbb{C}_n$.
- $\mathbb{T}$: the tetrahedral group with 12 elements consisting of the identity; three π rotations about the axes connecting midpoints of opposite sides; and four $\pi/3$ and $2\pi/3$ rotations about the axes connecting each vertex and the center of the opposite face.
- $\mathbb{O}$: the octahedral group with 24 elements consisting of the identity; three $\pi/2$, π, and $3\pi/2$ rotations about the axes connecting centers of opposite surfaces; four $2\pi/3$ and $4\pi/3$ rotations about the axes connecting opposite vertices; and six π rotations about the axes connecting midpoints of opposite sides.
- $\mathbb{I}$: the icosahedral group with 60 elements, consisting of the identity; six $2\pi/5$, $4\pi/5$, $6\pi/5$, and $8\pi/5$ rotations about the axes connecting opposite vertices; ten $2\pi/3$ and $4\pi/3$ rotations about the centers of opposite faces; and fifteen π rotations about the axes connecting the midpoints of opposite sides.

Many of these subgroups indeed have physical counterparts: $U(1)$, $\mathbb{C}_2$, $\mathbb{D}_2$, and $\mathbb{T}$ correspond to the ferromagnetic, antiferromagnetic, polar, and cyclic phases, respectively. We also expect that $\mathbb{O}$ finds a physical counterpart for a spin-3 BEC.

12.2.2 *Spin-1 BEC*

An element g of G in Eq. (12.30) corresponds to a transformation of the order parameter ψ as

$$g\psi = e^{i\theta} U(\alpha, \beta, \gamma)\psi, \tag{12.32}$$

where $e^{i\theta} \in U(1)_\phi$ and $U(\alpha, \beta, \gamma) = e^{if_z\alpha} e^{if_y\beta} e^{if_z\gamma} \in SO(3)_{\mathbf{S}}$. For the case of a spin-1 ferromagnetic phase with $\psi = (1, 0, 0)^T$, the general order parameter is given by Eq. (6.98). Here, we note that the gauge angle θ and the rotation angle γ appear as a linear superposition, so that elements of the isotropy group can be expressed as

$$h\psi = e^{i\gamma}U(0,0,\gamma)\psi = \psi. \tag{12.33}$$

Therefore, the isotropy group is given by

$$H^{\mathrm{F}} = U(1)_{\phi+S_\gamma}, \tag{12.34}$$

where the subscript $\phi + S_\gamma$ indicates that elements of the isotropy group describe the combined operations of the gauge transformation and the spin rotation about the direction of the spin. The order parameter manifold of the ferromagnetic phase is given by

$$M^{\mathrm{F}} = \frac{U(1)_\phi \times SO(3)_{\mathbf{S}}}{U(1)_{\phi+S_\gamma}} = SO(3)_{\phi,\mathbf{S}}. \tag{12.35}$$

For the case of the spin-1 polar phase, the general order parameter is given by Eq. (6.107), which does not depend on γ. Moreover, it is invariant under the combined transformation of gauge $\phi = \theta \to \theta + \pi$ and spin $\beta \to \beta + \pi$ that constitutes the nontrivial element of the two-element group $\mathbb{Z}_2$, the other element being the identity. Thus, the isotropy group is given by

$$H^{\mathrm{P}} = U(1)_{S_\gamma} \times (\mathbb{Z}_2)_{\phi,S_\beta}, \tag{12.36}$$

and therefore, the order parameter manifold is

$$M^{\mathrm{P}} = \frac{U(1)_\phi \times SO(3)_{\mathbf{S}}}{U(1)_{S_\gamma} \times (\mathbb{Z}_2)_{\phi,S_\beta}} = \frac{U(1)_\phi \times S^2_{\mathbf{S}}}{(\mathbb{Z}_2)_{\phi,S_\beta}}, \tag{12.37}$$

where we use the rotation $SO(3)/U(1) = S^2$.

12.2.3 *Spin-2 BEC*

Next, we consider the case of a spin-2 BEC. For the ferromagnetic phase, we again obtain Eq. (12.35). For the uniaxial nematic phase, the same argument as the case of the spin-1 polar phase applies and we obtain Eq. (12.37). For the biaxial nematic (or antiferromagnetic) phase, the continuous $U(1)$ symmetry in Eq. (12.36) is broken; however, the system possesses an additional inversion symmetry about the axis perpendicular to the quantization axis of the spin, and consequently, the isotropy group becomes the quaternion group $\mathbb{Q}$. Thus,

$$M^{\mathrm{biaxial}} = \frac{U(1)_\phi \times SO(3)_{\mathbf{S}}}{\mathbb{Q}}. \tag{12.38}$$

Because the quaternion group is non-Abelian, the biaxial nematic phase of a spin-2 BEC exhibits nontrivial topological excitations.

The isotropy group of the cyclic phase of a spin-2 BEC is the tetrahedral group:

$$H^{\text{cyclic}} = (\mathbb{T})_{\phi,\mathbf{S}}. \tag{12.39}$$

This can be understood as follows. The fundamental building block of the cyclic phase is a spin-singlet trio that geometrically forms a regular triangle [Ueda and Koashi (2002)]. The possible polyhedral subgroups are the tetrahedron and icosahedron. Now, the order parameter of the spin-2 BEC is described by five complex numbers. Because of the normalization condition and the global gauge invariance, only four complex numbers are free parameters that specify four vertices of the tetrahedron [Barnett, *et al.* (2006)]. The order parameter manifold is therefore given by

$$M^{\text{cyclic}} = \frac{U(1)_\phi \times SO(3)_{\mathbf{S}}}{(\mathbb{T})_{\phi,\mathbf{S}}}. \tag{12.40}$$

The tetrahedral group has 12 elements that are divided into three conjugacy classes. It is non-Abelian and gives rise to non-Abelian vortices [Kobayashi, *et al.* (2009)].

Table 12.1 lists order parameter manifolds M and their homotopy groups $\pi_n(M)$ for some typical systems.

12.3 Classification of Defects

12.3.1 *Domains*

By convention, $\pi_0(M)$ gives the number of "domain walls" that separate the order parameter manifold M. If $\pi_0(M) = 0$, M is said to be connected. If $\pi_0(M) = 1$, M is divided into two disconnected regions, and so on.

12.3.2 *Line defects*

12.3.2.1 $U(1)$ *vortex*

Line defects such as quantized vortices in superfluids or disclinations in liquid crystals are classified by the first homotopy group $\pi_1(M)$ that describes a mapping from a loop in real space onto the order parameter manifold M. Let us consider, for example, a scalar order parameter $\psi(\mathbf{r}) = |\psi(\mathbf{r})|e^{i\phi(\mathbf{r})}$ of a superfluid. If the amplitude is constant, the order parameter is specified by the phase $\phi(\mathbf{r})$ that ranges from 0 and 2π. The order parameter

Table 12.1 List of order parameter manifolds M and their homotopy groups π_n, where $\mathbb{Q}$ and $\mathbb{T}^*$ are the quaternion group and the binary tetrahedral group, respectively; $\mathbb{R}P^2$ and $\mathbb{R}P^3$ are the two- and three-dimensional real projective spaces, respectively.

	M	π_1	π_2	π_3	π_4
planar spin	$U(1)$	$\mathbb{Z}$	0	0	0
Heisenberg spin	S^2	0	$\mathbb{Z}$	$\mathbb{Z}$	$\mathbb{Z}_2$
nematics	$\mathbb{R}P^2 \cong \frac{S^2}{\mathbb{Z}_2}$	$\mathbb{Z}_2$	$\mathbb{Z}$	$\mathbb{Z}$	$\mathbb{Z}_2$
biaxial nematics	$\frac{SU(2)}{\mathbb{Q}}$	$\mathbb{Q}$	0	0	$\mathbb{Z}_2$
ferromagnetic BEC	$SO(3)_{\phi.S_\gamma} \cong \mathbb{R}P^3_{\phi,S_\gamma}$	$\mathbb{Z}_2$	0	$\mathbb{Z}$	$\mathbb{Z}_2$
spin-1 polar BEC	$\frac{S^2_{\mathbf{S}} \times U(1)_\phi}{(Z_2)_{\phi,\mathbf{S}}}$	$\mathbb{Z}$	$\mathbb{Z}$	$\mathbb{Z}$	$\mathbb{Z}_2$
cyclic BEC	$\frac{SO(3)_{\mathbf{S}} \times U(1)_\phi}{\mathbb{T}_{\phi,\mathbf{S}}}$	$\mathbb{T}^*$	0	$\mathbb{Z}$	$\mathbb{Z}_2$
^{3}He-A (dipole-free)	$\frac{S^2 \times SO(3)}{\mathbb{Z}_2}$	$\mathbb{Z}_4$	$\mathbb{Z}$	$\mathbb{Z} \times \mathbb{Z}$	$\mathbb{Z}_2 \times \mathbb{Z}_2$
^{3}He-A (dipole-locked)	$SO(3)$	$\mathbb{Z}_2$	0	$\mathbb{Z}$	$\mathbb{Z}_2$
^{3}He-B (dipole-free)	$S^1 \times SO(3)$	$\mathbb{Z} \times \mathbb{Z}_2$	0	$\mathbb{Z}$	$\mathbb{Z}_2$
^{3}He-B (dipole-locked)	$S^1 \times S^2$	$\mathbb{Z}$	$\mathbb{Z}$	$\mathbb{Z}$	$\mathbb{Z}_2$

manifold is therefore described by a unit circle called 1-sphere S^1. In this case, $\pi_1(S^1)$ describes a mapping from a loop in real space to S^1 according to the correspondence [see Fig. 12.3]

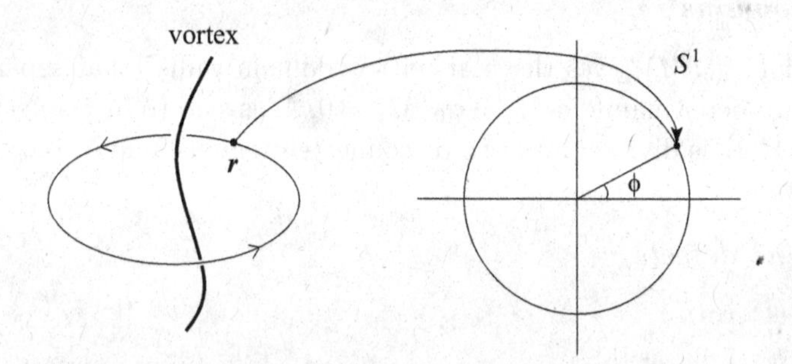

Fig. 12.3 Mapping from a loop in real space to the order parameter manifold S^1 according to the correspondence ψ:$\mathbf{r} \to \arg\psi(\mathbf{r}) = \phi(\mathbf{r})$. $\pi_1(S^1)$ classifies how many times this mapping covers S^1; therefore, $\pi_1(S^1) \cong \mathbb{Z}$

$$\psi : \mathbf{r} \to \arg\psi(\mathbf{r}) = \phi(\mathbf{r}). \tag{12.41}$$

If $\pi_1(S^1)$ gives n, the map covers S^1 n times. Physically, if $n = 0$, the loop contains no vortex. If $n = 1$, the loop contains a singly quantized vortex, and so on. Here, n is referred to as the winding number or topological charge of a vortex.

Thus, $\pi_1(S^1)$ classifies how many times the mapping (12.41) covers S^1; therefore, $\pi_1(S^1) \cong \mathbb{Z}$. The fact that $\pi_1(S^1)$ is isomorphic to the additive group of integers implies that two vortices with winding numbers m and n can coalesce into a single vortex with winding number $m + n$. Conversely, a vortex with winding number n can disintegrate into two vortices with winding numbers k and $n - k$. Whether these processes actually occur depends on the dynamics and energetics of an individual system.

Because any loop on a 2-sphere can be contracted to a point, we obtain $\pi_1(S^2) = 0$. More generally, $\pi_1(S^n) = 0$ for $n \geq 2$. Because the order parameter manifold of a three-dimensional ferromagnet is S^2, $\pi_1(S^2) = 0$, implying that this system cannot support a topologically stable line defect.

12.3.2.2 *SO(3) vortex*

The order parameter manifold of a ferromagnetic BEC is $SO(3)$, as shown in Eq. (12.35), and the fundamental group is

$$\pi_1(SO(3)) \cong \mathbb{Z}_2. \tag{12.42}$$

This can be understood if we represent an element of $SO(3)$ by a point within a sphere of radius π. A given point P in the sphere designates both the rotation axis $\overrightarrow{OP}$ and the rotation angle $\phi = \overline{OP}$, where the overbar denotes the length of the segment OP. If a loop in real space is mapped onto a loop, say C, within the sphere, it can be contracted to a point [see Fig. 12.4]. However, if the loop is mapped onto a curve that connects two diametrically opposite points, say A and A', on the sphere, the curve cannot be contracted to a point [see Fig. 12.4]. Thus, there exists only one type of nontrivial line defects in a ferromagnetic BEC, a singly quantized vortex called the polar-core vortex. A doubly quantized vortex can be continuously deformed to a uniform configuration, as schematically illustrated in Fig. 12.5. Mathematically, this is because $\pi_1(SO(3))$ is isomorphic to $\mathbb{Z}_2 = \{0, 1\}$—the additive group of integers modulo 2. Spin textures belonging to class 0 can continuously transform to a uniform spin configuration, whereas those belonging to class 1 involve singly quantized vortices that are topologically stable. Because $1 + 1 = 0 \pmod 2$, the coalescence of two singly quantized vortices is homotopic to a uniform spin configuration. The order parameter of a polar-core vortex can be obtained by setting $\theta = \gamma$ and $\alpha = \phi$ in Eq. (6.98):

$$\begin{pmatrix} \psi_1 \\ \psi_0 \\ \psi_{-1} \end{pmatrix} = \sqrt{n} \begin{pmatrix} e^{-i\phi} \cos^2 \frac{\beta}{2} \\ \frac{1}{\sqrt{2}} \sin\beta \\ e^{i\phi} \sin^2 \frac{\beta}{2} \end{pmatrix}. \tag{12.43}$$

The core of this vortex is filled by the "polar" (*i.e.*, $m = 0$) component; hence, it is called the polar-core vortex. It is a nonsingular vortex.

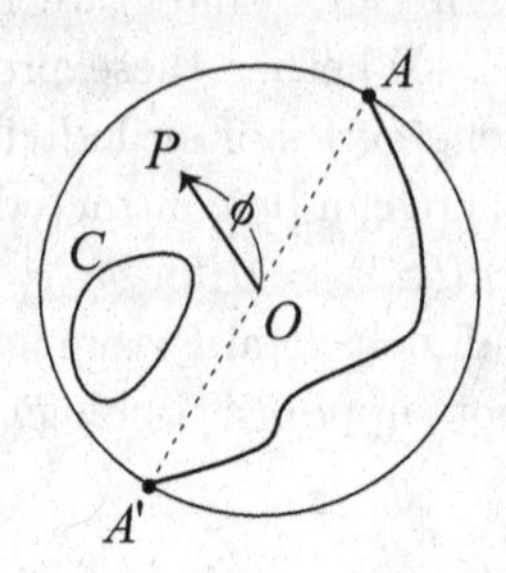

Fig. 12.4 Representation of an element of $SO(3)$ by a point P that lies within a sphere of radius π. The direction $\overrightarrow{OP}$ designates the rotation axis and the distance $\overline{OP}$ indicates the rotation angle ϕ. Note that two diametrically opposite points A and A' represent the same rotation because a π rotation about $\overrightarrow{OA}$ is equivalent to that about $\overrightarrow{OA'}$. A curve connecting A and A' represents a nonsingular polar-core vortex [see the text].

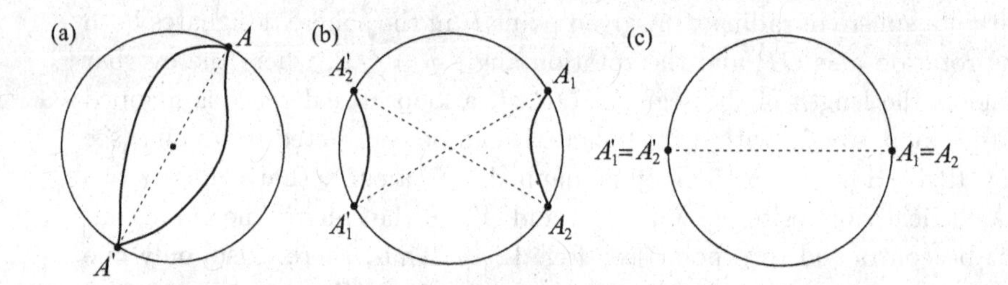

Fig. 12.5 A doubly quantized vortex (a) can be decomposed into a pair of singly quantized vortices that are diametrically connected (b); they can be continuously contracted to two equivalent single points (c).

12.3.2.3 *Half-quantum (Alice) vortex*

Let us next consider the polar phase of a spin-1 BEC, where the order parameter manifold is given by Eq. (12.37). Since S^2 makes a trivial contribution to π_1, we obtain from Eq. (12.21)

$$\pi_1(M^{\mathrm{P}}) \cong \mathbb{Z}. \tag{12.44}$$

Thus, the polar phase can have multiply quantized vortices. However, due to the combined $\mathbb{Z}_2$ symmetry of the spin and gauge shown in Eq. (12.37), the circulation is quantized in units of $h/2M$ rather than h/M. To show this, we consider the order parameter of the polar phase in Eq. (6.105):

$$\psi_{\mathrm{P}} = \sqrt{n}e^{i\theta} \begin{pmatrix} -\frac{e^{-i\alpha}}{\sqrt{2}} \sin\beta \\ \cos\beta \\ \frac{e^{i\alpha}}{\sqrt{2}} \sin\beta \end{pmatrix}. \tag{12.45}$$

As we circumnavigate the loop shown in Fig. 12.6, we let θ, α, and β vary

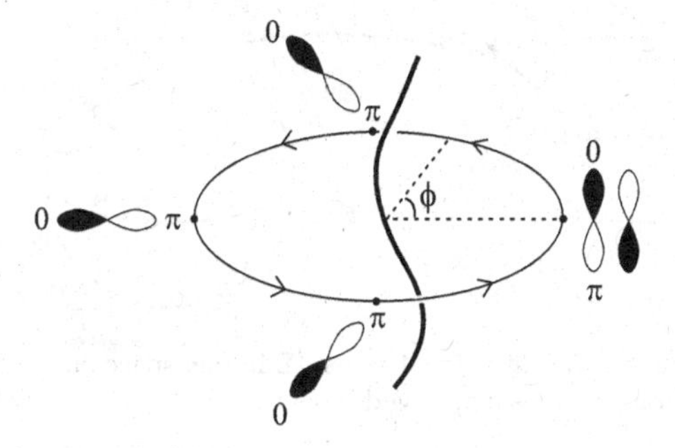

Fig. 12.6 Half-quantum (Alice) vortex of a spin-1 polar BEC. The order parameter rotates through π about a horizontal axis as one circumnavigates a loop. The single-valuedness of the order parameter is satisfied if the π rotation ($\beta \to \beta+\pi$) is accompanied by a gauge transformation by π ($\theta \to \theta + \pi$).

as

$$\theta = \frac{\phi}{2}, \quad \alpha = \text{constant}, \quad \beta = \frac{\phi}{2}. \tag{12.46}$$

Then, the superfluid velocity is calculated from Eq. (6.100) to give

$$\mathbf{v}_{\mathrm{s}}^{\mathrm{P}} = \frac{\hbar}{2M}\nabla\phi, \tag{12.47}$$

and thus, the circulation is quantized in units of $h/2M$ which is one half of the usual h/M. This vortex is therefore referred to as a half-quantum vortex or Alice vortex.

12.3.3 *Point defects*

12.3.3.1 *'t Hooft–Polyakov monopole*

Point defects such as monopoles and hedgehogs are characterized by $\pi_2(M)$ that classifies mappings from a sphere in real space onto an order parameter manifold M. Suppose that we enclose a point-like object O with a two-dimensional sphere Σ and consider a mapping from a point $\mathbf{r}$ on the sphere onto the order parameter manifold M. The second homotopy group $\pi_2(M)$ classifies such a map. For example, if $M = S^2$, $\pi_2(S^2)$ classifies the mapping according to the number of times the mapping covers S^2 [see Fig. 12.7]. Thus,

$$\pi_2(S^2) = \mathbb{Z}. \tag{12.48}$$

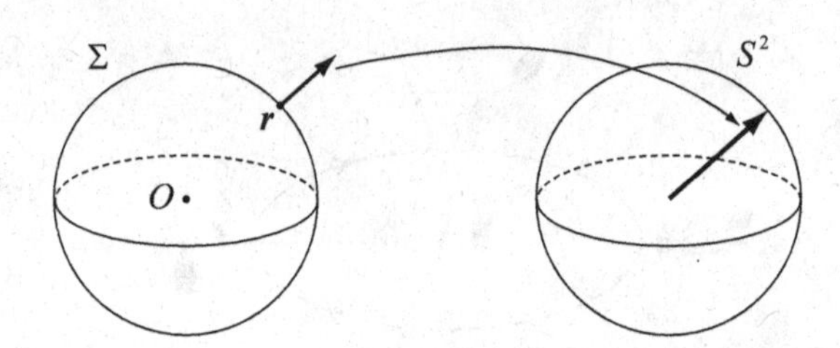

Fig. 12.7 Mapping from a two-dimensional sphere Σ in real space onto a 2-sphere S^2 in the order-parameter manifold. This mapping defines the second homotopy group $\pi_2(S^2)$.

The order parameter manifold M of the polar phase of a spin-1 BEC is given by Eq. (12.37) which involves the factor S^2 and gives

$$\pi_2(M^{\mathrm{P}}) = \mathbb{Z}.$$

Therefore, the system can possess a point defect. To understand the nature of this defect, we rewrite the order parameter (12.45) as

$$\psi_{\mathrm{P}} = \sqrt{\frac{n}{2}} e^{i\theta} \begin{pmatrix} -d_x + id_y \\ \sqrt{2} d_z \\ d_x + id_y \end{pmatrix}, \tag{12.49}$$

where

$$d_x = \sin\beta\cos\alpha, \quad d_y = \sin\beta\sin\alpha, \quad d_z = \cos\beta. \tag{12.50}$$

If we set $\theta = 0$ and

$$\mathbf{d}(\mathbf{r}) = \frac{\mathbf{r}}{r}, \tag{12.51}$$

the spin texture becomes spherical. The point defect of the type (12.50) is called a hedgehog or 't Hooft–Polyakov monopole with topological charge 1. This point defect, however, is unstable against deformation into the Alice ring [see Sec. 12.3.5.2].

12.3.3.2 *Dirac monopole*

Another type of point defect, the Dirac monopole, was originally envisaged as a magnetic analogue of the quantized electric charge. In analogy with the Poisson equation for an electric field, we assume that the magnetic field **B** obeys

$$\nabla \cdot \mathbf{B} = 4\pi g \delta(\mathbf{r}), \tag{12.52}$$

where g denotes the strength of the magnetic monopole. The solution of this equation is given by

$$\mathbf{B} = g\frac{\mathbf{r}}{r^3} \tag{12.53}$$

and the corresponding vector potential is

$$\mathbf{A} = \frac{g}{r(r-z)}(y, -x, 0) = -\frac{g(1+\cos\theta)}{r\sin\theta}\mathbf{e}_\varphi(\mathbf{r}), \tag{12.54}$$

where (r, θ, φ) are the polar coordinates and $\mathbf{e}_\varphi(\mathbf{r})$ is the unit vector along the φ direction at position $\mathbf{r}$. The vector potential in Eq. (12.54) reproduces the magnetic field in Eq. (12.53) except on the positive z-axis on which the magnetic field calculated from Eq. (12.54) exhibits a singularity that is called the Dirac string:

$$\mathrm{rot}\mathbf{A} = g\frac{\mathbf{r}}{r^3} - 4\pi g\delta(x)\delta(y)\theta(z)\mathbf{e}_z, \tag{12.55}$$

where $\theta(z)$ is the unit-step function and $\mathbf{e}_z$ is the unit vector along the z-direction.

The Dirac monopole can be created in the ferromagnetic phase of a spin-1 BEC [Ruostekoski and Anglin (2003)]. Substituting $\theta - \gamma \to -\varphi, \alpha \to \varphi$, and $\beta \to \theta$ in Eq. (6.98), we obtain

$$\boldsymbol{\xi}_\mathrm{F} \equiv \frac{1}{\sqrt{n}}\psi_\mathrm{F} = \begin{pmatrix} e^{-2i\varphi}\cos^2\frac{\theta}{2} \\ \frac{e^{-i\varphi}}{\sqrt{2}}\sin\theta \\ \sin^2\frac{\theta}{2} \end{pmatrix}. \tag{12.56}$$

The superfluid velocity

$$\mathbf{v}_s = -\frac{i\hbar}{M}\boldsymbol{\xi}_\mathrm{F}^\dagger\nabla\boldsymbol{\xi}_\mathrm{F} = -\frac{\hbar}{M}(1+\cos\theta)\nabla\varphi = -\frac{\hbar(1+\cos\theta)}{Mr\sin\theta}\mathbf{e}_\varphi \tag{12.57}$$

takes the same form as the vector potential (12.54) of the Dirac monopole with the identification $g = \hbar/M$. The magnetization is calculated to be

$$\langle \mathbf{F} \rangle = \sum_{\alpha,\beta} \xi_\alpha^* \mathbf{F}_{\alpha\beta} \xi_\beta = (\mathbf{e}_x \cos\varphi + \mathbf{e}_y \sin\varphi) \sin\theta + \mathbf{e}_z \cos\theta, \quad (12.58)$$

which looks like a hedgehog. However, when $\theta = 0$, φ cannot be determined from $\langle \mathbf{F} \rangle$. This is consistent with the fact that the $m = 1$ component of ξ_{F} is singular except for $\theta = \pi$. The Dirac monopole is attached to a line singularity along the positive z-axis. Equation (12.56) shows that this singularity is a doubly quantized vortex. In accordance with the general argument of the $SO(3)$ vortex shown in Fig. 12.5, the Dirac monopole can be continuously deformed into a nonsingular spin texture. In fact, if we replace θ by $\theta(1 - t) + \pi t$ in Eq. (12.56), the order parameter deforms continuously from the Dirac monopole at $t = 0$ to a nonsingular texture $\boldsymbol{\xi}_{\mathrm{F}} = (0, 0, 1)^{\mathrm{T}}$ at $t = 1$.

12.3.4 *Skyrmions*

The third homotoopy group $\pi_3(M)$ classifies topological objects that extend over the entire three-dimensional space. These objects are called Skyrmions or particle-like solitons. Examples include the Shankar Skyrmion[1] and various types of knots.

12.3.4.1 *Shankar Skyrmion*

The order parameter manifold of a ferromagnetic BEC is $SO(3)$ which supports a topological object called the Shankar Skyrmion because $\pi_3(SO(3)) = \mathbb{Z}$. [Shankar (1977)] The order parameter is characterized by the direction of the spin and the rotation angle about that direction. Because of the spin-gauge symmetry, the rotation angle is related to the overall phase of the order parameter. We can specify both of them with a single vector $\boldsymbol{\Omega}$ whose direction and magnitude give the direction of the spin and the rotation angle, respectively. Each element of $\pi_3(SO(3)) = \mathbb{Z}$ may be realized by rotating the order parameter $(1, 0, 0)^T$ at each position $\mathbf{r}$ through angle $f(r)n$ about the direction

$$\boldsymbol{\Omega} = \frac{\mathbf{r}}{r} f(r) n, \quad n \in \mathbb{Z}, \quad (12.59)$$

[1] Despite the widespread nomenclature of the Shankar monopole, it is, in fact, a skyrmion.

where $f(0) = 2\pi$ and $f(\infty) = 0$. The condition $f(\infty) = 0$ ensures that the order parameter is uniform at spatial infinity. By identifying all points at spatial infinity, the three-dimensional space is compactified to the 3-sphere S^3. The order parameter of the Shankar Skyrmion for the case of $n = 1$ is shown in Fig. 12.8. The unitary operation for the rotation specified in

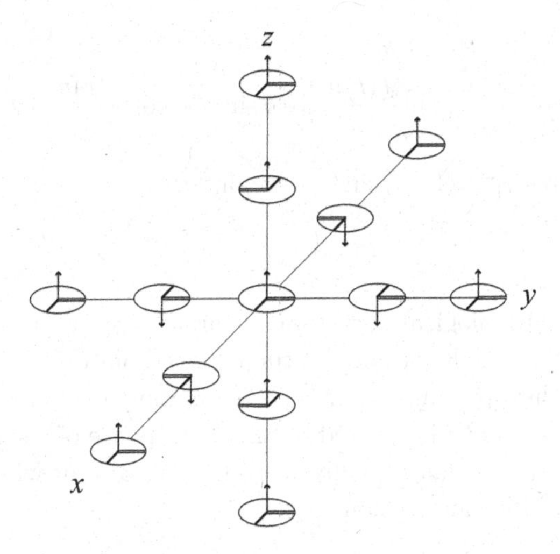

Fig. 12.8 Shankar Skyrmion with $n = 1$ in Eq. (12.59). In each circle, the arrow indicates the direction of the spin and the other two bars indicate how the overall phase of the order parameter changes over space. At spatial infinity, the direction of the spin and the phase of the order parameter are uniform; thus, the three-dimensional space is compactified to the 3-sphere S^3.

Eq. (12.59) is given by

$$\hat{U}(\mathbf{\Omega}) = \exp(-i\hat{\mathbf{f}} \cdot \mathbf{\Omega}), \tag{12.60}$$

where $\hat{\mathbf{f}} = (\hat{f}_x, \hat{f}_y, \hat{f}_z)$. For the spin-1 case, we obtain

$$\begin{aligned} \begin{pmatrix} \psi_1 \\ \psi_0 \\ \psi_{-1} \end{pmatrix} &= \hat{U}(\mathbf{\Omega}) \begin{pmatrix} 1 \\ 0 \\ 0 \end{pmatrix} \\ &= \begin{bmatrix} \left(\cos \frac{f(r)n}{2} - i \cos\theta \sin \frac{f(r)n}{2}\right)^2 \\ -\sqrt{2}i \left(\cos \frac{f(r)n}{2} - i\cos\theta \sin \frac{f(r)n}{2}\right) \sin \frac{f(r)n}{2} \sin\theta e^{i\phi} \\ -\sin^2 \frac{f(r)n}{2} \sin^2\theta e^{2i\phi} \end{bmatrix}, \end{aligned} \tag{12.61}$$

where (r, θ, ϕ) are the polar coordinates of $\mathbf{r}$. The spin vector (f_x, f_y, f_z) at $\mathbf{r}$ is therefore calculated to give

$$\begin{aligned}
f_x &= \sqrt{2}\mathrm{Re}\left\{(\psi_1 + \psi_{-1})\psi_0^*\right\} \\
&= 2\sin\frac{f(r)n}{2}\sin\theta\left(\sin\frac{f(r)n}{2}\cos\theta\cos\phi + \cos\frac{f(r)n}{2}\sin\phi\right), \\
f_y &= -\sqrt{2}\mathrm{Im}\left\{(\psi_1 - \psi_{-1})\psi_0^*\right\} \\
&= 2\sin\frac{f(r)n}{2}\sin\theta\left(\sin\frac{f(r)n}{2}\cos\theta\sin\phi - \cos\frac{f(r)n}{2}\cos\phi\right), \\
f_z &= |\psi_1|^2 - |\psi_{-1}|^2 = 1 - 2\sin^2\frac{f(r)n}{2}\sin^2\theta.
\end{aligned} \tag{12.62}$$

12.3.4.2 *Knots*

Another example of topological excitations characterized by the third homotopy group are knots. Knots are distinguished from other topological excitations in that they are characterized by the linking number rather than the winding number. Knots are hosted by the polar phase of a spin-1 BEC, whose order parameter manifold is given by Eq. (12.37), of which the S^2 part makes a nontrivial contribution to π_3:

$$\pi_3(S^2) = \mathbb{Z}. \tag{12.63}$$

The topological charge given by Eq. (12.63) is called the Hopf charge whose physical meaning can be understood as follows. Suppose that we trace a circle in a BEC along which the spinor order parameter points in one common direction $\mathbf{f}_1$ [see circle C_1 in Fig. 12.9]. We take another circle C_2 along which the order parameter points in another common direction $\mathbf{f}_2$ [see circle C_2]. If C_1 and C_2 link once, the linking number is 1. The linking number may be positive or negative depending on the relative orientation of the two loops.

A knot is a topological object in which a loop is embedded into itself in a nontrivial manner. The simplest example is the trefoil knot shown in Fig. 12.10. Technically speaking, the preimage of a single point on S^2 is called a simple unknotted loop, and therefore, the configuration in Fig. 12.9 is also called an unknot of a pair of closed loops with linking number 1. A topological object of the type shown in Fig. 12.9 is shown to be created in the polar phase of a spin-1 BEC [Kawaguchi, *et al.* (2008)].

The homotopy classification of topological excitations is summarized in Table 12.2.

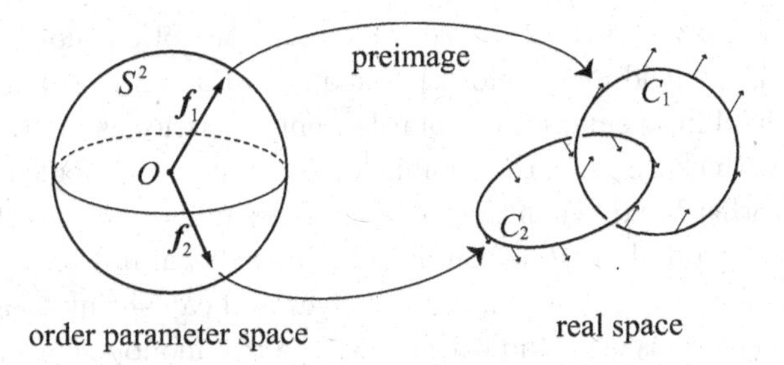

Fig. 12.9 Hopf charge (linking number). The preimage of each point in the order parameter space is a loop in real space. If C_1 and C_2 link once, the linking number is 1.

Fig. 12.10 Trefoil knot.

Table 12.2 Homotopy classification of defects and solitons.

π_n	defects	solitons
π_0	domain walls	
π_1	vortices	nonsingular domain walls
π_2	monopoles	2D Skyrmions
π_3	Skyrmions	Skyrmions, knots
π_4	instantons	$SO(3)$ instantons

12.3.5 *Influence of different types of defects*

12.3.5.1 *Influence of π_1 on π_2*

When two different types of topological objects coexist, they may influence each other in a nontrivial manner. As an example, consider a spin-1 polar phase in which two monopoles with charge one merge in the presence of a half-quantum vortex, as shown in Fig. 12.11 (a). The half-quantum vortex introduces a branch cut for field lines of monopoles, so that field lines on one side of the branch cut cannot merge with those on the other side by crossing it. Thus, if the two monopoles approach each other along path 1, the order-parameter field must deform, as shown in Fig. 12.11 (b). Because the number of field lines emanating from the center is twice as many as that of a monopole with charge one, the combined object has a topological charge of two. On the other hand, when the two monopoles merge along path 2, the order-parameter manifold can deform into the one shown in Fig. 12.11 (c). Because all the field lines close upon themselves and can shrink to a point, the topological charge is zero, indicating that the two monopoles annihilate in pairs. The latter example indicates that the charge of the monopole changes its sign as it moves around the half-quantum vortex along path 2. This example illustrates how the coalescence of topological objects is influenced by the presence of a different type of topological excitation.

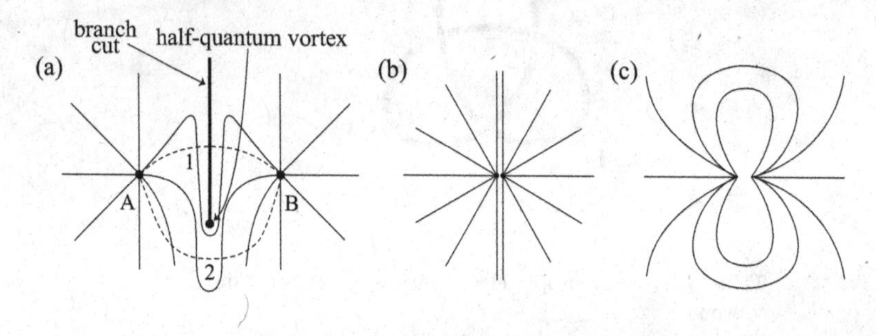

Fig. 12.11 (a) Coalescence of two monopoles with charge one in the presence of a half-quantum vortex. (b) If they merge along path 1, the topological charge of the combined object is two. (c) If they merge along path 2, the topological charge of the combined object is zero because the field lines can continuously shrink to a uniform configuration.

12.3.5.2 *Alice ring*

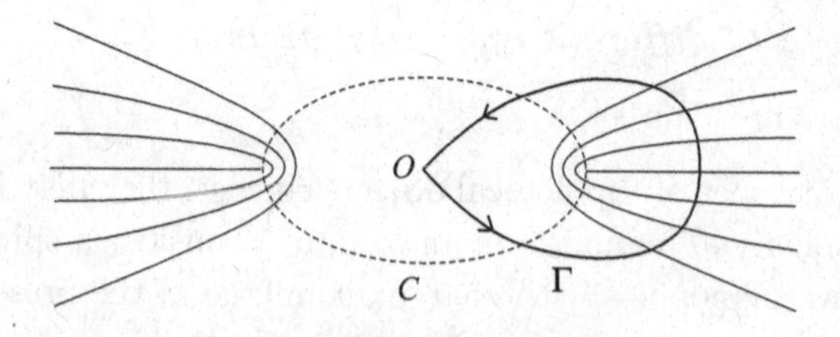

Fig. 12.12 Alice ring comprised of continuously distributed half-quantum vortices along a contour C. Far from the origin, it looks like a monopole.

The Alice ring is a combined object of two defects characterized by π_1 and π_2 [see Fig. 12.12]. Far from the origin it appears to be a point defect ($\pi_2 = 1$); however, along the ring C on which the order parameter is singular, it appears to be a continuous distribution of line defects with $\pi_1 = 1$. This composite topological object is called an Alice ring. The Alice ring was predicted to be realized in an optically trapped spin-1 ^{23}Na BEC [Ruostekoski and Anglin (2003)].

12.3.6 *Topological charges*

12.3.6.1 *Line defects: π_1*

The topological charge can be expressed algebraically as a function of the order-parameter field. The topological charge corresponding to $\pi_1(S^1)$ can be expressed as a line integral of the velocity field. The superfluid current density $\mathbf{j}$ is given by Eq. (6.100), and the corresponding expression for a spin-1 ferromagnetic BEC is given by Eq. (6.101).

Unlike a scalar BEC, where the superfluid velocity is irrotational, the superfluid velocity for the ferromagnetic BEC has nonvanishing $\nabla \times \mathbf{v}_s^F$ due to the Berry phase, as shown in Eq. (6.104). As a consequence, the circulation alone is not quantized; however, the difference between the circulation and the contribution from the Berry phase is quantized.

12.3.6.2 *Point defects: π_2*

The topological charge N_2 of a point defect can be calculated as follows. By definition, N_2 gives the number of times the unit vector $\mathbf{m}(\mathbf{r})$ representing the order parameter wraps S^2 [see Fig. 12.7]. Expressing the components of this vector as $m_x = \sin\alpha\cos\beta$, $m_y = \sin\alpha\sin\beta$, and $m_z = \cos\alpha$, N_2 is given by

$$N_2 = \frac{1}{4\pi}\int_\Sigma d\theta d\phi \sin\theta \left|\frac{\partial(\alpha,\beta)}{\partial(\theta,\phi)}\right|, \tag{12.64}$$

where θ and ϕ are the polar and azimuthal angles of $\mathbf{r}$ and the last term on the right-hand side is the Jacobian of the transformation of the coordinates. The right-hand side can be directly expressed in terms of $\mathbf{m}$ as

$$N_2 = \frac{1}{4\pi}\int_\Sigma d\theta d\phi \, \mathbf{m}\cdot\left(\frac{\partial \mathbf{m}}{\partial\theta} \times \frac{\partial \mathbf{m}}{\partial\phi}\right). \tag{12.65}$$

It follows from this that the topological charge of the hedgehog is $N_2 = 1$.

The topological charge N_2 corresponding to $\pi_2(S^2)$ can be expressed as a surface integral of a vector quantity $\mathbf{j}$:

$$N_2 = \frac{1}{4\pi}\int \mathbf{j}\cdot d\mathbf{S}, \tag{12.66}$$

where

$$\mathbf{j} = \frac{1}{2}\epsilon_{ijk} m_i (\nabla m_j \times \nabla m_k). \tag{12.67}$$

We can use Gauss' law to rewrite the right-hand side of Eq. (12.66) as a volume integral:

$$N_2 = \int \rho_{\mathrm{m}} d\mathbf{r}, \tag{12.68}$$

where

$$\rho_{\mathrm{m}}(\mathbf{r}) = \frac{1}{4\pi}\nabla\cdot\mathbf{j}(\mathbf{r}) \tag{12.69}$$

gives the density distribution of point singularities.

12.3.6.3 *Skyrmions:* π_3

The topological charge of the third homotopy group can be introduced in a manner similar to that of the second homotopy group. We consider a continuous map $U(\mathbf{r}) : \mathbf{R}^3 \to SU(2) \cong S^3$ that satisfies the boundary condition $U(\mathbf{r}) \to \hat{1}$ (identity map) as $|\mathbf{r}| \to \infty$. Then, $\mathbf{R}^3$ is compactified to S^3 and $U(\mathbf{r})$ describes mapping $S^3 \to S^3$. The topological charge of the third homotopy group can be calculated by using the Hopf map in which the $SU(2)$ matrix U is expressed in terms of two complex numbers z_1 and z_2 as

$$U(\mathbf{r}) = \begin{pmatrix} z_2 & -z_1^* \\ z_1 & z_2 \end{pmatrix}, \quad |z_1|^2 + |z_2|^2 = 1, \tag{12.70}$$

where z_1 and z_2 are related to $\mathbf{r}$ as follows. We first use $\mathbf{r}$ and $f(r)$ in Eq. (12.59) to define a four-component unit vector as

$$\mathbf{n} = \frac{\mathbf{r}}{r}\sin\frac{f(r)n}{2}, \quad n_4 = \cos\frac{f(r)n}{2}, \tag{12.71}$$

where $\mathbf{n} = (n_1, n_2, n_3)$. Then, $z_1 \equiv n_1 + in_2$ and $z_2 \equiv n_3 + in_4$ satisfy the normalization condition $|z_1|^2 + |z_2|^2 = 1$. Introducing $Z = (z_2, z_1)^T$, the spin vector is given by

$$\mathbf{f} = Z^\dagger \boldsymbol{\sigma} Z, \tag{12.72}$$

where $\boldsymbol{\sigma} = (\sigma_x, \sigma_y, \sigma_z)$ are the Pauli matrices. It can be shown that Eq. (12.72) reproduces Eq. (12.62). The fact that a local $U(1)$ transformation $Z \to e^{i\gamma} Z$ does not change $\mathbf{f}$ implies that S^3 is a $U(1)$ fibre bundle over S^2; physically, this means that the preimage of a point in the order parameter manifold is a closed loop [see Fig. 12.9]. We can use Z to introduce an $SU(2)$-valued gauge potential

$$A_j = \frac{i}{2}(Z^\dagger \partial_j Z - (\partial_j Z^\dagger) Z) \tag{12.73}$$

and the associated field tensor

$$F_{ij} = \partial_i A_j - \partial_j A_i. \tag{12.74}$$

Then, the Hopf charge is given by the integral of the Chern–Simons term over $\mathbf{R}^3$:

$$N_3 = \frac{1}{4\pi^2} \int d\mathbf{r} \epsilon_{ijk} F_{ij} A_k. \tag{12.75}$$

Expressing the right-hand side in terms of $(\mathbf{n}, n_4)$, we can rewrite the Hopf charge as the Jacobian of the transformations:

$$N_3 = \frac{1}{12\pi^2} \int d\mathbf{r} \epsilon_{ijk} \epsilon_{\alpha\beta\gamma\delta} n_\alpha \partial_i n_\beta \partial_j n_\gamma \partial_k n_\delta, \tag{12.76}$$

where ϵ_{ijk} and $\epsilon_{\alpha\beta\gamma\delta}$ are the completely antisymmetric tensors of the third and fourth orders, respectively, and the Roman letters run over x, y, z and the Greek letters run over $1, 2, 3, 4$. Substituting Eq. (12.71) in Eq. (12.76) and performing integration, we obtain $N_3 = n$. Thus, we find that the number of times the spin rotates as it goes from the origin to spatial infinity [see Eq. (12.59)] is equal to the Hopf charge.

Appendix A

Order of Phase Transition, Clausius–Clapeyron Formula, and Gibbs–Duhem Relation

When matter is physically and chemically uniform, it is said to form one phase. Each phase is described by a thermodynamic function of thermodynamic variables such as temperature and pressure. When two phases I and II are possible under given circumstances, the phase that has the lower chemical potential μ is the ground state. A change in the state of matter from one phase to another is called a phase transition; it is accompanied by singularities of physical quantities. A phase transition is said to be of the n-th order if up to the (n-1)-th derivatives of μ are continuous but the n-th derivative is discontinuous.

The first-order phase transition is defined as

$$\Delta\mu \equiv \mu_1 - \mu_2 = 0, \quad \frac{\partial \Delta\mu}{\partial T} \neq 0, \quad \frac{\partial \Delta\mu}{\partial P} \neq 0,$$

and it is accompanied by discontinuous changes in entropy and volume. The second-order phase transition is defined as

$$\Delta\mu = 0, \quad \frac{\partial \Delta\mu}{\partial T} = 0, \quad \frac{\partial \Delta\mu}{\partial P} = 0, \quad \frac{\partial^2 \Delta\mu}{\partial T^2} \neq 0, \quad \frac{\partial^2 \Delta\mu}{\partial P^2} \neq 0,$$

and it is accompanied by a cusp in entropy and a jump in specific heat as a function of temperature. The second-order phase transition is also accompanied by critical phenomena in which fluctuations of some physical quantities become anomalously large and correlation lengths of the fluctuations diverge.

At constant pressure, the Gibbs free energy must be continuous on the border line between different phases I and II, $G_I(P,T) = G_{II}(P,T)$. Along the phase boundary, we have

$$\frac{\partial(G_I - G_{II})}{\partial P} dP + \frac{\partial(G_I - G_{II})}{\partial T} dT = \Delta V \cdot dP - \Delta S \cdot dT = 0,$$

where

$$\Delta V \equiv V_I - V_{II}, \; \Delta S \equiv S_I - S_{II}.$$

Hence, we obtain the Clausius–Clapeyron formula

$$\frac{dP}{dT} = \frac{\Delta S}{\Delta V} = \frac{\Delta Q}{T\Delta V},$$

where $\Delta Q = T\Delta S$ is the latent heat of the transition per mole and ΔV, the change in molar volume at the phase transition.

The Gibbs free energy G is a function of temperature T, pressure P, and particle number N. Because the only extensive variable in G is N, we have

$$G(T, P, \alpha N) = \alpha G(T, P, N).$$

Differentiating both sides with respect to α and setting $\alpha = 1$, we obtain

$$G = \left(\frac{\partial G}{\partial N}\right)_{T,P} N = \mu N \quad \text{(Euler's relation)}.$$

Substituting this in $dG = -SdT + VdP + \mu dN$ gives the Gibbs–Duhem relation

$$SdT - VdP + Nd\mu = 0.$$

Appendix B

Bogoliubov Wave Functions in Coordinate Space

B.1 Ground-State Wave Function

The state vector of the ground state in the Bogoliubov approximation can be written from Eq. (2.81) as

$$|\psi_0\rangle = N \exp\left[\phi_0 \hat{a}_0^\dagger - \sum_{k_x>0} \alpha_k \hat{a}_{\mathbf{k}}^\dagger \hat{a}_{-\mathbf{k}}^\dagger\right] |\text{vac}\rangle = Z|\phi_0\rangle \prod_{k_x>0} \sum_{l_{\mathbf{k}}=0}^{\infty} (-\alpha_k)^{l_k} |l_{\mathbf{k}}\rangle, \tag{B.1}$$

where

$$Z = e^{\frac{|\phi|^2}{2}} N = \exp\left[-\frac{4N}{9}(3\pi - 8)\sqrt{\frac{na^3}{\pi}}\right], \tag{B.2}$$

$|\phi_0\rangle$ is the coherent state for the $\mathbf{k} = 0$ mode, and $|l_{\mathbf{k}}\rangle$ represents the state of $l_{\mathbf{k}}$ pairs of particles with wave vectors $\mathbf{k}$ and $-\mathbf{k}$. To find the coordinate representation of $|\psi_0\rangle$, we consider the completeness relation in the coordinate representation.

$$\frac{1}{N!} \int |\mathbf{r}_1, \mathbf{r}_2, \cdots, \mathbf{r}_N\rangle \langle \mathbf{r}_1, \mathbf{r}_2, \cdots, \mathbf{r}_N| d\mathbf{r}_1 d\mathbf{r}_2 \cdots d\mathbf{r}_N = \hat{1}, \tag{B.3}$$

where $|\mathbf{r}_1, \mathbf{r}_2, \cdots, \mathbf{r}_N\rangle / \sqrt{N!}$ is the completely symmetrized state vector. The coordinate representation of the wave function is given by

$$\psi_0(\mathbf{r}_1, \mathbf{r}_2, \cdots, \mathbf{r}_N) = \frac{1}{\sqrt{N!}} \langle \mathbf{r}_1, \mathbf{r}_2, \cdots, \mathbf{r}_N | \psi_0\rangle . \tag{B.4}$$

Operating the state vectors $|\psi_0\rangle$ from both sides of Eq. (B.3) gives

$$\int |\psi_0(\mathbf{r}_1, \mathbf{r}_2, \cdots, \mathbf{r}_N)|^2 d\mathbf{r}_1 d\mathbf{r}_2 \cdots d\mathbf{r}_N = 1. \tag{B.5}$$

Thus, the wave function ψ_0 is normalized to unity. The state vector $|\mathbf{r}_1, \mathbf{r}_2, \cdots, \mathbf{r}_N\rangle$ is expressed in terms of the field operator $\hat{\phi}^\dagger(\mathbf{r}) = \frac{1}{\sqrt{V}} \sum_{\mathbf{k}} \hat{a}_{\mathbf{k}}^\dagger e^{-i\mathbf{k}\mathbf{r}}$ as

$$\begin{aligned}|\mathbf{r}_1, \mathbf{r}_2, \cdots, \mathbf{r}_N\rangle &= \hat{\phi}^\dagger(\mathbf{r}_1)\hat{\phi}^\dagger(\mathbf{r}_2)\cdots\hat{\phi}^\dagger(\mathbf{r}_N)|\mathrm{vac}\rangle \\ &= \frac{1}{\sqrt{V^N}} \sum_{\mathbf{k}_1, \mathbf{k}_2, \cdots, \mathbf{k}_N} e^{-i\sum_{i=1}^N \mathbf{k}_i \mathbf{r}_i} \hat{a}_{\mathbf{k}_1}^\dagger \hat{a}_{\mathbf{k}_2}^\dagger \cdots \hat{a}_{\mathbf{k}_N}^\dagger |\mathrm{vac}\rangle. \qquad \text{(B.6)}\end{aligned}$$

Substituting this and Eq. (B.1) in Eq. (B.4) gives

$$\begin{aligned}\psi_0(\mathbf{r}_1, \mathbf{r}_2, \cdots, \mathbf{r}_N) = \frac{Z}{\sqrt{V^N N!}} \sum_{\mathbf{k}_1, \cdots, \mathbf{k}_N} e^{i\sum_{i=l}^N \mathbf{k}_i \mathbf{r}_i} \langle \mathrm{vac}|\hat{a}_{\mathbf{k}_1} \cdots \hat{a}_{\mathbf{k}_N}|\phi_0\rangle \\ \times \prod_{k_x > 0} \sum_{l_{\mathbf{k}}=0}^{\infty} (-\alpha_k)^{l_{\mathbf{k}}} |l_{\mathbf{k}}\rangle. \qquad \text{(B.7)}\end{aligned}$$

The matrix element on the right-hand side does not vanish only when $\hat{a}_{\mathbf{k}_1} \cdots \hat{a}_{\mathbf{k}_N}$ includes $l_{\mathbf{k}}$ $\hat{a}_{\mathbf{k}}$'s, $l_{\mathbf{k}}$ $\hat{a}_{-\mathbf{k}}$'s, and $N - 2l_{\mathbf{k}}$ $\hat{a}_{\mathbf{0}}$'s:

$$\begin{aligned}&\sum_{\mathbf{k}_1, \cdots, \mathbf{k}_N} e^{i\sum_{i=1}^N \mathbf{k}_i \mathbf{r}_i} \langle \mathrm{vac}|\hat{a}_{\mathbf{k}_1} \cdots \hat{a}_{\mathbf{k}_N}|\phi_0, l_{\mathbf{k}}\rangle \\ &= \sum_{(i_1, \cdots, i_{2l_{\mathbf{k}}})} e^{i\mathbf{k}\left[\left(\mathbf{r}_{i_1} - \mathbf{r}_{i_{l_{\mathbf{k}}+1}}\right) + \cdots + \left(\mathbf{r}_{i_{l_{\mathbf{k}}}} - \mathbf{r}_{i_{2l_{\mathbf{k}}}}\right)\right]} \times \langle \mathrm{vac}|\hat{a}_{\mathbf{k}}^{l_{\mathbf{k}}} \hat{a}_{-\mathbf{k}}^{l_{\mathbf{k}}} \hat{a}_0^{N-2l_{\mathbf{k}}}|\phi_0, l_{\mathbf{k}}\rangle \\ &= l_{\mathbf{k}}! \phi_0^{N-2l_{\mathbf{k}}} e^{-|\phi_0|^2/2} \sum_{(i_1, \cdots, i_{2l_{\mathbf{k}}})} e^{i\mathbf{k}\left[\left(\mathbf{r}_{i_1} - \mathbf{r}_{i_{l_{\mathbf{k}}+1}}\right) + \cdots + \left(\mathbf{r}_{l_{\mathbf{k}}} - \mathbf{r}_{2l_{\mathbf{k}}}\right)\right]} \langle \mathrm{vac}|\phi_0\rangle, \text{(B.8)}\end{aligned}$$

where $\langle \mathrm{vac}|\phi_0\rangle = e^{-|\phi_0|^2/2}$ is used, and the sum runs over all $2l_{\mathbf{k}}$'s combinations out of $1, 2, \cdots, N$. Substituting Eq. (B.8) in Eq. (B.7) gives

$$\begin{aligned}\psi_0(\mathbf{r}_1, \mathbf{r}_2, \cdots, \mathbf{r}_N) = \frac{Ze^{-\frac{|\phi_0|^2}{2}} \phi_0^N}{\sqrt{V^N N!}} \prod_{k_x > 0} \sum_{l_{\mathbf{k}}=0}^{\infty} l_k! \\ \times \sum_{(i_1, \cdots, i_{2l_{\mathbf{k}}})} \prod_{j=1}^{l_{\mathbf{k}}} \left[-\frac{\alpha_k}{\phi_0^2} e^{i\mathbf{k}\sum_{j=1}^{l_{\mathbf{k}}}\left(\mathbf{r}_{i_j} - \mathbf{r}_{i_{l_{\mathbf{k}}+j}}\right)}\right], \text{(B.9)}\end{aligned}$$

where it is understood that for the case of $l_{\mathbf{k}} = 0$, the last factor is equal to 1. Taking ϕ_0 to be real, from Eq. (2.79), we have

$$\phi_0^N = N^{\frac{N}{2}} \left(1 - \frac{8}{3}A\right)^{\frac{N}{2}} \simeq N^{\frac{N}{2}} e^{-\frac{4}{3}NA}, \qquad \text{(B.10)}$$

where $A = \sqrt{na^3/\pi}$. These results combined with Stirling's formula $N! \simeq \sqrt{2\pi N}(N/e)^N$ yield $e^{-|\phi_0|^2/2}\phi_0^N/\sqrt{N!} \simeq 1/4\sqrt{2\pi N}$. Separating out the term with $l_{\mathbf{k}} = 0$ in Eq. (B.9) gives

$$\psi_0(\mathbf{r}_1, \cdots, \mathbf{r}_N) \simeq \frac{Z}{\sqrt[4]{2\pi N} V^{N/2}} \times \prod_{k_x>0} \left[1 + \sum_{l_k \neq 0}^{\infty} l_k! \sum_{(i_1, \cdots, i_{2l_k})} \prod_{j=1}^{l_{\mathbf{k}}} g^{\mathbf{k}}(i_j, i_{l_{\mathbf{k}}+j}) \right], \tag{B.11}$$

where $g^{\mathbf{k}}(i,j) \equiv -(\alpha_k/\phi_0^2)e^{i\mathbf{k}(\mathbf{r}_i - \mathbf{r}_j)} \propto 1/N$, because $\phi_0^2 \propto N$. Writing down the terms of the product $\prod_{k_x>0}$ and the sum $\sum_{l_{\mathbf{k}}}$ in Eq. (B.11) explicitly, we have

$$\begin{aligned}
&\prod_{k_x>0} \left[1 + \sum_{(i_1,i_2)} g^{\mathbf{k}}(i_1,i_2) + 2! \sum_{(i_1,\cdots,i_4)} g^{\mathbf{k}}(i_1,i_3)\, g^{\mathbf{k}}(i_2,i_4) \right. \\
&\left. \quad + 3! \sum_{(i_1,\cdots,i_6)} g^{\mathbf{k}}(i_1,i_4)\, g^{\mathbf{k}}(i_2,i_5)\, g^{\mathbf{k}}(i_3,i_6) + \cdots \right] \\
&= 1 + \sum_{k_x>0} \sum_{(i_1,i_2)} g^{\mathbf{k}}(i_1,i_2) + \left[\frac{1}{2!} \sum_{k_x,k'_x>0} \sum_{(i_1,i_2),(i'_1,i'_2)} g^{\mathbf{k}}(i_j,i_2)\, g^{\mathbf{k}'}(i'_1,i'_2) \right. \\
&\left. \quad + 2! \sum_{k_x>0} \sum_{(i_1,\cdots,i_4)} g^{\mathbf{k}}(i_1,i_3)\, g^{\mathbf{k}}(i_2,i_4) \right] \\
&\quad + \left[\frac{1}{3!} \sum_{k_x,k'_x,k''_x>0} \sum_{(i_1,i_2),(i'_1,i'_2),(i''_1,i''_2)} g^{\mathbf{k}}(i_1,i_2)\, g^{\mathbf{k}'}(i'_1,i'_2)\, g^{\mathbf{k}'}(i''_1,i''_2) \right. \\
&\quad + \frac{1}{2!} \sum_{k_x,k'_x>0} g^{\mathbf{k}}(i_1,i_2)\, 2! \sum_{(i'_1,\cdots,i'_4)} g^{\mathbf{k}'}(i'_1,i'_3)\, g^{\mathbf{k}'}(i'_2,i'_4) \\
&\left. \quad + 3! \sum_{(i_1,\cdots,i_6)} g^{\mathbf{k}}(i_1,i_4) g^{\mathbf{k}}(i_2,i_5) g^{\mathbf{k}}(i_3,i_6) \right] + \cdots .
\end{aligned} \tag{B.12}$$

Because each $g^{\mathbf{k}}$ is of the order of $1/N$, only the first term in each square bracket [ ] remains nonvanishing in the thermodynamic limit. Thus, the right-hand side of Eq. (B.12) becomes

$$1+\sum_{k_x>0}\sum_{(i_1,i_2)} g^{\mathbf{k}}(i_1,i_2)+\frac{1}{2!}\left[\sum_{k_x>0}\sum_{(i_1,i_2)} g^{\mathbf{k}}(i_1,i_2)\right]^2+\cdots$$

$$=\exp\left[\sum_{k_x>0}\sum_{(i_1,i_2)} g^{\mathbf{k}}(i_1,i_2)\right].$$

Thus, we obtain

$$\psi_0(\mathbf{r}_1,\cdots,\mathbf{r}_N)\simeq\frac{Z}{\sqrt[4]{2\pi N}V^{N/2}}\exp\left[\sum_{k_x>0}\sum_{(i_1,i_2)} g^{\mathbf{k}}(i_1,i_2)\right]. \quad \text{(B.13)}$$

We may replace $\sum_{k_x>0}\sum_{(i_1,i_2)} g^{\mathbf{k}}(i_1,i_2)$ with $\sum_{\mathbf{k}}\sum_{i<j} g^{\mathbf{k}}(i,j)$ because the condition $k_x>0$ is equivalent to $i<j$. Furthermore, we have

$$\sum_{\mathbf{k}} g^{\mathbf{k}}(i,j)\simeq-\frac{1}{8\pi^3 n}\int d^3k\alpha_k e^{i\mathbf{k}(\mathbf{r}_i-\mathbf{r}_j)}.$$

Because α_k is independent of the angle, we have

$$\int d^3k\alpha_k e^{i\mathbf{k}(\mathbf{r}_i-\mathbf{r}_i)}=2\pi\int_0^\infty k^2dk\alpha_k\int_0^\pi d\theta\sin\theta e^{ik|\mathbf{r}_i-\mathbf{r}_j|}\cos\theta$$

$$=2\pi\int_0^\infty k^2dk\alpha_k\frac{2\sin kr_{ij}}{kr_{ij}}=\frac{4\pi}{r_{ij}}\int_0^\infty dk\alpha_k k\sin kr_{ij},$$

where $r_{ij}\equiv|\mathbf{r}_i-\mathbf{r}_j|$. Thus,

$$\sum_{\mathbf{k}} g^{\mathbf{k}}(i,j)=-\frac{1}{2\pi^2 nr_{ij}}\int_0^\infty dk\alpha_k k\sin kr_{ij}$$

$$=-\frac{4a}{\pi r_{ij}}\int_0^\infty dx\frac{x\sin\sqrt{8\pi anr_{ij}}x}{1+x^2+x\sqrt{x^2+2}}.$$

We therefore obtain the following wave function in the coordinate space:

$$\psi_0(\mathbf{r}_1,\cdots,\mathbf{r}_N)\simeq\frac{Z}{\sqrt[4]{2\pi N}V^{N/2}}\exp\left[-\frac{4}{\pi}\sum_{i<j}\frac{a}{r_{ij}}\int_0^\infty dx\frac{x\sin\left(\sqrt{8\pi anr_{ij}}x\right)}{1+x^2+x\sqrt{x^2+2}}\right]. \quad \text{(B.14)}$$

Substituting $\alpha\equiv\sqrt{8\pi anr_{ij}}$ and $t\equiv\alpha x$, we obtain

$$\int_0^\infty dx\frac{x\sin\alpha x}{1+x^2+x\sqrt{x^2+2}}\simeq\frac{1}{2}\int_0^\infty dt\frac{t\sin t}{t^2+\alpha^2}=\frac{\pi}{4}e^{-\alpha}. \quad \text{(B.15)}$$

Thus, for $\alpha\equiv\sqrt{8\pi anr_{ij}}\lesssim 1$, Eq. (B.14) can be approximated as

$$\psi_0(\mathbf{r}_1,\cdots,\mathbf{r}_N)\simeq\frac{Z}{4\sqrt{2\pi N}V^{N/2}}\exp\left(-\sum_{i<j}\frac{a}{r_{ij}}e^{-\sqrt{8\pi anr_{ij}}}\right). \quad \text{(B.16)}$$

B.2 One-Phonon State

The state vector for a one-phonon state is given from Eq. (2.74) by

$$|\psi_{\mathbf{k}}\rangle = \sqrt{1-\alpha_k^2}\hat{a}_{\mathbf{k}}^{\dagger}|\psi_0\rangle. \tag{B.17}$$

Substituting Eq. (B.2) in this equation gives

$$|\psi_{\mathbf{k}}\rangle = Z\sqrt{1-\alpha_k^2}\prod_{k'_x>0}\sum_{n_{\mathbf{k}'}=0}^{\infty}(-\alpha_{k'})^{n_{k'}}\,\hat{a}_{\mathbf{k}}^{\dagger}|\phi_0, n_{\mathbf{k}'}\rangle. \tag{B.18}$$

The equation corresponding to Eq. (B.7) is

$$\begin{aligned}\psi_{\mathbf{k}}(\mathbf{r}_1,\cdots,\mathbf{r}_N) = {} & Z\sqrt{\frac{1-\alpha_k^2}{V^N N!}}\prod_{k'_x>0}\sum_{n_{\mathbf{k}'}=0}^{\infty}(-\alpha_{k'})^{n_{\mathbf{k}'}}\sum_{\mathbf{k}_1,\cdots,\mathbf{k}_N}e^{i\sum_{i=1}^{N}\mathbf{k}_i\mathbf{r}_i}\\ & \times\left\langle \text{vac}|\hat{a}_{\mathbf{k}_1}\cdots\hat{a}_{\mathbf{k}_N}\hat{a}_{\mathbf{k}}^{\dagger}|\phi_0, n_{\mathbf{k}'}\right\rangle.\end{aligned} \tag{B.19}$$

For $\mathbf{k}_1,\cdots,\mathbf{k}_N$, the matrix element on the right-hand side of Eq. (B.19) remains nonvanishing only when it has $n_{\mathbf{k}'}$'s terms equal to $\mathbf{k}'$, $n_{\mathbf{k}'}$'s terms equal to $-\mathbf{k}'$, one term equal to $\mathbf{k}$, and the remaining $N-2n_{\mathbf{k}'}-1$ wave numbers vanish. Hence,

$$\begin{aligned}&\sum_{\mathbf{k}_1,\cdots,\mathbf{k}_N}e^{i\sum_{i=1}^{N}\mathbf{k}_i\mathbf{r}_i}\left\langle \text{vac}|\hat{a}_{\mathbf{k}_1}\cdots\hat{a}_{\mathbf{k}_N}\hat{a}_{\mathbf{k}}^{\dagger}|\phi_0, n_{\mathbf{k}'}\right\rangle = \sum_{(i_0,i_1,\cdots,i_{2n_{\mathbf{k}}})}\\ &\times e^{i\mathbf{k}\mathbf{r}_{i_0}}e^{i\mathbf{k}'\sum_{j=1}^{n_k}\left(\mathbf{r}_{i_j}-\mathbf{r}_{i_{j+n_{\mathbf{k}}}}\right)}\left\langle \text{vac}|\hat{a}_{\mathbf{k}'}^{n_{\mathbf{k}'}}\hat{a}_{-\mathbf{k}'}^{n'_{\mathbf{k}}}\hat{a}_{\mathbf{k}}\hat{a}_0^{N-2n_{\mathbf{k}'}-1}\hat{a}_{\mathbf{k}}^{\dagger}|\phi_0, n'_{\mathbf{k}}\right\rangle,\end{aligned}$$

where

$$\begin{aligned}&\left\langle \text{vac}|\hat{a}_{\mathbf{k}'}^{n_{\mathbf{k}'}}\hat{a}_{-\mathbf{k}'}^{n'_{\mathbf{k}}}\hat{a}_{\mathbf{k}}\hat{a}_0^{N-2n_{\mathbf{k}'}-1}\hat{a}_{\mathbf{k}}^{\dagger}|\phi_0, n'_{\mathbf{k}}\right\rangle\\ &= \left[(1-\delta_{\mathbf{k},\mathbf{k}'}-\delta_{\mathbf{k},-\mathbf{k}'})\,n_{\mathbf{k}'}! + (\delta_{\mathbf{k},\mathbf{k}'}+\delta_{\mathbf{k},-\mathbf{k}'})\,(n_{k'}+1)!\right]\phi_0^{N-2n_{\mathbf{k}'}-1}e^{-\frac{|\phi_0|^2}{2}}\\ &= \left[n_{\mathbf{k}'}! + (\delta_{\mathbf{k},\mathbf{k}'}+\delta_{\mathbf{k},-\mathbf{k}'})\,n_{\mathbf{k}'}\cdot n_{\mathbf{k}'}!\right]\phi_0^{N-2n_{\mathbf{k}'}-1}e^{-\frac{|\phi_0|^2}{2}}.\end{aligned}$$

Here, because the terms proportional to $n_{\mathbf{k}'}!$ are equivalent to those in Eq. (B.8) except for the factor $\sum_{i_0}e^{i\mathbf{k}\mathbf{r}_0}$, substituting the terms in Eq. (B.19) gives $\sum_{i_0}e^{i\mathbf{k}\mathbf{r}_0}\psi_0(\mathbf{r}_1,\cdots,\mathbf{r}_N)$. The remaining terms, which are proportional to $\delta_{\mathbf{k},\mathbf{k}'}+\delta_{\mathbf{k},-\mathbf{k}'}$, do not vanish only when either $\mathbf{k}'$ or $-\mathbf{k}'$ is equal to $\mathbf{k}$; otherwise, they vanish in the thermodynamic limit. Therefore, we obtain

$$\psi_{\mathbf{k}}(\mathbf{r}_1,\mathbf{r}_2,\cdots,\mathbf{r}_N) = \sum_{n=1}^{N}e^{i\mathbf{k}\mathbf{r}_n}\psi_0(\mathbf{r}_1,\mathbf{r}_2,\cdots,\mathbf{r}_N). \tag{B.20}$$

Appendix C

Effective Mass, Sound Velocity, and Spin Susceptibility of Fermi Liquid

The formula for the effective mass—Eq. (8.41)—can be derived from Galileo's relativity principle. Because of the assumption of adiabatic continuity, the number of quasiparticles is equal to that of real particles. The mass flux density, which is the flux of particles multiplied by the mass of the particles, is therefore given by

$$\sum_{\mathbf{p}\sigma} M \frac{\partial \varepsilon_{\mathbf{p}\sigma}}{\partial \mathbf{p}} n_{\mathbf{p}\sigma}. \tag{C.1}$$

According to Galileo's relativity principle[2] , this is equal to the total momentum of the fluid $\sum_{\sigma} \mathbf{p} n_{\mathbf{p}\sigma}$. Thus,

$$\sum_{\mathbf{p}\sigma} \mathbf{p} n_{\mathbf{p}\sigma} = \sum_{\mathbf{p}\sigma} M \frac{\partial \varepsilon_{\mathbf{p}\sigma}}{\partial \mathbf{p}} n_{\mathbf{p}\sigma}. \tag{C.2}$$

Dividing both sides by m and taking the variation with respect to $n_{\mathbf{p}\sigma}$, we obtain

$$\begin{aligned}\sum_{\mathbf{p}\sigma} \frac{\mathbf{p}}{M} \delta n_{\mathbf{p}\sigma} &= \sum_{\mathbf{p}\sigma} \frac{\partial \varepsilon_{\mathbf{p}\sigma}}{\partial \mathbf{p}} \delta n_{\mathbf{p}\sigma} + \frac{1}{\Omega} \sum_{\mathbf{p}\mathbf{p}'\sigma\sigma'} \frac{\partial f_{\sigma\sigma'}(\mathbf{p}, \mathbf{p}')}{\partial \mathbf{p}} n_{\mathbf{p}\sigma} \delta n_{\mathbf{p}'\sigma'} \\ &= \sum_{\mathbf{p}\sigma} \frac{\partial \varepsilon_{\mathbf{p}\sigma}}{\partial \mathbf{p}} \delta n_{\mathbf{p}\sigma} - \frac{1}{\Omega} \sum_{\mathbf{p}\mathbf{p}'\sigma\sigma'} f_{\sigma\sigma'}(\mathbf{p}, \mathbf{p}') \frac{\partial n_{\mathbf{p}\sigma}}{\partial \mathbf{p}} \delta n_{\mathbf{p}'\sigma'}.\end{aligned}$$

[2]There is no net flow of particles if we observe the system from the frame of reference moving with velocity $(1/N) \sum_{\mathbf{p}\sigma} \frac{\partial \varepsilon_{\mathbf{p}\sigma}}{\partial \mathbf{p}} n_{\mathbf{p}\sigma}$. The total momentum of the system viewed from the rest frame is therefore given by

$$\frac{1}{N} \sum_{\mathbf{p}\sigma} \frac{\partial \varepsilon_{\mathbf{p}\sigma}}{\partial \mathbf{p}} n_{\mathbf{p}\sigma} \times MN = \sum_{\mathbf{p}\sigma} M \frac{\partial \varepsilon_{\mathbf{p}\sigma}}{\partial \mathbf{p}} n_{\mathbf{p}\sigma}.$$

Because this equation holds for arbitrary $\delta n_{\mathbf{p}\sigma}$, we obtain

$$\frac{\mathbf{p}}{M} = \frac{\partial \varepsilon_{\mathbf{p}\sigma}}{\partial \mathbf{p}} - \frac{1}{\Omega}\sum_{\mathbf{p}'\sigma'} f_{\sigma'\sigma}(\mathbf{p}', \mathbf{p})\frac{\partial n_{\mathbf{p}'\sigma'}}{\partial \mathbf{p}'}.$$

Near the Fermi surface, we have

$$\frac{\partial \varepsilon_{\mathbf{p}\sigma}}{\partial \mathbf{p}} = \frac{\mathbf{p}_{\mathrm{F}}}{M^*}, \quad \frac{\partial n_{\mathbf{p}'\sigma'}}{\partial \mathbf{p}'} \simeq -\frac{\mathbf{p}_{\mathrm{F}}'}{p_{\mathrm{F}}}\delta(p - p_{\mathrm{F}}), \quad \frac{1}{\Omega}\sum_{\mathbf{p}'} \simeq \frac{p_{\mathrm{F}}^2 dp' d\Omega'}{(2\pi\hbar)^3},$$

where we ignore the spin dependence and $d\Omega'$ is the solid angle. We thus obtain

$$\frac{\mathbf{p}_{\mathrm{F}}}{M} = \frac{\mathbf{p}_{\mathrm{F}}}{M^*} + \frac{2p_{\mathrm{F}}}{(2\pi\hbar)^3}\int f(\theta)\mathbf{p}_{\mathrm{F}}' d\Omega',$$

where a factor of 2 results from the spin degrees of freedom. Thus,

$$\begin{aligned}\frac{1}{M} &= \frac{1}{M^*}\left[1 + \frac{1}{4\pi}\int \frac{M^* p_{\mathrm{F}}}{\pi^2\hbar^3} f(\theta)\cos\theta d\Omega\right] \\ &= \frac{1}{M^*}\left[1 + \overline{F^{\mathrm{s}}(\theta)\cos\theta}\right] \\ &= \frac{1}{M^*}\left[1 + \frac{1}{3}F^{\mathrm{s}}\right].\end{aligned} \tag{C.3}$$

This proves Eq. (8.41).

To prove Eq. (8.42) for sound velocity, we start by considering a change in the chemical potential associated with a change in the total number of particles:

$$\delta\mu = \int f(\theta)\delta n(\mathbf{p}')\frac{d^3p'}{(2\pi\hbar)^3} + \frac{\partial \varepsilon_{\mathrm{F}}}{\partial p_{\mathrm{F}}}\delta p_{\mathrm{F}}, \tag{C.4}$$

where the first term on the right-hand side describes the contribution due to a change in the distribution function with θ being the polar angle of $\mathbf{p}'$, and the second term gives the contribution from a change in the Fermi momentum. The first term is approximated as

$$\int f(\theta)\delta n(\mathbf{p}')\frac{d^3p'}{(2\pi\hbar)^3} \simeq \int f(\theta)d\Omega \cdot \frac{1}{4\pi}\int \delta n(\mathbf{p}')\frac{d^3p'}{(2\pi\hbar)^3} = \frac{\delta N}{8\pi\Omega}\int f(\theta)d\Omega,$$

whereas the second term can be rewritten using Eq. (8.5) as

$$\frac{\partial \varepsilon_{\mathrm{F}}}{\partial p_{\mathrm{F}}}\delta p_{\mathrm{F}} \equiv \frac{\pi^2\hbar^3}{M^* p_{\mathrm{F}}\Omega}\delta N.$$

Thus, Eq. (C.4) leads to

$$\left(\frac{\partial \mu}{\partial N}\right)_{\Omega} = \frac{\pi^2\hbar^3}{M^* p_{\mathrm{F}}\Omega} + \frac{1}{8\pi\Omega}\int f(\theta)d\Omega. \tag{C.5}$$

The sound velocity is

$$u = \sqrt{\frac{N}{M}\left(\frac{\partial \mu}{\partial N}\right)_\Omega}. \tag{C.6}$$

Substituting Eq. (C.5) into Eq. (C.6), we obtain

$$u^2 = \frac{p_F^2}{3MM^*} + \frac{1}{3M}\left(\frac{p_F}{2\pi\hbar}\right)^3 \int f(\theta)d\Omega = \frac{p_F^2}{3MM^*}(1+F_0^s). \tag{C.7}$$

Substituting Eq. (8.41) in Eq. (C.7), we obtain Eq. (8.42).

To prove Eq. (8.43), we have to take into account the spin dependence of the quasiparticle energy:

$$\varepsilon_{\mathbf{p}\uparrow} = \varepsilon_{\mathbf{p}\uparrow}{}^0 - \mu_B H + \frac{1}{\Omega}\sum_{\mathbf{p}} [f_{\uparrow\uparrow}(\mathbf{p},\mathbf{p}')\delta n_{\mathbf{p}'\uparrow} + f_{\uparrow\downarrow}(\mathbf{p},\mathbf{p}')\delta n_{\mathbf{p}'\downarrow}], \tag{C.8}$$

where the second term on the right-hand side is the Zeeman energy and the last two terms give the contributions from variations in the distribution function. Because $n_{\mathbf{p}'\uparrow} + n_{\mathbf{p}'\downarrow} = 0$, Eq. (C.8) can be rewritten as

$$\delta\varepsilon_{\mathbf{p}\uparrow} \equiv \varepsilon_{\mathbf{p}\uparrow} - \varepsilon_{\mathbf{p}\uparrow}{}^0 = -\mu_{\mathbf{B}} H + \frac{2}{\Omega}\sum_{\mathbf{p}'} f^a(\mathbf{p},\mathbf{p}')\delta n_{\mathbf{p}'\uparrow}. \tag{C.9}$$

Here, $\delta n_{\mathbf{p}'\uparrow}$ deviates significantly from zero only near the Fermi surface. Assuming the isotropy of $\delta n_{\mathbf{p}'\uparrow}$, we may substitute $\delta n_{\mathbf{p}'\uparrow} = \delta n_{p_F\uparrow} \equiv \delta n_\uparrow$ in Eq. (C.9). Then,

$$\delta\varepsilon_{\mathbf{p}\uparrow} = -\mu_B\left(H - \frac{2}{\mu_B} f_0^a \delta n_\uparrow\right), \tag{C.10}$$

where $f_0^a \equiv (1/\Omega)\sum_{\mathbf{p}'} f^a(\mathbf{p},\mathbf{p}')$. Equation (C.10) shows that the effect of interactions can be represented by a molecular field $H_m = -(2/\mu_B) f_0^a \delta n_\uparrow$. The energy shift (C.10), in turn, changes the quasiparticle distribution by

$$\delta n_\uparrow = \frac{1}{2} D_F |\delta\varepsilon_{\mathbf{p}\uparrow}| = \frac{1}{2} D_F \mu_B \left(H - \frac{2}{\mu_B} f_0^a \delta n_\uparrow\right). \tag{C.11}$$

We can solve Eqs. (C.10) and (C.11), obtaining

$$\delta n_\uparrow = \frac{D_F \mu_B H}{2(1 + D_F f_0^a)} = \frac{D_F \mu_B H}{2(1+F_0^a)}.$$

This gives rise to magnetization

$$M = \mu H(\delta n_\uparrow - \delta n_\downarrow) - 2\mu H \delta n_\uparrow = \frac{D_F {\mu_B}^2}{1+F_0^a} H, \tag{C.12}$$

which proves Eq. (8.43).

Appendix D

Derivation of Eq. (8.155)

Equation (8.155) can be derived by evaluating Eq. (8.154) with

$$\hat{H}_1(\tau) = -\frac{V}{\Omega} \sum_{\mathbf{kk}'\mathbf{k}''} \hat{c}^\dagger_{\mathbf{k}+\mathbf{k}''\uparrow}(\tau)\hat{c}^\dagger_{\mathbf{k}'-\mathbf{k}''\downarrow}(\tau)\hat{c}_{\mathbf{k}'\downarrow}(\tau)\hat{c}_{\mathbf{k}\uparrow}(\tau), \tag{D.1}$$

where

$$\hat{c}_{\mathbf{k}\sigma}(\tau) \equiv e^{\frac{1}{\hbar}\hat{H}_0\tau}\hat{c}_{\mathbf{k}\sigma}e^{-\frac{1}{\hbar}\hat{H}_0\tau} = \hat{c}_{\mathbf{k}\sigma}e^{-\frac{1}{\hbar}\xi_k\tau}. \tag{D.2}$$

Using Wick's theorem and spin conservation, the $n = 1$ term in Eq. (8.155) is decomposed into

$$\begin{aligned}\Omega_1^{(1)} = \frac{V}{\Omega}\sum_{\mathbf{k},\mathbf{q}} \int_0^{\beta\hbar} d\tau \left\langle T_\tau\left\{\hat{c}_{\mathbf{k}+\frac{\mathbf{q}}{2}\uparrow}(\tau)\hat{c}^\dagger_{\mathbf{k}+\frac{\mathbf{q}}{2}\uparrow}(\tau)\right\}\right\rangle_0 \\ \times \left\langle T_\tau\left\{\hat{c}_{-\mathbf{k}+\frac{\mathbf{q}}{2}\downarrow}(\tau)\hat{c}^\dagger_{-\mathbf{k}+\frac{\mathbf{q}}{2}\downarrow}(\tau)\right\}\right\rangle_0 .\end{aligned} \tag{D.3}$$

Here, we introduce the single-particle temperature Green's function

$$\begin{aligned}G_{\mathbf{k}\sigma}(\tau-\tau') &\equiv -\left\langle T_\tau\left\{\hat{c}_{\mathbf{k}\sigma}(\tau)\hat{c}^\dagger_{\mathbf{k}\sigma}(\tau')\right\}\right\rangle_0 \\ &\equiv e^{-\xi_k(\tau-\tau')}\left[\theta(\tau'-\tau)f(\xi_k) - \theta(\tau-\tau')(1-f(\xi_k))\right],\end{aligned} \tag{D.4}$$

and its Fourier transform

$$G_{\mathbf{k}\sigma}(i\nu_n) = \int_0^{\beta\hbar} d\tau G_{\mathbf{k}\sigma}(\tau)e^{i\mathbf{k}\tau} = \frac{1}{i\nu_n - \xi_k}, \tag{D.5}$$

$$G_{\mathbf{k}\sigma}(\tau) = \frac{1}{\beta\hbar}\sum_{n=-\infty}^{\infty} G_{\mathbf{k}\sigma}(i\nu_n)e^{-i\nu_n\tau}, \tag{D.6}$$

where $\nu_n = \pi(2n+1)/(\beta\hbar)$ is the Matsubara frequency for fermions. In terms of Green's functions, $\Omega_1^{(1)}$ can be written as

$$\Omega_1^{(1)} = V\sum_{\mathbf{kq}} \int_0^{\beta\hbar} d\tau G_{\mathbf{k}+\frac{\mathbf{q}}{2}\uparrow}(\tau-\tau^+)G_{-\mathbf{k}+\frac{\mathbf{q}}{2}\downarrow}(\tau-\tau^+), \tag{D.7}$$

where $\tau^+ = \tau + 0^+$. Substituting Eq. (D.6) into Eq. (D.7) and using the formula

$$\sum_{n=-\infty}^{\infty} \frac{1}{\omega_n^2 - \omega_n a + b} = \frac{\beta\hbar}{2\sqrt{a^2-4b}} \left(\cot \frac{\beta\hbar(a-\sqrt{a^2-4b})}{4} - \cot \frac{\beta\hbar(a+\sqrt{a^2-4b})}{4} \right), \quad (D.8)$$

where $\omega_n = 2\pi n/(\beta\hbar)$ is the Matsubara frequency for bosons, we obtain

$$\Omega_1^{(1)} = V \sum_{\mathbf{k},n} e^{i\omega_n 0^+} \Pi(\mathbf{k}, i\omega_n), \quad (D.9)$$

where

$$\Pi(\mathbf{k}, i\omega_n) = -\sum_{\mathbf{q}} \frac{1 - f\left(\xi_{\mathbf{q}+\frac{\mathbf{k}}{2}}\right) - f\left(\xi_{\mathbf{q}-\frac{\mathbf{k}}{2}}\right)}{i\omega_n - \xi_{\mathbf{q}+\frac{\mathbf{k}}{2}} - \xi_{\mathbf{q}-\frac{\mathbf{k}}{2}}}. \quad (D.10)$$

Next, we consider $\Omega_1^{(2)}$. In terms of Green's functions, it is expressed as

$$\Omega_1^{(2)} = \frac{V^2}{2} \int_0^{\beta\hbar} d\tau_1 \int_0^{\beta\hbar} d\tau_2 \sum_{\mathbf{k},\mathbf{k}',\mathbf{q}} G_{\mathbf{k}+\frac{\mathbf{q}}{2}\uparrow}(\tau_2 - \tau_1^+) G_{-\mathbf{k}'+\frac{\mathbf{q}}{2}\uparrow}(\tau_1 - \tau_2^+)$$
$$\times G_{-\mathbf{k}+\frac{\mathbf{q}}{2}\downarrow}(\tau_2 - \tau_1^+) G_{\mathbf{k}'+\frac{\mathbf{q}}{2}\downarrow}(\tau_1 - \tau_2^+). \quad (D.11)$$

Substituting Eq. (D.6) into this equation, we obtain

$$\Omega_1^{(2)} = -\frac{V^2}{2} \sum_{\mathbf{k},n} \left[\Pi(\mathbf{k}, i\omega_n)\right]^2 e^{i\omega_n 0^+}. \quad (D.12)$$

For $m \geq 3$, there exist $(m-1)!$ possible nonzero combinations in the Wick expansion and the coefficient becomes $m! \times 1/(m-1)! = m$. Thus, we obtain

$$\Omega_1^{(m)} = \frac{V^m}{m} \sum_{\mathbf{k},n} \left[\Pi(\mathbf{k}, i\omega_n)\right]^m e^{i\omega_n 0^+}, \quad (D.13)$$

which proves Eq. (8.155).

Appendix E

f-Sum Rule

The *f*-sum rule holds for an arbitrary system that conserves the number of particles, and it applies for both bosons and fermions. To show this, we consider the Hamiltonian

$$\begin{aligned}\hat{H} &= \sum_{\mathbf{p}} \epsilon_{\mathbf{p}} \hat{a}_{\mathbf{p}}^\dagger \hat{a}_{\mathbf{p}} + \frac{1}{2} \sum_{\mathbf{k},\mathbf{p},\mathbf{q}} V_{\mathbf{k}} \hat{a}_{\mathbf{p}+\mathbf{k}}^\dagger \hat{a}_{\mathbf{q}-\mathbf{k}}^\dagger \hat{a}_{\mathbf{q}} \hat{a}_{\mathbf{p}} \\ &= \sum_{\mathbf{p}} \epsilon_{\mathbf{p}} \hat{a}_{\mathbf{p}}^\dagger \hat{a}_{\mathbf{p}} + \frac{1}{2} \sum_{\mathbf{k}} V_{\mathbf{k}} \hat{\rho}_{\mathbf{k}}^\dagger \hat{\rho}_{\mathbf{k}} - \frac{1}{2} N V(\mathbf{r}=0), \end{aligned} \tag{E.1}$$

where $V(\mathbf{r}=0) = \sum_{\mathbf{p}} V_{\mathbf{p}}$ and $\hat{\rho}_{\mathbf{k}} = \sum_{\mathbf{p}} \hat{a}_{\mathbf{p}}^\dagger \hat{a}_{\mathbf{p}+\mathbf{k}}$. By straightforward algebraic manipulation, we have

$$[\hat{H}, \hat{\rho}_{\mathbf{p}}] = \sum_{\mathbf{k}} (\epsilon_{\mathbf{k}} - \epsilon_{\mathbf{k}+\mathbf{p}}) \hat{a}_{\mathbf{k}}^\dagger \hat{a}_{\mathbf{k}+\mathbf{p}}, \tag{E.2}$$

$$\left[\hat{\rho}_{\mathbf{p}}^\dagger, [\hat{H}, \hat{\rho}_{\mathbf{p}}]\right] = \sum_{\mathbf{k}} (\epsilon_{\mathbf{k}+\mathbf{p}} + \epsilon_{\mathbf{k}-\mathbf{p}} - 2\epsilon_{\mathbf{k}}) \hat{a}_{\mathbf{k}}^\dagger \hat{a}_{\mathbf{k}}. \tag{E.3}$$

Substituting $\epsilon_{\mathbf{k}} = \mathbf{k}^2/(2m)$ into $(E.3)$, we obtain

$$\left[\hat{\rho}_{\mathbf{p}}, [\hat{H}, \hat{\rho}_{\mathbf{p}}]\right] = \frac{\mathbf{p}^2}{M} \hat{N}, \tag{E.4}$$

where $\hat{N} \equiv \sum_{\mathbf{k}} \hat{a}_{\mathbf{k}}^\dagger \hat{a}_{\mathbf{k}}$. Multiplying both sides of $(E.4)$ by $\hat{\rho} = Z^{-1} e^{-\beta \hat{H}} (Z \equiv \mathrm{Tr} e^{-\beta \hat{H}})$ and taking the trace, we obtain

$$\int_0^\infty \hbar\omega S(\mathbf{p}, \omega) d\omega = \epsilon_{\mathbf{p}} N, \quad (f\text{-sum rule}) \tag{E.5}$$

where

$$S(\mathbf{p}, \omega) \equiv \frac{1}{Z} \sum_{m,n} e^{-\beta E_m} \left(|\langle m|\hat{\rho}_{\mathbf{p}}^\dagger|n\rangle|^2 + |\langle m|\hat{\rho}_{\mathbf{p}}|n\rangle|^2 \right) \delta(\hbar\omega - \hbar\omega_{nm}) \tag{E.6}$$

is the dynamic structure factor. From the Schwartz inequality, we have

$$S(\mathbf{p}) = \int_{-\infty}^{\infty} S(\mathbf{p},\omega)d\omega \leq \sqrt{\int_{-\infty}^{\infty} \hbar\omega S(\mathbf{p},\omega)d\omega \int_{-\infty}^{\infty} \frac{1}{\hbar\omega} S(\mathbf{p},\omega)d\omega}, \quad \text{(E.7)}$$

where $S(\mathbf{p})$ is the static structure factor. Substituting $(E.5)$, we obtain

$$S(\mathbf{p}) \leq \sqrt{2\epsilon_{\mathbf{p}} N G(\mathbf{p})}, \quad \text{(E.8)}$$

where

$$G(\mathbf{p}) \equiv \int_{-\infty}^{\infty} \frac{1}{\hbar\omega} S(\mathbf{p},\omega)d\omega \quad \text{(E.9)}$$

is the inverse-energy-weighted moment that satisfies the compressibility sum rule [see Sec. 4.2.2]:

$$\lim_{p\to 0} G(\mathbf{p}) = \frac{\hbar N}{Mc^2}, \quad \text{(E.10)}$$

where c is the sound velocity. It follows from $(E.8)$ and $(E.10)$ that

$$S(\mathbf{p}) \leq \frac{p}{Mc} N \ (p \to 0). \quad \text{(E.11)}$$

Bibliography

J. R. Abo-Shaeer, C. Raman, J. M. Vogels, and W. Ketterle, Science **292**, 476 (2001).

A. A. Abrikosov, Zh. Eksp. Teor. Fiz. **32**, 1442 (1957) [Sov. Phys. JETP **5**, 1174 (1959)].

A. Aftalion, X. Blanc, and F. Nier, Phys. Rev. A **71**, 023611 (2005).

A. F. Andreev and I. M. Lifshitz, Sov. Phys. JETP **29**, 1107 (1969).

C. D. Andereck and W. I. Glaberson, J. Low Temp. Phys. **48**, 257 (1982).

M. Anderlini, P. Lee, B. Brown, J. Sebby-Strabley, W. Phillips, J. Porto, Nature **448**, 452 (2007).

M. H. Anderson, J. R. Ensher, M. R. Mathews, C. E. Wieman, and E. A. Cornell, Science **269**, 198 (1995).

P. W. Anderson and W. F. Brinkman, Phys. Rev. Lett. **30**, 1108 (1973).

P. W. Anderson and P. Morel, Phys. Rev. **123**, 1911 (1961).

M. R. Andrews, C. G. Townsend, H.-J. Miesner, D. S. Durfee, D. M. Kurn, and W. Ketterle, Science **275**, 637 (1997).

M. R. Andrews, D. M. Kurn, H. -J. Miesner, D. S. Durfee, C. G. Townsend, S. Inouye, and W. Ketterle, Phys. Rev. Lett. **79**, 553 (1997); **80**, 2967 (1998).

N. W. Ashcroft and N. D. Mermin, *Solid State Physics* (Holt, Rinehardt and Winston, New York, 1976).

G. E. Astrakharchik, J. Boronat, J. Casulleras, and and S. Giorgini, Phys. Rev. Lett. **93**, 200404 (2004).

V. Bagnato and D. Kleppner, Phys. Rev. A **44**, 7439 (1991).

R. Balian and N. R. Werthamer, hys. Rev. **131**, 1553 (1963).

M. A. Baranov, Phys. Rep. **464**, 71 (2008).

J. Bardeen, L. N. Cooper, and J. R. Schrieffer, Phys. Rev. **106**, 162 (1957); **108**, 1175.

R. Barnett, A. Turner, and E. Demler, Phys. Rev. Lett. **97**, 180412 (2006).

G. Baym, Phys. Rev. Lett. **91**, 110402 (2003).

G. Baym, J. -P. Blaizot, M. Holzmann, F. Laloë, and D. Vautherin, Phys. Rev. Lett. **83**, 1703 (1999); Eur. Phys. J. B **24**, 107 (2001).

G. Baym and E. Chandler, J. Low Temp. Phys. **50**, 57 (1983).

G. Baym and C. J. Pethick, Phys. Rev. Lett. **76**, 6 (1996).

G. Baym and C. J. Pethick, Phys. Rev. A **69**, 0437619 (2004).
C. Becker, P. Soltan-Panahi, J. Kronjager, S. Dorscher, K. Bongs, K. Sengstock, arXiv:0912.3646.
P. F. Bedaque, H. Caldas, and G. Rupak, Phys. Rev. Lett. **91**, 247002 (2003).
V. L. Berezinskii, Zh. Eksp. Teor. Fiz. **59**, 907 (1970) [Sov. Phys. JETP **32**, 493 (1971)]; **61**, 1144 (1971) [**34**, 610 (1972)].
G. F. Bertsch and T. Papenbrock, Phys. Rev. Lett. **83**, 5412 (1999).
H. A. Bethe and R. Peierls, Proc. R. Soc. London, Ser. A **148**, 146 (1935).
A. Bijl, Physica **7**, 860 (1940).
D. J. Bishop and J. D. Reppy, Phys. Rev. Lett. **40**, 1727 (1978); Phys. Rev. B **22**, 5171 (1980).
I. Bloch, Nature **453**, 1016 (2008).
I. Bloch, J. Dalibard, and W. Zwerger, Rev. Mod. Phys. **80**, 885 (2008).
N. N. Bogoliubov, Phys. USSR **11**, 23 (1947).
N. N. Bogoliubov, Nuovo Cimento **7**, 6 (1958); **7**, 794 (1958); J. Valatin, *ibid.* **7**, 843 (1958).
N. N. Bogoliubov, Physik. Adhandl. Sowjetunion **6**, 1, 113, 229 (1962).
M. Boninsegni and N. Prokof'ev, Phys. Rev. Lett. **95**, 237204 (2005).
E. Braaten, H. -W. Hammer, and S. Hermans, Phys. Rev. A **63**, 063609 (2001).
C. C. Bradley, C. A. Sackett, J. J. Tollet, and R. G. Hulet, Phys. Rev. Lett. **75**, 1687 (1995); **79**, 1170 (1997); C. C. Bradley, C. A. Sackett, and R. G. Hulet, *ibid.* **78**, 985 (1997).
A. Bulgac, J. E. Drut, and P. Magierski, Phys. Rev. Lett. **96**, 090404 (2006).
A. Bulgac, J. E. Drut, and P. Magierski, Phys. Rev. A **78**, 023625 (2008).
E. Burovski, N. Prokof'ev, B. Svistunov, and M. Troyer, Phys. Rev. Lett. **96**, 160402 (2006).
D. A. Butts and D. S. Rokhsar, Nature **397**, 327 (1999).
B. Capogrosso-Sansone, N. V. Prokof'ev, and B. V. Svistunov, Phys. Rev. B **75**, 134302 (2007).
F. Calogero and A. Degasperis, Phys. Rev. A **11**, 265 (1975).
J. Carlson, S.-Y. Chang, V. R. Pandharipande, and K. E. Schmidt, Phys. Rev. Lett. **91**, 050401 (2003).
J. Carlson and S. Reddy, Phys. Rev. Lett. **95**, 060401 (2005).
J. Carlson and S. Reddy, Phys. Rev. Lett. **100**, 150403 (2008).
Y. Castin and J. Dalibard, Phys. Rev. A **55**, 4330 (1997).
Y. Castin and C. Herzog, C. R. Acad. Sci. Ser. 4 **2**, 419 (2001).
B. S. Chandrasekhar, App. Phys. Lett. **1**, 7 (1962).
C. Chin, R. Grimm, P. Julienne, and E. Tiesinga, arXiv:0812.1496v1.
S. Choi, S. A. Morgan, and K. Burnett, Phys. Rev. A **57**, 4057 (1998).
P. Cladé, C. Ryu, A. Ramanathan, K. Helmerson, and W. D. Phillips, Phys. Rev. Lett. **102**, 170401 (2009).
A. M. Clogston, Phys. Rev. Lett. **9**, 266 (1962).
I. Coddington, P. Engels, V. Schweikhard, and E. A. Cornell, Phys. Rev. Lett. **91**, 100402 (2003).
L. N. Cooper, Phys. Rev. **104**, 1189 (1956).

N. R. Cooper, N. K. Wilkin, and J. M. F. Gunn, Phys. Rev. Lett. **87**, 120405 (2001).
N. R. Cooper, Adv. Phys. **57**, 539 (2008).
S. L. Cornish, N. R. Claussen, J. L. Roberts, E. A. Cornell, and C. E. Wieman, Phys. Rev. Lett. **85**, 1795 (2000).
Ph. Courteille, R. S. Freeland, D. J. Heinzen, F. A. van Abeelen, and B. J. Verhaar, Phys. Rev. Lett. **81**, 69 (1998).
M. Cozzini1, L. P. Pitaevskii, and S. Stringari, Phys. Rev. Lett. **92**, 220401 (2004).
K. B. Davis, M. -O. Mewes, M. R. Andrews, N. J. van Druten, D. S. Durfee, D. M. Kurn, and W. Ketterle, Phys. Rev. Lett. **75**, 3969 (1995).
P. G. de Gennes, *Superconductivity of Metals and Alloys* (Addison Wesley, Reading, MA, 1989).
S. R. de Groot, G. J. Hooyman, and C. A. ten Seldam, Proc. R. Soc. London A **203**, 266 (1950).
L. Deng, E. W. Hagley, J. Wen, M. Trippenbach, Y. Band, P. S. Julienne, J. E. Simsarian, K. Helmerson, S. L. Rolston, and W. D. Phillips, Nature **398**, 218 (1999).
J. Denschlag, J. E. Simsarian, D. L. Feder, C. W. Clark, L. A. Collins, J. Cubizolles, L. Deng, E. W. Hagley, K. Helmerson, W. P. Reinhardt, S. L. Rolston, B. I. Schneider, and W. D. Phillips, Science **287**, 97 (2000).
R. J. Dodd, Mark Edwards, C. J. Williams, C. W. Clark, M. J. Holland, P. A. Ruprecht, and K. Burnett, Phys. Rev. A **54**, 661 (1996).
E. A. Donley, N. R. Claussen, S. L. Cornish, J. L. Roberts, E. A. Cornell, and C. E. Wieman, Nature (London) **412**, 295 (2001).
R. J. Donnelly, *Quantized Vortices in Helium II* (Cambridge University Press, 1991).
T. Donner, S. Ritter, T. Bourdel, A. öttl, M. Köhl, and T. Esslinger, Science **315**, 1556 (2007).
T. C. Dorlas, Commun. Math. Phys. **154**, 347 (1993).
D. M. Eagles, Phys. Rev. **186**, 456 (1969).
C. Eberlein, S. Giovanazzi, and D. H. J. O'Dell, Phys. Rev. A **71**, 033618 (2005).
A. Einstein, S. B. Preuss. Akad. Wiss. Phys. -math. Klasse **13**, 3 (1925).
T. Ellis and P. V. E. McClintock, Phil. Trans. R. Soc. A **315**, 259 (1985).
J. R. Ensher, D. S. Jin, M. R. Matthews, C. E. Wieman, and E. A. Cornell, Phys. Rev. Lett. **77**, 4984 (1996).
U. Fano, Phys. Rev. **124**, 1866 (1961).
H. Feshbach, Ann. Phys. (N.Y.) **5**, 357 (1958).
A. L. Fetter, Ann. Phys. (N.Y.) **70**, 67 (1972).
A. L. Fetter and A. A. Svidzinsky, J. Phys. Condens. Matter **13**, R135 (2001).
R. P. Feynman, Phys. Rev. **94**, 262 (1954).
U. Fischer and G. Baym, Phys. Rev. Lett. **90**, 140402 (2003).
M. Fisher, P. Weichman, G. Gristein, and D. Fisher, Phys. Rev. B **40**, 546 (1989).
S. Fölling, S. Trotzky, P. Cheinet, M. Feld, R. Saers, A. Widera, T. Muller, I. Bloch, Nature **448**, 1029 (2007).
P. J. Forrester, N. E. Frankel, T. M. Garoni, and N. S. Witte, Phys. Rev. A **67**, 043607 (2003).

J. K. Freericks and H. Monien, Phys. Rev. B **53**, 2691 (1996).
P. Fulde and R. A. Ferrell, Phys. Rev. **135**, A550 (1964).
F. Gerbier, J. H. Thywissen, S. Richard, M. Hugbart, P. Bouyer, and A. Aspect, Phys. Rev. Lett. **92**, 030405 (2004).
J. M. Gerton, D. Strekalov, I. Prodan, and R. G. Hulet, Nature (London) **408**, 692 (2000).
V. L. Ginzburg and L. P. Pitaevskii, Zh. Eksp. Teor. Fiz. **34**, 1240 (1958) [Sov. Phys. JETP **7**, 858 (1958)].
S. Giorgini, L. P. Pitaevskii, and S. Stringari, Phys. Rev. A **54**, R4633 (1996).
S. Giorgini, L. P. Pitaevskii, and S. Stringari, Rev. Mod. Phys. **80**, 1215 (2008).
S. Giovanazzi, A. Görlitz, and T. Pfau, Phys. Rev. Lett. **89**, 130401 (2002).
M. Girardeau, J. Math Phys. **1**, 516 (1960).
M. D. Girardeau, Phys. Rev. B **139**, 500 (1965).
M. D. Girardeau and E. M. Wright, Phys. Rev. Lett. **87**, 050403 (2001).
L. P. Gor'kov, Zh. Eksp. Teor. Fiz. **34**, 735 (1958) [Sov. Phys. JETP **7**, 505 (1958)].
L. P. Gor'kov and T. K. Melik-Barkhudarov, Zh. Eksp. Teor. Fiz. **40**, 1452 (1961) [Sov. Phys. JETP **13**, 1018 (1961)].
I. S. Gradshteyn and I. M. Ryzhik, *Table of Integrals, Series, and Products* (Academic, New York, 1980) Formula 3, p. 774.
M. Greiner, I. Bloch, O. Mandel, T. W. Hänsch, and T. Esslinger, Phys. Rev. Lett. **87**, 160405 (2001).
M. Greiner, M. O Mandel, T. Esslinger, T. Hänsch, and I. Bloch, Nature **415**, 39 (2002).
M. Greiner, C. A. Regal, and D. S. Jin, Nature **426**, 537 (2003).
G. F. Gribakin and V. V. Flambaum, Phys. Rev. A **48**, 546 (1993).
A. Griesmaier, J. Werner, S. Hensler, J. Stuhler, and T. Pfau, Phys. Rev. Lett. **94**, 160401 (2005).
E. P. Gross, Nuovo Cimento **20**, 454 (1961).
S. Grossmann and M. Holthaus, Z. Naturforsch. **50a**, 921 (1995).
D. Guéry-Odelin and S. Stringari, Phys. Rev. Lett. **83**, 4452 (1999).
R. Haag, *Local Quantum Physics: Fields Particles, Algebras*, 2nd Ed. (Springer-Verlag, Berlin, 1996).
Z. Hadzibabic, P. Krüger, M. Cheneau, B. Battelier, and J. Dalibard, Nature **441**, 1118 (2006).
Z. Hadzibabic, P. Krüger, M. Cheneau, S. P. Rath, and J. Dalibard, New J. Phys. **10**, 045006 (2008).
Z. Hadzibabic, S. Stock, B. Battelier, V. Bretin, and J. Dalibard, Phys. Rev. Lett. **93**, 180403 (2004).
R. Hanbury Brown and R. Q. Twiss, Nature **177**, 27 (1956).
G. B. Hess and W. M. Fairbank, Phys. Rev. Lett. **19**, 216 (1967).
T. -L. Ho, Phys. Rev. Lett. **87**, 060403 (2001).
T. -L. Ho, Phys. Rev. Lett. **92**, 090402 (2004).
T. -L. Ho and E. J. Mueller, Phys. Rev. Lett. **92**, 160404 (2004).
T. -L. Ho and S. K. Yip, Phys. Rev. Lett. **84**, 4031 (2000).

E. Hodby, G. Hechenblaikner, S. A. Hopkins, O. M. Marag, and C. J. Foot, Phys. Rev. Lett. **88**, 010405 (2002).
P. C. Hohenberg, Phys. Rev. **158**, 383 (1967).
M. Horikoshi, S. Nakajima, M. Ueda, and T. Mukaiyama, Science (to be published).
K. Huang and C. N. Yang, Phys. Rev. **105**, 767 (1957).
N. M. Hugenholtz and D. Pines, Phys. Rev. **116**, 489 (1959).
M. Inguscio, W. Ketterle, and C. Salomon (eds.) *Ultra-Cold Fermi Gases: Proceedings of the International School of Physics "Enrico Fermi", Course CLXIV, Varenna, 20–30 June 2006* (IOS Press, Amsterdam, 2008).
S. Inouye, M. R. Andrews, J. Stenger, H. -J. Miesner, D. M. Stamper-Kurn, and W. Ketterle, Nature **392**, 151 (1998).
M. Ishikawa and H. Takayama, J. Phys. Soc. Jpn. **49**, 1242 (1980).
D. Jaksch1, V. Venturi, J. I. Cirac, C. J. Williams, and P. Zoller Phys. Rev. Lett. **89**, 040402 (2002).
D. Jaksch, C. Bruder, J. I. Cirac, C. W. Gardiner, and P. Zoller, Phys. Rev. Lett. **81**, 3108 (1998).
J. Javanainen and S. M. Yoo, Phys. Rev. Lett. **76**, 161 (1996).
B. D. Josephson, Phys. Lett. **21**, 608 (1966); G. Baym, in *Mathematical Methods in Solid State and Superfluid Theory*, edited by R. C. Clark and G. H. Derrick (Oliver and Boyd, Edinburgh, 1969), p. 121.
B. Kahn and G. E. Uhlenbeck, Physica **5**, 399 (1938).
R. Kanamoto, L. D. Carr, and M. Ueda, arXiv:0910.2805.
K. Kasamatsu and M. Tsubota, Phys. Rev. A **67**, 033610 (2003).
Y. Kawaguchi, M. Nitta, and M. Ueda, Phys. Rev. Lett. **100**, 180403 (2008).
Y. Kawaguchi, H. Saito, and M. Ueda, Phys. Rev. Lett. **96**, 080405 (2006).
Y. Kawaguchi, H. Saito, and M. Ueda, Phys. Rev. Lett. **97**, 130404 (2006).
Y. Kawaguchi, H. Saito, and M. Ueda, Phys. Rev. Lett. **98**, 110406 (2007).
W. Ketterle and N. J. van Druten, Phys. Rev. A **54**, 656 (1996).
T. Kinoshita, T. Wenger, and D. S. Weiss, Science **305**, 1125 (2004).
W. H. Kleiner, L. M. Roth, and S. H. Autler, Phys. Rev. **133**, A1226 (1964).
M. Koashi and M. Ueda, Phys. Rev. Lett. **84**, 1066 (2000).
M. Kobayashi, Y. Kawaguchi, M. Nitta, and M. Ueda, Phys. Rev. Lett. **103**, 115301 (2009).
T. Köhler, K. Góral, and P. S. Julienne, Rev. Mod. Phys. **78**, 1311 (2006).
C. Kollath, U. Schollwöck, J. von Delft, and W. Zwerger, Phys. Rev. A **69**, 031601(R) (2004).
J. M. Kosterlitz, J. Phys. C **7**, 1046 (1974).
J. M. Kosterlitz and D. J. Thouless, J. Phys. C **6**, 1181 (1973).
R. Kubo, M. Toda, and N. Hashitume, *Statistical Physics II. Nonequilibrium Statistical Mechanics* (Springer, Berlin, 1992).
T. D. Kuhner, S. R. White, and H. Monien, Phys. Rev. B **61**, 12474 (2000).
T. Lahaye, T. Koch, B. Fröhlich, M. Fattori, J. Metz, A. Griesmaier, S. Giovanazzi, and T. Pfau, Nature **448**, 672 (2007).
T. Lahaye, J. Metz, B. Frhlich, T. Koch, M. Meister, A. Griesmaier, T. Pfau, H. Saito, Y. Kawaguchi, and M. Ueda, Phys. Rev. Lett. **101**, 080401 (2008).

T. Lahaye, C. Menotti, L. Santos, M. Lewenstein, and T. Pfau, Rep. Prog. Phys. **72**, 126401 (2009).
L. D. Landau, J. Phys. USSR **5**, 71 (1941); **11**, 91 (1947).
L. D. Landau, Zh. Eksp. Teor. Fiz. **32**, 59 (1957) [Sov. Phys. JETP **5**, 101 (1957)].
L. D. Landau and E. M. Lifshitz, *Fluid Mechanics* (Butterworth-Heinemann, Oxford, 1987).
L. D. Landau and E. M. Lifshitz, *Quantum Mechanics* (Butterworth-Heinemann, Oxford, 1999).
G.J. Lapeyre, Jr., M.D. Girardeau, and E.M. Wright, Phys. Rev. A **66**, 023606 (2002).
A. I. Larkin and Yu. N. Ovchinnikov, Zh. Eksp. Teor. Fiz. **47**, 1136 (1964) [Sov. Phys. JETP **20**, 762 (1965)].
C. K. Law, H. Pu, and N. P. Bigelow, Phys. Rev. Lett. **81**, 5257 (1998).
A. E. Leanhardt, A. Grlitz, A. P. Chikkatur, D. Kielpinski, Y. Shin, D. E. Pritchard, and W. Ketterle, Phys. Rev. Lett. **89**, 190403 (2002).
T. D. Lee, K. Huang, and C. N. Yang. Phys. Rev. **106**, 1135 (1957).
T. D. Lee and C. N. Yang, Phys. Rev. **105**, 1119 (1957).
A. J. Leggett, J. Phys. Colloques. **41**, C7-19 (1980).
A. J. Leggett, in *Low Temperature Physics*, edited by M. J. R. Hoch and R. H. Lemmer (Springer-Verlag, Berlin, 1991), pp. 1-92.
A. J. Leggett, Rev. Mod. Phys. **73**, 307 (2001).
E. H. Lieb, Phys. Rev. **130**, 1616 (1963).
E. H. Lieb and W. Liniger, Phys. Rev. **130**, 1605 (1963).
E. H. Lieb, R. Seiringer, and J. Yngvason, Phys. Rev. A **61**, 043602 (2000).
E. M. Lifshitz and L. P. Pitaevskii, *Statistical Physics, Part 2* (Butterworth-Heinemann, Oxford, 1991).
K.-S. Liu and M. E. Fisher, J. Low Temp. Phys. **10**, 655 (1973).
W. V. Liu and F. Wilczek, Phys. Rev. Lett. **90**, 047002 (2003).
C. Lobo, A. Recati, S. Giorgini, and S. Stringari, Phys. Rev. Lett. **97**, 200403 (2006).
F. London, Nature (London) **141**, 643 (1938).
K.W. Madison, F. Chevy, W. Wohlleben, and J. Dalibard, Phys. Rev. Lett. **84**, 806 (2000).
H. Mäkelä, J. Phys. A.: Math. Gen. **39**, 7423 (2006).
O. M. Maragò, S. A. Hopkins, J. Arlt, E. Hodby, G. Hechenblaikner, and C. J. Foot, Phys. Rev. Lett. **84**, 2056 (2000).
H. Matsuda and T. Tsuneto, Prog. Theor. Phys. **46**, 411 (1970).
M. R. Matthews, B. P. Anderson, P. C. Haljan, D. S. Hall, C. E. Wieman, and E. A. Cornell, Phys. Rev. Lett. **83**, 2498 (1999).
G. Moore and N. Reed, Nucl. Phys. B **360**, 362 (1991).
J. B. McGuire, J. Math. Phys. **5**, 622 (1964).
N. D. Mermin, Phys. Rev. **176**, 250 (1968).
N. D. Mermin and H. Wagner, Phys. Rev. Lett. **17**, 1133 (1966).
T. Mizushima, Y. Kawaguchi, K. Machida, T. Ohmi, T. Isoshima, and M. M. Salomaa, Phys. Rev. Lett. **92**, 060407 (2004).

H. Moritz, T. Stöferle, K. Günter, M. Köhl, and T, Esslinger, Phys. Rev. Lett. **94**, 210401 (2005).
H. Moritz, T. Stöferle, M. Köhl, and T, Esslinger, Phys. Rev. Lett. **91**, 250402 (2003).
O. Morsch, J. H. Muller, M. Cristiani, D. Ciampini, and E. Arimondo, Phys. Rev. Lett. **87**, 140402 (2001).
E. J. Mueller, T.-L. Ho, M. Ueda, and G. Baym, Phys. Rev. A **74**, 033612 (2006).
W. J. Mullin, J. Low. Temp. Phys. **106**, 615 (1997).
K. Murata, H. Saito, and M. Ueda, Phys. Rev. A **75**, 013607 (2007).
G. Murthy, D. Arovas, and A. Auerbach, Phys. Rev. B **55**, 3104 (1997).
M. Nakahara, T. Isoshima, K. Machida, S. -I. Ogawa, and T. Ohmi, Physica. B **284-288**, 17 (2000).
T. Nakajima and M. Ueda, Phys. Rev. Lett. **91**, 140401 (2003).
M. Naraschewski, H. Wallis, and A. Schenzle, J. I. Cirac, P. Zoller, Phys. Rev. A **54**, 2185 (1996).
S. Nascimbène, N. Navon, K. Jiang, F. Chevy, and C. Salomon, arXiv:0907.3032.
D. R. Nelson and J. M. Kosterlitz, Phys. Rev. Lett. **39**, 1201 (1977).
P. Nozières, in *Bose–Einstein Condensation*, edited by A. Griffin, D. W. Snoke, and S. Stringari (Cambridge University Press, New York, 1995), p. 15.
P. Nozières and D. Saint James, J. Phys. (Paris) **43**, 1133 (1982).
P. Nozières and S. Schmitt-Rink, J. Low. Temp. Phys. **59**, 195 (1985).
D. H. J. O'Dell, S. Giovanazzi, and C. Eberlein, Phys. Rev. Lett. **92**, 250401 (2004).
M. Olshanii, Phys. Rev. Lett. **81**, 938 (1998).
L. Onsager, Nuovo Cimento **6**, Suppl. 2, 249 (1949).
S. B. Papp, J. M. Pino, R. J. Wild, S. Ronen, C. E. Wieman, D. S. Jin, and E. A. Cornell, Phys. Rev. Lett. **101**, 135301 (2008).
B. Paredes, P. Fedichev, J. I. Cirac, and P. Zoller, Phys. Rev. Lett. **87**, 010402 (2001).
B. Paredes, A. Widera, V. Murg, O. Mandel, S. Folling, I. Cirac, G. V. Shlyapnikov, T. W. Hansch, and I. Bloch, Nature (London) **429**, 277 (2004).
P. Pedri and L. Santos, Phys. Rev. Lett. **95**, 200404 (2005).
S. Peil, J. V. Porto, B. L. Tolra, J. M. Obrecht, B. E. King, M. Subbotin, S. L. Rolston, and W. D. Phillips, Phys. Rev. A **67**, 051603 (2003).
O. Penrose and L. Onsager, Phys. Rev. **104**, 576 (1956).
Víctor M. Pérez–García, H. Michinel, J. I. Cirac, M. Lewenstein, and P. Zoller, Phys. Rev. Lett. **77**, 5320 (1996); Phys. Rev. A **56**, 1424 (1997).
C. J. Pethick and H. Smith, *Bose-Einstein Condensation in Dilute Gases* (Cambridge University Press, Cambridge, 2002).
D. S. Petrov, D. M. Gangardt, and G. V. Shlyapnikov, J. Phys. IV (France) **116**, 5 (2004).
D. S. Petrov, C. Salomon, and G. V. Shlyapnikov, Phys. Rev. Lett. **93**, 090404 (2004).
L. P. Pitaevskii, Zh. Eksp. Teor. Fiz. **40**, 646 (1961) [Sov. Phys. JETP **13**, 451 (1961)].
L. Pitaevskii and S. Stringari, J. Low. Temp. Phys. **85**, 377 (1991).

S. E. Pollack, D. Dries, M. Junker, Y. P. Chen, T. A. Corcovilos, and R. G. Hulet, Phys. Rev. Lett. **102**, 090402 (2009).
R. E. Prange and S. M. Girvin (eds.), *The Quantum Hall Effect* (Springer, Berlin, 1990).
N. Prokof'ev, O. Ruebenacker, and B. Svistunov, Phys. Rev. Lett. **87**, 270402 (2001).
N. Prokof'ev and B. Svistunov, Phys. Rev. A **66**, 043608 (2002).
C. Raman, J. R. Abo-Shaeer, J. M. Vogels, K. Xu, and W. Ketterle, Phys. Rev. Lett. **87**, 210402 (2001).
C. Raman, M. Köhl, R. Onofrio, D. S. Durfee, C. E. Kuklewicz, Z. Hadzibabic, and W. Ketterle, Phys. Rev. Lett. **83**, 8502 (1999).
A. Recati, F. Zambelli, and S. Stringari, Phys. Rev. Lett. **86**, 377 (2001).
C. A. Regal, M. Greiner, and D. S. Jin, Phys. Rev. Lett. **92**, 040403 (2004).
N. Reed and E. H. Rezayi, Phys. Rev. B **59**, 8084 (1999).
N. Regnault and Th. Jolicoeur, Phys. Rev. Lett. **91**, 030402 (2003).
D. S. Rokhsar and B. G. Kotliar, Phys. Rev. B **44**, 10328 (1991).
S. L. Rolston, and W. D. Phillips, Nature **416**, 219 (2002).
S. Ronen, D. C. E. Bortolotti, and J. L. Bohn, Phys. Rev. Lett. **98**, 030406 (2007).
I. Rudnick, Phys. Rev. Lett. **40**, 1454 (1978).
J. Ruostekoski and J. R. Anglin, Phys. Rev. Lett. **91**, 190402 (2003).
H. Saito, Y. Kawaguchi, and M. Ueda, Phys. Rev. Lett. **102**, 230403 (2009).
H. Saito and M. Ueda, Phys. Rev. A **63**, 043601 (2001).
H. Saito and M. Ueda, Phys. Rev. A **72**, 053628 (2005).
L. Santos, G. V. Shlyapnikov, and M. Lewenstein, Phys. Rev. Lett. **90**, 250403 (2003).
L. Santos, G. V. Shlyapnikov, P. Zoller, and M. Lewenstein, Phys. Rev. Lett. **85**, 1791 (2000).
L. Santos and T. Pfau, Phys. Rev. Lett. **96**, 190404 (2006).
G. Sarma, J. Phys. Chem. Solids **24**, 1029 (1963).
K. Sawada, Phys. Rev. **116**, 1344 (1959).
C. H. Schunck, M. W. Zwierlein, C. A. Stan, S. M. F. Raupauch, W. Ketterle, A. Simoni, E. Teisinga, C. J. Williams, and P. S. Julienne, Phys. Rev. A **71**, 045601 (2005).
J. Sebby-Strabley, M. Anderlini, P. S. Jessen, and J. V Porto, Phys. Rev. A **73**, 033605 (2006).
G. W. Semenoff and F. Zhou, Phys. Rev. Lett. **98**, 100401 (2007).
R. Shankar, J. Phys. (Paris) **38**, 1405 (1977).
K. Sheshadri, H. R. Krishnamurty, R. Pandit, and T V. Ramarkrishnan, Europhys. Lett. **22**, 257 (1993).
Y. Shin, M. W. Zwierlein, C. H. Schunck, A. Schirotzek, and W. Ketterle, Phys. Rev. Lett. **97**, 030401 (2006).
S. Sinha and Y. Castin, Phys. Rev. Lett. **87**, 190402 (2001).
J. Sinova, C. B. Hanna, and A. H. MacDonald, Phys. Rev. Lett. **89**, 030403 (2002).
R. A. Smith and N. K. Wilkin, Phys. Rev. A **62**, 061602 (2000).

J. Stenger, S. Inouye, D. M. Stamper-Kurn, H.-J. Miesner, A. P. Chikkatur, and W. Ketterle, Nature **396**, 345 (1998).

S. Stringari, in *Bose–Einstein Condensation*, edited by A. Griffin, D. W. Snoke, and S. Stringari (Cambridge University Press, Cambridge, 1995), p. 86.

S. Stringari, Phys. Rev. Lett. **77**, 2360 (1996).

J. Stuhler, A. Griesmaier, T. Koch, M. Fattori, T. Pfau, S. Giovanazzi, P. Pedri, and L. Santos, Phys. Rev. Lett. **95**, 150406 (2005).

A. A. Svidzinsky and A. L. Fetter, Phys. Rev. A **62**, 063617 (2000).

J. E. Thomas, J. Kinast, and A. Turlapov, Phys. Rev. Lett. **95**, 120402 (2005).

D. J. Thouless, *Topological Quantum Numbers in Nonrelativistic Physics* (World Scientific, Singapore, 1998).

C. Ticknor, C. A. Regal, D. S. Jin, and J. L. Bohn, Phys. Rev. A **69**, 042712 (2004).

V. K. Tkachenko, Zh. Eksp. Teor. Fiz. **49**, 1875 (1965) [Sov. Phys. JETP **22**, 1282 (1966)], ; **56**, 1763 (1969) [**29**, 245 (1969)].

B. L. Tolra, K. M. O'Hara, J. H. Huckans, W. D. Phillips, S. L. Rolston, and J. V. Porto, Phys. Rev. Lett. **92**, 190401 (2004).

L. Tonks, Phys. Rev. **50**, 955 (1936).

M. Tsubota, K. Kasamatsu, and M. Ueda, Phys. Rev. A **65**, 023603 (2002).

M. Ueda and Y. Kawaguchi, Phy. Rep. (to be published).

M. Ueda and M. Koashi, Phys. Rev. A **65**, 063602 (2002).

M. Ueda and A. J. Leggett, Phys. Rev. Lett. **80**, 1576 (1998).

M. Ueda and T. Nakajima, Phys. Rev. A **73**, 043603 (2006).

M. Ueda and H. Saito, in *Emergent Nonlinear Phenomena in Bose-Einstein Condensates: Theory and Experiment*, edited by P. G. Kevrekidis, D. J. Frantzeskakis, and R. C-. Gonzalez (Springer, 2007), pp. 211-227.

D. van Oosten, P. van der Straten, and H. T. C. Stoof, Phys. Rev. A **63**, 053601 (2001).

D. Vollhardt and P. Wölfle, *The Superfluid Phases of Helium 3* (Taylor & Francis, London, 1990).

H. Wagner, Z. Physik **195**, 273 (1966).

G. Watanabe, G. Baym, and C. J. Pethick, Phys. Rev. Lett. **93**, 190401 (2004).

S. Wessel and M. Troyer, Phys. Rev. Lett. **95**, 127205 (2005).

A. Widom, Phys. Rev. **168**, 150 (1968).

N. K. Wilkin, J. M. F. Gunn, and R. A. Smith, Phys. Rev. Lett. **80**, 2265 (1998).

T. T. Wu, Phys. Rev. **115**, 1390 (1959).

E. J. Yarmchuk, M. J. V. Gordon, and R. E. Packard, Phys. Rev. Lett. **43**, 214 (1979).

C. N. Yang, Rev. Mod. Phys. **34**, 694 (1962).

C. N. Yang and C. P. Yang, J. Math. Phys. **10**, 1115 (1969).

V. I. Yukalov, Laser Phys. Lett. **1**, 435 (2004).

V. I. Yukalov and M. D. Girardeau, Laser Phys. Lett. **2**, 375 (2005).

Y. Zhang, H. Mäkelä, and K. -A. Suominen, in *Bosons, Ferromagnetism And Crystal Growth Research*, eduted by E. D. Seifer (Nova Science Publishers, New York, 2006), cond-mat/0404138.

J. Zhang, E. G. M. van Kempen, T. Bourdel, L. Khaykovich, J. Cubizolles, F. Chevy, M. Teichmann, L. Tarruell, S. J. J. M. F. Kokkelmans, and C. Salomon, Phys. Rev. A **70**, 030702(R) (2004).

F. Zhou, Phys. Rev. Lett. **87**, 080401 (2001).

M.W. Zwierlein, C. A. Stan, C. H. Schunck, S. M. F. Raupach, S. Gupta, Z. Hadzibabic, and W. Ketterle, Phys. Rev. Lett. **91**, 250401 (2003).

M.W. Zwierlein, C. A. Stan, C. H. Schunck, S. M. F. Raupach, A. J. Kerman, and W. Ketterle, Phys. Rev. Lett. **92**, 120403 (2004).

Index